Natural and Social Sciences of Patagonia

Despite being an underpopulated region, Patagonia has attracted the attention of scientists since the very beginning of its settlement. From classical explorers such as Darwin or D´Orbigny, to modern science including nuclear and satellite developments, several disciplines have focused their efforts on unraveling Patagonia's natural and social history. Today, scientific and technological research is shifting from being shaped by northern agendas, towards more locally oriented objectives, such as the management of natural resources, the modernization of energy production and distribution, and the coexistence of rural and cosmopolitan social lifestyles. At the intersection of all these topics, new conflicts concerning the economy, human development, population, and the proper and long-standing planification and management of the landscape and its natural resources have emerged. These conflicts, of course, have also caught the attention of many interdisciplinary research groups.

This series is aimed at describing and discussing various aspects of this complex reality, but also at bridging the gaps between the scientific community and governments, policymakers, and society in general. The respective volumes will analyze and synthesize our knowledge of Patagonian biodiversity at different scales, from alleles, genes and species, to ecosystems and the biosphere, including its multilevel interactions. As humans cannot be viewed as being separate from biodiversity, the series' volumes will also share anthropological, archaeological, sociological and historical views of humanity, and highlight the wide range of benefits that ecosystems provide to humanity including provisioning, regulating and cultural services.

Trace Gale-Detrich • Andrea Ednie
Keith Bosak

Editors

Tourism and Conservation-based Development in the Periphery

Lessons from Patagonia for a Rapidly Changing World

 Springer

Editors
Trace Gale-Detrich
Centro de Investigación en Ecosistemas de
la Patagonia (CIEP), Sustainable Tourism
Research Line, Human-Environmental
Interactions Group, Coyhaique, Chile

Cape Horn International Center (CHIC)
Puerto Williams, Magallanes and Chilean
Antarctica Region, Chile

Andrea Ednie
College of Education &
Professional Studies
University of Wisconsin–Whitewater
Whitewater, WI, USA

Keith Bosak
Department of Society and Conservation
University of Montana, College of Forestry
and Conservation
Missoula, MT, USA

ISSN 2662-3463   ISSN 2662-3471   (electronic)
Natural and Social Sciences of Patagonia
ISBN 978-3-031-38047-1   ISBN 978-3-031-38048-8   (eBook)
https://doi.org/10.1007/978-3-031-38048-8

This Springer imprint is published by the registered company Springer Nature Switzerland AG
The registered company address is: Gewerbestrasse 11, 6330 Cham, Switzerland

Paper in this product is recyclable.

Foreword

Change comes to all places and peoples, usually at inconvenient times, and change affects everyone. Patagonia is no exception. For 500 years, Patagonia has seen and experienced change, all brought by colonial powers, first violently and then more subtlety as governing, land tenure, and religious systems became embedded in a progression of social systems. Colonialization brought globalization and modernization to an indigenous culture which resulted in many other conflicts and fundamentally transformed what had been a traditional, sustainable set of land use practices. That transformation brought uncertainty, with various geographic imaginaries and meanings of landscapes successively laid on the landscape. Each resulted in a dominant paradigm of resource development that has led to the current one based on sustainable use of landscapes, primarily through commodification of them by tourism. International non-governmental organizations collude with the national governments and lead much of the discussion of sustainable development. In distinct contrast to local initiatives which typify the world's interest in sustainable development in other places, this conversation seems to begin in national capitals.

Understanding the roots of change, how it has occurred, what human and ecological responses have ensued, and how to address them are the nuclei of this book. And it focuses on efforts toward conservation based on sustainable resource development—a relatively new land use for the region, but itself a change agent that coincidentally seeks to institutionalize actions to maintain an ecology that is relatively variable.

Editors Gale-Detrich, Ednie, and Bosak bring us important contributions to understanding these changes in Patagonia: how and why Patagonia has been undergoing change for over 500 years. Those changes are dramatic and continue to provide surprises as unintended consequences pop up and force additional change in response. The volume shows the interconnectedness between various change agents and consequences, often lasting centuries.

The volume begins with the editors describing the chapters and terminology and is followed by a section utilizing critical history to show us the principal agent of change here—colonialization. Colonization had several immediate effects including the near complete extirpation of indigenous people and the resulting transformation of land use to more commercial extractive activities. Then the volume turns to an extensive discussion of the evolution of tourism as a use of the landscape. Here various authors point out, from different points of view, the development of nature-based tourism as an economic activity. Tourism in the region is almost totally dependent on a high-quality natural environment and formalized protection of landscapes through both governmental and private reserves. This section also documents the growing recognition that many landscape-dependent resources and processes are transboundary in character, crossing both regional and international boundaries. Their protection requires international level collaboration. Readers will find a chapter that examines the "stories" about the Tompkins donation to the Chilean government of a sizeable land donation particularly valuable given its contemporary public interest.

The final section focuses on the future and innovative ways of managing land use, particularly tourism development and activities. The chapters provide value with alternatives to conventional planning based on the highly reductionistic rational comprehensive planning approach that dominates conservation planning worldwide. Here we learn more about collaboration and imagining the future. Enhancing resilience is also brought more sharply into focus. This goal is extremely difficult to achieve, yet necessary to conservation-based resource development in the uncertain and changing world in which we live and work. The discussions provide hints and ideas about alternative ways that we can apply policy and learn from our mistakes.

Patagonia is an ideal "natural laboratory" to study conservation-based development: it is rather remote, in the sense of being hard to get to and generally out of people's day-to-day thinking; yet it's connected to some of the most important world events with respect to privately influenced conservation in recent history. While external pressures and amenity migration are on the rise, population density levels and industrial development are still relatively low, so the economic and political setting is not as complex and multifaceted as Yellowstone or Kruger, the Rocky Mountains of North America, or the Great Dividing Range of Australia. This means that in Patagonia, the effects of different geographic imaginaries are relatively easy to see.

What I like most about this book is that every chapter shows a people-environment link, instead of treating them as independent entities. People's actions do affect the environment and vice versa. To enhance resilience, we need to find the leverage points in this inextricable relationship. The search for sustainability is neither easy nor simple. An answer may lie just beyond the horizon. Works like this book move us forward in this search.

Professor Emeritus   Stephen F. McCool
University of Montana,
Missoula, MT, USA

Preface

This book is for researchers, graduate students, and professionals who are interested in better understanding conservation-based development, and its complexities with relation to core and peripheral areas, neoliberalism, sustainable development, governance, and territorial transitions. This book documents the evolution of strategic initiatives to foster conservation-based development in the southern reaches of Argentina and Chile. It examines the approaches and challenges associated with contemporary conservation-based development, and offers strategies for building resilience and sustainability, contextualized through Patagonia's varied and volatile circumstances of climate change, environmental and human health, and geographic periphery.

The foundations of contemporary Patagonian social-ecological systems (SES) evolved during the nineteenth century, when Argentina and Chile emerged as independent states, with formal colonization and frontier expansion efforts that extended the length of the southern cone periphery, including claims to strategic points of control for the Strait of Magellan. Recent human settlement, emergent national development strategies, formal colonization, and frontier expansion efforts designed to exploit natural resource-rich territories have led to the manifestation of Patagonia as an ideal region for conservation-based development.

This work applies an SES lens to conservation-based development in Patagonia, bringing together authors with historical, contemporary, and future-oriented perspectives, to increase understanding of the social and environmental implications of nature-based tourism and other forms of conservation-based territorial development. By focusing on Patagonia (as a region) and its various forms of conservation-based development, this book contributes one of the first collections of peer-reviewed South American-based lessons and will be valuable to researchers and practitioners, both locally and around the world, seeking to better understand complex interconnections between social and ecological environments, and pursue a similar path to resilience and sustainability.

Reviewers for the book shared,

- "Patagonia offers a fascinating regional perspective on conservation-based development, yet literature is limited. Further, much of what exists is written by scholars outside of the region, and this [book] integrates local researchers, which enhances the scholarly integrity of the proposal and contributes importantly to the field overall."
- "The shared emphasis on human well-being, social resiliency, economic viability and biodiversity conservation is notable and has been called for in numerous academic and governmental forums."
- "As a former graduate student researcher in the region, I scoured available materials in English and Spanish on this topic, and found scattered pieces, but no synthesis book of this kind. I imagine any graduate student entering the field would be thrilled to find this book."

Coyhaique, Chile   Trace Gale-Detrich
Whitewater, WI, USA   Andrea Ednie
Missoula, MT, USA   Keith Bosak

Acknowledgements

The proposal for this book went through a rigorous process of peer review before it was accepted by Springer and the series editors, and we would like to acknowledge that work and our appreciation of the reviewer suggestions. During the development of this book, each of the chapters also underwent several rounds of single-anonymous peer review and editor feedback that focused on the validity, quality, and originality. During this process, scientists working in the same fields as the authors reviewed each chapter submission, offering feedback on ways in which the authors could improve their acknowledgement and incorporation of other relevant works, the design and validity of the methods, the presentation of their data, and conclusions that were supported by the data. While some of the chapters could still contain inaccuracies, we believe our peer review process has greatly improved the overall quality of the book and its 18 chapters, ensuring that the valuable research shared by the authors complies with scientific standards and includes the necessary information richness to enable replication in other contexts and settings. We truly appreciate the hard work of all whom participated in this process, including the reviewers, the editors, and of course, the authors.

We are especially grateful to Andrés Adiego, of the Centro de Investigación en Ecosistemas de la Patagonia (CIEP) and the Department of Geography and Territorial Planning and Management of the Universidad de Zaragoza, for his elaboration of the cartography for Chaps. 1, 3, 6, 8, 11, 12, 13, and 18, and his general editorial and coordination support. We also appreciate Zac Hummel, of the University of Montana, for his dedicated editorial and coordination support. And we would like to acknowledge our family, friends, and colleagues who have supported the process of this book's development, and in particular, Dr. Steve McCool, Ms. Lee Ann Gale, Dr. Karen Beeftink, and Dr. Cesar Mendéz.

This work was supported by Chile's National Research and Development Agency (ANID) under ANID's Regional Program R17A10002; the CIEP R20F0002 project; the NODOSLN0002 project; ANID FONDECYT Regular 1230020; and the CHIC-ANID PIA/BASAL PFB210018.

About This Book

Tourism and conservation-based development in the periphery: Lessons from Patagonia for a Rapidly Changing World is part of the Natural and Social Sciences of Patagonia Springer book series. This work applies a social-ecological lens to conservation-based development in Patagonia, bringing together authors with historical, contemporary, and future-oriented perspectives, to increase the understanding of the social and environmental implications of nature-based tourism and other forms of conservation-based development. It provides researchers and graduate students with a robust collection of theoretical constructs and methodological tools that can be applied to their own place-based research interests.

Part I: Evolution of the *Green Economy* in Patagonia documents the evolution of strategic initiatives to foster conservation-based territorial development in the southern reaches of Argentina and Chile. Chapter 2 draws on political ecology scholarship, contributing to an emerging critical assessment of Francisco Moreno's land donation to the Argentine federal government in 1903, that one an initiating point for Patagonian conservation. Chapter 3 explores evolutions in environmental values, utilizing case study methods to examine the contemporary social reassessment of Chilean Patagonia landscapes, in which social, cultural, and ancestral forces are reasserting meaning and value to nature, and orienting territories around the conservation of ecosystems that are currently highly threatened by predominant (neo) extractivist regimes of capital accumulation. Chapter 4 employs the social imaginary framework, discourse, and institutional analysis, to conduct an historical analysis of local, national, and international influences regarding the way nature and tourism are conceived and managed in Argentine national protected areas. Chapter 5 combines critical, post-structural, and posthumanist social science perspectives with geographic imaginaries theory to develop a critical analysis of land privatization in western Patagonia. Historical data, contemporary secondary data, and land tenure transactions facilitated the production of narratives, land privatization trends, and contemporary tourism industry practices. Chapter 6 reflects extensive bibliographical review, document analysis, and fieldwork that was conducted through numerous periodic stays in Pewenche communities of northwestern Patagonia, applying geographical imaginaries theory to analyze the dual processes of exploitation/protection of the Araucaria, in the mountainous communes of Lonquimay and Alto Bío-Bío, where 54% and 83% of the population identify themselves as Pewenche, respectively.

Part II: Contemporary *Conservation-Based Development: Challenges for Green Integration* examines the potential, approaches, and challenges associated with contemporary conservation-based development in Patagonia. Informed by relational spatial theory for the production of space, Chap. 7 employed intrinsic case study methods, including interviews, participant observation, field visits, and documentary analysis, to understand how sustainability was being considered within a cross-border nature-based tourism circuit of capital accumulation. Sensitized by the literature on supporting and inhibiting factors of transboundary collaboration and conservation, Chap. 8 employs semi-structured interviews to explore stakeholder perspectives on key factors influencing transboundary conservation collaboration. Chapter 9 employs narrative analysis grounded in social constructivism to trace the formation of narratives developed by researchers, conservation entities, politicians, and other actors who have given meaning to Douglas and Kristine Tompkins effort to create protected areas in Chilean Patagonia and Argentina. Chapter 10 employs a collective case study approach to explore how *Modernization, Transformation,* and *Control* Sustainable Development imaginaries and trajectories interacted, nuanced, and mediated the approaches of neoliberal development initiatives and the conflicts which surrounded them. Chapter 11 examines sustainable tourism development causal chains within the Chilean Cerro Castillo and Torres del Paine National Parks, using Ante Mandić's (2020) conception of the *Drivers, Pressures, State, Impact, and Response (DPSIR) model.* Chapter 12 employs natural landscape concepts within contemporary landscape theory to examine how prospective national tourists might perceive visible salmon aquaculture infrastructure within Puyuhuapi's landscapes, through online A/B testing of two hypothetical experience scenarios ($n = 804$ responses).

Part III: Building Resilience and Sustainability offers strategies for building resilience and sustainability, contextualized through Patagonia's varied and volatile circumstances of climate change, environmental and human health, and geographic periphery. Chapter 13 presents a recent regional project developed to improve tourism governance in and around the protected areas, including in-depth description of the process employed to develop and test Local Tourism Councils. Chapter 14 applies a measurement model, incorporating natural risk assessment, to assess three fundamental pillars of resilience (capabilities, ownership, and connections), as a mechanism to relate resilience capacity to the territorial context. Chapter 15 presents an innovative process, including stakeholder identification, semi-structured survey interviews, and document analysis, to develop a matrix of weighted criteria to assess the potential for sustainable ST development within a destination. Chapter 16 discusses the merits of harmonious relationships between people, society, and nature, and their potential to help address increasing societal vulnerabilities, through the nature bathing initiative, which has been developed by the Chilean National Forestry Corporation (CONAF) within its Nature for Everyone program. Chapter 17 compares Indigenous and Western paradigms for sustainability as a framework to evaluate different alternative living projects. Informed by biocultural philosophy and ethics and neoliberal/post-neoliberal paradigms, Chapter 18 employs a mixed-methods intrinsic case study of three initiatives taking place in Chilean Patagonia to understand they might be integrated to achieve holistic conservation-based development through ethical travel experiences rooted in subantarctic bioculture. Methods combine secondary data analysis, open-ended interviews, participant observation, field visits, and a systematized geo-literature review and geobibliometric analysis, which evaluated and situated the scientific production that has occurred in two subantarctic natural laboratories during recent decades.

Contents

Contributors

Andrés Adiego Centro de Investigación en Ecosistemas de la Patagonia (CIEP), Sustainable Tourism Research Line, Human-Environmental Interactions Group, Coyhaique, Chile

Universidad de Zaragoza, Department of Geography and Territorial Planning, Zaragoza, Spain

Christopher B. Anderson Universidad Nacional de Tierra del Fuego (UNTDF) Instituto de Ciencias Polares, Ambiente y Recursos Naturales (ICPA) and Consejo Nacional de Investigaciones Científicas y Técnicas (CONICET) Centro Austral de Investigaciones Científicas (CADIC), Ushuaia, Tierra del Fuego, Argentina

Jessica L. Archibald Northern Arizona University, School of Earth and Sustainability, Flagstaff, AZ, USA

Pamela Bachmann-Vargas Wageningen University & Research, Environmental Policy Group, Wageningen, Netherlands

Andrea Baéz Montenegro Universidad Austral de Chile, Institute of Statistics, Valdivia, Chile

Centro de Investigación en Ecosistemas de la Patagonia (CIEP), Human-Environmental Interactions Group, Coyhaique, Chile

Heidi Blair University of Montana, College of Forestry and Conservation, Missoula, MT, USA

Keith Bosak University of Montana, College of Forestry and Conservation, Department of Society and Conservation, Missoula, MT, USA

Fabien Bourlon Centro de Investigación en Ecosistemas de la Patagonia (CIEP), Sustainable Tourism Research Line, Human-Environmental Interactions Group, Coyhaique, Chile

Université Grenoble Alpes, Institute of Urban Planning and Alpine Geography - UMR 5194, Grenoble, France

Ronald Cancino Salas Universidad de la Frontera, Department of Social Sciences, Temuco, Chile

José Coloma Zapata Universidad de la Frontera, Center for Social Research of the South, Temuco, Chile

Mara Dicenta William & Mary, Department of Anthropology and Institute for Integrative Conservation, Williamsburg, VA, USA

Andrea Ednie University of Wisconsin – Whitewater, College of Education and Professional Studies, Whitewater, WI, USA

Trace Gale-Detrich Centro de Investigación en Ecosistemas de la Patagonia (CIEP), Sustainable Tourism Research Line, Human-Environmental Interactions Group, Coyhaique, Chile

Cape Horn International Center (CHIC), Puerto Williams, Chile

Cecilia Gutiérrez Vega Universidad Austral de Chile, Faculty of Economics and Administrative Sciences - Institute of Tourism, Valdivia, Chile

Carla Henríquez V. Universidad de Magallanes, GAIA Antarctica Investigation Center, Punta Arenas, Chile

Diego Hernández Soto Universidad de la Frontera, Department of Social Sciences, Temuco, Chile

Gabriel Inostroza Villanueva Universidad Austral de Chile, Patagonia Campus, Coyhaique, Chile

Angel Custodio Lazo Álvarez Chilean National Forestry Corporation (CONAF), Management of Protected Wildlife Areas, Santiago, Chile

Pamela Maldonado Universidad de Magallanes, GAIA Antarctica Investigation Center, Punta Arenas, Chile

Nelson Martínez-Berríos Corporación Municipal de La Florida, Education Area, Santiago, Chile

Claudia Matus Pontificia Universidad Católica de Chile, Educational Justice Center, Santiago, Chile

Marcos Mendoza University of Mississippi, Department of Sociology & Anthropology, University, MS, USA

Sanober R. Mirza University of Montana, College of Forestry and Conservation, Department of Society and Conservation, Missoula, MT, USA

Lorna Moldenhauer Ortega Universidad de Magallanes, Coyhaique University Center, Coyhaique, Chile

Manuel Mora Chepo Universidad de la Frontera, Department of Social Sciences, Temuco, Chile

María Dolores Muñoz Rebolledo   Department of Urban Planning, Faculty of Architecture, Urbanism, and Geography, Universidad de Concepción, Concepción, Biobío Region, Chile

Matías Navarrete Almonacid   Universidad de la Frontera, Department of Social Sciences, Temuco, Chile

Jorge Olea-Peñaloza   Universidad Católica de Temuco, Environmental Science Department, Temuco, Chile

Guillermo Sebastián Pacheco Habert   Universidad Austral de Chile, Faculty of Economics and Administrative Sciences - Institute of Tourism, Valdivia, Chile

Sabrina Elizabeth Picone   Consejo Nacional de Investigaciones Científicas y Técnicas (CONICET), Santa Cruz Research and Transfer Center, Río Gallegos, Argentina

Brenda Sofía Ponzi   Consejo Nacional de Investigaciones Científicas y Técnicas (CONICET), Santa Cruz Research and Transfer Center, Río Gallegos, Argentina

Fiorella Repetto Giavelli   Torres del Paine Legacy Fund, Punta Arenas, Chile

Laura Rodríguez   Universidad Austral de Chile, Architecture and Urban Planning, Faculty of Architecture and Arts, Valdivia, Chile

Fulvio Rossetti   Faculty of Engineering, School of Architecture & Centro de Innovación de Ingeniería Aplicada (CIIA), Universidad Católica del Maule, Talca, Región del Maule, Chile

Adriano Rovira   Universidad Austral de Chile, Faculty of Economics and Administrative Sciences - Institute of Tourism, Valdivia, Chile

Ricardo Rozzi   University of North Texas, Department of Philosophy & Religion Studies, Denton, TX, USA

Universidad de Magallanes, Sub-Antarctic Biocultural Conservation Program, Puerto Williams, Chile

Cape Horn International Center (CHIC), Puerto Williams, Chile

Alejandro Salazar-Burrows   Pontificia Universidad Católica de Chile, Institute of Geography. UC Patagonia Station of Interdisciplinary Research, Santiago, Chile

Laura Sánchez Jardón   Universidad de Magallanes, Coyhaique University Center, Coyhaique, Chile

Cape Horn International Center, Puerto Williams, Chile

Alejandro F. Schweitzer   Consejo Nacional de Investigaciones Científicas y Técnicas (CONICET), Santa Cruz Research and Transfer Center, Río Gallegos, Argentina

Bastien Sepúlveda   Research affiliate at UMR SENS, Laboratory, Montpellier, France

Christopher Serenari Texas State University; Department of Biology, San Marcos, TX, USA

Florencia Spirito Universidad de Aysén, Department of Natural Sciences and Technology, Coyhaique, Chile

Pablo Szmulewicz Universidad Austral de Chile, Faculty of Economics and Administrative Sciences - Institute of Tourism, Valdivia, Chile

Jennifer M. Thomsen University of Montana, College of Forestry and Conservation, Department of Society and Conservation, Missoula, MT, USA

Robinson Torres-Salinas Department of Sociology, Faculty of Social Sciences & Department of Territorial Planning, Faculty of Environmental Sciences, Universidad de Concepción, Concepción, Biobío Region, Chile

Santiago Urrutia Reveco Instituto de Geografía, Universidad de Buenos Aires (UBA)-CONICET, Buenos Aires, Argentina

Alejandro E. J. Valenzuela Universidad Nacional de Tierra del Fuego (UNTDF) Instituto de Ciencias Polares, Ambiente y Recursos Naturales (ICPA) and Consejo Nacional de Investigaciones Científicas y Técnicas (CONICET) Instituto de Ciencias Polares, Ambiente y Recursos Naturales (ICPA), Ushuaia, Tierra del Fuego, Argentina

Germaynee Vela-Ruiz Figueroa Vela-Ruiz Consultorías Ambientales SpA, Punta Arenas, Chile

Katerina Veloso Universidad Austral de Chile, Austral Patagonia Program & Faculty of Economics and Administrative Sciences - Institute of Tourism, Valdivia, Chile

Rodrigo Villa-Martínez Universidad de Magallanes, GAIA Antarctica Investigation Center, Punta Arenas, Chile

Cape Horn International Center (CHIC), Puerto Williams, Chile

Hugo Marcelo Zunino Universidad de La Frontera, Social Sciences Department / Social Sciences Nucleus, Temuco, Chile

About the Editors

Trace Gale-Detrich is a senior researcher in CIEP (Center for Investigation in Ecosystems of Patagonia), where she has worked since 2009 and is currently serving as the coordinator for both the Sustainable Tourism Research Line and the Human-Environmental Interactions group (HEI). Her research interests focus on human/human and human/non-human interactions within the ecosystems of Patagonia, with most of her work addressing the intersection of conservation and development. Her areas of focus include human values, perceptions, affect, and experiences, with the goal of understanding how these human dynamics converge with regards to environmental stewardship, transdisciplinary governance, human development/wellbeing, and the integration of protected areas with their bordering lands and communities; especially in contexts of territorial transition, involving social and climate change.

Andrea Ednie is currently interim associate dean of the College of Education and Professional Studies at the University of Wisconsin-Whitewater, where she has worked as a faculty member since 2014. Her research examines protected area planning and stakeholder values, the role of soundscapes within Healthy Parks Healthy People processes, and wellness benefits associated with connections to nature and protected area experiences within Chilean Patagonia and protected areas within the Midwestern US. She also studies motivations, sense of place, experiences, and behavior choices within the contexts of outdoor recreation management and outdoor exercise. Dr. Ednie is currently the Young Scholars program coordinator for the Western Society for Kinesiology & Wellness (WSKW).

Keith Bosak currently serves as a professor in the Department of Society and Conservation at the University of Montana, USA. Bosak's research interests are broadly centered on the intersection of conservation and development, and as such, he often studies nature-based tourism and sustainable tourism in the context of development and protected areas. He has conducted research on ecotourism and environmental justice in India, scientific tourism in Chile, and Geotourism in Montana. Aside from tourism, Dr. Bosak has conducted research on climate change impacts and adaptations among tribal populations in the Himalaya, private protected areas in Chile, and conservation and development initiatives in Montana. Dr. Bosak also conducts workshops on protected area planning and management around the world.

Abbreviations

ABB	Alto Bío Bío
ADI	Indigenous Development Areas [Áreas de Desarrollo Indígenas]
APN	Argentine National Park Administration
AR	Argentina
CA	Chilean Longitudinal Austral Highway or Southern Highway or Route 7 [Carretera Austral]
CBD	Conservation-Based Development
CCNP	Cerro Castillo National Park
CFL	Community for Life
CH	Cape Horn
CIEP	Center for Investigation in Ecosystems of Patagonia [Centro de Investigación en Ecosistemas de la Patagonia]
CL	Chile
CONADI	Chilean National Indigenous Development Corporation [Corporación Nacional de Desarrollo Indígena]
CONAF	Chilean National Forestry Corporation [Corporación Nacional Forestal de Chile]
CRI	Chilean Indian Registration Committee [Comisión Radicadora de Indígenas de Chile]
DMO	Destination Management Organization
DPN	Argentine National Park Directorate
DPSIR	Drivers, Pressures, State, Impact, and Response model
EIA	Environmental Impact Assessment
EIS	Environmental Impact Study
FDE	Foundation for Deep Ecology
FEP	Field Environmental Philosophy
GEF	Global Environmental Facility
INGO	International Non-governmental Organization
IRP	Iberá Rewilding Program
ISTN	International Network for Scientific Tourism Research and Development

IUCN	International Union for the Conservation of Nature
LTC	Local Tourism Councils
MCIP	Marine-Coastal Interjurisdictional Parks
MIP	Marine Interjurisdictional parks
MPA	Marine Protected Areas
NBT	Nature-Based Tourism
ND	Argentine National Decree
NGOs	Non-governmental organizations
NL	Argentine National Law
NP	National Park
OEP	Omora Ethnobotanical Park
PAs	Protected Areas
PPAs	Private Protected Areas
PUP	Public Use Plan
QRE	Quick Risk Estimation
SBAP	Chilean Biodiversity and Protected Areas Service
SD	Sustainable Development
SEA	Chilean Environmental Assessment Service [*Servicio de Evaluación Ambiental de Chile*]
SEIA	Chilean Environmental Impact Assessment System [*Sistema de Evaluación de Impacto Ambiental de Chile*]
SERNATUR	Chilean National Tourism Service [*Servicio Nacional de Turismo de Chile*]
SES	Social-ecological systems
SNAP	Chilean National System of Protected Areas (SNAP)
SNASPE	Chilean National System of Protected Wildlife Areas [*Sistema Nacional de Áreas Silvestre Protegidas por el Estado de Chile*]
SNL	Subantarctic Natural Laboratories
SRL	San Rafael Lagoon
ST	Scientific Tourism
TM	Titles of Merced
TPNP	Torres del Paine National Park
UNESCO	United Nations Educational, Scientific and Cultural Organization
US	United States of America
WST	World System Theory
ZOIT	Chilean Tourism Interest Zone [*Zona de Interés Turística de Chile*]

List of Figures

List of Tables

Chapter 1
Tourism and Conservation-Based Development in the Periphery

Keith Bosak, Trace Gale-Detrich, and Andrea Ednie

Abstract This overview introduces the major concepts and themes that are addressed within the three parts of this book (Part I: Evolution of the green economy in Patagonia; Part II: Contemporary conservation-based development: challenges for green integration; and Part III: Building resilience and sustainability). Fundamental concepts, including core– periphery interactions, conservation-based development, and imaginaries, are described, and we explain how they are prominent themes within this book's chapters. A geographical overview of the studies represented within this book is provided, along with an outline of the geopolitical and historical contexts of the imaginary region of Patagonia. The three book parts are contextualized within the context of some of the major challenges facing nature-based tourism in Patagonia, including recent and upcoming initiatives that may contribute to sustainability and resilience. This chapter concludes with an overview of this book's contributors.

Keywords Patagonia · Conservation-based development · Tourism · Sustainability · Core-periphery

The original version of the chapter has been revised. A correction to this chapter can be found at https://doi.org/10.1007/978-3-031-38048-8_19

K. Bosak
Department of Society and Conservation, College of Forestry and Conservation, University of Montana, Missoula, MT, USA
e-mail: keith.bosak@umontana.edu

T. Gale-Detrich (✉)
Centro de Investigación en Ecosistemas de la Patagonia (CIEP), Sustainable Tourism Research Line, Human-Environmental Interactions Group, Coyhaique, Chile

Cape Horn International Center (CHIC),
Puerto Williams, Magallanes and Chilean Antarctica Region, Chile
e-mail: tracegale@ciep.cl

A. Ednie
University of Wisconsin – Whitewater, College of Education & Professional Studies, Whitewater, WI, USA
e-mail: edniea@uww.edu

1

1.1   Geographic Context for "Tourism and Conservation-Based Development in the Periphery: Lessons from Patagonia for a Rapidly Changing World"

This chapter offers an introduction to the collection of peer-reviewed essays and research projects presented within this book, Tourism and conservation-based development in the periphery: Lessons from Patagonia for a rapidly changing world. We would like to begin by clarifying some essential aspects of Patagonia, which is an imagined territory with boundaries that are subject to ongoing interpretation and dispute. Although humans have inhabited the remote, peripheral region of Patagonia since approximately 13,000 BP (Borrero et al. 2019), this isolated, peripheral zone in southern South America was essentially undeveloped with no fixed political borders until the last few centuries.

Patagonia represents the last region of the Americas to be settled by humans. Due to dramatic differences in the geographic and environmental characteristics of the territory, it was likely settled in a dispersed manner, through a complex spatial-temporal ranking (Borrero and Franco 1997). The earliest human settlement sites in Patagonia are located within the Santa Cruz Province of Argentina, along the central plateau. Sites along the Andes in western Chilean Patagonia developed much later (Méndez 2013; Méndez et al. 2018). Perhaps, these patterns resulted from glacial retreat by the end of the last ice age, which determined the timeframe for the formation of viable ecosystems to support biodiversity and human life (Méndez et al. 2018).

This relatively recent peopling defines the region as one of the most recently developed territories for human–nonhuman interactions and as such, an extremely interesting part of the world in which to work and live. Today, Patagonia is generally accepted as encompassing the southernmost regions of South America, from around the 38° latitude south to 56° latitude south. The limits of what is, and what is not Patagonia, continue to evolve through a series of social and political iterations that are subject to ongoing interpretation and dispute (Carte and Zunino 2022; Navarro Floria 2007; Warren 2013). This book does not attempt to define the limits of Patagonia, nor to posit an argument for how that should occur. Nevertheless, for the purposes of orienting the works contained in this book within the Patagonian geographic imaginary, we have chosen an interpretation of how Patagonia is often defined in contemporary conversations and debates (Fig. 1.1). Today, the eastern and western divisions of Patagonia are clearer as a result of ongoing geopolitical processes of border definition that use the southern Andes' continental divide, and the flow of rivers to the Pacific and Atlantic Oceans, to mark the Chilean–Argentine border. This same border defines eastern and western Patagonia.

Figure 1.1 divides contemporary Patagonia into three imaginary zones: north, central, and south (black dashed lines). In accordance, northern Patagonia would extend from 38° latitude south to 42° latitude south, including the Argentine provinces of Río Negro and Neuquén, most of the Chilean Araucanía Region and most of the Los Lagos Region, to Hornopirén. Central Patagonia would extend from 42° latitude south to 46° latitude south, encompassing the Argentine province of Chubut

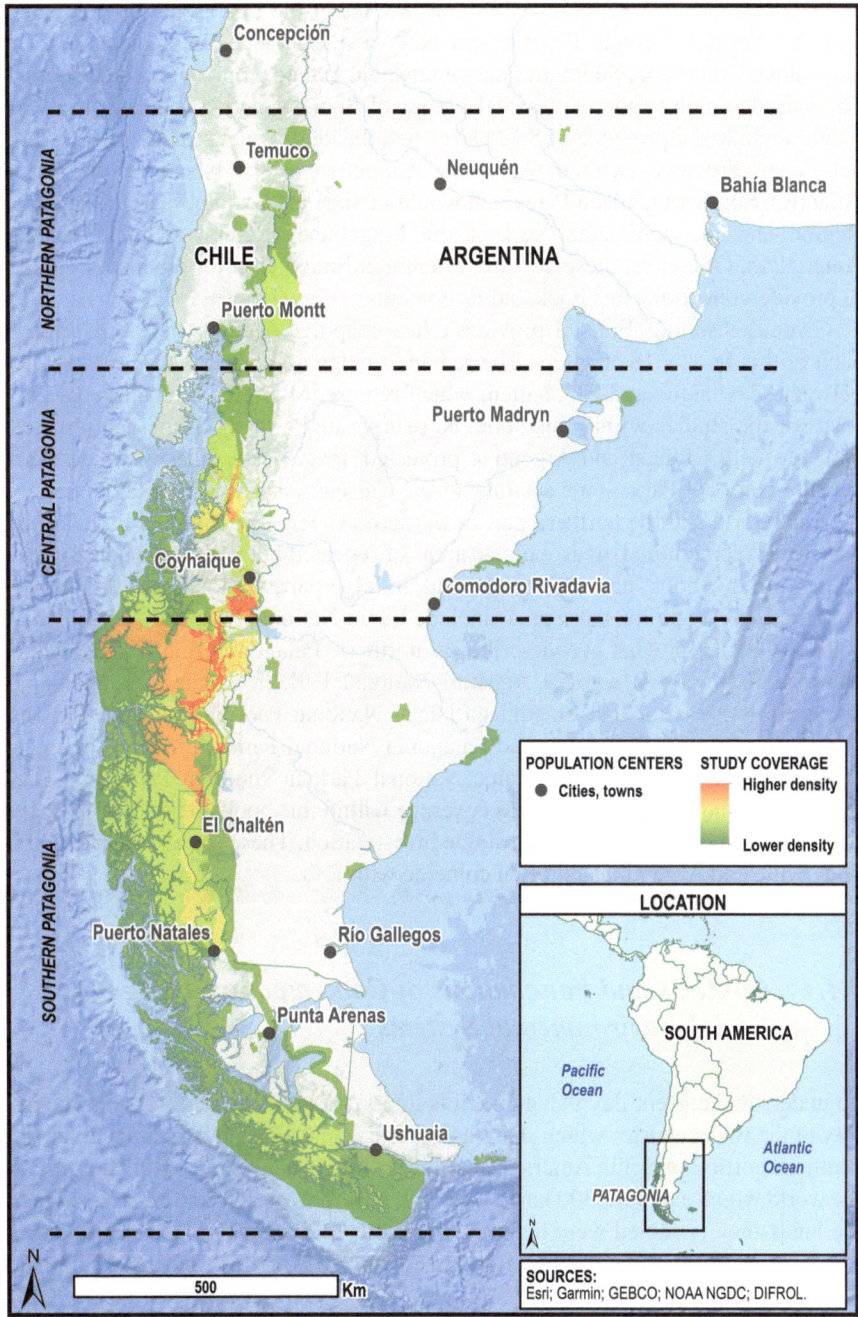

Fig. 1.1 Geographic dispersion of the research presented in this book

and Chilean territory within the southern reaches of the Los Lagos Region, south from Hornopirén through the northern parts of the Aysén Region, including the Coyhaique Commune. Southern Patagonia would extend from 46° latitude south to 56° latitude south to the continental border of South America. In Argentina, this would include the province of Santa Cruz and the continental reaches of the Tierra del Fuego Province (without Argentine Antarctica or the Islands of the South Atlantic). Southern Chilean Patagonia would include the southern part of the Aysén Region and the continental part of the Magallanes Region (without Chilean Antarctica). Of course, these are only referential limits, used for practical purposes to provide context for this book and its contents.

Given this setting, Fig. 1.1 provides a heat map that overlaps the study areas of each of this book's 18 chapters. The red and orange colors show the areas located between Coyhaique and El Chaltén, which represent the predominant geographic focus of the studies within this book. Not surprisingly, many of these study areas coincide with Chilean and Argentine protected areas (PAs). These include Cerro Castillo National Park, some sections of the Carretera Austral, San Rafael Lagoon National Park, and the northern part of Bernardo O'Higgins National Park. Yellow and light green colored areas indicate areas of repeated but with less intense coverage within this book, and also correspond, in large part, with national PAs in both countries. In Argentina, these areas include Lanín National Park and Nahuel Huapi National Park, in what we describe as northern Patagonia, and Los Glaciares National Park and Alberto de Agostini National Park in southern Patagonia. In Chile, these areas include Magdalena Island National Park and Queulat National Park within central Patagonia, and Patagonia National Park, Bernardo O'Higgins National Park, and Torres del Paine National Park in southern Patagonia. Dark green zones indicate areas with less coverage within this book: generally relating to the study area designated within a single investigation. These are also distributed in both Chile and Argentina and often coincide with PAs.

1.1.1   Contexts and Foundations of Contemporary Patagonian Social-Environmental Systems

To understand current-day Patagonia, it is important to understand some of the history of the two countries which share its governance. Chile and Argentina divide the southern portion of South America, sharing the third longest international border in the world which extends 5300 km (Thies 2001). Before these countries were formed, the lands they occupied were home to numerous Indigenous groups whose ancestors migrated from Asia, through Beringia, down the Americas, reaching Chile and Argentina sometime between 18,500 and 15,000 years ago (Prates et al. 2013). During the sixteenth century, Chile and Argentina were colonized by Europe through distinct processes (Cervantes 2020). The Chilean conquest began in 1541 as part of a southern extension of the Peruvian conquest. Argentina was initially settled overland from Peru, Chile, and the Atlantic, with a permanent Spanish colony being established in Buenos Aires in 1580. Both Chile and Argentina were colonies of Spain until the intervention of Napoleon in 1808 which left Spain without a clear ruler; thus, leaving the colonies to forge their own provisional governments and begin moving toward autonomy. Officially, Argentine independence was declared on July 9, 1816, while Chile's official declaration was issued on February 12, 1818 (Cervantes 2020).

The foundations of contemporary Patagonian social-environmental systems evolved during the nineteenth century when Argentina and Chile emerged as independent states with their own formal colonization and frontier expansion efforts that extended the length of the southern cone periphery, including claims to strategic points of control for the Strait of Magellan (Thies 2001). Although the Patagonia boundary was formally agreed to in 1881, conflicts and tensions about the territory have persisted and taken on new forms (Thies 2001). Extractive industries like mining, forestry, and ranching have been used by both countries to establish, occupy, and develop their respective territories and borderlands. More recently, *green*, or *conservation-based development* (CBD), has emerged as an alternative development mechanism that is used for similar ends. This mechanism has manifested through purposeful protected area designations and an increasing emphasis on rural transitions, positioning nature-based tourism as a central, and often controversial element, within territorial strategies.

1.1.2 Patagonia as a Peripheral Territory

Immanuel Wallerstein's critical World System Theory (WST), developed during the 1970s, broke away from functionalist tradition by introducing a new conceptual understanding of capitalism, industrialization, and world order through a Marxist-inspired lens (Hier 2001; Lennerfors et al. 2015). WST rejected the division of the world's countries within first, second, and third worlds, instead approaching social analysis and change from the standpoint that, "there is only one world connected by a complex network of economic exchange relationships" (Sorinel 2010, p. 220). Drawing on dependency theory concepts (Frank 1969), WST focused on the place-based social and economic inequalities that arise in the world as a result of capitalism's need to expand geographically in order to survive (Lennerfors et al. 2015). He described capitalist exploitation in terms of three geographic concepts: core (i.e., developed, rich countries), periphery (i.e., poor countries), and in-between—semi-periphery—positing that the wealth of core zones depends on the raw materials, goods, and labor of the periphery (Lennerfors et al. 2015).

In the years since its inception, WST has been extended and applied to various scales and contexts. For example, Hornborg connected ecology with WST in order to conceptualize the environmental effects and implications of capitalism for core, periphery, and in-between territories (Hornborg 1998, 2001, 2009). Responding to increased social-spatial inequality, spatial research has re-engaged with the concept of peripheralization in recent years (Bachmann-Vargas and van Koppen 2020; Kühn

2015; Kühn et al. 2017). Kühn (2015) described modern consideration of peripheralization as related to process, rather than place. As such, peripheralization can apply to any spatial process in which political, social, economic, or communicative processes lead to marginalization from a structural perspective (i.e., social marginality and poverty, political dependency or exclusion, economic polarization).

Both Chile and Argentina can be considered as semi-peripheral territories (Cairói-Céspedes and Palacios Cívico 2022; Önis 2006). They are generally considered industrialized and contribute to the manufacture and export of a spectrum of goods. While they offer their inhabitants diverse economic opportunities, they also exhibit wide gaps between social conditions, depending on economic situations. Patagonia (both Argentina and Chile) can be characterized as a periphery within a semi-periphery (Núñez 2015; see Chap. 7, Schweitzer et al.). The region is rich in natural resources, including an abundance of public land, freshwater reserves, forests, minerals, fjords, fisheries, biodiversity, pampas, and grasslands, among others. Patagonia has low population densities and few urban centers (Gale and Ednie 2019; Núñez 2015). It is less developed than the central regions of both Chile and Argentina, including its social support systems (e.g., education, health care, arts), and is dependent on (and some would say, exploited by) the services and funding provided by the core areas of Chilean and Argentina semi-peripheries, and foreign interests of the international core (Gale et al. 2013). Although these conditions manifest differently in Argentina and Chile as a result of differences in federal and republic systems of government, similarities in lived experiences have been observed (Gale et al. 2013; Lambert and Scribner 2021).

Recent human settlement, emergent national development strategies, formal colonization, and frontier expansion efforts designed to exploit natural resource-rich territories have led to the manifestation of Patagonia as an ideal region for CBD. In Patagonia and other peripheral territories around the world, CBD often marginalizes peripheral areas as it is an outcome of the interaction between the resources of the periphery and the needs, wants, and power of the core (Abrams et al. 2012; Blair et al. 2019; Gale et al. 2013; Núñez et al. 2020). In many ways, CBD is another mechanism by which the core exploits the periphery (Gale et al. 2013; Núñez et al. 2020). Placed within the current context of climate change, global biodiversity crisis, as well as the complex network of international organizations and agreements that oversee and manage these crises; however, CBD becomes much more complex and dynamic, as myriad relationships and flows of power emerge and evolve (Louder and Bosak 2022; Núñez et al. 2020).

1.1.3   Government, Power, and Neoliberalism in Chilean and Argentine Patagonia

Both Argentina and Chile have experienced long periods of democratic rule and multiple periods of dictatorship. During the twentieth century, Argentina experienced six coup d'états, in 1930, 1943, 1955, 1962, 1966, and 1976 (Ormaechea

2021; Vitale 2009). While the first four involved interim rule, during the 17-year period from 1966 to 1983, the dictatorships ruled through a bureaucratic-authoritarian state (Vitale 2009). During the twentieth century, Chile experienced two coup d'états: one in 1924 that resulted in a new constitution the following year, and the other involving a violent takeover and prolonged authoritarian military dictatorship that lasted 17 years, between 1973 and 1990 (Lambert and Scribner 2021; Thies 2001; Warren 2013).

Current systems of government within Argentina and Chile are very distinct. Chile is governed through a representative democratic republic, as a unitary state in which most of the governing power resides centrally with its president. Presidents are elected to a 4-year term and may be re-elected twice, though their terms must be noncontinuous. The president appoints and leads the Cabinet of Ministers, who in turn manage the administration of the country. The country is divided into 15 regions, 54 provinces, and 345 communes, which are governed at the local level through municipal governments. Legislative power is shared by the government and the bi-chamber National Congress, composed of the Chamber of Deputies and the Senate. The Chamber of Deputies is formed by 155 directly elected members who are appointed for 4 years. The Senate is formed by 50 directly elected members who are appointed for alternating four-year terms, with half of its members renewed every 4 years (Antía 2019; Niedzwiecki and Pribble 2017).

In contrast, Argentina has a representative, republican, federal union. Its constitution was implemented in 1853 and modeled on the United States' constitution. In 1994, it was amended to provide for consecutive presidential terms, though the 1853 document remains largely intact. Argentina is divided into 23 provinces with Buenos Aires as its federal capital district. While executive power resides with the president, who is elected to terms of 4 years (with a maximum of two consecutive terms), their role is much more limited than in Chile. They lead the armed forces, make civil and judicial appointments, and lead the Cabinet of Ministers, who administer federal public services. The legislature in Argentina consists of two houses, including a 257-seat Chamber of Deputies who serve four-year terms and a 72-seat Senate, who are elected for six-year terms. Each province has its own constitution, government, legislative, and judicial branches, and retains all powers not specifically set apart for the federal government (Warren 2013).

Chile has had several constitutions since establishing its independence (Kennedy 2017). The Constitution of 1833 was in place until 1925, followed by another constitution that remained in effect until 1973 when the military dictatorship of Augusto Pinochet suspended it, along with Congress. In 1980, a constitutional referendum replaced the 1925 constitution with a new constitution, granting significant overarching powers to Augusto Pinochet as President of the Republic. Some of these powers were modified or eliminated after 1990 in the more than 70 amendments since Chile's return to democracy and the re-establishment of Congress (Kennedy 2017). Nevertheless, increasing recognition of the long-standing limitations of the 1980 constitution led to growing social unrest in Chile, culminating in a national plebiscite in 2020 in which 78% of voters called for a new Constitution. Subsequently, a diverse, democratically elected assembly drafted a new Constitution which was submitted for popular vote and overwhelmingly rejected by Chile's eligible voters. Thus, a new process is currently underway, involving two popular votes in 2023. The first will take place in May and will elect a 50-person Constitutional Council to draft a new Constitutional document, which will be submitted to a vote during a December referendum.

These historical differences have a great impact on the format and processes for conservation funding and governance in the two countries, particularly in the context of privately driven conservation. For example, in Argentina, the 1994 constitutional reform established the environment as a collective and legal good, defining that the people have the right and the obligation to preserve a healthy environment (Argentine National Ministry of the Environment and Sustainable Development 2017). Furthermore, Article 124 of this Constitution provided the provinces with legislative power over their natural resources, regardless of ownership. Protected area management and legality are regulated both at a national and at a provincial level depending on their jurisdiction (Myron et al. 2019a). Legal restrictions impede foreign land purchase, especially with respect to lands close to the national border (Myron et al. 2019b; Ponzi 2020). Civil code in Argentina does not define private protected areas or consider private conservation protection in perpetuity (Myron et al. 2019b). Therefore, to establish a new national park or amend an existing one, the Provincial government must agree to cede lands and control through its legislature, even when the lands are donated by private landholders (Ponzi 2020; see Chap. 7, Schweitzer et al.).

In Chile, the 1931 Forest Law provided the president with the power to establish national parks and reserves (Myron et al. 2019b). Land rights are disaggregated (e.g., freshwater, subsoil minerals, geothermal water and energy, coastal intertidal zone, real property), so there can be overlaps between parties and claims, in which some of the disaggregated rights supercede real property rights (Myron et al. 2019b). That said, private entities can hold legal title over land in Chile, and to date, when private lands are purchased in Chile (which is legal) and then donated back to the State, negotiations can, and have, occurred at a national level with the executive branch, without the necessity of local territorial involvement (Blair et al. 2019; Borrie et al. 2020; Gale and Ednie 2019; Myron et al. 2019b; see Chap. 10, Inostroza et al.).

One of the more complex issues that have arisen in Chile and Argentina with respect to CBD has involved perceptions and realities related to neoliberalism, especially as they pertain to government oversight of the environment and the inclusion of local communities and territories in protected area governance (Borrie et al. 2020; De Matheus e Silva et al. 2018; Gentes and Policzer 2022; Grugel and Riggirozzi 2012; Jones 2012; Latta and Aguayo 2012; McCarthy 2012; Önis 2006; Tabbush and Caminotti 2015). Neoliberal restructuring has played a major role in the modern development of both Argentine and Chilean Patagonia. Argentina was a rich agricultural exporter during the 1930s but did not maintain those trends after World War II. During authoritarian rule from 1966 to 1983, the country experienced a prolonged period of low growth and hyperinflation. The 1976 coup d'état was motivated as an attempt to overcome these economic obstacles and began to roll back high levels of state intervention and protectionism, to open up the Argentine economy (Ormaechea 2021; Undurraga 2015). Argentina returned to democracy with Raúl Alfonsin's election in 1983, and subsequently underwent extreme neoliberal reforms designed and implemented in the final years of the 1980s.

During most of the 1990s, Argentina experienced a prolonged period of high economic growth, with reduced levels of inflation. This period ended with recession from 1998 to 2002, punctuated by a major economic crisis in 2001. In the aftermath of this crisis, the country experienced massive social and political unrest which criticized populism and corruption within the Federal structure, exposing the risks of Argentina's neoliberal policies that emphasized heavy international borrowing and economic growth driven by short-term capital inflows (Önis 2006; Ormaechea 2021). As an example of the severity of unrest and distrust, there were five presidents between December 20, 2001, and January 2, 2002 (Ormaechea 2021). Subsequent economic policy in Argentina, led by the Kirchner governments, moved away from neoliberal tactics, with self-described post-neoliberal politics that promised social justice delivered through government-led and -controlled economic and social intervention (Ormaechea 2021; Undurraga 2015; Villalón 2007). During this time, researchers began to understand that neoliberalism and post-neoliberalism extended well beyond economic and political policy in the mind of the public (Grugel and Riggirozzi 2012; Ormaechea 2021; Undurraga 2015). For example, Grugel and Riggirozzi (2012) expanded post-neoliberalism to involve more than just economic aspects, saying, "It is also a call for a new kind of politics, rooted in and responsive to local traditions and communities, and an attempt to forge a new pact between society and the state" (p. 3). Nevertheless, since the death of Néstor Kirchner in 2010, Argentina has remained a country in dispute, marked by frequent social discord, crisis, political corruption, and uncertainty, suggesting that this form of post-neoliberal development is still very much in evolution (Elbert and Pérez 2018; Lublin 2021; Tabbush and Caminotti 2015; Undurraga 2015; Wylde 2016).

Much has been written about Chile's deployment of neoliberal policies and structures during the dictatorship of Augusto Pinochet and the relatively few changes that have occurred in these policies since Chile's return to democracy more than three decades ago (Gentes and Policzer 2022; Harvey 2005; Latta and Aguayo 2012). Chile's environmental oversight and controls remain diffuse, spread among at least 14 ministries and 28 public services at a national level, 16 regional governments, and 345 municipalities, resulting in a complex institutional scenario. Perhaps the biggest change to date for conservation and environmental concern relates to the implementation of Chile's Environmental Assessment System (SEIA) in 1997, though this system has also been met with ongoing question and debate (Latta and Aguayo 2012; OECD 2016).

In Chile, there is no single institution overseeing biodiversity conservation priorities nor a single public service dedicated to strengthening protected areas in an integrated manner (BIOFIN Chile 2017; Chilean Ministry of the Environment 2019). In addition to governance complexities, funding deficits are a long-standing issue for biodiversity conservation. In 2010, a joint report of the United Nations Food and Agriculture Organization (FAO) and Spain's national parks agency

(Organismo Autonomo Parques Nacionales, OAPN) identified significant protected area funding risks throughout Latin America. The report noted that, at the time, Chile invested only around US$0.95 per 10,000 m^2 in protected areas, representing one of the lowest rates of investment in Latin America (FAO/OAPN 2010).

During the decade that followed this influential report, work has focused on development of legal instruments, management systems, financial tools, and governance approaches that comprise the current institutional structure and financing system for conservation. In 2011, a law was presented to the Chilean congress to create the Biodiversity and Protected Areas Service (SBAP) and an integrated National System of Protected Areas (SNAP) to include all of Chile's public and private protected areas, both terrestrial and marine (Donoso 2019). Subsequent debate and modifications have focused on improving the proposed law, resulting in an amended submission in 2014 (Donoso 2019). The 2014 modification proposed funding based on a combination of private resources, public support, and income-generating mechanisms, relying heavily on the contributions of private entities and initiatives that will require modifications in law related to the tax treatment of charitable donations (Walker 2018).

Although much time has passed, the bill remains in active debate. It was approved by the Environment Committee in 2017, by the Finance Committee and the Senate Chamber in 2019, and also has the support of the Labor and Social Welfare Committee. In 2021, the Agriculture, Forestry, and Rural Development Committee of the Chamber of Deputies approved the bill. Next steps include the Chamber's vote and subsequent consideration by the Senate. As such, while big changes seem imminent for Chile's conservation funding and governance, it seems likely the country will continue to endorse neoliberal approaches as necessary for the successful designation and financing of Chile's protected areas, including accelerated development of protected area nature-based tourism concessions.

1.1.4 Conservation-Based Development

CBD is a key theme of this book and as such we begin with a discussion of this phenomenon and its relationship to peripheral areas. CBD is a process by which conservation serves as a catalyst for development. Oftentimes, conservation is embodied through PAs whether they be public or private. These PAs attract visitors who in turn spend money both at the site and in the surrounding areas. This spending spurs further development, particularly in the tourism and recreation sectors. As conservation-related tourism and recreation grow, local livelihoods change as do the character of communities. Eventually, the natural amenities of these regions attract new residents, be they full time or part time, that buy real estate and begin to change the very demographics of these areas. Central governments around the world have keyed into this phenomenon and are now promoting CBD in their peripheral areas as a way to boost gross domestic product. The relationship between CBD and peripheral areas is predicated on the pattern of conservation and in particular PAs.

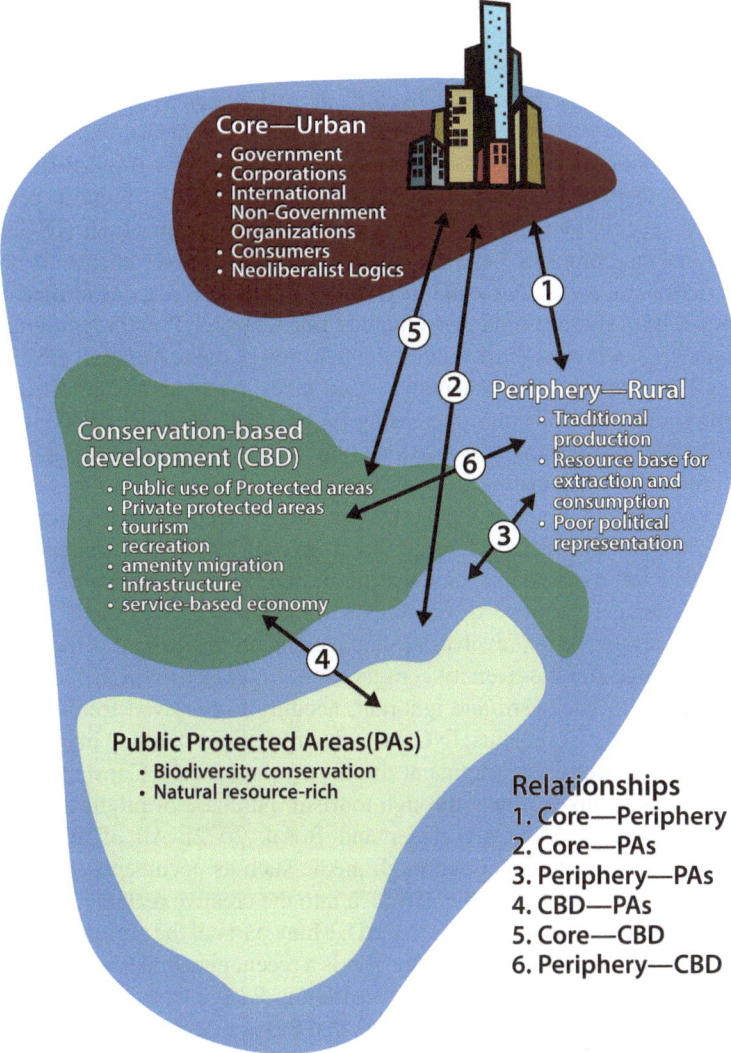

Fig. 1.2 Conceptual map of conservation-based development relationships

PAs are largely located in peripheral areas for the simple fact that it is easier to conserve large tracts of land where there are fewer people and less demand for the resources from agriculture or extraction (Fig. 1.2).

Therefore, CBD exists primarily in peripheral areas. This is evidenced in the proliferation of PAs in mountain regions which tend to exist in the periphery. These peripheral areas are mostly rural, and often have a history of resource extraction such as logging, mining, and/or agricultural production. Most of the products from peripheral areas go to meeting the needs of the core. With CBD, the change from extraction and production-based economies to conservation-based economies often produces conflict. Policy shifts from central governments toward developing or enlarging PAs and PA systems are often an additional source of conflict. The development of PAs often limits local livelihoods and threatens traditional cultures. Tensions can arise between local people, national governments, and PAs because conservation policies are made in the centers of power and implemented in the periphery, in many cases with little or no input from people living in peripheral areas. More recently, International Non-Governmental Organizations (INGOs) and multinational corporations have aligned with national governments to promote CBD in peripheral areas. One outcome of this alignment is a series of mechanisms to finance CBD. These mechanisms include financing for PAs through grants and other transfers, and return-based investments, among others. National governments support CBD through environmentally based taxes, penalties for industrial pollution, carbon offsetting schemes, increasing PA coverage, and financing of biodiversity and habitat restoration. Corporations play a role in financing INGOs, buying carbon credits or participating in other environmentally conscious business practices, and working with national governments to gain access to the resources of the periphery. This might include foreign direct investment in infrastructure and services needed for tourism and amenity migration. The imaginary of CBD then coalesces around a narrative of capitalism for conservation (Louder and Bosak 2022). This narrative often revolves around a crisis. In this case, it is the dual crisis of climate change and biodiversity conservation. Local people in the periphery are often portrayed as backward and ignorant, needing to be saved from themselves. The heroes are the governments, INGOs, and corporations that step in to protect the land and offer economic development to the wayward locals. Consumers are then invited to support this narrative through tourism, purchase of products, and donations to conservation INGOs (Louder and Bosak 2022). All of this is further cemented in the global zeitgeist through media such as documentaries, photography, art, and books that extend the narrative into the creative realm.

Patagonia is a relative newcomer to CBD. Many parts of the region only received road access within the last 30 years. Tourism is a recent phenomenon as is the arrival of the energy industry. The process of transition in Patagonia is happening at a frenetic rate and while it shares some parallels to CBD in places like the United States, there are also differences.

CBD in peripheral areas has been well documented in the shift in the United States from the *Old West* to the *New West*. This shift is often characterized as one where working (productive) landscapes become conservation and recreation landscapes. This shift is accompanied by the changing role of public lands from places that once provided timber, minerals, and grass for grazing to places of recreation, leisure, and tourism (Gosnell and Jesse 2011; Shumway and Otterstrom 2001). The changing role of public lands is a reflection of changing economic trends and changes in government policy; both of which are moving away from extractive and productive activities to emphasize more consumptive activities like outdoor recreation. Modernity and capitalism are implicated as drivers of this shift from productive to consumptive economies as people's identities become intertwined with their lifestyles, culture, and recreational activities. These consumers are increasingly attracted to rural and mountain areas in the periphery because they offer the amenities that align with their values. These amenities include access to public lands including national parks, forests and wildlife refuges, as well as proximity to rural communities and the recreational activities they seek to pursue (Shumway and Otterstrom 2001; Stewart 2002).

In the United States, this shift has focused on the immigration of people from the core, urban areas to peripheral and mountain areas in the West in search of scenic beauty and recreational opportunities. However, the rural transition in the United States began with tourism based on the conservation of natural areas. The US government has long promoted tourism on public lands, and tourism has served as an economic driver for many gateway communities adjacent to these lands. Tourists to peripheral and mountain areas of the US West eventually began to move from short stays in temporary accommodations to seasonal residences and finally permanent homes (Stewart 2002). These gentrifiers of the US West brought with them their own set of values and ideas of how to relate to nature. These values oftentimes conflicted with those of long-time residents who saw their culture as threatened by the newcomers (Ooi et al. 2016). CBD in the US West was predicated on the existence of public lands, the willingness of the government to shift policies for these lands from production to consumption (tourism and recreation), the consumers who had an interest in the landscapes, and the local communities who were more than willing to embrace the economic transition (Abrams and Gosnell 2012). This situation in Patagonia is much the same as it was in the US West in the late twentieth century. There are large swaths of public lands that governments are promoting for recreation and tourism, willing consumers who are eager to leave crowded urban areas for a better life, and communities who are either willing or have no other choice but to shift their economies to cater to these new consumers.

While many of the patterns seen in the United States with the shift from the *old west* to the *new west* are evident in Patagonia, there are some important differences. The shift to CBD in Patagonia is fueled by the expansion of PAs both public and private, and the dual crises of climate change and biodiversity loss provide the justification for this expansion. Furthermore, corporations, INGOs, and national governments are working in conjunction to develop a narrative around the need for CBD based on these crises. They then collude to provide funding mechanisms for conservation and development. The coordination of national governments, INGOs, and corporations has accelerated the pace of CBD in Patagonia to the point where local people are often caught unaware of the changes and powerless to influence the direction or character of the development.

1.2   Overview of the Book Sections

This book is divided into three parts: each with a collection of peer-reviewed studies grounded in the imaginary of Patagonia. Table 1.1 and Fig. 1.3 present an overview of the geographic emphasis for these studies. The sections that follow introduce some of the important theoretical concepts within each part of this book, with an overview of chapter contributions.

The studies that consider Argentine Patagonia mainly contemplate areas bordering Chile, although scattered points were also recorded toward the central-eastern sector of the country. Chapters 2, 4, and 6 offer perspectives rooted in northern Argentine Patagonia. Chapter 4 considers the central Argentine area of Patagonia. Chapters 2, 4, 7, and 8 focus on spaces within southern Argentine Patagonia. Chapter 9 also considers Argentina but concentrates on Iberá National Park in the northeastern reaches of the country, near Uruguay and Paraguay. Nevertheless, the majority of the chapters' study areas are concentrated in Chilean Patagonia. Chapters 6, 9, 16, and 17 consider settings within the northern sectors of Chilean Patagonia. Eleven of the chapters considered central Chilean Patagonia within their study areas, including Chaps. 3, 5, 9, 10, 11, 12, 13, 14, 15, 16, and 18. Southern Chilean Patagonia was also studied within 11 of this book's chapters, including Chaps. 3, 5, 7, 8, 9, 10, 11, 14, 15, 16, and 18.

Figure 1.3 specifies the area of study for each of these chapters, dividing them between the three parts of this book. Part I of this book: *Evolution of the green economy in Patagonia* features research from all three zones of Argentine and Chilean Patagonia. Part II of this book: *Contemporary conservation-based development: Challenges for green integration* emphasizes research from all three zones of Chilean Patagonia and the southern zone of Argentine Patagonia. Part III of this book: *Resilience and sustainability* shares lessons from all three zones of Chilean Patagonia.

1.2.1   Part I: Evolution of the Green Economy in Patagonia

While social imaginaries can be conceptualized as the different ways in which people imagine their social whole (e.g., combinations of sociocultural practices, meaning and materiality, and human agency), environmental imaginaries can be conceptualized as the "constellation of ideas that groups of humans develop about a given landscape" (Davis and Burke 2011, p.134). Environmental governance researchers have found it useful to consider these imaginaries to better understand how relations with the environment are created and staged (Chhetri et al. 2022).

During the twenty-first century, the *geography of imaginaries* concept has expanded within human geography research and thinking (Chhetri et al. 2022). This term builds on foundations that began around 1947 when John Wright posited that geographic imaginaries (cultural and ideological constructions of place) were

Table 1.1 Geographic overview of study areas within northern, central, and southern Patagonia, for each book chapter

Chapter name/geographic focus (N=northern; C=central; S=southern)	Argentina			Chile		
	N	C	S	N	C	S
Part I: Evolution of the Green Economy in Patagonia						
Chapter 2: *Territorializing Capital: Moreno's Gift and the Political Economy of Nature in Argentine Patagonia*	X		X			
Chapter 3: *Connectivity, Tourism, and Conservation: From Extractive Appropriation to Socio- Environmental Reappropriation of Nature in Aysén*				X	X	
Chapter 4: *How Changing Imaginaries of Nature and Tourism Shape National Protected Area Creation in Argentine Patagonia*	X	X	X			
Chapter 5: *Western Patagonia: From Frontiers of Exploration to the Commodification of Nature*				X	X	
Chapter 6: *Geographic Imaginaries in Dispute in Northern Patagonia: Tourism, Environmental Conservation, and Indigenous Territorial Rights in Quinquén, Chile*	X			X		
Part II: Contemporary Conservation-Based Development: Challenges for Green Integration						
Chapter 7: *The Production of Space in the Frontiers of Tourism: Critical Analysis of the Huella de Glaciares Circuit Between El Chaltén, Argentina, and Villa O'Higgins, Chile*		X				X
Chapter 8: *Beyond the Border: Understanding Freshwater Resources, Shared Identity, and Transboundary Cooperation in Southern Patagonia*		X				X
Chapter 9: *Values, Conflicts, and Narratives of Private Protected Areas: The Case of Tompkins Conservation in Chilean Patagonia and Argentina*				X	X	X
Chapter 10: *Exploring Social Representations of Nature-Based Tourism, Development Conflict, and Sustainable Development Futures in Chilean Patagonia*				X	X	
Chapter 11: *Identification of Causal Chains for Sustainable Tourism Development Within Two Chilean Patagonia National Parks: Cerro Castillo and Torres del Paine*				X	X	
Chapter 12: *Visual Dimensions of Conservation Landscapes: An Exploration of Patagonian Fjordic Landscapes from the Perspective of Prospective Chilean Tourists*				X		
Part III: Building Resilience and Sustainability						
Chapter 13: *Employing Local Tourism Councils to Improve Protected Area Tourism Development and Governance in the Aysén Region of Chile*				X		
Chapter 14: *Key Resilience Factors in Four Patagonia Nature-Based Tourism Destinations in the Aysén Region of Chile*				X	X	
Chapter 15: *Evaluating Scientific Tourism Potential for Nature-Based Destinations: Expert Validation and Field Testing of Criteria and Indicators in the Aysén Región of Chilean Patagonia*				X	X	
Chapter 16: *Contributions of Nature Bathing to Resilience and Sustainability*				X	X	X
Chapter 17: *(Re) imagining the Relationship Between Society and Nature in Northern Chilean Patagonia: Encounters and (Mis)encounters with the Modern World*				X		
Chapter 18: *Catalyzing Holistic Conservation-Based Development Through Ethical Travel Experiences Rooted in the Biocultural of Patagonia's Subantarctic Natural Laboratories*				X	X	

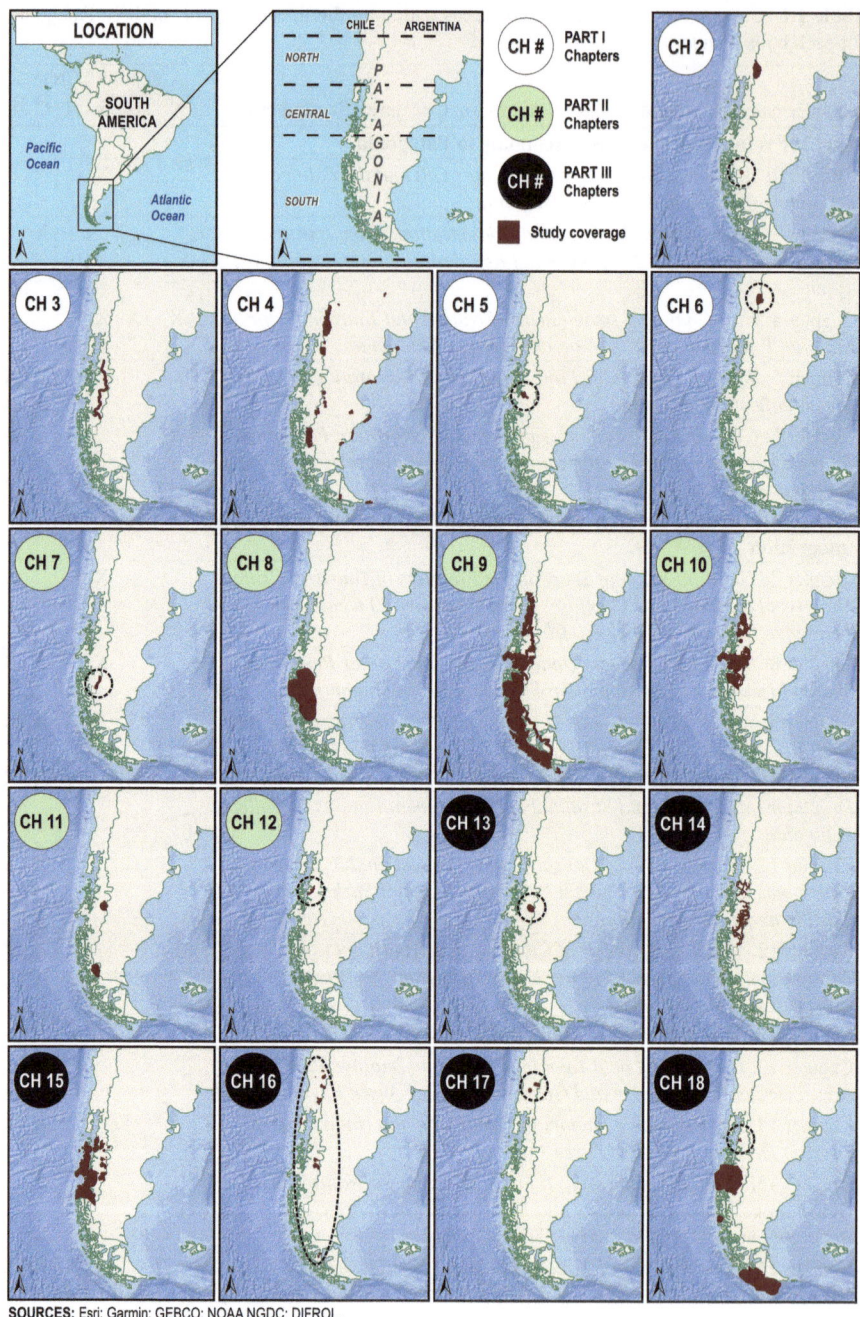

Fig. 1.3 Geographical representation of the study areas for each chapter

created by all sorts of people, from geographers to farmers and painters, tourists, and conservationists. These observations set the course for the contemporary idea of geography as a pluriverse of different imaginations, many of which are shared between social groups. The concepts of environmental and social imaginaries were integrated in the 1980s when geographers suggested that geographic imaginaries influenced social constructions of nature and society (e.g., Duncan and Duncan 1988; Massey 1984). In the 1990s, the concept of geographic imaginaries continued to evolve and integrate with critical social theory, as described by Howie and Lewis (2014):

> Geographers began to see popular, institutional, political and technical representations of the world as structured by more or less fixed, distinctive and discernible framings of relations between people, place and territory. These framings may be intuited, discursive, textual or institutionalized, but they shape and frame how people understand their worlds and those of others. (p. 133)

Today, the term geography of imaginaries is used to express that there are multiple geographic imaginations within the world which can be used, politically, to frame our understandings about the world and influence social discourse and action (Howie and Lewis 2014). As geographers have come to understand the deep embeddedness of social imaginaries within specific territories (Taussing 1997), they have observed an increasing plurality of environmental imaginaries. These imaginaries can be applied across tenses, affecting a territory's current state, framing how a territory came to be, and allowing stakeholders to project a future trajectory. In fact, some describe that environmental change can either be the cause, or effect of, environmental imaginaries (Davis and Burke 2011). Several authors have noted how impactful environmental imaginaries can be, and how their construction, through discourse and narratives, can have widespread and lasting effects on the ways people think about territories (e.g., Bachmann-Vargas and van Koppen 2020; Bachmann-Vargas et al. 2021; Chhetri et al. 2022; Davis and Burke 2011). While these imaginaries can unite groups or people, they can also provoke conflicts stemming from differences in envisioned societal outcomes (Davis and Burke 2011; Chhetri et al. 2022).

Thus, for environmental governance researchers, whose work is intimately connected with place, it is critical to understand and consider the range of environmental imaginaries that have shaped collective action (Chhetri et al. 2022). Who tells the stories that shape, and frame imaginaries is important, since power plays a role in their trajectories. Davis and Burke (2011) pointed out this dynamic in reference to imperial and colonial settings, emphasizing that when environmental representations are constructed from outside of a territory, the imaginaries may be imported and imposed rather than representative of the lived experiences within the territory.

Part I of this book explores the imaginaries that have arisen in Chile and Argentina seeking to shape development in Patagonia. Mendoza (Chap. 2) presents an examination of the political economy of nature and the Moreno-centric imaginary which dominates Argentine history, positing that this imaginary has rendered invisible state violence and Indigenous dispossession, and that such invisibility is a precondition of national conservation. His argument focuses on two historic aspects of capitalist territorialization in Patagonia: the clearing-out strategy pursued by the Argentine government to open Patagonia for colonization and agrarian capitalism; the re-territorialization of space through the creation of national parks and the promotion of leisure capitalism. Muñoz Rebolledo et al. (Chap. 3) explore how the Longitudinal Austral Highway (Carretera Austral, in Spanish), which facilitates connectivity and access to natural environments throughout the Aysén region of Chilean Patagonia, has operated as the mobilizing axis for different processes of imagining, valuing, and appropriating nature. The authors posit that the links between connectivity, tourism, and conservation have changed over time as material and symbolic appropriations of nature in Patagonia have evolved.

Anderson et al. (Chap. 4) propose Patagonia as a dynamic social-ecological system that has been imagined and reimagined over time, through discourses, practices, and institutions that are connected and interrelated along local-to-international scales. Their research employed the social imaginary framework to conduct an historical analysis of local, national, and international influences regarding the way nature and tourism are conceived and managed in national PAs. Salazar-Burrows et al. (Chap. 5) discuss the impact of normative notions of nature and culture in the production of narratives about western Patagonia, exploring how these imaginaries underscore the practices of contemporary CBD and tourism, especially within the Exploradores Valley. They reflect on two major transformations in western Patagonia: first, the shift in historical and environmental configurations of western Patagonia, including the continuity of the colonization processes up to the present day, and the continuities and ruptures of economic activities, mainly tourism. Second, they posit that the current occupation strategies, which mainly focus on the touristification of western Patagonia, are best understood as a process of commodification of nature. Sepúlveda and Martínez-Berríos (Chap. 6) discuss tourism both as a window revealing the tensions between environmental conservation and Indigenous territorial rights and as a sociopolitical process that could resolve them. They explore the dual processes of exploitation/protection of the Araucaria, in northern Chilean Patagonia, contextualizing these processes within the framework of the territorial dispossession that has affected (and continues to affect) the Pewenche in the upper basin of the Bío-Bío River.

1.2.2 Part II: Contemporary Conservation-Based Development: Challenges for Green Integration

Part II of this book explores the relationship between the neoliberal model of CBD that is widespread in Chilean Patagonia and advancing in Argentine Patagonia, to better understand actual and potential impacts that this model has on people, cultures, economies, and ecosystems. For example, in Borrie et al.'s (2020) recent case study, the authors identified three vulnerabilities arising from neoliberal approaches to conservation: a loss of the social embeddedness of nature; an imposition of global, capital dynamics; and conflicting discourses and assumptions. The authors call for greater attention toward social equity and justice, emphasizing processes to build social capital around PAs (public and private), and more generally, citizen-led conservation. Specifically, they recommended, "regional and [private protected area]-specific land-use planning needs to incorporate greater public engagement, cross-jurisdictional coordination, and transparent and inclusive decision-making" (Borrie et al. 2020, p. 1). They also emphasize the importance of supporting local communities with the preservation of their histories and identities. Several similar challenges arose in the chapters presented within this part of this book: both in Chile and in Argentina.

Chapters 7 and 8 focus on Patagonia as a transboundary region, evaluating how cross-border disputes and collaboration processes occur for Argentina and Chile. Specifically, Schweitzer et al. (Chap. 7) focus on cross-border nature-based tourism strategies along the border, presenting a case study of the *Huella de Glaciares* (Trail of the Glaciers), situated in the Santa Cruz Province of Argentina and the Aysén Region of Chile. According to the authors' research, the Huella de Glaciares and several other overland routes that connect with the larger Patagonia territory (e.g., Route of the Parks including the Carretera Austral in Chile, and Scenic Route 41 in Argentina) respond to renewed cross-border processes. The authors warn that these products are being promoted by both the Argentine and Chilean governments and private tourism agents without sufficient consideration of their sustainability. Mirza et al. (Chap. 8) explored transboundary conservation across political and spatial scales in the Southern Patagonian Ice Field shared between Chile and Argentina. The authors underscored the role of freshwater resources in disputed, transboundary landscapes and found that local community collaboration, rooted in shared identity, was the basis of existing transboundary collaboration in southern Patagonia.

Chapters 9 and 10 explore some of the development values and conflicts that have arisen in recent decades in Patagonia. Specifically, Serenari and Bachmann-Vargas (Chap. 9) analyze values and conflicts within the narratives that arose around Tompkins Conservation in Chilean Patagonia and Argentina. Their case study traces the development of meaning regarding Douglas and Kristine Tompkins' effort to create private protected areas in Chilean Patagonia and Argentina. Narratives developed by researchers, conservation entities, politicians, and other actors reveal connections, tensions, and contradictions produced by the broader Tompkins project. Inostroza Villanueva et al. (Chap. 10) explore how *Modernization, Transformation*, and *Control* sustainable development (SD) imaginaries and trajectories interacted with three large-scale development proposal: the Patagonia National Parks network, the HidroAysén hydroelectric project, and the Río Cuervo hydroelectric project. Their case study identified six themes: a desire for greater proactiveness around transparency, a binding participation process of governance, bottom-up decision-making, re-empowerment of local groups, decentralization, and improved oversight practices. They suggest that SD agility, or "the strategic ability to maneuver between

SD imaginaries and trajectories to achieve strategic SD outcomes," may present an important capacity for SD futures trajectories.

The final two chapters of Part II seek to better understand tourism development and sustainability in Chilean Patagonia. Adiego et al. (Chap. 11) use Ante Mandić's (2020) conception of *Drivers, Pressures, State, Impacts,* and *Responses* (DPSIR) to identify causal chains for sustainable tourism development within two Chilean Patagonia national parks: Cerro Castillo and Torres del Paine. Outcomes of the study represent an important first step for developing a better understanding of the causal chains related to the economic, social, and environmental dynamics of tourism in PAs within Chilean Patagonia. Báez Montenegro et al. (Chap. 12) use two hypothetical visual experience scenarios to explore the effect of salmon aquaculture infrastructure on how Chilean tourists' value tourism experiences in and around the Chilean village of Puyuhuapi. Results support current nature-based tourism experience positioning but find several interactions between tourism experience attributes and socio-demographic characteristics, including population density, level of education, and sex. They define a series of hypotheses to expand our current understanding of Chilean perspectives and imaginaries of Patagonia.

1.2.3   Part III: Redefining and Evolving Conservation-Based Development Toward Locally Led Resilience and Sustainability

Part III of this book explores contemporary efforts that are occurring in Patagonia in an effort to move beyond the current governance models to forge new stewardship, governance, and relational models for locally led CBD. Chapter authors provide numerous case studies that illustrate the complexities of integrating resilience and sustainability approaches in current CBD models. Throughout Patagonia, there is a realization that any movement toward resilience and sustainable development must be led by empowered locals. However, that does not mean that national and global actors have no place: national governments can provide policy direction, legal protection, resources, and systems. However, it must be noted that political changes at the national level in Argentina and Chile can mean that progress toward local governance in the CBD space can be reversed quickly. For their part, local communities can work with national governments and INGOs to provide opportunities for ethical and deliberate decision-making, transdisciplinary science, quality travel experiences, and co-creation of management plans within the region.

Chapters 13, 14, and 15 focus on tourism in and around protected areas of the Aysén Region to better understand the potential for shared governance, resilience planning, and the development of sustainable tourism through science. Rovira et al. (Chap. 13) describe a case study into the development of a local tourism council for Cerro Castillo National Park as part of a recent regional tourism governance project. The project seeks to develop a participatory multi-scale governance system that would allow local communities to collaborate with SNASPE PAs (PAs) to improve tourism services, both within PAs and surrounding communities. They observed enabling factors that may inform the creation of local tourism councils in other areas and help stimulate local economies, thereby improving the potential for tourism development to be compatible with the conservation of natural and cultural heritage. Gutiérrez-Vega et al. (Chap. 14) identified key resilience factors (capabilities, ownership, and connections) in four nature-based tourism destinations in the Aysén Region of Chile: Aysén Patagonia Queulat, Coyhaique-Puerto Aysén-Cerro Castillo, Lago General Carrera, and Provincia de los Glaciares (Province of the Glaciers). They report positive evaluations for several resilience factors in three of the four tourism destinations; however, all four destinations presented high levels of natural risks, with the Provincia de los Glaciares destination as the most vulnerable. They suggest destinations with higher levels of natural risks should focus on strengthening their resilience factors. Veloso et al. (Chap. 15) evaluated Scientific Tourism potential in the Aysén Region of Chilean Patagonia. They developed a matrix of weighted criteria to assess the potential for sustainable Scientific Tourism and found that involving travelers in research initiatives taking place in Patagonian destinations allows them to develop lasting connections with the heritage and institutions of these territories.

Chapters 16, 17, and 18 conclude Part III with a series of experiments that are occurring in Chilean Patagonia to redefine human–human and human–nonhuman relationships and their connections with stewardship and conservation. Lazo Álvarez et al. (Chap. 16) explore the potential for the Nature Bathing initiative, developed within the Chilean National Forestry Corporation's (CONAF) Nature for Everyone program, to contribute to resilience and sustainability. CONAF's Nature Bathing program integrates elements of forest bathing (*Shinrin Yoku*, in Japanese), grounding, and Andean Indigenous and popular culture. They discuss how such programs can help to strengthen the role of PAs in supporting public health and helping visitors build resilience while connecting with nature. Zunino and Spirito (Chap. 17) describe three alternative ways for inhabiting the territory of Southern Chile that are being developed as local community projects in the mountainous area of the Araucanía region: (1) a community project recreating Mesoamerican Indigenous practices, (2) the Waldorf Educational Project that represents a pedagogical counterproposal developed by the European spiritual thinker Rudolf Steiner (1861–1925), and (3) permaculture projects that seek new forms of food production through a close link with nature. These projects have raised interest through the profound transformation in how locals interrelate with nature. They suggest new forms of living will be needed as society progresses through crises that break away from dualistic ways of thinking about humans and nature. Finally, Gale et al. (Chap. 18) close this book, by bringing together some of the challenges that arose in Part II with an integrative proposal for moving forward through locally led approaches and programs that promise new policies and mechanisms for oversight. The authors explored three initiatives underway in the Aysén and Magallanes regions of Chile that foster biocultural reawakening, democratize science, and catalyze sustainable development: (1) Subantarctic Natural Laboratories; (2) 3-Hs Biocultural Ethic and

Field Environmental Philosophy Cycle Approach; and (3) Scientific Tourism Collaborative Learning Networks. These project methods were studied to explore how their integration might strengthen CBD in Patagonia through ethical travel experiences rooted in the biocultural of local communities. Results suggest promise for a combined approach; thus, additional research and consideration is merited.

1.3   A Bit About the Authors Involved in This Project

We would like to close this chapter by sharing a bit about the authors involved in this project. We sought to include a range of voices and perspectives, made up of a broad diversity of gender and nationality, with a particular emphasis on the research perspectives of Argentinian and Chilean authors. There are a total of 51 authors involved in the 18 chapters of this book. Our group of authors are affiliated with institutions in Argentina (5), Chile (39), France (2), the Netherlands (1), Spain (1), and the United States (10). Twenty-three (45%) are women and 28 are men (55%). There are six chapters (33%) with a female first author and twelve (67%) with a female second author. There are 12 chapters (67%) with a male first author and six with a male second author (33%). The first author for 11 chapters (61%) and the second author for 12 chapters (67%) are from either Chile or from Argentina. Finally, 29 (57%) of the authors live and work in Patagonia. Eight of the chapters (44%) feature a first author who lives and works in Patagonia, and another 10 chapters (56%) have a second author who lives and works in this special region of the world.

Acknowledgments This work was supported by Chile's National Research and Development Agency (ANID) under ANID's Regional Program R17A10002; the CIEP R20F0002 project; the NODOSLN0002 project; ANID FONDECYT Regular 1230020; and the CHIC-ANID PIA/ BASAL PFB210018. We are grateful to Andrés Adiego, of the Centro de Investigación en Ecosistemas de la Patagonia (CIEP), and the Department of Geography and Territorial Planning and Management of the Universidad de Zaragoza, for his support with the elaboration of the cartography in Figs. 1.1 and 1.3.

References

J.B. Abrams, H. Gosnell, The politics of marginality in Wallowa County, Oregon: Contesting the production of landscapes of consumption. J. Rural. Stud. **28**(1), 30–37 (2012). https://doi. org/10.1016/j.jrurstud.2011.09.004
J.B. Abrams, H. Gosnell, N.J. Gill, P.J. Klepeis, Re-creating the rural, reconstructing nature: An international literature review of the environmental implications of amenity migration. Conserv. Soc. **10**(3), 270–285 (2012). https://doi.org/10.4103/0972-4923.101837
F. Antía, The political dynamic of redistribution in unequal democracies: The center-left governments of Chile and Uruguay in comparative perspective. Lat. Am. Perspect. **46**(1), 152–166 (2019). https://doi.org/10.1177/0094582X18806827

Argentine National Ministry of the Environment and Sustainable Development, *Estrategia nacio-nal sobre la biodiversidad y plan de acción 2016–2020 (ENBPA)* (National Biodiversity Strategy and Action Plan 2016–2020 (ENBPA)) (Buenos Aires, AR, 2017)

P. Bachmann-Vargas, C.S.A. van Koppen, Disentangling environmental and development discourses in a peripheral spatial context: The case of the Aysén Region, Patagonia, Chile. J. Environ. Dev. **29**(3), 366–390 (2020). https://doi.org/10.1177/1070496520937041

P. Bachmann-Vargas, C.S.A.(.K.). van Koppen, M. Lamers, A social practice approach to nature-based tours: The case of the Marble Caves in northern Patagonia, Chile. J. Ecotour. **21**(1), 1–17 (2021). https://doi.org/10.1080/14724049.2021.1913176

BIOFIN Chile, *Policy Brief: Biodiversidad en Chile. Propuestas para financiar su conservación y uso sostenible* (Policy Brief: Biodiversity in Chile. Proposals for Financing Conservation and Sustainable Use) (PNUD, Chile, 2017). Retrieved January 7, 2023, from https://www.undp.org/es/chile/publications/biodiversidad-en-chile-propuestas-para-financiar-su-conservaci%C3%B3n-y-uso-sostenible

H. Blair, K. Bosak, T. Gale, Protected areas, tourism, and rural transition in Aysén, Chile. Sustainability (Switzerland) **11**(24), Article 7087 (2019). https://doi.org/10.3390/su11247087

L.A. Borrero, N.V. Franco, Early Patagonian hunter-gatherers: Subsistence and technology. J. Anthropol. Res. **53**(2), 219–239 (1997)

L.A. Borrero, A. Nuevo Delaunay, C. Méndez, Ethnographical and historical accounts for understanding the exploration of new lands: The case of central western Patagonia, southernmost South America. J. Anthropol. Archaeol. **54**, 1–16 (2019). https://doi.org/10.1016/j.jaa.2019.02.001

B. Borrie, T. Gale, K. Bosak, Privately protected areas in increasingly turbulent social contexts: Strategic roles, extent, and governance. J. Sustain. Tour. **30**(11), 2631–2648 (2020). https://doi.org/10.1080/09669582.2020.1845709

G. Cairó-i-Céspedes, J.C. Palacios Cívico, Beyond core and periphery: The role of the semi-periphery in global capitalism. Third World Q. **43**(8), 1950–1969 (2022). https://doi.org/10.1080/01436597.2022.2079488

L. Carte, H. Zunino, Chilean Patagonia. Geogr. Rev. **112**(5), 615–621 (2022). https://doi.org/10.1080/00167428.2022.2121652

F. Cervantes, *Conquistadores: A New History of Spanish Discovery and Conquest* (Penguin Random House LLC, 2020)

N. Chhetri, R. Ghimire, D.C. Eisenhauer, Geographies of imaginaries and environmental governance. Prof. Geogr. **75**(2), 263–268 (2022). https://doi.org/10.1080/00330124.2022.2087698

Chilean Ministry of the Environment, *Sexto informe nacional de biodiversidad de Chile ante el Convenio sobre la Diversidad Biológica (CDB)* (Sixth National Biodiversity Report of Chile to the Convention on Biological Diversity (CBD)) (Government of Chile, 2019)

D.K. Davis, E. Burke, Imperialism, orientalism, and the environment in the Middle East: History, policy, power, and practice, in *Environmental Imaginaries of the Middle East and North Africa*, ed. by D.K. Davis, E. Burke, (Ohio University Press, 2011), pp. 1–22

L.F. De Matheus e Silva, H.M. Zunino, V. Huiliñir Curío, El negocio de la conservación ambiental: Cómo la naturaleza se ha convertido en una nueva estrategia de acumulación capitalista en la zona andino-lacustre de Los Ríos, sur de Chile (The business of environmental conservation: How nature has become a new strategy of capitalist accumulation in the Andean-lacustrine zone of Los Ríos, Southern Chile). Scripta Nova **22**(583), 1–25 (2018). https://doi.org/10.1344/sn2018.22.19021

R.J.J. Donoso, *Servicio de biodiversidad y áreas protegidas, Boletín N° 9404-12. (National Service of Biodiversity and Protected Areas, Bulletin N° 9404-12)*, Presentation, August 1, 2019 (Ministry of the Environment, Coyhaique, 2019)

J. Duncan, N. Duncan, (Re)reading the land- scape. Environ. Plann. D Soc. Space **6**, 117–126 (1988). https://doi.org/10.1068/d060117

R. Elbert, P. Pérez, The identity of class in Latin America: Objective class position and subjective class identification in Argentina and Chile (2009). Curr. Sociol. **66**(5), 724–747 (2018). https://doi.org/10.1177/0011392117749685

FAO/OAPN Program, *Sostenibilidad financiera para áreas protegidas en América Latina. Financiamiento sustentable en áreas protegidas* (Financial Sustainability for Protected Areas in Latin America. Sustainable Financing in Protected Areas) (FAO, 2010)

A.G. Frank, *Latin America: Underdevelopment or Revolution. Essays on the Development and Underdevelopment and the Immediate Enemy* (Monthly Review Press, 1969)

T. Gale, A. Ednie, Can intrinsic, instrumental, and relational value assignments inform more integrative methods of protected area conflict resolution? Exploratory findings from Aysén, Chile. J. Tour. Cult. Chang. **18**(6), 690–710 (2019). https://doi.org/10.1080/14766825.2019.1633336

T. Gale, K. Bosak, L. Caplins, Moving beyond tourists' concepts of authenticity: Place-based tourism differentiation within rural zones of Chilean Patagonia. J. Tour. Cult. Chang. **11**(4), 264–286 (2013). https://doi.org/10.1080/14766825.2013.851201

I. Gentes, P. Policzer, Weakness by design: Neoliberal governance over mining and water in Chile, in *Territory, Politics, Governance*, (Advance Online Publication, 2022). https://doi.org/10.1080/21622671.2022.2134196

H. Gosnell, A. Jesse, Amenity migration: Diverse conceptualizations of drivers, socioeconomic dimensions, and emerging challenges. GeoJournal **76**, 303–322 (2011). https://doi.org/10.1007/s10708-009-9295-4

J. Grugel, P. Riggirozzi, Post-neoliberalism in Latin America: Rebuilding and reclaiming the state after crisis. Dev. Chang. **43**(1), 1–21 (2012). https://doi.org/10.1111/j.1467-7660.2011.01746.x

D.L. Harvey, *A Brief History of Neoliberalism* (Oxford University Press, 2005)

S.P. Hier, The forgotten architect: Cox, Wallerstein and world-system theory. Race Class **42**(3), 69–86 (2001). https://doi.org/10.1177/0306396801423004

A. Hornborg, Towards an ecological theory of unequal exchange: Articulating world system theory and ecological economics. Ecol. Econ. **25**(1), 127–136 (1998). https://doi.org/10.1016/S0921-8009(97)00100-6

A. Hornborg, *The Power of the Machine: Global Inequalities of Economy, Technology, and Environment* (Rowman & Littlefield, Lanham, 2001)

A. Hornborg, Zero-sum world, challenges in conceptualizing environmental load displacement and ecologically unequal exchange in the world-system. Int. J. Comp. Sociol. **50**(3–4), 237–262 (2009). https://doi.org/10.1177/0020715209105141

B. Howie, N. Lewis, Geographical imaginaries: Articulating the values of geography. N. Z. Geogr. **70**(2), 131–139 (2014). https://doi.org/10.1111/nzg.12051

C. Jones, Ecophilanthropy, neoliberal conservation, and the transformation of Chilean Patagonia's Chacabuco Valley. Oceania **82**(3), 250–263 (2012). https://doi.org/10.1002/j.1834-4461.2012.tb00132.x

R. Kennedy, Replacing the Chilean constitution. Constellations **24**(3), 456–469 (2017). https://doi.org/10.1111/1467-8675.12303

M. Kühn, Peripheralization: Theoretical concepts explaining socio-spatial inequalities. Eur. Plan. Stud. **23**(2), 367–378 (2015). https://doi.org/10.1080/09654313.2013.862518

M. Kühn, M. Bernt, L. Colini, Power, politics and peripheralization: Two eastern German cities. Eur. Urban Region. Stud. **24**(3), 258–273 (2017). https://doi.org/10.1177/0969776416637207

P. Lambert, D. Scribner, Constitutions and gender equality in Chile and Argentina. J. Polit. Latin Am. **13**(2), 219–242 (2021). https://doi.org/10.1177/1866802X211024245

A. Latta, B.E.C. Aguayo, Testing the limits neoliberal ecologies from Pinochet to Bachelet. Latin Am. Perspect. **39**(185), 163–180 (2012). https://doi.org/10.1177/0094582X12439050

T.T. Lennerfors, P. Fors, J. van Rooijen, ICT and environmental sustainability in a changing society: The view of ecological World Systems Theory. Inf. Technol. People **28**(4), 758–774 (2015). https://doi.org/10.1108/ITP-09-2014-0219

E. Louder, K. Bosak, Spectacle of nature 2.0: The (re)production of Patagonia National Park. Ann. Am. Assoc. Geogr. **113**(2), 331–345 (2022). https://doi.org/10.1080/24694452.2022.2106176

G. Lublin, Adjusting the focus: Looking at Patagonia and the wider Argentine state through the lens of settler colonial theory. Settler Colonial Stud. **11**(3), 386–409 (2021). https://doi.org/1 0.1080/2201473X.2021.2001961

A. Mandić, Structuring challenges of sustainable tourism development in protected natural areas with driving force–pressure–state–impact–response (DPSIR) framework. Environ. Syst. Decis. **40**(4), 560–576 (2020). https://doi.org/10.1007/s10669-020-09759-y

D. Massey, *Spatial Divisions of Labour: Social Structures and the Geography of Production* (Macmillan, 1984)

J. Mccarthy, The financial crisis and environmental governance "after" neoliberalism. Tijdschr. Econ. Soc. Geogr. **103**(2), 180–195 (2012). https://doi.org/10.1111/j.1467-9663.2012.00711.x

C. Méndez, Terminal Pleistocene/early Holocene 14C dates form archaeological sites in Chile: Critical chronological issues for the initial peopling of the region. Quat. Int. **301**, 60–73 (2013). https://doi.org/10.1016/j.quaint.2012.04.003

C. Méndez, A. Nuevo-Delaunay, O. Reyes, I.L. Ozán, C. Belmar, P. López-Mendoza, The initial peopling of central western Patagonia (southernmost South America): Late Pleistocene through Holocene site context and archaeological assemblages from Cueva de la Vieja site. Quat. Int. **473**, 261–277 (2018). https://doi.org/10.1016/j.quaint.2017.07.014

E. Myron, C. Fabiano, H. Ahmed, *International Outlook for Privately Protected Areas: Argentina Country Profile* (International Land Conservation Network (a project of the Lincoln Institute of Land Policy) – United Nations Development Programme, 2019a). From https://www.landcon-servationnetwork.org/sites/default/files/pictures/Argentina%20Country%20Profile%20on%20 Privately%20Protected%20Areas_7_24_2019.pdf

E. Myron, L. Gloss, C. Fabiano, H. Ahmed, *International Outlook for Privately Protected Areas: Chile Country Profile* (International Land Conservation Network (a project of the Lincoln Institute of Land Policy) - United Nations Development Programme, 2019b). Retrieved January 16, 2023, from https://www.landconservationnetwork.org/sites/default/files/pictures/Chile%20 Country%20Profile%20on%20Privately%20Protected%20Areas_7_24_2019.pdf

P. Navarro Floria, Landscapes of an uncertain progress: Northern Patagonia in Argentine scientific journals (1876–1909). J. Lat. Am. Cult. Stud. **16**(3), 261–283 (2007). https://doi. org/10.1080/13569320701682476

S. Niedzwiecki, J. Pribble, Social policies and center-right governments in Argentina and Chile. Latin Am. Polit. Soc. **59**(3), 72–97 (2017). https://doi.org/10.1111/laps.12027

P.G. Núñez, The 'She-Land,' social consequences of the sexualized construction of landscape in North Patagonia. Gend. Place Cult. **22**(10), 1445–1462 (2015). https://doi.org/10.108 0/0966369X.2014.991695

A. Núñez, M.C. Benwell, E. Aliste, Interrogating green discourses in Patagonia-Aysén (Chile): Green grabbing and eco-extractivism as a new strategy of capitalism? Geogr. Rev. **112**(5), 688–706 (2020). https://doi.org/10.1080/00167428.2020.1798764

OCDE. Regulatory Policy in Chile: Government Capacity to Ensure High-Quality Regulation (OECD Review). OECD Publishing. (2016). https://www.oecd-ilibrary.org/governance/regulatory-policy-in-croatia_b1c44413-en

Z. Önis, Varieties and crises of neoliberal globalisation: Argentina, Turkey and the IMF. Third World Q. **27**(2), 239–263 (2006). https://doi.org/10.1080/01436590500432366

N. Ooi, J. Laing, J. Mair, Sociocultural change facing ranchers in the Rocky Mountain West as a result of mountain resort tourism and amenity migration. J. Rural. Stud. **41**, 59–71 (2016). https://doi.org/10.1016/j.jrurstud.2015.07.005

E. Ormaechea, The failures of neoliberalism in Argentina. J. Econ. Issues **55**(2), 318–324 (2021). https://doi.org/10.1080/00213624.2021.1907155

B.S. Ponzi, Han tomado la parte del fondo: La territorialización del Parque Nacional Patagonia, Santa Cruz (Argentina) (They have taken the background part: The territorialization of Patagonia National Park, Santa Cruz (Argentina)). AMBIENTES: Revista de Geografía e Ecología Política **2**(1), 228 (2020). https://doi.org/10.48075/amb.v2i1.24284

L. Prates, G. Politis, J. Steele, Radiocarbon chronology of the early human occupation of Argentina. Quat. Int. **301**, 104–122 (2013). https://doi.org/10.1016/j.quaint.2013.03.011

M. Shumway, S. Otterstrom, Spatial patterns of migration and income change in the mountain west: The dominance of service-based, amenity-rich counties. Prof. Geogr. **53**(4), 492–502 (2001). https://doi.org/10.1111/0033-0124.00299

C. Sorinel, Immanuel Wallerstein's world system theory. Ann. Fac. Econ. **1**(2), 220–225 (2010)

S. Stewart, *Amenity Migration in Trends 2000s: Shaping the Future–5th Outdoor Recreation & Tourism Trends Symposium* (Michigan State University, 2002)

C. Tabbush, M. Caminotti, Igualdad de género y movimientos sociales en la Argentina posneo-liberal: La organización barrial Tupac Amaru (Gender equality and social movements in post-neoliberal Argentina: The neighborhood organization Tupac Amaru). Perfiles Latinoam. **23**(46), 147–171 (2015). https://doi.org/10.18504/pl2346-147-2015

K.S. Taussing, Calvinism and chromosomes: Religion, the geographical imaginary, and medical genetics in The Netherlands. Sci. Cult. **6**(4), 495–524 (1997). https://doi.org/10.1080/09505439709526483

C.G. Thies, Territorial nationalism in spatial rivalries: An institutionalist account of the Argentine-Chilean rivalry. Int. Interact. **27**(4), 399–431 (2001). https://doi.org/10.1080/03050620108434992

T. Undurraga, Neoliberalism in Argentina and Chile: Common antecedents, divergent paths. Revista de Sociología e Política **23**(55), 11–34 (2015). https://doi.org/10.1590/1678-987315235502

R. Villalón, Neoliberalism, corruption and legacies of contention: Argentina's social movements, 1993-2006. Lat. Am. Perspect. **34**(2), 139–156 (2007)

M.A. Vitale, La dimensión argumentativa de las memorias discursivas: El caso de los discursos golpistas de la prensa escrita Argentina (1930-1976) (The argumentative dimension in discursive memoirs: The case of the coup d'etat discourses in the Argentinian press (1930–1976)). Forma y Función **22**(1), 125–144 (2009)

P. Walker, Legislación ambiental: Desafíos para el Chile futuro (Environmental legislation: Future challenges for Chile), in *La vía medioambiental: Desafíos y proyecciones para un Chile futuro* (The Environmental Way: Challenges and Projections for a Future Chile), ed. by P.S. Quintana, (Ministerio del Medio Ambiente, 2018), pp. 105–113

S. Warren, A nation divided: Building the cross-border Mapuche nation in Chile and Argentina. J. Lat. Am. Stud. **45**(2), 235–264 (2013). https://doi.org/10.1017/S0022216X13000023

C. Wylde, Post-neoliberal developmental regimes in Latin America: Argentina under Cristina Fernandez de Kirchner. New Polit. Econ. **21**(3), 322–341 (2016). https://doi.org/10.1080/13563467.2016.111394

Part I
Evolution of the Green Economy in Patagonia

Part I documents the evolution of strategic initiatives to foster conservation-based territorial development in the southern reaches of Argentina and Chile.

Part I
Evolution of the Green Economy in Patagonia

Chapter 2
Territorializing Capital: Moreno's Gift and the Political Economy of Nature in Argentine Patagonia

Marcos Mendoza

Abstract This chapter examines the political economy of nature and the legacy of Francisco Moreno, scientist and explorer, within Argentine Patagonia. Moreno is institutionally recognized for a land donation he made to the federal government in 1903, which is celebrated for inaugurating the national park conservation movement. This Moreno-centric official history, however, has rendered invisible state violence and Indigenous dispossession as preconditions of national conservation. Moving beyond this official history of conservation, the discussion highlights two histories of capitalist territorialization. The first focuses on the clearing-out strategy pursued by the Argentine government to open Patagonia for colonization and agrarian capitalism. The second attends to the re-territorialization of space through the creation of national parks and the promotion of leisure capitalism. Using the concept of "the gift" to assess Moreno's legacy, this chapter shows that the "spirit of the gift"—heralded by the Argentine federal government—is chained to these two projects of capitalist territorialization. These territorialization histories challenge the halcyon representation of Moreno's gift promoted by the state. Drawing upon scholarship in political ecology, this study is a contribution to an emerging critical assessment of "the gift" within Patagonian conservation.

Keywords Patagonia · Political ecology · Francisco Moreno · National park · Conservation

2.1 Introduction

In Argentina, National Parks Day is celebrated on November 6 to commemorate the date in 1903 when Francisco Moreno, explorer and scientist, donated 75 km² of land to the federal government to create a public nature park (*parque público natural*).

M. Mendoza (✉)
University of Mississippi, Department of Sociology & Anthropology, University, MS, USA
e-mail: mendoza@olemiss.edu

© The Author(s) 2023
T. Gale-Detrich et al. (eds.), *Tourism and Conservation-based Development in the Periphery*, Natural and Social Sciences of Patagonia,
https://doi.org/10.1007/978-3-031-38048-8_2

Moreno's donation of territory near Lake Nahuel Huapi in northern Patagonia has become recognized by the Argentine National Park Administration (APN) as a foundational act that created the first national park within the country and ignited the conservation movement. This gift was legally incorporated into the new Park of the South established by presidential decree in 1922 and redesignated as Nahuel Huapi National Park in 1934 (Freitas 2021). This baptismal moment positioned Argentina at the cutting edge of international conservation, making it—according to the APN—only the "third country in América" and "the fifth country in the world" to have created a national park (National Parks Administration of Argentina 2012, p. 13). This language extols Argentina as a conservation leader with deep commitments to its biophysical environments and nonhuman populations. Scholars have criticized this Moreno-centric narrative—propagated by the APN—for marginalizing conservation histories that might foreground Iguazú Falls, the Atlantic-forest biome, and the work of Carlos Thays (Freitas 2021; Kaltmeier 2021). The Moreno-centric narrative also renders invisible state violence and Indigenous dispossession as conditions of possibility for national conservation.

Annual celebrations of National Parks Day retell this founding story, highlighting how Moreno's gift established a public domain of "inalienable wealth" (Weiner 1985) that has expanded over the generations into the present-day system of protected areas (PAs). I was fortunate to attend one of these anniversary events in 2009 while conducting ethnographic research in the mountain village of El Chaltén adjacent to Glaciers National Park (*Parque Nacional Los Glaciares*). The event was held in the elementary school auditorium. The audience, approximately 140 people, included schoolchildren, gendarmes, public officials, town residents, and the rangers from the Viedma Lake Section (*Seccional Lago Viedma*), the station in charge of the northern sector of the park. The event kicked off with the presentation of the flags (national and provincial) and the singing of the national anthem. The seniormost ranger gave an opening address focused on the children. A charismatic speaker, Enrique recounted the importance of Francisco Moreno to the park system and explained the value of conservation to the El Chaltén community, which was expanded beyond humans to encompass forests, wildlife, and glaciers. This was followed by a series of humorous skits that explained rules for park conservation, such as not lighting campfires and refraining from littering. The person littering in the park was depicted as a foreign tourist, while the person intervening was represented as a resident of El Chaltén. The resident addressed and sought to educate the tourist—in English—about the rules of park conservation. The skits then segued into a PowerPoint presentation focused on the adults, which briefly explained the history and key features of Glaciers National Park. The event concluded by naming and celebrating each ranger—dressed in their signature tan and green uniforms—as they stood at the front of the room. Then everyone ate cake.

This anniversary event in El Chaltén raises a number of key issues. It indicates how Moreno's legacy remains central to the historical imagination of national conservation. Celebrating National Parks Day is a way to recount a story told concerning a selfless act of gift-giving that would create an expanding patrimony of nature. The ritual singing of the anthem, saluting of the flags, and assembling of state personnel affirm the official significance of this legacy, but also direct attention onto protected nature. This highlights how the nation recognizes itself not just through flags, anthems, and uniformed officials, but also through the narrative of protection of vulnerable environments threatened by anthropogenic forces such as ranching, mining, and deforestation. This institutional history of Moreno is also a call to an ethics of conservation. Moreno's gift to the nation only means something—the APN suggests—if everyday citizens take up the duty of environmental care and the "greening" of Argentina. To care for nature is to perform an act of citizenship that implies a will to protect and cherish Argentina's national heritage. This ethical injunction is incorporated into the presentation given by rangers, calling upon children and adults alike to be patriotic stewards of the gift.

To apply the scholarship of Marcel Mauss (2000), there is an ethical "spirit" of the gift at work within the Moreno legacy presented by the APN. Mauss's theory of the gift highlights the obligations to give, to receive, and to reciprocate. There are various gifts and countergifts to consider. The first entails the gift logic linking Moreno and the Argentine government. Within the APN's institutional history of Moreno, the initial gift of land was received by the state as a foundational donation that was generative of the new category of the national patrimony of protected nature. By receiving the donation, the state had to reciprocate or repay this gift. This act of repayment involved the commitment to honor Moreno's legacy by expanding the system of protected nature. This repayment consists of two sub-gifts: (1) an institutional effort to lionize Moreno as the founding father of Argentine conservation; and (2) the creation of an expanding system of protected nature that indexes the initial gift. However, this first moment of gift exchange opens up a second moment that places the two initial actors (Moreno and the state) into the category of gift-giver and locates the citizenry in the category of recipient. Taken together, Moreno and Argentine state have given the citizenry the gift of the founding idea and institutional realization of the national park system. This sets up the injunction or call for the citizenry to reciprocate. This repayment is fulfilled by practicing an ethic of conservation. The ethic of conservation is an open field of action: volunteering time for trail restoration; creating monitoring groups for endangered species; establishing nongovernmental organizations (NGOs) to tap foreign donor networks; or supporting the sustainable development protocols that govern the green economy (Mendoza 2018). This logic calls the citizenry to valorize the first gift (Moreno's donation), and the second gift (the APN's expanding park system), by becoming the third party to this expanding circuit of reciprocity.

The spirit of the gift materializes through the connections of people, organizations, institutions, and nonhumans whose actions are folded into the ethic of conservation. This opens up the gift logic beyond the Moreno–state–citizenry triad to enfold a fourth figure: the foreigner. Foreigners include tourists visiting national parks, paying money to the APN for entrance fees, and contributing to the green economy. The APN calls upon visitors—many involved in leisure pursuits such as trekking, kayaking, birding, climbing, and sightseeing—to practice conservation ethics inside parks (Mendoza 2016). Nevertheless, there are other foreigners in Argentina with questionable motives, such as land barons who have consolidated massive estates (Sánchez 2007). Facing intense public criticism, such foreigners can work to transform their identities through participation in the spirit of the gift. As discussed later, eco-philanthropic organizations such as Tompkins Conservation have participated in the call to conservation through the buying of properties and the gifting of these to the state to become national parkland.

This institutionalized discourse about Moreno is significant for creating and circulating a conservation imaginary organized around the spirit of the gift and the call to "green" Argentina, which is open to citizens and noncitizens alike. It is also significant for what it does not reveal. This chapter explores the political economy of nature that undergirds the legacy of Moreno's gift. The discussion takes its point of departure from this routine celebration of National Parks Day but probes the histories of appropriation and extraction that undergird this institutional legacy. I demonstrate that the spirit of the gift is chained to projects of capitalist territorialization in Patagonia. The discussion focuses on two territorialization projects: rangeland farming and Andean conservation. Dispossession and violence are integral to capitalist territorialization, challenging the halcyon representation of Moreno's gift operative within the APN's official history. Following a discussion of capitalist territorialization, this chapter explores settler colonialism and livestock farming in Patagonia. It then focuses on the creation of the national park administration, the genesis of Andean border parks, and the efforts of eco-philanthropists.

2.2   Territorializing Capital

Capitalist territorialization refers to the production of space for capital accumulation (Brenner 1999; Lefebvre 2004). Capital formation requires the construction of spaces of appropriation that facilitate value extraction from humans, nonhumans, and the environment, adapting to the shifting dynamics of profit seeking, rent capture, and market creation (Harvey 2001; Lefebvre 2004; Moore 2015). As it gains fixity or traction in these produced spaces, capital is established through "systems of resource control—rights, authorities, jurisdictions, and their spatial representations" (Rasmussen and Lund 2018, p. 388). States are integral to the fashioning of systems of resource control that seek to include or exclude certain populations (Vandergeest and Peluso 1995). This is a particularly salient point for Patagonia and the strategy of Indigenous annihilation and the clearing of the region for capitalism and settler colonialism pursued by the Argentine and Chilean governments (Bandieri 2005; Di Giminiani 2018; Edwards 2022; Klubock 2014; Navarro-Floria 1999; Ogden 2021; Rasmussen 2021).

I conceptualize capitalist territorialization projects as sites of gathering that assemble actors around the production of space through appropriation and extraction. These gathering sites are open, thus capable of recruiting expected and unexpected human and nonhuman actors (Blanco et al. 2015; Dicenta and Correa 2021; Ogden 2021). Territorialization projects produce tensions and contradictions as they unfold, which may provoke new strategies to re-territorialize spaces and their market connections (Brenner 1999; Edwards 2022; Rasmussen and Lund 2018). Territorialization projects also generate imaginaries: shared understandings and interpretive frameworks that shape the practical orientations and value designations that actors ascribe to a particular space (Taylor 2004). Mendoza et al. (2017) discuss the rise of a Patagonian territorial imaginary that—since the 1990s—has framed the Southern Andes as an eco-region earmarked for "green development," which in and of itself is a contested term within the territory. This eco-region is constituted through the images, representations, and values produced by the conjoining of tourism markets, the outdoor industry, and environmentalism. Though contested, this Patagonian imaginary has enrolled various state, corporate, and civil society actors within an emerging project to extend green capitalism.

This chapter contributes to scholarship by highlighting the intersection of gift legacies and capitalist territorialization. Rather than conceptualizing the "gift economy" as antagonistic and external to the "market economy," scholars have highlighted how capitalism depends upon various economic logics ranging from markets to sharing to reciprocity to redistribution, as well as appropriating unpaid work and raiding the environmental commons (Gibson-Graham 2006; Moore 2015; Mendoza et al. 2021). My discussion scrutinizes two histories of capitalist territorialization and the production of space in Argentine Patagonia to situate the spirit of the gift and Moreno's legacy. The first focuses on the clearing-out strategy pursued by the Argentine government to open the Patagonian "desert" for colonization and agrarian capitalism. This produced a space of rangeland extraction that—though greatly diminished—continues to structure Patagonian social life in the present. The second focuses on the perceived failure of this first strategy in the Andean borderlands and the re-territorialization of space through the creation of national parks and the promotion of leisure capitalism. In the 1990s, this second strategy began to attract unexpected actors: foreign eco-philanthropists.

2.2.1 Agrarian Capitalism on the Rangelands

The first aspect of the Moreno gift to consider is the land he received that was then donated to the national government. Moreno's (2002) letter to the Ministry of Agriculture in 1903, explaining his motives, identified the park systems in the United States and other countries as the inspiration. Moreno was given land by the Argentine government in recognition of his services as a scientist exploring Patagonia and contributing to the scientific diplomacy that defended national sovereignty in the Andean borderlands (Wakild 2017). Having received this land grant, Moreno wished to contribute to the founding of a nature park that would benefit "current and future generations" and inspire the Argentine government to set aside "magnificent unspoiled parkland" that would become a "catalyst for human advancement" (Moreno 2002). Moreno gestured toward the benefits of the park for scientific study, visitation by tourists who might marvel at its beauty and appreciate its serenity, and for peaceful transborder coexistence and international conviviality.

What remained unsaid within Moreno's letter was that the land grant was anchored within a history of capitalist territorialization tied to settler colonialism and livestock farming.

In the second half of the nineteenth century, the Argentine state initiated a territorialization project that sought to open Patagonia for capitalist investment and market formation. This territorialization project was spurred by the Argentine military and the prosecution of the so-called Conquest of the Desert (1878–1885). Social elites viewed Indigenous controlled lands as a "desert" in which civilization was absent, barbarism thrived, and racialized others prevented the exercise of integral sovereignty by the state over its claimed territories (Gordillo 2004; Nouzeilles 1999). The Conquest of the Desert was executed by military forces that first established control over the Pampas region before moving south of the Río Negro into Patagonia. Indigenous groups fought an "asymmetric war" in which they used their superior knowledge of the terrain to avoid engaging in large battles on open ground (Vezub and Healey 2020). The military, however, continued to push south as it succeeded in killing combatants and unarmed groups. Larson (2020b) notes that the "official" narrative of the military conquest ends with the surrender of the cacique Saygüeque in January 1885. Scholars have demonstrated that the "conquest" did not end in 1885 but instead took on new forms of internal colonization that sought to contain ongoing Indigenous resistance (Larson 2020a). Indeed, the war inflicted genocidal violence on Indigenous peoples, as the military established concentration camps for survivors and conscripted Indigenous men into the armed forces (Delrio and Pérez 2020; Larson 2020b; Vezub and Healey 2020). The war machine sought to clear out the "desert" and prepare the way for "progress" based on private property, agriculture, and white settlement.

The Conquest of the Desert had enormous implications for the Indigenous peoples of Santa Cruz. The Tehuelches were a nomadic foraging society integrated into the wider Indigenous networks in Patagonia that stretched across both sides of the Andes and included Tierra del Fuego (Pero 2002). Various Indigenous populations, fleeing violence to the north, took refuge in southern Patagonia. The government created six reservations between 1898 and 1927, including one near Lake Viedma, called Reserva del Lago Viedma, totaling some 200 km^2 (Nuevo-Delaunay et al. 2020; Rodríguez 2014). The Indigenous population was largely confined to reservations where they were expected to "disappear" or "go extinct" based on prevailing assumptions deriving from evolutionary anthropology (Argentine Ministry of Education and Sports 2016). Sited on lands with precarious rights, the reservations were populated by Tehuelche, Mapuche, and Mapuche-Tehuelche families (Argentine Ministry of Education and Sports 2016). In some cases, families were able to establish marginal holdings in the hinterlands away from white settlers where they could engage in hunting, livestock rearing, and horticulture (Nuevo-Delaunay 2012). However, individual holdings and reservations were subject to land grabs by settlers and, later, the state. In 1966, the Santa Cruz provincial government eliminated three reservations altogether, including the Lago Viedma Reserve (*Reserva del Lago Viedma*), and shrank the size of the remaining ones (Argentine Ministry of Education and Sports 2016; Rodríguez 2014). The reservation lands were sold off, and the government sought to push the resident populations toward urbanized areas where they would be assimilated into the Santacruceño working class (Nuevo-Delaunay 2012).

Territorial dispossession opened Patagonia to white settler colonialism (Bandieri 2005; Gott 2007). The Argentine government supported European immigration as a way to populate the frontier, perceiving white settlers as a biopolitical tool to bring civilization to the desert. European settlers—along with initial waves of Argentine and Chilean migrants—traveled to the region and founded or expanded coastal settlements in Punta Arenas, Ushuaia, Río Gallegos, and Puerto Madryn (Bandieri 2005; Edwards 2022). From the 1880s to the 1910s, southern Patagonia was established as a region integrated into transnational circuits of capital investment tied to the Falkland Islands (Islas Malvinas) and European markets (Barbería 1994; Harambour-Ross 2016). A landowning class established latifundios (large agricultural estates) on the most fertile and accessible spaces (Oliva et al. 2016). Landowners consolidated territory and created rural leagues (*sociedades rurales*) that pressed for favorable public policies and defended the prerogatives and impunity of elites. This impunity extended to the violent repression of farmworker unions pressing for better working conditions (Bayer 2008; Coronato and Tourrand 2020). This included the notorious *Patagonia rebelde* massacre of hundreds of workers and anarchist organizers in 1920–1921. Intermarriage and business alliances helped stitch together a regional oligarchy that developed companies and consortia that controlled the import-export trade, shipping, banking, and commerce (Banderi 2005). Subsequent waves of settlers pushed inland into the steppe and sub-Andean zones where they established small- or medium-sized holdings on more marginal territories, sometimes by way of formal leasing contracts and in other cases by informal land occupations (Barbieri 1994).

Territorial dispossession and white settler colonialism facilitated the transformation of Patagonia into a region dominated by agrarian rangeland capitalism. This landscape of capitalist opportunity took material form as a rangeland to be exploited primarily through sheep farming (supplemented by cattle and goat herds). Argentine Patagonia was colonized not just by white settlers but also by the sheep herds driven in from both the south and the north. In 1876, some 300 sheep were transported from the Falkland Islands to southern Patagonia to establish an initial stock (Aagesen 2000). Sheep were also introduced by way of the Pampas (Coronato and Tourrand 2020). Landholders created production systems on estancias initially oriented toward wool exports. The advent of refrigerated ships in 1894 allowed the industry to export meat as well (Aagesen 2000). However, this productive model—imposed on Patagonia—had taken shape within the environments of the Falkland Islands and Pampas with higher levels of precipitation (Coronato and Tourrand 2020). In the short run, livestock multiplied and expanded across environments now conceived as "rangelands," a process that was completed in the late 1930s (Coronato and Tourrand 2020). The sheep population grew to a high point of 22 million in the 1950s before declining over the next 50 years (Coronato et al. 2016; Oliva et al. 2016). Livestock farming began to stagnate as a result of environmental degradation.

Agrarian rangeland capitalism generated significant environmental degradation across Patagonia. Capital accumulation emerges through processes of appropriation that degrade landscapes, destroy plant and animal life (i.e., biota), strip the soil of nutrients, pollute the atmosphere and oceans, and create environmental externalities not incorporated into the valuation systems through which capital grows (Foster and Clark 2020). Extra-Andean Patagonia is dominated by steppe ecologies. Livestock farming—as it was implemented by settler colonists—precipitated environmental damage as this productive system fostered overgrazing, overstocking, erosion, and unsustainable land use (Del Valle et al. 1998; Aagesen 2000; Oliva et al. 2016). Despite warnings from scientists beginning in the early 1900s, the livestock sector continued to spread and exert pressures that outstripped the regenerative capacity of plants and soils (Andrade 2002). This process of deterioration—what environmental scientists refer to as "desertification"—entails plant, soil, and water resource degradation (Mazzonia and Vazquez 2009). Ironically, the very productive model that sought to remake the Indigenous Patagonian "desert" into a landscape of capitalist progress unleashed productive forces that undermined the ecological conditions of accumulation.

After peaking in the 1950s, sheep farming began a long decline. This led to a crisis of the Patagonian estancia that was defined by shrinking stocks, jobs, and profits, which depopulated the interior of the region (Andrade 2002; Coronato et al. 2016). This had serious economic effects on provincial governments trying to maintain the viability of rural society and deal with enormous inequality in land tenure and the abandonment of estancias (Aagesen 2000; Mazzonia and Vazquez 2009). By the end of the twentieth century, there were 500 abandoned estancias out of 1260 total in Santa Cruz alone (Coronato and Tourrand 2020). Del Valle et al. (1998) found that within the 785,000 km^2 that comprise Argentine Patagonia, 93.6% of the region exhibited at least some degree of desertification: slight (9.3%), moderate (17.1%), moderate to severe (35.4%), severe (23.3%), and very severe (8.5%). They argue that the "severe" and "very severe" categories, comprising some 31.8% of Argentine Patagonia, represent zones that most likely are irreversibly degraded (Del Valle et al. 1998). These environments can no longer support livestock. However, land degradation presents an opportunity for new waves of capital accumulation layered atop and alongside agrarian rangeland capitalism and a regional sheep population that still numbers nearly ten million (Coronato and Tourrand 2020). The mining and hydrocarbon industries would soon begin to dominate and convert the interior and coastlines into spaces of subterranean extraction (Shever 2012).

Moreno donated the plot adjacent to Lake Nahuel Huapi with the hope that it would become a "catalyst for human advancement" as a PA (Moreno 2002). This donation was made possible by military violence, the destruction of Indigenous lifeworlds, land dispossession, and white settler colonialism. This territorialization project sought to convert Patagonia into a capitalist rangeland for livestock farming. Moreno had long championed the development and populating of Patagonia with settlers who would help secure Argentine sovereignty. In a letter to General Julio

Roca in 1899, Moreno extols the military leader of the Conquest of the Desert for paving the way for "civilization to take hold" thus allowing Moreno to see his "dreams coming true" sooner than anticipated (Moreno 2002). Moreno talks up the economic potential of the "Patagonian territories," adding that "extraordinary things could be accomplished there" (Moreno 2002, p. 225).

2.2.2 Andean Conservation, Leisure Capitalism, and Eco-philanthropy

The second aspect to consider regarding Moreno's gift is how the APN institutionalized his vision for conservation. In the 1930s, a new territorialization project sought to create a national park system and strengthen sovereignty within the Andes. This vision reorganized the Andean borderlands around leisure capitalism based on tourism markets and the selling of an alpine landscape aesthetic. In the late twentieth century, this territorialization project was extended by eco-philanthropic investment.

Agrarian rangeland capitalism emerged against the backdrop of border disputes between Chile and Argentina and lingering concerns about the British Empire. In 1833, the British gained control over the Falkland Islands, adding to their South Atlantic holdings (Dodds and Benwell 2010). An imperial outpost for knowledge production (Blair 2019), the Falklands were also a base for ships to be repaired and resupplied, and a commercial hub linking mainland Patagonia to European markets (Bandieri 2005; Minchinton 1985). Over subsequent decades, there was a back-and-forth diplomatic messaging war by the Chilean and Argentine governments to establish sovereignty claims over Tierra del Fuego and eastern Patagonia (Perry 1980). The 1881 Border Treaty created a legal framework for the division of space along the "most elevated crests" of the Andean cordillera that "may divide the waters" (United Nations 1902). This legal rule proved difficult to apply in practice along stretches of the Andes where the orographic and hydrological lines diverged (Bandieri 2005). The 1902 arbitration case—overseen by the British Crown—more precisely established the international boundary line, with Francisco Moreno serving on the border commission (Wakild 2017). Diplomatic conflicts persisted, particularly in relation to the Southern Patagonian Ice Field (Sopeña 2008). In 1965 there was a small engagement between Chilean and Argentine armed forces that transpired just north of the Chaltén Massif and resulted in the death of one soldier. Ongoing border disputes at times pushed the two countries to the brink of war.

Conservative elites took advantage of the ousting of President Yrigoyen in 1930 to implement a new vision for Patagonia linking conservation, colonialism, and capitalism. This territorializing project sought to facilitate tourism and create a series of national parks in geopolitically sensitive zones. This began with the birth of the Argentine National Park Directorate (DPN) in 1934 (Law 12,103) and the formation of Iguazú National Park and Nahuel Huapi National Park (Kaltmeier 2021). Exequiel Bustillo, the first DPN president, highlighted the intersection of colonization and conservation in his memoir, "The Awakening of Bariloche: A Patagonian Strategy." Bustillo viewed the DPN and borderland parks as the means to "carry ahead our civilizing business in the lakes of the South and Iguazú" (Bustillo 1999, p. 123). As signaled by the title of his memoir, Bustillo's central concern was for Patagonia—not Iguazú and the far north (Freitas 2021). Bustillo (1999) lamented the perceived failure of the government to secure control over Andean Patagonia and lionized the conservation state—the park administration and the PA system under its management—as a heroic agency that had facilitated "the accumulation of capital, population, and all those elements essential for progress" (p. 11). For Bustillo (1999) the conservation state was an "integral tool for colonization" that activated tourism and thus "prepared the ground to complete the conquest" (p. 15). As noted above, the Conquest of the Desert did not end in 1885 (Larson 2020a). Indeed, this conquest logic was repurposed to legitimate and institutionalize an ethic of conservation.

The conservation state promulgated a landscape ideology of alpine aesthetics. This aesthetic took the European Alps as a geophysical referent and landscape category. The designation "alpine" entailed a specific way of understanding this environment—replete with mountains, forests, glaciers, and lakes—that reflected prevailing understandings of the Alps as a site for exploration and tourism for the leisured class (Navarro-Floria 2008). In an 1883 letter, Francisco Moreno (2002) described the Lakes Region as the "Switzerland of South America" and compared Lake Nahuel Huapi to Lake Geneva, imagining a future city to be built by the army, a "New Geneva" that would be "even more majestic than its Alpine counterpart" (p. 218). This alpine vision of the Patagonian Andes was affirmed and extended by Bustillo and the DPN (Navarro-Floria 2008). As Bustillo (1999) writes, the goal was to build the town of Bariloche into a picturesque mountain city as exists in "Switzerland" or "Tyrol." Bustillo (1999) eagerly quotes the French ambassador's impressions: "It is at once the Engadine, the Alps of Savoie, Italian and Swiss lakes, l'Esterel, and even the landscapes of Tuscany and Umbria, plus *le grand décor* of the Andes which is unto itself. Truly it is one of the most beautiful corners of the world" (p. 174). Ambassador D'Ormesson's letter reaffirmed the Lakes Region to be a stunning compendium of Italian, Swiss, and French alpine environments. This alpine aesthetic was enacted through an architectonic style elaborated with respect to public buildings in Bariloche and park infrastructure in Nahuel Huapi (Frischknecht 2006; Picone 2022). The first decade of the DPN (later renamed the APN) began the project of commodifying the Patagonian Andes as landscape experiences to be sold for upper-class leisure activities such as hiking, boating, and sightseeing. The Peronist administration would later open up leisure spaces for the working and middle classes. Beginning with Bariloche and Nahuel Huapi, the conservation state labored to produce space for leisure capitalism through selling aesthetic grandeur and recreational activities.

The conservation state expanded its territorialization project down the length of the Patagonian Andes. This involved the addition of new PAs in 1937, such as Glaciers, Lanín, Los Alerces, Lago Puelo, and Perito Moreno (National Parks

Administration of Argentina 2012). Other PAs followed, such as Bosque Petrificado, Laguna Blanca, Arrayanes, and Tierra del Fuego. Almost all of these parks were strategically placed along the Chilean border to consolidate territorial sovereignty, establish stable population bases, generate communities of Argentine nationals, attract capital investments to build tourism-based economies, and sell the alpine aesthetic to visitors. Until the late twentieth century, however, many PAs in southern Patagonia had minimal tourism infrastructure, fielded small ranger corps to enforce conservation rules, and remained closely connected to estancia-based livestock farming (Mendoza 2018). In Santa Cruz, Glaciers National Park is a case in point. The nascent park service recruited settlers to become the first park rangers. They patrolled on horseback to enforce park rules. Tourism was still incipient, though a trail to Perito Moreno Glacier was established for visitors on horses. A road to the Moreno Glacier was finally finished in the early 1960s, allowing tourists to reach it by automobile (Cousido 2003, p. 163). Not until the end of the twentieth century did Glaciers National Park become a booming tourism destination for sightseers and trekkers to complement the mountaineering expeditions that had long visited (Mendoza 2020).

Alongside the conservation state, this territorialization project gathered together entrepreneurs, workers, and visitors within tourism destinations along the cordillera. Capitalist territorialization projects are contingent sites of gathering and can attract unexpected actors. In the late twentieth century, a new generation of elites began purchasing large tracts of land in Chilean and Argentine Patagonia (Holmes 2015; Sánchez 2007; Tecklin and Sepulveda 2014). One eco-philanthropic organization in particular, Tompkins Conservation, has amassed holdings to create PAs and pursue the goals of rewilding ecosystems, building tourism infrastructure, and donating properties to the Argentine and Chilean governments to be converted into national parks (Beer 2022; Gale and Ednie 2019; García and Mulrennan 2020; Louder and Bosak 2019; Louder and Bosak 2022; Mendoza et al. 2017; Wakild 2009). Tompkins Conservation—especially its late founder Doug Tompkins—has faced accusations of green grabbing and land dispossession (Busscher et al. 2018; Louder and Bosak 2019; Wakild 2009). For Doug Tompkins, Francisco Moreno was a "mythical" person—a founding figure of conservation who should be recognized as Argentina's "John Muir" (Tompkins 2013, p. 127–129). Through its participation in the spirit of the gift, Tompkins Conservation has sought to construct an eco-philanthropic image in Argentina (and Chile) that advances Moreno's legacy and the institutionalized ethic of conservation.

Tompkins Conservation has primarily targeted the Patagonian Andes for its land acquisition and conservation efforts. Most significantly, Tompkins Conservation worked with the Chilean government to create the *Ruta de los Parques de la Patagonia* (The Route of Parks of Patagonia), establishing PAs from Puerto Montt to Cape Horn that contain some 17 national parks and cover 115,000 km^2. As Clare Beer (2022) has shown, Tompkins Conservation attached financing conditionality arrangements to their gift as a way to influence conservation governance outcomes and secure funding from the Chilean state to PAs within the Ruta de los Parques. In Argentina, Tompkins Conservation has sought to expand existing Andean parks or to create new PAs. The organization donated 150 km² to expand Perito Moreno National Park (Butler 2016, p. 69) and has worked to create the new Patagonia National Park as part of a transboundary protected zone that includes Chile's Patagonia National Park—the latter also spearheaded by Tompkins Conservation. In Tierra del Fuego, the organization has endeavored to create a park in Peninsula Mitre to complement the existing Tierra del Fuego National Park (Rewilding Argentina Foundation 2020). Eco-philanthropy has thus contributed to capitalist territorialization based on the expansion of the national park system in the Andes. In southern Patagonia, there is a binational eco-tourism circuit that connects the towns of Ushuaia, Puerto Natales, El Calafate, and El Chaltén and their respective national parks (Mendoza et al. 2017). There is also an important tourism circuit in northern Patagonia linking Bariloche, San Martín de los Andes, Pucón, and Puerto Varas, among other destinations. The goal—for Tompkins Conservation—is to create an eco-tourism circuit that traverses the Ruta de los Parques, generating revenue for Andean communities and building local support for parks, tourism, and green development. The hope is that this Chilean route will fuse synergistically with the existing northern and southern Patagonian corridors, thereby creating a massive park-tourism complex that dominates the Patagonian Andes.

Tompkins Conservation has also targeted coastal and marine environments in Argentine Patagonia. This began with the creation of the Monte León National Park in 2004 in concert with Fundación Vida Silvestre Argentina (Butler 2016). It has also sought to build upon the existing Interjurisdictional Coastal Marine Park of Southern Patagonia created in 2008 to which surrounding terrestrial and marine biosphere reserves were added in 2015 (Rewilding Argentina Foundation 2020). Tompkins Conservation has worked to establish a South Atlantic eco-tourism circuit selling the sublime experiences of coastal and marine environments.

2.3   Territorializing Moreno's Gift

The year before his death, Francisco Moreno reflected on the significance of his life. Moreno wrote that he had given land to create

> a National Park for the benefit of future citizens, so that they may find solace and renewed strength to serve this country. Yet I have nothing to give my children, not even a tiny plot in which to bury my ashes. (Moreno 2002, p. 13)

Moreno's narrative identifies the foundational gift of the national park that he has offered to citizens of Argentina. Moreno frames this gift around its potential to inspire a countergift: that citizens will find "renewed strength to serve this country" (Moreno 2002, p. 13). In contemporary Argentina, the APN has organized the history of the conservation state around the spirit of the gift. It has repaid Moreno by creating an institutional legacy in which he occupies a heroic role as the founding figure of conservation. National Parks Day in Argentina celebrates this legacy and continues to use the gift as a political resource to teach and inspire children and adults in Andean settlements like El Chaltén, exhorting the citizenry to answer the call to an ethic of conservation. In addition to the discourse of the gift revisited every sixth of November, Moreno's name now adorns libraries, towns, schools, roads, and parks throughout Argentina.

I have argued that this spirit of the gift is chained to projects of capitalist territorialization involving rangeland farming and Andean conservation. The first territorializing project employed state violence to clear out Indigenous populations in Patagonia for white settler colonialism and the creation of estancias for sheep farming. This produced a space for agrarian rangeland capitalism. Over generations, however, livestock farming facilitated widespread environmental change that undermined the ecological conditions of accumulation. Though it endures, sheep farming is significantly reduced compared to its zenith in the mid-twentieth century. New industries—mining and hydrocarbon extraction—have spread throughout the interior and coastal zones, generating expanding circuits of accumulation. A second territorialization project emerged in the 1930s that sought to respond to the perceived failures of the state to establish territorial sovereignty in the Andean borderlands. The conservation state promoted tourism markets and created a series of Andean parks to establish population centers and facilitate renewed colonization efforts. The conservation state enlisted an alpine landscape ideology to sell recreational activities and the aesthetic grandeur of the cordillera, at least initially, to wealthy tourists. The production of Andean space for leisure capitalism was unevenly accomplished, with parks in the south lagging far behind in terms of the tourism infrastructure that existed in Bariloche and Nahuel Huapi National Park. In the late twentieth century, eco-philanthropy organizations like Tompkins Conservation engaged in massive land purchases to create private estates that could then be donated to the Argentine and Chilean governments to expand the national park systems. Eco-philanthropists have thus sought to make (and perhaps launder) their reputations through participation in the spirit of the gift.

One consequence of these histories of capitalist territorialization is the division of Patagonian space into two sections. The Andean domain has been slotted for conservation, parks, and green development. By contrast, the extra-Andean domain (steppe, monte, and coastal zones) has been earmarked for resource extraction industries and infrastructure that intersect and overlap with rangeland farming. Eco-philanthropists—acting in concert with park administration allies—have sought to defend Andean ecosystems through land purchases and deal-making with federal governments, thereby legally placing these outside the realm of livestock farming, mining, and energy extraction. With the creation of the Ruta de los Parques in Chile, there is now a vast set of Andean PAs. There is a mega-park system that links the south (Bernado O'Higgins, Glaciers, Kawésqar, Torres del Paine, Alberto de Agostini, Yendegaia, and Tierra del Fuego) and a well-established network of the parks in the north (Lanín, Nahuel Huapi, Vincente Perez Rosales, Pumalín, and Corcovado, among others). This legal expansion of national parks has contributed to, reinforced, and deepened the Patagonian territorial imaginary that represents the Southern Andes as an eco-region.

The legacy of Moreno's gift is grounded within histories of capitalist territorialization. The official discourse of the gift is tethered to the progressive creation of an expanding national park system. National parks are symbolically resonant elements of the inalienable wealth protected by the state. In the first two decades of the twenty-first century, center-left and center-right administrations have contributed significantly to the federal system. However, this institutional legacy is equally important for what is not revealed in official discourse. This APN history—retold annually under the auspices of state ritual—has fundamentally failed to recognize and reckon with histories of extraction, violence, and dispossession. This chapter is a contribution to an emerging critical assessment of the gift within Patagonian conservation.

References

D. Aagesen, Crisis and conservation at the end of the world: Sheep ranching in Argentine Patagonia. Environ. Conserv. **27**(2), 208–215 (2000). https://doi.org/10.1017/S0376892900000229

L. Andrade, Territorio y ganadería en la Patagonia Argentina: Desertificación y rentabilidad en la Meseta Central de Santa Cruz [Territory and livestock farming in Argentine Patagonia: Desertification and profitability in the Central Plateau of Santa Cruz]. Econ. Soc. Territ. **3**(12), 675–706 (2002)

Argentine Ministry of Education and Sports, *Tehuelches y Selk'nam (Santa Cruz y Tierra del Fuego): No desaparecimos [Tehuelches and Selk'nam (Santa Cruz and Tierra del Fuego): We Did Not Disappear]* (Ministerio de Educación y Deportes, 2016)

S. Bandieri, *Historia de la Patagonia [History of Patagonia]* (Editorial Sudamericana, 2005)

E. Barbieri, El extremo austral Sudamericano. Ocupación y relaciones de los territorios argentinos y chilenos, 1880–1920 [The South American south. Occupation and relations of Argentine and Chilean territories, 1880–1920]. Estud. Front. **33**, 185–212 (1994)

O. Bayer, *La Patagonia rebelde [Rebel Patagonia]* (Booket, 2008)

C.M. Beer, Bankrolling biodiversity: The politics of philanthropic conservation finance in Chile. Environ. Plann. E: Nat. Space. Advance online publication (2022). https://doi.org/10.1177/25148486221108171

J.A. Blair, South Atlantic universals: Science, sovereignty, and self-determination in the Falkland Islands (Malvinas). Tapuya: Lat. Am. Sci. Technol. Soc. **2**(1), 220–236 (2019). https://doi.org/10.1080/25729861.2019.1633225

G. Blanco, A. Arce, E. Fisher, Becoming a region, becoming global, becoming imperceptible: Territorialising salmon in Chilean Patagonia. J. Rural. Stud. **42**, 179–190 (2015). https://doi.org/10.1016/j.jrurstud.2015.10.007

N. Brenner, Beyond state-centrism? Space, territoriality, and geographical scale in globalization studies. Theory Soc. **28**(1), 39–78 (1999). https://doi.org/10.1023/A:1006996806674

N. Busscher, C. Parra, F. Vanclay, Land grabbing within a protected area: The experience of local communities with conservation and forestry activities in Los Esteros del Iberá, Argentina. Land Use Policy **78**, 572–582 (2018). https://doi.org/10.1016/j.landusepol.2018.07.024

E. Bustillo, *El despertar de Bariloche: Una estrategia patagónica [Bariloche's Awakening: A Patagonian Strategy]* (Editorial Sudamericana, 1999)

T. Butler, *Tompkins Conservation 25: A Quarter Century of Work to Save the Wild* (Tompkins Conservation, 2016)

F.R. Coronato, J.F. Tourrand, Sheep policy in the colonization of Argentine Patagonia, in *Livestock Policy*, ed. by J.F. Tourrand, P.D. Waquil, M.C. Maraval, M.T. Sraïri, L.G. Duarte, G.V. Kozloski, (CIRAD, 2020), pp. 67–78

F. Coronato, E. Fasioli, A. Schweitzer, J.F. Tourrand, Repenser le rôle des moutons dans le développement local de la Patagonie en Argentine [Rethinking the role of sheep in the local development of Patagonia, Argentina]. Rev. Elev. Med. Vet. Pays Trop. **68**(2–3), 129–133 (2016). https://doi.org/10.19182/remvt.20599

F.J. Cousido, *Siglo parques [Century Parks]* (Editorial Amalevi, 2003)

H.F. Del Valle, N.O. Elissalde, D.A. Gagliardini, J. Milovich, Status of desertification in the Patagonian region: Assessment and mapping from satellite imagery. Arid Land Res. Manag. **12**(2), 95–121 (1998). https://doi.org/10.1080/15324989809381502

W. Delrio, P. Pérez, Beyond the 'desert': Indigenous genocide as a structuring event in northern Patagonia, in *The Conquest of the Desert: Argentina's Indigenous Peoples and the Battle for History*, ed. by C.R. Larson, (University of New Mexico Press, Albuquerque, 2020), pp. 122–145

P. Di Giminiani, *Sentient Lands: Indigeneity, Property, and Political Imagination in Neoliberal Chile* (University of Arizona Press, 2018)

M. Dicenta, G. Correa, Worlding the end: A story about colonial and scientific anxieties over beaver vitalities in the Castorcene. Tapuya: Lat. Am. Sci. Technol. Soc. **4**(1). Advance online publication (2021). https://doi.org/10.1080/25729861.2021.1973290

K. Dodds, M. Benwell, More unfinished business: The Falklands/Malvinas, maritime claims, and the spectre of oil in the South Atlantic. Environ. Plann. D: Soc. Space **28**(4), 571–580 (2010). https://doi.org/10.1068/d2804cm

R.C. Edwards, *A Carceral Ecology: Ushuaia and the History of Landscape and Punishment in Argentina* (University of California Press, Oakland, 2022)

J.B. Foster, B. Clark, *The Robbery of Nature: Capitalism and the Ecological Rift* (Monthly Review Press, New York, 2020)

F. Freitas, *Nationalizing Nature: Iguazu Falls and National Parks at the Brazil-Argentina Border* (Cambridge University Press, Cambridge, 2021)

M. Frischknecht, Placemaking: From colonisation to architectonic style in Andean Patagonia. J. Archit. **11**(2), 209–223 (2006). https://doi.org/10.1080/13602360600786142

T. Gale, A. Ednie, Can intrinsic, instrumental, and relational value assignments inform more integrative methods of protected area conflict resolution? Exploratory findings from Aysén, Chile. J. Tour. Cult. Chang. **18**(6), 690–710 (2019). https://doi.org/10.1080/14766825.2019.1633336

M. García, M.E. Mulrennan, Tracking the history of protected areas in Chile: Territorialization strategies and shifting state rationalities. J. Lat. Am. Geogr. **19**(4), 199–233 (2020). https://doi.org/10.1353/lag.2020.0085

J.K. Gibson-Graham, *The End of Capitalism (As We Knew it): A Feminist Critique of Political Economy* (University of Minnesota Press, Minneapolis, 2006)

G. Gordillo, *Landscapes of Devils: Tensions of Place and Memory in the Argentinean Chaco* (Duke University Press, Durham, 2004)

R. Gott, Latin America as a White settler society. Bull. Lat. Am. Res. **26**(2), 269–289 (2007). https://doi.org/10.1111/j.1470-9856.2007.00224.x

A. Harambour-Ross, Sheep sovereignties: The colonization of The Falkland Islands/Malvinas, Patagonia, and Tierra del Fuego, 1830s–1910s, in *Oxford Research Encyclopedia of Latin American History*, (2016), pp. 1–21. https://doi.org/10.1093/acrefore/9780199366439.013.351

D. Harvey, *Spaces of Capital: Towards a Critical Geography* (Routledge, 2001)

G. Holmes, Markets, nature, neoliberalism, and conservation through private protected areas in southern Chile. Environ. Plann. A: Econ. Space **47**(4), 850–866 (2015). https://doi.org/10.1068/a140194p

O. Kaltmeier, *National Parks from North to South: An Entangled History of Conservation and Colonization in Argentina* (University of New Orleans Press, 2021)

T. Klubock, *La Frontera: Forests and Ecological Conflict in Chile's Frontier Territory* (Duke University Press, Durham, 2014)

C. Larson, *The Conquest of the Desert: Argentina's Indigenous Peoples and the Battle for History* (University of New Mexico Press, Albuquerque, 2020a)

C. Larson, The conquest of the desert: The official story, in *The Conquest of the Desert: Argentina's Indigenous Peoples and the Battle for History*, ed. by C.R. Larson, (University of New Mexico Press, Albuquerque, 2020b), pp. 17–42

H. Lefebvre, *The Production of Space* (Blackwell Publishing, Oxford, 2004)

E. Louder, K. Bosak, What the gringos brought: Local perspectives on a private protected area in Chilean Patagonia. Conserv. Soc. **17**(2), 161–172 (2019). https://doi.org/10.4103/cs.cs_17_169

E. Louder, K. Bosak, Spectacle of nature 2.0: The (re)production of Patagonia National Park. Ann. Am. Assoc. Geogr. **113**(2), 331–345 (2022). https://doi.org/10.1080/24694452.2022.2106176

M. Mauss, *The Gift: The Form and Reason for Exchange in Archaic Societies* (W.W. Norton, New York, 2000)

E. Mazzonia, M. Vázquez, Desertification in Patagonia. Dev. Earth Surf. Process. **13**, 351–377 (2009). https://doi.org/10.1016/S0928-2025(08)10017-7

M. Mendoza, Educational policing: Park rangers and the politics of the green (e)state in Patagonia. J. Lat. Am. Caribb. Anthropol. **21**(1), 173–192 (2016). https://doi.org/10.1111/jlca.12195

M. Mendoza, *The Patagonian Sublime: The Green Economy and Post-neoliberal Politics* (Rutgers University Press, New Brunswick, 2018)

M. Mendoza, Alpine masculinity: A gendered figuration of capital in the Patagonian Andes. Bull. Lat. Am. Res. **39**(2), 208–222 (2020). https://doi.org/10.1111/blar.12839

M. Mendoza, R. Fletcher, G. Holmes, L.A. Ogden, C. Schaeffer, The Patagonian imaginary: Natural resources and global capitalism as the far end of the world. J. Lat. Am. Geogr. **16**(2), 93–116 (2017). https://doi.org/10.1353/lag.2017.0023

M. Mendoza, M. Greenleaf, E.H. Thomas, Green distributive politics: Legitimizing green capitalism and environmental protection in Latin America. Geoforum **126**, 1–12 (2021). https://doi.org/10.1016/j.geoforum.2021.07.012

W. Minchinton, The role of the British South Atlantic Islands in sea-borne commerce in the nineteenth century. Actas del IV Coloq. Hist. Canar.-Am. **4**, 543–576 (1985)

J.W. Moore, *Capitalism in the Web of Life: Ecology and the Accumulation of Capital* (Verso, New York, 2015)

F. Moreno, *Perito Moreno's Travel Journal: A Personal Reminiscence* (El Elefante Blanco, Buenos Aires, 2002)

National Parks Administration of Argentina, *Guía visual parques nacionales de la Argentina [Visual Guide to National Parks in Argentina]* (Administración de Parques Nacionales–Ministerio de Medio Ambiente, 2012)

P. Navarro-Floria, *Historia de la Patagonia [History of Patagonia]* (Ciudad Argentina, 1999)

P. Navarro-Floria, El proceso de construcción social de la región del Nahuel Huapi en la práctica simbólica y material de Exequiel Bustillo (1934-1944) [The process of social construction of the Nahuel Huapi region in the symbolic and material practice of Exequiel Bustillo 1934–1944]. Rev. Pilquen **10**(1), 1–14 (2008). https://revele.uncoma.edu.ar/index.php/Sociales/article/view/2065/58559

G. Nouzeilles, Patagonia as borderland: Nature, capital, and the idea of the state. J. Lat. Am. Cult. Stud. **8**(1), 35–48 (1999). https://doi.org/10.1080/13569329909361947

A. Nuevo-Delaunay, Disarticulation of Aónikenk hunter-gatherer lifeways during the late nineteenth and early twentieth centuries: Two case studies from Argentinean Patagonia. Hist. Archaeol. **46**(3), 149–164 (2012). https://doi.org/10.1007/BF03376875

A. Nuevo-Delaunay, J. Belardi, F. Carballo, Nuevas evidencias de sitios arqueológicos Tehuelche/Aoniken-Mapuche (siglo XX) en Santa Cruz, Patagonia (Argentina) [New evidence of Tehuelche/Aoniken-Mapuche archaeological sites (20th century) in Santa Cruz, Patagonia (Argentina)]. Magallania **48**(1), 161–172 (2020). https://doi.org/10.4067/S0718-22442020000100161

L. Ogden, *Loss and Wonder at the World's End* (Duke University Press, Durham, 2021)

G. Oliva, J. Gaitan, D. Ferrante, Humans cause deserts: Evidence of irreversible changes in Argentinian Patagonia rangelands, in *The End of Desertification?* ed. by R.H. Behnke, M. Mortimore, (Springer Earth System Sciences, Heidelberg, 2016), pp. 363–386

A. Pero, The Tehuelches of Patagonia as chronicled by travelers and explorers in the nineteenth century, in *Archaeological and Anthropological Perspectives on the Native Peoples of Pampa, Patagonia, and Tierra del Fuego to the Nineteenth Century*, ed. by C. Briones, J.L. Lantana, (Bergin & Garvey, Westport, 2002), pp. 103–119

R. Perry, Argentina and Chile: The struggle for Patagonia, 1843–1881. Americas **36**(3), 347–363 (1980)

M. Picone, Settling Bariloche: Explorations, violence, and tourism in the Argentine frontier, in *Oxford Research Encyclopedia of Latin American History*, (2022), pp. 1–31. https://doi.org/10.1093/acrefore/9780199366439.013.1060

M. Rasmussen, Institutionalizing precarity: Settler identities, national parks and the containment of political spaces in Patagonia. Geoforum **119**, 289–297 (2021). https://doi.org/10.1016/j.geoforum.2019.06.005

M. Rasmussen, C. Lund, Reconfiguring frontier spaces: The territorialization of resource control. World Dev. **101**, 188–399 (2018). https://doi.org/10.1016/j.worlddev.2017.01.018

Rewilding Argentina Foundation, *Annual Report 2019* (2020). Retrieved 20 Jan 2023, from https://ww2.rewildingargentina.org/library/boletines/ra-2019-annual-report.pdf

M.E. Rodríguez, Trayectorias de una recuperación en suspenso (ex Reserva Lago Viedma) [Trajectories of a recovery in suspense (former Lago Viedma Reserve)]. Ava: Rev. Antropol. **14**, 1–22 (2014). http://argos.fhycs.unam.edu.ar/handle/123456789/304

G. Sánchez, *La Patagonia vendida: Los nuevos dueños de la tierra [Patagonia Sold: The New Owners of the Land]* (Marea, 2007)

E. Shever, *Resources for Reform: Oil and Neoliberalism in Argentina* (Stanford University Press, Stanford, 2012)

G. Sopeña, *Memorias de Patagonia: Crónicas, escenarios, personajes [Memories of Patagonia: Chronicles, Scenarios, Characters]* (Booket, 2008)

C. Taylor, *Modern Social Imaginaries* (Duke University Press, Durham, 2004)

D. Tecklin, C. Sepúlveda, The diverse properties of private land conservation in Chile: Growth and barriers to private protected areas in a market-friendly context. Conserv. Soc. **12**(2), 202–217 (2014). https://doi.org/10.4103/0972-4923.138422

D. Tompkins, Adventures of the irresponsible, in *Climbing Fitz Roy 1968: Reflections on the Lost Photos of the Third Ascent*, ed. by Y. Chouinard, D. Dorworth, C. Jones, L. Tejada-Flores, D. Tompkins, (Patagonia Books, New York, 2013)

United Nations, *The Cordillera of the Andes Boundary Case (Argentina and Chile). Reports of the International Arbitral Awards*, vol IX (United Nations, 1902)

P. Vandergeest, N. Peluso, Territorialization and state power in Thailand. Theory Soc. **24**(3), 385–426 (1995). https://doi.org/10.1007/BF00993352

J. Vezub, M. Healey, Occupy every road and prepare for combat': Mapuche and Tehuelche leaders face the war in Patagonia, in *The Conquest of the Desert: Argentina's Indigenous Peoples and the Battle for History*, ed. by C.R. Larson, (University of New Mexico Press, Albuquerque, 2020), pp. 43–70

E. Wakild, Purchasing Patagonia: The contradictions of conservation in free market Chile, in *Lost in the Long Transition: Struggle for Social Justice in Neoliberal Chile*, ed. by W.L. Alexander, (Lexington Books, Lanham, 2009), pp. 112–125

E. Wakild, Protecting Patagonia: Science, conservation and the pre-history of the nature state on a South American frontier, 1903–1934, in *The Nature State: Rethinking the History of Conservation*, ed. by W.G.V. Hardenberg, M. Kelly, C. Leal, E. Wakild, (Routledge, 2017), pp. 37–54

A. Weiner, Inalienable wealth. Am. Ethnol. **12**(2), 210–227 (1985). https://doi.org/10.1525/ae.1985.12.2.02a00020

Chapter 3
Connectivity, Tourism, and Conservation: From Extractive Appropriation to Socio-Environmental Reappropriation of Nature in Aysén

María Dolores Muñoz Rebolledo, Fulvio Rossetti, Robinson Torres-Salinas, and Santiago Urrutia Reveco

Abstract This chapter explores how connectivity has interacted with tourism and nature conservation in the Aysén region of Chilean Patagonia, and some of the social, environmental, and territorial issues that have arisen from this articulation. We posit that the Longitudinal Austral Highway (*Carretera Austral*, in Spanish), as a road that facilitates connectivity and access to natural environments throughout the Aysén region, has operated as the mobilizing axis for different processes of valuing and appropriating nature. We argue that the links between connectivity, tourism, and conservation have changed over time, as actors' material and symbolic appropriations of nature in Patagonia have evolved. Our goal with this chapter is to inform ongoing Chilean debate about the ways in which nature should be valued as the country moves forward. Our case study research identifies and discusses the new social, environmental, and territorial processes brought about by the *Carretera Austral* and the consequent resignification and reappropriation of nature that has

M. D. Muñoz Rebolledo (✉)
Department of Urban Planning, Faculty of Architecture, Urbanism, and Geography, Universidad de Concepción, Concepción, Biobío Region, Chile
e-mail: marmunoz@udec.cl

F. Rossetti
Faculty of Engineering, School of Architecture & Centro de Innovación de Ingeniería Aplicada (CIIA), Universidad Católica del Maule, Talca, Región del Maule, Chile
e-mail: frossett@uc.cl

R. Torres-Salinas
Department of Sociology, Faculty of Social Sciences & Department of Territorial Planning, Faculty of Environmental Sciences, Universidad de Concepción, Concepción, Biobío Region, Chile
e-mail: robtorre@udec.cl

S. Urrutia Reveco
Instituto de Geografía, Universidad de Buenos Aires (UBA)-CONICET, Buenos Aires, Argentina
e-mail: surrutiareveco@gmail.com

© The Author(s) 2023   47
T. Gale-Detrich et al. (eds.), *Tourism and Conservation-based Development in the Periphery*, Natural and Social Sciences of Patagonia,
https://doi.org/10.1007/978-3-031-38048-8_3

occurred in Aysén. We base our case study on review of secondary data sources including public reports, plans, and policies related to the Carretera Austral and territorial development, as well as scholarly research on related topics, including values, identity, territorial transformation, and heritage.

Keywords Patagonia · Connectivity · Tourism · Conservation · Reappropriation of nature

3.1 Introduction

This chapter explores how connectivity has interacted with tourism and nature conservation in the Aysén region of Chilean Patagonia, and some of the social, environmental, and territorial issues that have arisen from this articulation. We argue that the links between connectivity, tourism, and conservation have changed over time, as actors' material and symbolic appropriations of nature in Patagonia have developed and evolved. Our thesis is built on the assumption that historical processes of government appropriation and social reappropriation have evolved through a modernist, or instrumental view of the environment, that is, a resource-oriented view of the world that views nature in terms of economic production and accumulation (Leff 2004, 2019).

Specifically, capitalist extractivist oriented appropriation of nature has guided development in the Aysén region of Chilean Patagonia since the time of modern colonization, during the early twentieth century (Blair et al. 2019; Núñez et al. 2018; Rossetti 2018a, b). Diamanti (2018) defined this as *extractivism*, saying, "Extractivism names a given economic form of organizing natural and social resources in which sustained profitability depends on the extraction, over time, of an increasing amount of natural resources from the earth" (p. 54). Gudynas (2015) placed extractivism within the Latin American context, describing, "a type of extraction of natural resources, in large volumes or high intensity, that are essentially oriented to be exported as raw materials without or with minimal processing" (p. 13).

Since the construction of the Longitudinal Austral Highway (*Carretera Austral*, in Spanish) began in 1976, there has been a fragmentation in social values associated with the appropriation of nature in Aysén. This new social appropriation of nature is based on environmental imaginaries linked to protection and conservation, and with the development of a local economy increasingly centered around nature-based tourism (Blair et al. 2019; Núñez et al. 2018; Rossetti 2018a, b). Thus, we argue that the dominant twentieth century modernist or instrumental worldviews are in the process of shifting for a considerable part of the population.

Our goal with this chapter is to inform ongoing debate within Chile about how nature should be valued as the country moves forward. In this context, we posit that the Carretera Austral (CA), as a road that facilitates connectivity and access to natural environments throughout the Aysén region, has operated as the articulating and mobilizing axis for these different processes of appropriating nature. Thus, our case study research identifies and discusses the new social, environmental, and territorial processes brought about by increasing connectivity (i.e., the CA), and the consequent resignification and reappropriation of nature that has occurred in Aysén. We base our case study on review of secondary data sources including public reports, plans, and policies related to the CA and territorial development, as well as scholarly research on related topics, including values, identity, territorial transformation, and heritage.

3.2 Theoretical Framework

Mexican socio-environmental thinker Enrique Leff (2004, 2019) has argued that the capitalist appropriation of nature has manifested through a mobilization and application of modern scientific knowledge, that, in its eagerness to control and dominate nature to make it productive, has generated entropic processes of environmental degradation, biodiversity loss, and sociocultural fracture that are not sustainable for life on the planet. He has described the capitalist appropriation of nature as a socio-environmental crisis, at a planetary level, and a catalyst for a growing ecological culture that is revaluing nature as sustenance for all forms of life among diverse communities and Indigenous *peoples of the Earth*. During this process of social redefinition of the value of the environment, sociocultural perceptions change. For example, seeing and/or experiencing the impacts associated with capitalist and extractive appropriation of nature might provoke a revaluation of nature and the environment, accompanied by demands for more sustainable strategies for the use of nature, or even, the recognition that nature also has rights that should be recognized and respected.

Leff has attributed the growth of these movements to a revaluation of nature, writing, "socio-environmental demands are revaluing nature, [which] entails a process of resignification and revaluation of nature by different social groups, in different ecological and cultural contexts" (2019, p. 31). Martinez-Alier (2011) described various environmental and conservation movements that have gained force during the late twentieth century, like deep ecology, eco-efficiency (an ecological modernization), and social movements that fight for environmental justice, which he described as *ecology of the poor*. Moore (2020) described similar tendencies, noting a revaluation of biodiversity rich areas, as *pristine nature*, the emergence of communities with a growing ecological culture, and new socio-environmental movements that are defending nature within territories.

Landscape researchers, namely, Maderuelo (2006) and Rodríguez (2021), have described landscapes within territories as being both physical spaces and cultural constructions. Landscapes represent a series of ideas, sensations, and feelings. They are socially elaborated and mobilized cultural projections that are motivated by the sensitive contemplation of a place, and they are dynamic over time. Landscapes within territories are experienced both through contemplation, and also through human action; thus, humans can be both landscape spectators and protagonists (Marchán 2006). In synthesis, landscape is simultaneously an expression of geographic, cultural, and temporal realities that is endowed with meanings and affective values (Silvestri 2011). Enrique Leff's work describes a contemporary social reassessment of landscapes, in Latin America and Chile, which is driven by social, cultural, and ancestral forces that are reassigning, or perhaps, reasserting meaning and value to nature, and orienting territories around the conservation of ecosystems that are currently highly threatened by predominant (neo)extractivist regimes of capital accumulation (Svampa 2019). Through these mutual relationships between the symbolic and the material of the landscape, this chapter will explore the different manners in which the nature of Aysén has been valued, signified, and appropriated. Special attention will be placed on nature-based tourism and different kinds of conservation projects, as well as the catalyzing role of evolving terrestrial connectivity infrastructure.

3.3 Methodology

3.3.1 Study Area

The Aysén Región of Chile is situated 1650 km south of the country's capital city of Santiago, in the Patagonia cultural region of the southern cone, which extends through southern Chile and Argentina. The continental territory of Aysén represents around 14% of Chile, making it the third largest region in the country; yet, its population, of around 103,000, represents just 1% of the country's total (Gale and Ednie 2019; Chilean National Statistics Institute 2017; Muñoz and Torres Salinas 2010). In contrast to the grasslands, pampas, and Atlantic shoreline that dominate Argentine Patagonia, the terrain in Aysén offers enormous diversity, with lush native forests, grasslands, steppe, lakes, and rivers, towering peaks, glacial fields, islands, fjords, and Pacific shoreline (Blair et al. 2019). For centuries, long before these lands were Chilean, and well into the nineteenth century, this territory was inhabited by various hunter-gatherer groups, including the Aonikenk (Tehuelche), Chono, and Kawéskar (Martinic 2014; Méndez and Nuevo-Delaunay 2019; Méndez et al. 2020; Reyes et al. 2019).

The complexity of the relief and the lack of articulation with important production circuits explain why even at the beginning of the twentieth century, some three and a half centuries after Chile was first colonized by Spain, the Aysén was still one of the most uninhabited and unknown regions for modern Chilean settlers (Martinic 2014). Modern Chilean occupation of the Aysén territory, and the subsequent creation of population centers, began in the first half of the twentieth century, based on a series of extractive activities, including logging, ranching, mining, and fisheries (Martinic 2014). Early settlement during this period occurred in two parallel manners. First, the Chilean government granted a number of large land concessions that brought in large companies and their workers. Second, spontaneous migration occurred with individual families and/or small groups of settlers who arrived from other parts of Chile, Argentina, and other parts of the world, all establishing claims within the territory, and establishing a range of livelihoods, based on farming, ranching, fishing, logging, and other extractive pursuits (Martinic 2014).

Twentieth century colonization of the region constituted a complex sociocultural process in a territory with strong challenges for permanent settlement. During the early decades of this century, Aysén was only accessible by land from Argentina because the highly fragmented formations in Chile, including the fjords, effectively separated it from the rest of the Chilean mainland (Martinic 2014; Muñoz and Torres Salinas 2010). By the middle of the twentieth century, the shores of General Carrera Lake, the area along the Baker, Cisnes, and Palena Rivers, and the Puyuhuapi Fjord had begun to transform with the establishment of settlements connected by lake and river navigation routes. From the 1970s onward, processes of territorial occupation accelerated in Aysén, with the construction and gradual opening of the CA and other land and maritime routes (Fig. 3.1).

Currently, the CA, more than 1270 km in length, begins in Puerto Montt (Los Lagos Region) and ends in the small town of Villa O'Higgins within the southern reaches of the Aysén Region, providing connectivity to a territory of some 128,000 km^2 (Adiego et al. 2018). The CA serves the region as the north to south artery for overland transit and is crossed by secondary roads that link the eastern areas of the region with the fiords to the west, paralleling the course of the main river basins.

3.3.2 Methods

This chapter examines impacts of the increased connectivity created by the CA for local actors of the Aysén Region, with the goal of exploring a territorial example of changing conceptualizations and relationships between humans and nature. Our methods focused on the review of secondary data sources, including public reports, plans, and policies related to the CA and territorial development, as well as scholarly research on related topics, including values, identity, territorial transformation, and heritage. Documents were analyzed through a qualitative, inductive process, focused on extracting individual meaning within passages and quotes, that would enable a subsequent process of open coding, triangulation, and an interpretation of results that adequately captured the complexity of the situation in terms of perspectives and influences (Creswell 2014; Schutt and Chambliss 2013; Stake 2003; Yin 2011). Our analysis focused on the identification of underlying factors and decisions that occurred with respect to access and connectivity in the Aysén region and of associated cultural signals that materialized in attitudes and values around nature. Cultural signals were coded in terms of their association with an extractive appropriation of nature, or a conservation-based appropriation of nature oriented toward sustainable use and the care of nature in Aysén.

Fig. 3.1 Map of the Carretera Austral and other connectivity infrastructure in the Aysén Region of Chile

3.4   Results

3.4.1   The History of Aysén's Austral Road Network and Nature Protection

The resulting narrative reconstructs the factors and signals we observed in the data, tracing the transition we observed to be occurring in Aysén, and the changes we have identified in the ways in which nature is being appropriated. Results are organized around three primary periods of change: each with their own overarching theme. First, we present the period between 1902 and 1929, which we have characterized as being a time of *Opening Routes*. Next, we focus on the period between 1930 and 1989, which we have characterized under the theme of *The Carretera Austral and nature as economic-geopolitical resources*. We conclude our results with the period between 1990 and today, which we view as a time in which *Connectivity has become a platform for the resignification of nature*.

3.4.2   Opening Routes (1902–1949)

Today, the CA is considered the main road within the Aysén regional road system, providing north-south connectivity. It is complemented by east-west routes, to form the Austral Road Network. Arguably, its origin dates back to trails that were developed during the expeditions that began in 1902, to settle the border dispute with Argentina along the courses of the Simpson, Aysén, Cisnes, Palena, and Baker rivers which transverse the region (Rossetti 2018a). Once the international boundary was agreed, population and development policies began to be defined. Chile's central government delegated many of the responsibilities for integration of central-western Patagonia into the Chilean society to private companies (mainly backed by foreign capital), through a series of long-term ranching concessions (e.g., Sociedad Industrial de Aysén, Compañía Explotadora del Báker, Sociedad Ganadera Río Cisnes). The terms of the concessions included that these companies facilitate the settlement of the region by constructing roads, housing, schools, and other basic infrastructure. However, at the end of the 1920s, little had changed, especially in terms of transportation infrastructure and settlement. These failures led to the generation of new policies characterized by a greater role of the State.

At that time, Chile's strategy for connectivity between the region and the rest of the country depended almost entirely on maritime routes; thus, transportation routes were planned around a series of ports and included transversal roads to be built along the same routes as the existing trails alongside the region's rivers. Nevertheless, by the end of the 1920s, President Carlos Ibáñez del Campo (1927–1931) had only managed to realize the first of these envisioned roads, which connected the region's main port in Puerto Aysén, the town of Baquedano (later Coyhaique), and the small border town of Balmaceda. After crossing the border, Argentine roads continued to the Atlantic (Rossetti 2018a).

At the end of the 1930s, this same maritime-focused connectivity strategy led President Pedro Aguirre Cerda (1938–1941) to inaugurate the first official tourist cruises to the San Rafael Lagoon and glacier. The Ofqui Canal project was initiated, which intended to build a canal across the Isthmus of Ofqui (Taitao peninsula) to improve maritime connectivity between Magallanes, Aysén, and the rest of Chilean Territory (Martinic 2013; Rossetti 2018b). In parallel, the ranching companies' concessions were reduced, and settlement was encouraged through a program of land grants designed to balance the population distribution, which was considered to be too heavily concentrated in areas that bordered Argentina. In fact, several documents of the time (Pomar 1923; Oportus Mena 1928; Monge 1944) testified about the lack of transportation connections between Aysén's cattle-raising pampas to the east, and the Pacific, to the west. These conditions were perceived as contributors to geopolitical fragility, as they created a dependence by settlers on the trans-Andean territory; specifically, on Argentine road systems, for connectivity and access to markets and services.

During this period of opening routes, both authorities and settlers in Aysén fundamentally viewed nature as a resource to be dominated and extracted. Thus, its native Lenga forests and other native species were viewed as obstacles that must be overcome to advance the hopes of progress and sovereignty, which, at the time, rested on livestock and ranching. These ideologies are exemplified by the snapshots taken of the recently created province of Aysén included in the book, *Chile: 280 Copper Engravings* (Gerstmann 1932). In his sequences, a cattle-raising future was extolled and the burning of forests to build roads was celebrated (Fig. 3.2). In this context, it is also not surprising that in 1940 the Hotel de la Laguna San Rafael was erected in front of the homonymous glacier, and that its selection as the main tourist destination was not considered incompatible with the transformation of the environment left by the construction works of the Ofqui canal (Rossetti 2016).

3.4.3 The Carretera Austral and Nature as Economic-Geopolitical Resources (1950–1989)

For most of the twentieth century, transversal roads in Patagonia were much more developed in Argentina than Chile; various routes reached from the Chilean border to the Atlantic; yet only a very few reached west to the Pacific. Thus, many of the spontaneous Chilean settlers in Aysén migrated through Argentina to reach the region, and subsequently maintained their dependence on Argentine territory as a route for communication, transport, and trade (Rossetti 2018a, b). Their settlement tended to be fluid; they would move back and forth across the borderlands regardless of the international boundary when grazing lands were no longer productive on one side of the border. As historian Adolfo Ibáñez Santa María (1973) pointed out,

Fig. 3.2 Historic photo portraying the burning of native forests along the Simpson River to pave the way for roads in Aysén. (Gerstmann 1932)

at this point in history, "occupying land on the Chilean side is nothing more than an accident" (p. 312). The geopolitical problems created by this prolonged lack of control over the flow of Chilean settlers were compounded by a continuous burning of forests in Chile, to prepare the territory for livestock grazing. Fires often burned out of control, spreading and affecting the neighboring territory of Argentina. The ability to control these fires was greatly hindered by wind, the absence of roads, and access difficulties caused by the prolific density of forests and marshes. By the 1950s, increasing cross-border tensions fostered urgency and the will to concentrate Chilean government planning and development efforts on the eastern interior of Aysén, in closer proximity to the Argentine border. These relatively populated areas were considered to be more conflictive and isolated, in comparison with the less inhabited and supposedly unproductive Pacific coastline and maritime area. Thus, the Chilean government began to concentrate connectivity efforts on road infrastructure, rather than continued overdependence on maritime routes (Rossetti 2018a, b).

As the 1950s ended, tourism was still emerging and the lack of roads limited tourism activity to maritime cruises within the remote Pacific archipelagos, channels, and fjords. This territory was perceived as being uninhabitable; thus, in 1959, it was designated as San Rafael Lagoon National Park, an immense PA that currently spans 17,420 km². As exemplified by the photographs in Gerstmann's 1959 edition, increasing knowledge of this PA led to greater awareness of Aysén's eminent mountainous, aquatic, and glacial landscapes. With increasing levels of awareness, appreciation of the region's landscape began to grow, and new discourses began to emerge, no longer conceptualizing Aysén's nature as being hostile or an obstacle to development; rather, representing the region's incredible natural landscapes as common goods to be preserved. Figure 3.3 demonstrates the contrast in how the region's landscape was portrayed pre- and post- this conceptual change. In comparison to the 1932 first edition of this book (Fig. 3.2), in which Patagonia was presented as a territory in the process of a radical anthropogenic transformation of nature, this reprint presented a completely different image, representing Patagonia through a series of pristine maritime and glacial landscapes.

Nevertheless, these subtle shifts were largely occurring for landscapes visible from the sea, as terrestrial tourism in the continental interior of the region was extremely limited due to the lack of infrastructure. Thus, it would be necessary to wait for the development of the CA for a greater articulation of this type of social revaluation of nature, which would emerge in parallel to the development of tourism in the continental interior.

During the 1960s, under the presidency of Eduardo Frei Montalva, a series of efforts were implemented to alleviate tensions with Argentina and to bring Chilean Patagonia into geopolitical symmetry with the more advanced economic and territorial infrastructure present within Argentine Patagonia. Efforts focused on reducing transhumance dynamics by improving the quality of eroded soil and facilitating the foundation of new, permanent Chilean settlements (Martinic 2014; Rossetti 2018a, b). Specific initiatives included the implementation of a new fire control and prevention policy in 1962 (Law 15.066), a bi-national agreement to regulate and monitor fires on the international border in 1967 (Decree 254), and the establishment of several new national PAs. The state also developed tree nurseries to facilitate

Fig. 3.3 Historic photography of nature in San Rafael Lagoon National Park. (Gerstmann 1959)

mechanisms that would increase timber exports and established pre-cooperative ranching settlements, composed of groups of families who worked together, as part of the agrarian reform (Martinic 2014; Rossetti 2018b). In parallel, a number of expeditions were made to determine feasible overland routes for Aysén, and the 1968 proposal, *Carretera Longitudinal Austral, Puerto Montt—Aisén, Antecedentes del proyecto*, was published by the Chilean Roads Administration. This document proposed, "to build a longitudinal road that, starting from Puerto Montt in the province of Llanquihue, would develop through the areas with the greatest possibilities of exploitation and the population centers of these provinces, up to the city of Aysén" (Chilean Highway Administration of the Chilean Ministry of Public Works 1968, p. 11). As the proposal stated, "The fundamental purpose of the road submitted for consideration is to incorporate a large territorial area into the Nation, which today is physically disconnected and marginalized from economic and social development" (Chilean Highway Administration of the Chilean Ministry of Public Works 1968, p. 38). From this proposal came the plans that were finally implemented in the 1970s, 1980s, and 1990s, through the Army Engineers Command and the Military Labor Corps, in coordination with the Ministry of Public Works and several private companies, which provided work for unemployed workers through the Minimum Employment Program (PEM) (Adiego et al. 2018; Ruiz 2016).

Although the landscape's tourism potential was not a motivator within the road proposal, it can easily be identified within the types of economic development the project supported. The most notable advances in the construction of the longitudinal road took place between 1982 and 1988, in a context marked by the authoritarian regime and border tensions between the Chilean and Argentine dictatorships. In 1982, the first section of the road was inaugurated between Chaitén—a town until then accessible only by sea—and Coyhaique, the regional capital of Aysén. In 1988 the road was extended northward to Puerto Montt, joining the Pan-American Highway and south from Coyhaique to Cochrane (Van Schouwen 2003).

The CA was understood as one of the most representative works of the dictatorship because of its supposed geopolitical importance (García 1989). The military regime viewed the CA as being a core element in achieving territorial integrity, national sovereignty, and local and national economic development. Between longitudinal and transversal sections, 2400 kilometers of roads were built, with 118 permanent bridges, three ferries (e.g., Reloncaví, Comau, Riñihue), and an investment of US$198 million (Chilean Ministry of Public Works 1988). Within this framework, nature continued to be interpreted from a strategic, economic, and extractive point of view (Pinochet 1993; Von Chrismar 1986).

Although it is designated as a highway (*Carretera* in Spanish), its original construction did not follow international standards for roads categorized as highways. In order to reduce costs, it was designed and developed as a *penetration road*, with the objective of connecting locations through the shortest routes (Adiego et al. 2018). The road layout was curvilinear, responding to the natural slope of the terrain, thus avoiding costly civil works like bridges or viaducts, whenever possible. Roads were built with gravel and of minimum widths, under the assumption that widening to two lanes and asphalt paving would occur in subsequent projects

(Horvath 1992). Some of the assumptions made in the original design of the carretera set maximum speeds of between 30 km and 40 km per hour, minimum curve
radii of between 20 and 30 m, and maximum slopes of between 10% and 12%. Its
width was between three and four meters, with culverts and minor bridges (up to
15 m) of native wood, often harvested in the process of opening up the corridors
(Adiego et al. 2018). Larger bridges that could not be avoided were between 15 and
100 m in length, designed in double arch, with metal beams. Ferry systems were
developed for watercourse crossing of over 100 meters (Ruiz 2016).

In spite of its original geopolitical and economic-extractive motivation, the highway's operational standards and restrictions meant that the road was developed
within the landscape it crossed in such a way that largely blended in with the existing natural surroundings. As such, the CA ultimately became a scenic route, providing value to the territory through increased accessibility to the region's natural
surroundings that did not exist before.

As this road work progressed in the 1980s and 1990s, the need for lodging and
food was generated for workers of the contracting companies. These circumstances
triggered a spontaneous tourism service offering by the inhabitants of various towns
that gradually resulted in the production of basic lodging and food services. Over
time, more specialized services were added as more tourists arrived. For example,
during an interview conducted in the summer of 2007, a local camping area operator
described:

> I started in the seventies to provide board and lodging, I had a lot of people because at that
> time nobody else was working, but then more people moved in, so fewer people started to
> come to my place. And then it occurred to me, because one day we found a Japanese man
> sleeping in the square, and we took him there and he pitched a tent, in the [current] camping
> area, and that's when it dawned on me and I said I'm going to make a camping area, and I
> started to make little tables, stoves, and then I installed lighting, and there I am, and then I
> built a bathroom. (Torres and Rojas 2011)

In a sense, this example of an emerging business illustrates how different tourism
services were created and adapted as demand grew, from contractor workers and
public officials to national and international tourists. Particularly in rural communities, the growing interaction between locals and tourists has played an important
role in the creation of new tourism businesses, as tourist recommendations became
implemented as new services (Torres and Rojas 2011). In this sense, we believe the
southern road has also served as a springboard for the configuration of what Doreen
Massey calls "a global sense of place," that is, a place where global and local ideas
interact fluidly (Albet and Benach 2012). Thus, over the years, new forms of tourism appeared as a result of the opening and territorial connectivity brought about by
the CA (Muñoz and Torres Salinas 2010; Torres and Rojas 2011).

Increased accessibility facilitated the flow of tourists, making it possible, toward
the end of the dictatorship, to consider Aysén as the *great tourist reserve of Chile*
(Diario de Aysén 1988). Popular press and discourse began to shift messaging
around the road, emphasizing:

> special relevance that the work of the Longitudinal Austral Road has for tourism in Aysén..
> [as]...an integrating axis that has allowed the regional tourism to offer to incorporate a great

variety and quantity of tourist resources and to take advantage of a natural space free of contamination in which there are scenic views of rivers, lakes, mountains, glaciers, islands, channels and fjords, wildlife, and exuberant vegetation. (Diario de Aysén 1988, p. 3)

The media contributed to a growing tourist imaginary and soon, new practices emerged. For example, caravans, or rallies, were organized by tourism companies and automobile associations, around experiences designed to get to know the road (Urrutia et al. 2019).

Between 1980 and 1989, annual visitation increased from 5000 to 10,000 (Central Bank of Chile 1991), and along the first stretches of the road, tourist service infrastructure was constructed including the Ralún Inn (1977) and the Termas de Puyuhuapi Hotel (1991). Also, national PAs were created or reclassified along the new road network; for example, the Queulat National Park was created in October 1983 to preserve and protect the resources and places of scenic beauty around the CA.

We posit that a culture of care and respect for nature in Aysén has quietly and gradually emerged through the development process of the CA. Thus, if its construction was based on an ideology that saw the region's nature as a source of strategic economic resources to be extracted, it also catalyzed an appreciation of nature's scenic and cultural values. Traveling exhibitions, images, television documentaries (Urrutia 2019, 2021), and photographic books, like the Chilean National Tourism Service's (1987), *Carretera Austral, Presidente Pinochet: The Austral Road*, record the coexistence of these values. Especially demonstrative are George Munro's books of photographs which demonstrate the shift taking place. Specifically, his 1982 work, *Carretera Austral, integración de Chile* (Munro 1982), extolled sovereignty and transformation of the territory (Fig. 3.4); however, less than a decade later, in his 1989 work, *Los ecos del silencio: Carretera Austral* (Munro 1989), these topics took a back seat to a new narrative that prioritized natural landscapes and denounced the ecological damage and devastation of the colonization (Fig. 3.4, photo d). We believe the CA catalyzed a new image of the landscape and a general shift in values, including a call for protection, from a nostalgic perspective that connoted the region as a sort of lost paradise (Rossetti 2018a).

Fig. 3.4 Photos (**a**), (**b**), and (**c**) are from George Munro's (1982) work, *Carretera Austral, integración de Chile*. Photo (**d**) is from George Munro's (1989) work, *Los ecos del silencio: Carretera Austral*

3.4.4   Connectivity as a Platform for the Resignification of Nature (1990–Today)

Since the 1990s, Aysén's road network has continued to expand southward, reaching the towns of Puerto Yungay (1996), Villa O'Higgins (1999), and Caleta Tortel (2003) by land. Since 2016, it has been further extended, connecting Aysén with the southernmost Chilean region of Magallanes through a ferry route, named the *route of the glaciers*. These recent extensions of the CA have occurred in parallel to a changing general context about the role and value of nature in central Patagonia; both from within the Aysén region and from outside (nationally and internationally). During recent decades, Aysén's positioning has evolved from being a region that offered Chile an abundance of resources for extraction, to a region whose greatest characteristic is to have remained *pristine*, and (apparently) at the margin of capitalist development (Schweitzer 2014; Núñez and Aliste 2017). For example, the 2009 Aysén Regional Development Strategy defined the importance of sustainability for Aysén, saying,

> …the environmental quality of the Aysén region is a competitive advantage that must be safeguarded in order to sustain the production of goods and services of all kinds, but particularly those related to the special interest tourism industry. Consequently, the region has adopted the slogan "Aysén reserve of life", which invites its citizens to create a sustainable society that can persist through generations and that is able to achieve the welfare of its population, relating harmoniously with the natural environment, thus satisfying the present material needs and establishing the basis for every individual to deploy their human potential, without compromising the development capacity of future generations. (Aysén Regional Government 2009, p. 15)

In parallel, the CA has evolved to become one of the most popular tourism experiences in Aysén (Adiego et al. 2018), increasingly positioned as a scenic route connecting tourists with the region's natural landscapes; described as the most spectacular road in South America (Linde 2013; Sinclair and Houlbrooke 2015).

Massive rewilding efforts and socio-environmental campaigns in Aysén provide other examples of evolving environmental resignifications and valuations of nature. For example, in 1995, a few years after the extension of the CA reached Cochrane in 1988, Douglas and Kris Tompkins made their first visit to Valle Chacabuco to climb and hike (Rewilding Chile 2022). At the time, they noted the spectacular natural beauty of the valley and its importance for biodiversity. This valley has a significant history within colonial Aysén. Early in the twentieth century, Valle Chacabuco was the site for one of the largest ranching estancia concessions granted by the Chilean government. Later, concession lands were redistributed to local peasant families under the Chilean Agrarian Reform Act of 1964. Then in 1980, these peasant lands were dispossessed and appropriated by the military government of Augusto Pinochet and sold in 1980 to Belgian immigrant, Françoise de Smet (Rewilding Chile 2022). The Tompkins visited the de Smet estancia repeatedly over the next two decades and acquired the estancia in 2004, through their NGO, Tompkins Conservation, subsequently initiating one of the largest private rewilding projects in history (Rewilding Chile 2022). In 2018, the Tompkins Conservation successfully transferred their private park within the Chacabuco valley, Parque Patagonia, to the Chilean State in an agreement signed between then president, Michelle Bachelet and Kristine McDivitt Tompkins (*Tompkins Conservation*), to create Patagonia National Park. This historic agreement also created the *Route of the Parks of Patagonia* (rutadelosparques.org), a conceptual 2800 km *route* of terrestrial and maritime segments that extends between Puerto Montt and Cape Horn, Chile, connecting visitors with 17 National Parks and more than 60 local communities. With the entire length of the CA included within the Route of the Parks of Patagonia, the Route of Parks of Patagonia effectively consolidated earlier positioning of the route as a globally important scenic and structuring road for tourism.

Based on these developments, we propose that transportation infrastructure, tourism, and conservation policies have effectively structured the evolution of the Aysén region since the beginning of the twentieth century, albeit with historically changing objectives. In general, there has been a shift from a largely geopolitical interest in the region that centered around the protection of nature in strategic areas and the extraction of nature via *comparative advantages*, to a valuation of the landscape for its apparent pristineness and biodiversity. The socio-territorial impact of these shifts has not been free of tensions. The purchase of the valley, which was home to the Baker *Estancia*, one of the first cattle ranches in Aysén, and the subsequent *rewilding* of Patagonia National Park, has been accompanied by controversies around the capitalist ethics of foreign investors modifying traditional production patterns of the region, such as sheep ranching (Borrie et al. 2020; Gale and Ednie 2019). At the community level, certain sectors perceive these activities to be features of local identity and memory.

3.5 Discussion. Social Reappropriation of Nature: Emergence of Tourism and Environmental Protection in Aysén

The preceding narrative illustrates the factors we observed in the data with respect to how nature appropriation has transitioned within the Aysén region of Patagonia. Increasingly, we observed a high valuation of nature based on aesthetic, recreation, instrumental, non-use, and even intrinsic values (Gale and Ednie 2019). Key examples from the tourism sector and socio-environmental groups include *Patagonia sin Represas* (Patagonia without Dams), the Tehuelche Youth group, which is dedicated to the defense of territory and waters, and, more recently, the *Chile Sustentable* (Sustainable Chile) program (chilesustentable.net), which is self-described as a citizen proposal for change (Torres et al. 2017). Considering the proposed HidroAysén megaproject as an example, many in local communities, particularly those in the Baker River basin, had economic and employment expectations around the project, yet after learning about and weighing the destruction of nature that the dams would bring, many became convinced that it was better to opt for protecting and conserving Patagonia. This is when the narrative of *Aysén, Reserva de Vida* (Aysén, Reserve of Life) gained strength and became a regional, national, and global slogan to defend the region from extractive hydroelectric projects. Concretely, in 2007, 36% of Chileans were against dams in Aysén; and that percentage increased to 74% by 2011 (Torres et al. 2017, p. 157). We believe that the above may highlight a growing regional preference, largely mobilized by the presence of the CA, of the need to focus the local economy around activities that are socially and ecologically sustainable. This would illustrate Leff's (2019) thesis that territorial actors are currently building new ecological cultures and strategies for sustainable use of natural resources, in this case, based on nature tourism.

Nevertheless, evidence of evolving natural values facilitated by the construction of the CA must be contrasted with evidence of values that prioritize use and extraction of nature. For example, the recent extension of the CA, via the route of the glaciers ferry route, that connects Aysén with Magallanes, has facilitated the extraction and exportation of sphagnum moss on a concerning scale (León et al. 2021). And, increased access has led to real estate development and increased anthropogenic impacts, from amenity and climate migration that has not been accompanied by sufficient planning, to the lack of basic infrastructure to protect natural resources from associated pressures (Núñez et al. 2019). For example, of the 33 regional population centers, only 9 have territory planning documents in place, and some of these documents are more than 30 years old.

Many have expressed concern about the impacts and externalities that accompany nature-based tourism in remote and natural areas like Aysén (e.g., Butler 2018; Séraphin et al. 2019). Recent studies have identified local perceptions of concern regarding the growing tourism presence and conservation ethics that have erased much of the pioneer heritage and ways of life that existed in the region (Blair et al. 2019; Borrie et al. 2020; Gale and Ednie 2019). Others have observed that Aysén tourism has developed through two distinct market segments: local tourism service providers with low levels of specialization that appeal to backpackers and other independent tourists; and global tourism companies, based in Santiago or in other countries, that are more specialized and tend to focus on a more elite and specialized small-group clientele (Aysén Regional Government 2009). These dynamics have created some inequities within the tourism sector that are important to consider. First, over the past several decades several specialized national and global tourism services have acquired large tracts of land within Aysén in strategic locations, around PAs and along waterways. This *privatization* of nature has created and augmented access problems for local residents, tourists, and tourism providers (Blair et al. 2019; Borrie et al. 2020). Some construction projects linked to these acquisitions have introduced and imported design solutions that compete with the traditional development forms and scale. The specialized tourism services that these companies provide respond to high-scale, foreign demands with products that reflect a high level of innovation, diversification, and prices. This demand has primarily been met by extra-regional operators (national and global), who have the knowledge and resources to undertake costly development projects. The context described above leads to the question of whether locally rooted service operators, accustomed to responding with fewer resources to a less specialized tourism demand, will be able to compete given their own means in a market that is moving toward greater sophistication and diversification.

Multiple analyses have identified that the major limiting factors for the competitiveness of tourism as a driver of local development are the weak reception capacity of communities, the small size of local businesses, the irregular distribution of supply, the seasonality of tourism activities, and, above all, the lack of technical support at the community, regional, and national levels (Chilean National Tourism Service 2014, 2017; Chilean National Tourism Service and Guazzinni Consultores 2017; Chilean Subsecretary of Tourism 2017). These factors should inform the development of adequate policies for governance to allow full integration of local urban and rural inhabitants in the opportunities and benefits afforded by Aysén's growing tourism sector. The risk of not devising these strategies is that the valuation of nature could provoke a process of proletarianization of local residents whose communities have inhabited the region for over a century. The well-being of the very citizens whose emergence necessitated the CA might be sacrificed in the name of *wilderness*, *remoteness*, and these new discourses of conservationism and development.

Recent programs like the *Strategy of Gateway Communities of the Protected Areas of Chilean Patagonia*, funded by a public-private fund to ensure the future conservation of Chilean Patagonia, developed through an agreement between Tompkins Conservation, the Pew Charitable Trusts, and the Chilean government, are working to address the current and potential social inequities. Successful implementation of this program would contribute to a more integrated system of PA management and conservation, and to the local development and well-being for regional inhabitants. With a time horizon set for the year 2030, the gradual implementation of the gateway communities strategy strives to strengthen citizen participation, identify infrastructure needs, provide education, support public policy advocacy, and leverage public and private financing to make the initiative sustainable. The plan seeks to generate a relevant action framework linking the 26 municipalities that make up Chilean Patagonia, the Regional Governments of Los Lagos, Aysén, and Magallanes, the Chilean National Forestry Corporation (CONAF), the Ministry of the Environment, and the Ministry of Public Lands. The initiative is supported technically by the Austral Patagonia Program of the Austral University of Chile, Balloon Latam, Round River Conservation Studies, and the Pew Charitable Trusts (Pew Charitable Trusts 2021).

3.6 Conclusions

Historically, the links between connectivity, infrastructure, tourism, and nature protection in Aysén have been so close that the trajectory of one cannot be understood without looking into the others. While historically, and in the inaugural years of the CA the predominant geopolitical vision and state participation in the region promoted a capitalist and extractive appropriation of nature, contemporary practices and valuations of nature have shifted. On the one hand, our data suggests a social and community reappropriation of nature. As well, we have observed conservation processes underlying a growing valorization and commodification of nature. Thus, the CA appears not only as a platform that has mobilized spatial resignifications; it also appears to have facilitated new territorial and socio-environmental practices and policies.

Our study supports the hypothesis that development of the CA has catalyzed new ways of valuing the natural landscape, which include aesthetic, non-use, and intrinsic rights. These values seem to have motivated a range of local, community, and outsider efforts to protect the territory, through new PAs that extend *rewilding* operations within former ranching sectors, tourism development, amenity migration, and a prioritization of tourism within regional development strategies.

3.6.1 But Beware of Risks of Extractivisms, Real Estate Development, and Over-Tourism

It is vital to stress that increased connectivity has also facilitated the emergence of extractivisms of Patagonian natural resources (Gudynas 2015), including salmon aquiculture and mining. These extractivisms have been associated with several environmental disasters, including the contamination of continental and ocean waters. There has been a growing socio-environmental opposition to mining extractivism, as illustrated by the shifting focus of the Patagonia sin Represas movement as a new coalition called, *Patagonia sin Mineras* (Patagonia without Mines). These processes are parallel but closely associated with real estate development, the commodification of nature as a tourism resource, and the proletarianization of local communities. We worry that these dynamics may be propelled by regional development strategies that are overdependent on tourism, and, if poorly managed, may deteriorate the main resources that sustain them: the landscape.

Thus, more research is required to understand and avoid the potential negative impacts associated with current uses and extractivisms of nature in Aysén. Research foci should include potential proletarianization of local communities, possible environmentalism of the poor, and environmental injustices in Patagonia that could lead to local community losses or diminished territorial access and control due to the arrival of extra-local investments (Blair et al. 2019; Borrie et al. 2020; Martinez-Alier 2011). The challenge, therefore, is that the *valorization* of nature does not lead to new forms of colonization and dispossession analogous to those experienced in the first decades of the twentieth century. At an applied level, Aysén's great challenge involves knowing how to measure and plan for the desired development of transportation infrastructure, the expansion and maintenance of PAs, and an appropriate, equitable, and sustainable tourism exploitation, while controlling the pressures that these generate on a territory that is valued and perceived as pristine and remote. The risks are significant: if the valuation of remoteness and pristine nature in Aysén catalyzes an increase of visitors, tourism investments, real estate development, and new forms of exploitation without proper monitoring and controls, there could come a time when Aysén will cease to be perceived for these values and mass tourism will replace the conservation-based tourism strategies that currently exist.

Acknowledgments We would like to thank Andrés Adiego, of the Centro de Investigación en Ecosistemas de la Patagonia (CIEP), and the Department of Geography and Territorial Planning/Management of the Universidad de Zaragoza, for his assistance with the development of the map for our study area (Fig. 3.1). María Muñoz and Robinson Torres received support from Chile's National Agency for Research and Development (ANID) under the ANID Regional Program R17A10002; CIEP project R20F0002 and the ANID/FONDAP/15130015 program.

References

A. Adiego, A. Antas, T. Gale, M. Grossman, C. Mardones, M.J. May, R. Merino, G. Orellana, M.J. Oyarzo, *Carretera austral: Catalizador de turismo en Aysén Patagonia* (Carretera Austral: Catalyst for Tourism in Aysén Patagonia) (CIEP Centro de Investigación en Ecosistemas de la Patagonia, 2018). Retrieved January 21, 2023, from https://drive.google.com/file/d/0ByuKTEJ GEnPWTWptVWNlSk9ubVpuNnJsWURTZ2FBVGZhb0Nj/view?usp=sharing

A. Albet, N. Benach, *Doreen Massey. Un sentido global del lugar* (Doreen Massey. A Global Sense of Place) (Icaria, 2012)

Aysén Regional Government, *Estrategia regional de desarrollo de Aysén* (Regional Development Strategy of Aysén) (2009), Retrieved January 3, 2023, from https://www.goreaysen.cl/controls/ neochannels/neo_ch112/appinstances/media42/EDR_AYSEN.pdf

H. Blair, K. Bosak, T. Gale, Protected areas, tourism, and rural transition in Aysén, Chile. Sustainability **11**(24), Article 7087 (2019). https://doi.org/10.3390/su11247087

B. Borrie, T. Gale, K. Bosak, Privately protected areas in increasingly turbulent social contexts: Strategic roles, extent, and governance. J. Sustain. Tour. **30**(11), 2631–2648 (2020). https://doi. org/10.1080/09669582.2020.1845709

R. Butler, Sustainable tourism in sensitive environments: A wolf in sheep's clothing? Sustainability **10**(6), Article 1789 (2018). https://doi.org/10.3390/su10061789

Central Bank of Chile, *Indicadores económicos y sociales regionales, 1980–1989* (Regional Economic and Social Indicators, 1980–1989) (Banco Central de Chile, 1991)

Chilean Highway Administration of the Chilean Ministry of Public Works, *Carretera Longitudinal Austral. Puerto Montt – Aysén. Antecedentes del proyecto.* (Longitudinal Austral Highway. Puerto Montt – Aysén. Project Background) (Ministerio de Obras Públicas, 1968)

Chilean Ministry of Public Works, *Ministerio de Obras Públicas: Factor de integración nacional: 15 años de progreso sostenido: 1973–1988* (Ministry of Public Works: Factor of National Integration: 15 Years of Sustained Progress: 1973–1988) (Ministerio de Obras Públicas, 1988)

Chilean National Statistics Institute, *Resultados censo 2017, por país, regiones y comunas* (2017 National, Regional, Communal Census Results) (2017), Retrieved February 16, 2023, from http://resultados.censo2017.cl/

Chilean National Tourism Service, *Carretera Austral, presidente Pinochet* (Carretera Austral, President Pinochet) (Sernatur, 1987)

Chilean National Tourism Service, *Plan de acción Región de Aysén del General Carlos Ibáñez del Campo sector turismo 2014–2018* (Action Plan for the Aysén Region of General Carlos Ibáñez del Campo Tourism Sector 2014–2018) (2014), Retrieved January 10, 2023, from https://www. sernatur.cl/wp-content/uploads/2018/10/Plan-de-Accio%CC%81n-Aysen.pdf

Chilean National Tourism Service, *Plan de acción para la gestión participativa de Zonas de Interés Turístico (ZOIT): Aysén Patagonia Queulat* (Action Plan for Participatory Management of Zones of Tourist Interest (ZOIT): Aysén Patagonia Queulat) (2017), Retrieved January 19, 2023, from http://www.subturismo.gob.cl/wp-content/uploads/2015/10/Plan-Accion-ZOIT-Patagonia-Queulat.pdf

Chilean National Tourism Service, & Guazzinni Consultores, *Plan de acción para la gestión participativa de las Zonas de Interés Turístico (ZOIT), zona de interés turístico Chelenko* (Action Plan for the Participatory Management of Zones of Tourist Interest (ZOIT), Chelenko Zone of Tourist Interest) (2017), Retrieved January 19, 2023, from http://www.subturismo.gob.cl/wp-content/uploads/2015/10/Plan-Accion-ZOIT-Chelenko.pdf

Chilean Subsecretary of Tourism, *Plan de acción para la gestión participativa de las zonas de interés turístico (ZOIT), zona de interés turístico Provincia de los Glaciares* (Action Plan for the Participatory Management of the Zones of Tourist Interest (ZOIT), Glacier Province Zone of Tourist Interest) (2017), Retrieved January 19, 2023, from http://www.subturismo.gob.cl/wp-content/uploads/2015/10/PLAN-ACCI%C3%93N-PROVINCIA-DE-LOS-GLACIARES.pdf

J. Creswell, *Research Design: Qualitative, Quantitative, and Mixed Methods Approaches* (SAGE Publications Inc, 2014)

J. Diamanti, Extractivism. Krisis **2018**(2), 54–57 (2018)

Diario de Aysén, *El despegue turístico de Aysén* (Aysén's Tourism Takeoff) (Diario de Aysén, 1988)

T. Gale, A. Ednie, Can intrinsic, instrumental, and relational value assignments inform more integrative methods of protected area conflict resolution? Exploratory findings from Aysén, Chile. J. Tour. Cult. Chang. **18**(6), 690–710 (2019). https://doi.org/10.1080/14766825.2019.1633336

G. García, Carretera Longitudinal Austral. La respuesta a un desafío (Carretera Longitudinal Austral. The response to a challenge). Revista Chilena de Geopolítica **5**(3), 51–69 (1989)

R. Gerstmann, *Chile: 280 grabados en cobre* (Chile: 280 Copperplate Engravings) (Braun & Cie, 1932)

R. Gerstmann, *Chile en 235 cuadros* (Chile in 235 Pictures). (Hub Hoch-Düsserldorf, 1959)

E. Gudynas, *Extractivismos: Ecología, economía y política de un modo de entender el desarrollo y la naturaleza* (Extractivism: Ecology, Economics and Politics of a Way of Understanding Development and Nature) (Sagitario SRL Artes Gráficas, 2015)

A. Horvath, *Integración y desarrollo de la zona austral* (Integration and Development of the Southern Zone) (Revista CA 70, 1992)

A. Ibáñez Santa María, La incorporación de Aisén a la vida nacional, 1902–1936 (The incorporation of Aisén to the national life, 1902–1936). Historia **11**, 259–378 (1973). http://www.memoriachilena.gob.cl/602/w3-article-71193.html

E. Leff, *Racionalidad ambiental. La reapropiación social de la naturaleza* (Environmental Rationality. The Social Reappropriation of Nature) (Siglo XXI Editores, 2004)

E. Leff, *Ecología política: De la deconstrucción del capital a la territorialización de la vida* (Political Ecology: From the Deconstruction of Capital to the Territorialization of Life) (Siglo XXI Editores, 2019)

C.A. León, M. Gabriel, C. Rodríguez, R. Iturraspe, A. Savoretti, V. Pancotto, et al., Peatlands of southern South America: A review. Mires and Peat **27**, 1–29 (2021). https://doi.org/10.19189/MaP.2020.SNPG.StA.2021

W. Linde, *La Carretera Austral. El camino más espectacular de Sudamérica* (The Carretera Austral. The most Spectacular Road in South America) (Ocho Libros, 2013)

J. Maderuelo, *El paisaje. Génesis de un concepto* (The landscape. Genesis of a Concept) (Abada Editores, 2006)

S. Marchán, La experiencia estética de la naturaleza y la construcción del paisaje (The aesthetic experience of nature and the construction of landscape), in *Paisaje y pensamiento* (Landscape and Thought), ed. by S. Marchán, J. Maderuelo, (Abada Editores, 2006), pp. 11–54

J. Martínez-Alier, *El ecologismo de los pobres: Conflictos ambientales y lenguajes de valoración* (The Environmentalism of the Poor: Environmental Conflicts and Languages of Valuation) (Icaria, 2011)

B.M. Martinic, Apertura del istmo de Ofqui: Historia de una quimera, consideraciones sobre la vigencia de sus razones (Opening of the Ofqui isthmus: History of a dream, reflections on the validity of its motives). Magallania (Punta Arenas) **41**(2), 5–50 (2013). https://doi.org/10.4067/s0718-22442013000200001

M. Martinic, *De Trapananda al Aysén. Una mirada reflexiva sobre el acontecer de la Región de Aysén desde la prehistoria hasta nuestros días* (From Trapananda to Aysén. A Reflective Look on the Events of the Aysén Region from Prehistory to the Present Day) (Fundación Río Baker, 2014)

C. Méndez, A. Nuevo-Delaunay, Evidencias a cielo abierto para discutir superficies potenciales de actividad temprana en Patagonia centro occidental (44-45° S) (The open-air evidences for discussing potential early human activity surfaces in central-western Patagonia (44-45°S)). Magallania (Punta Arenas) **47**(1), 105–116 (2019). https://doi.org/10.4067/s0718-22442019000100105

C. Méndez, A. Nuevo-Delaunay, O. Reyes, J.B. Belardi, B. Thompson, J. Carranza, Ocupación humana del bosque caducifolio de Aisén durante el Holoceno medio: Nuevos datos de la localidad de Altos del Moro (río Cisnes) (Human occupation of the deciduous forest of Aisen during the middle Holocene: New data from the locality of Altos del Moro (Cisnes river)). Boletín de La Sociedad Chilena de Arqueología **50**, 65–73 (2020)

J. Monge, *El istmo de Ofqui. II. El proyecto del Canal* (The Ofqui Isthmus. II. The Canal Project) (La Sociedad, 1944)

J. Moore, *El capitalismo en la trama de la vida: Ecología y acumulación de capital* (Capitalism in the Web of Life: Ecology and Capital Accumulation) (Traficantes de Sueños, 2020)

M. Muñoz, R. Torres Salinas, Conectividad, apertura territorial y formación de un destino turístico de naturaleza. El caso de Aysén (Patagonia chilena) (Connectivity, territorial openness and the formation of a nature tourism destination. The case of Aysén (Chilean Patagonia)). Estudios y Perspectivas en Turismo **19**(4), 447–470 (2010)

G. Munro, *Carretera Austral, integración de Chile* (Carretera Austral, Chile's Integration) (Ediciones Servicios Promocionales, 1982)

G. Munro, *Los ecos del silencio, Carretera Austral* (The Echoes of Silence, Carretera Austral) (Editorial Ediciones y Publicidad, 1989)

A. Núñez, E. Aliste, Discursos ambientales y procesos de fronterización en Patagonia-Aysén (Chile): De los paisajes de la mala hierba a los del bosque sagrado. (Environmental discourses and frontier processes in Patagonia-Aysén (Chile): From weed-filled landscapes to sacred forests). Front. J. Soc. Technol. Environ. Sci. **6**(1), 198–218 (2017)

A. Núñez, E. Aliste, A. Bello, J. Astaburuaga, Eco-extractivismo y los discursos de la naturaleza en Patagonia-Aysén: Nuevos imaginarios geográficos y renovados procesos de control territorial (Eco-extractivism and nature discourses in Patagonia-Aysén: New geographical imaginaries and renewed processes of territorial control). Revista Austral de Ciencias Sociales **35**, 133–153 (2018). https://doi.org/10.4206/rev.austral.cienc.soc.2018.n35-09

A. Núñez, F. Miranda, E. Aliste, S. Urrutia, Conservacionismo y desarrollo sustentable en la geografía del capitalismo: Negocio ambiental y nuevas formas de colonialidad en Patagonia-Aysén (Conservationism and sustainable development in the geography of capitalism: Environmental business and new forms of coloniality in Patagonia-Aysén), in *(Las) Otras geografías en Chile. Perspectivas sociales y enfoques críticos* ((The) Other Geographies in Chile. Social Perspectives and Critical Approaches), ed. by A. Núñez, E. Aliste, R. Molina, (Lom, 2019), pp. 23–46

C. Oportus Mena. Informe sobre el problema de colonización de la Zona del Río Baker (Report about the colonization problem in the Baker River zone) (Issue 81, Santiago, 1928).

Pew Charitable Trusts, *Las comunidades portal y su rol en la protección de la Patagonia chilena* (Portal Communities and Their Role in Protecting Chilean Patagonia) (2021), Retrieved January 16, 2023, from https://www.pewtrusts.org/-/media/assets/2021/10/gatewaycommunitiestranslation_v3.pdf

A. Pinochet, *Ejército de Chile: Posibles elementos a considerar en su proyección futura* (Chilean Army: Possible Elements to Consider in its Future Projection) (FASOC, No. 4. Conferencia dictada en Santiago de Chile el 19 de agosto de 1993, 1993)

Pomar, J. (1923). *La concesión del Aisén y el Valle Simpson (notas y recuerdos de inspección de un viaje en mayo y junio de 1920)* (The Aisén and Simpson Valley concessions (Notes and Inspection Recollections of a Trip in May and June 1920)). (Imprenta Cervantes, 1923)

Rewilding Chile, *Parque Nacional Patagonia* (2022), Retrieved January 16, 2023, from https://www.rewildingchile.org/proyectos/parque-nacional-patagonia/

O. Reyes, C. Méndez, M. San Román, Cronología de la ocupación humana en los canales septentrionales de Patagonia occidental, Chile (Chronology of the human occupation of the Northern water courses of western Patagonian, Chile). Intersecciones En Antropología **20**(2), 195–209 (2019). https://doi.org/10.37176/IEA.20.2.2019.449

L. Rodríguez, Paisajes del trauma: Una cinta de Moebius (Landscapes of trauma: A Moebius tape), in *Historia, trauma, memoria* (History, Trauma, Memory), ed. by C. Balbontín, L. Rodríguez, (Libros del amanecer, 2021), pp. 173–196

F. Rossetti, El hotel de la Laguna San Rafael, el canal de Ofqui y la apertura de la frontera centro patagónica occidental: Ciudad, arquitectura y paisaje en el discurso estatal (The San Rafael Lagoon hotel, the Ofqui channel and the opening of the western central Patagonian frontier: City, architecture and landscape in the state discourse). NODO **11**(21), 21–33 (2016). https://doi.org/10.54104/nodo.v11n21.790

F. Rossetti, De infraestructura a paisaje. La Carretera Austral como motor de resignificación (From infrastructure to landscape. The Carretera Austral as an engine of resignification). ARQ (Santiago) **99**, 86–95 (2018a). https://doi.org/10.4067/S0717-69962018000200086

F. Rossetti, *Entre Trapananda e HidroAysén. Territorio y nación en la conformación de las figuras culturales de Aysén, Patagonia centro occidental* (Between Trapananda and HidroAysén. Territory and Nation in the Conformation of the Cultural Figures of Aysén, Central-Western Patagonia) (Pontificia Universidad Católica de Chile, 2018b)

M. Ruiz, *Acerca del problema de la conectividad en la zona austral de Chile: El caso de la Carretera Austral 1976–1996* (On the Problem of Connectivity in the Southern Zone of Chile: The Case of the Carretera Austral 1976–1996) (Universidad de Chile, 2016)

R.K. Schutt, D.F. Chambliss, Chapter 10: Qualitative data analysis, in *Making Sense of the Social World: Methods of Investigation*, ed. by R.K. Schutt, D.F. Chambliss, (SAGE Publications Inc, 2013), pp. 320–357

A.F. Schweitzer, Patagonia, naturaleza y territorios (Patagonia, nature and territories). Aust. Geogr. **10**(2), 1–24 (2014). https://memoria.fahce.unlp.edu.ar/art_revistas/pr.6468/pr.6468.pdf

H. Séraphin, M. Zaman, S. Olver, S. Bourliataux-Lajoinie, F. Dosquet, Destination branding and overtourism. J. Hosp. Tour. Manag. **38**, 1–4 (2019). https://doi.org/10.1016/j.jhtm.2018.11.003

G. Silvestri, *El lugar común. Una historia de las figuras de paisaje en el río de la Plata* (The Common Place. A History of Landscape Figures in the río de la Plata) (Edhasa, 2011)

H. Sinclair, W. Houlbrooke, *Chile: Carretera Austral: A Guide to One of the world's most Scenic Road Trips*, 1st edn. (Bradt Travel Guide, 2015)

R.E. Stake, Case studies, in *Strategies of Qualitative Inquiry*, ed. by N.K. Denzin, Y.S. Lincoln, 2nd edn., (SAGE Publications Inc, 2003), pp. 134–164

M. Svampa, *Las fronteras del neoextractivismo en América Latina: Conflictos socioambientales, giro ecoterritorial y nuevas dependencias* (The Frontiers of Neo-Extractivism in Latin America: Socio-environmental Conflicts, Eco-territorial Turn and New Dependencies) (Editorial UCR, 2019)

R. Torres, J. Rojas, Naturaleza, cultura, y formas turísticas de vida en Aysén (Nature, culture, and tourist ways of life in Aysén). Sociedad Hoy **20**, 77–109 (2011)

R. Torres, A. García, J. Rojas, Privatizando el agua, produciendo sujetos hídricos: Análisis de las políticas de escala en la movilización socio-hídrica contra Pascua Lama e HidroAysén en Chile (Privatizing water, producing water subjects: Analysis of the politics of scale in the socio-water mobilization against Pascua Lama and HidroAysén in Chile). Agua y Territorio **10**, 149–166 (2017)

S. Urrutia, Dispositivo, imagen y transparencia. El Camino Longitudinal Austral y la Patagonia-Aysén durante el régimen de Pinochet (1973–1990) (Device, image and transparency. The Southern Longitudinal Road and Patagonia-Aysén during the Pinochet regime (1973–1990)). Aust. Geogr. **9**(2), 138–142 (2019). https://ri.conicet.gov.ar/handle/11336/174766

S. Urrutia, Paisaje, ensamblaje, movimiento. La imagen en movimiento en la construcción de la Carretera Austral durante la dictadura cívico militar (Patagonia, Chile) (Landscape, assemblage, movement. The moving image in the construction of the Carretera Austral during the civil-military dictatorship (Patagonia, Chile)). Diálogo Andino **66**, 41–51 (2021). https://doi.org/10.4067/S0719-26812021000300041

S. Urrutia, A. Núñez, E. Aliste, Naturaleza salvaje y agreste: Los imaginarios de la naturaleza en la construcción del Camino Longitudinal Austral, Chile 1976-1990 (Wild and rugged nature: Imaginaries of nature in the construction of the Camino Longitudinal Austral, Chile 1976-1990). Magallania (Punta Arenas) **47**(2), 55–72 (2019). https://doi.org/10.4067/S0718-22442019000200055

G. Van Schouwen, *Ejército de Chile: Historia del Cuerpo Militar del Trabajo: 50 años* (Army of Chile: History of the Military Labor Corps: 50 Years) (IGM, 2003)

J. Von Chrismar, Trascendencia geopolítica de la Carretera Austral 'presidente Pinochet' (Geopolitical significance of the Carretera Austral 'President Pinochet'). Revista Chilena de Geopolítica **3**(1), 35–43 (1986)

R. Yin, *Qualitative Research from Start to Finish* (SAGE Publications Inc., 2011)

Chapter 4
How Changing Imaginaries of Nature and Tourism Have Shaped National Protected Area Creation in Argentine Patagonia

Christopher B. Anderson, Mara Dicenta, Jessica L. Archibald, and Alejandro E. J. Valenzuela

Abstract Even regions of the planet widely considered to be "remote" or "pristine" like Patagonia are actually dynamic social-ecological systems with interrelated local-international connections of discourses, practices, and institutions. Yet, their study and management often do not consider this complexity. In Argentine Patagonia's iconic landscapes, protected areas (PAs) represent a major human-nature relationship, and PA creation has been motivated by objectives ranging from geopolitical interests to biodiversity conservation. In this chapter, we employed the social imaginary framework to conduct an historical analysis of local, national, and international influences regarding the way nature and tourism are conceived and managed in national PAs. We evaluated the discourses (ideals, values, beliefs) and institutions (norms, rules, structures, stakeholders) involved in creating these PAs in Argentine Patagonia. The national PA system was legally formed in the 1930s, but initial efforts reach back as far as the early 1900s. We found that while the globalization of Patagonian conservation-based development has consolidated since the 1980s, local-international relationships extended over more than a century to co-produce these social-ecological systems.

Keywords Patagonia · Human-nature relationships · Social imaginary · Tourism · Conservation discourse · Protected areas

C. B. Anderson (✉)
Universidad Nacional de Tierra del Fuego (UNTDF), Instituto de Ciencias Polares, Ambiente y Recursos Naturales (ICPA), Ushuaia, Tierra del Fuego, Argentina

Consejo Nacional de Investigaciones Científicas y Técnicas (CONICET), Centro Austral de Investigaciones Científicas (CADIC), Ushuaia, Tierra del Fuego, Argentina
e-mail: canderson@alumni.unc.edu

M. Dicenta
William & Mary, Department of Anthropology & Institute for Integrative Conservation (IIC), Williamsburg, VA, USA
e-mail: mdicentavilker@wm.edu

71

T. Gale-Detrich et al. (eds.), *Tourism and Conservation-based Development in the Periphery*, Natural and Social Sciences of Patagonia,
https://doi.org/10.1007/978-3-031-38048-8_4

4.1 Introduction

The current magnitude and extent of global environmental change have led some natural scientists to posit a new geological epoch, the Anthropocene, whereby humans dominate planetary biogeochemical processes (Steffan et al. 2007). Social scientists and humanities scholars, however, have pointed out that the human species does not drive these novel conditions; rather, they are caused by specific social practices (e.g., burning fossil fuels) and actors (e.g., high consuming socio-economic groups; Dicenta and Correa 2021; Malm and Hornborg 2014). In this context, it has become increasingly evident that understanding and managing the world's multiple "environmental" crises (e.g., climate change, biodiversity loss, unequitable exposure to pollution, pandemic emergence) require better integration of these human and natural dimensions operating from local to global scales (IPBES 2019; United Nations 2015). As such, traditional environmental management models like protected areas (PAs) are being challenged, shifting away from conserving "nature for nature" or "nature despite people" toward more holistic versions of "nature and people" (Mace 2014; West et al. 2006).

These shifting academic and social-political paradigms also affect Patagonia. While in Western thought the region has been conceived largely as the "uttermost ends of the Earth" (Bridges 1948), in fact, it has been inhabited by diverse humans for at least 7,000 years (Morello et al. 2012) and non-human species for much longer (McEwan et al. 1997). Furthermore, in the 500 years since Europeans first reached the area, it has increasingly become a space of encounter, friction, and circulation of local to global ideas about the origins and futures of humans and nature (Moss 2008; Wakild 2017). In this sense, Patagonia can be approached both as a tangible location that has its own history and evolution and as a symbolic space that has powerfully produced meanings and representations beyond its geographic limits; together, these realms co-construct a dynamic and multi-faceted social-ecological territory.

As an ecological system, Patagonia displays stark biophysical gradients of elevation, topography, temperature, and precipitation that lead to sharply contrasting biotic communities (Oliva et al. 2020). In turn, this vast geographic expanse has

J. L. Archibald
Northern Arizona University, School of Earth and Sustainability, Flagstaff, AZ, USA
e-mail: jla396@nau.edu

A. E. J. Valenzuela
Universidad Nacional de Tierra del Fuego (UNTDF), Instituto de Ciencias Polares, Ambiente y Recursos Naturales (ICPA), Ushuaia, Tierra del Fuego, Argentina

Consejo Nacional de Investigaciones Científicas y Técnicas (CONICET), Instituto de Ciencias Polares, Ambiente y Recursos Naturales (ICPA),
Ushuaia, Tierra del Fuego, Argentina
e-mail: avalenzuela@untdf.edu.ar

occupied a privileged position in Western thought since the emergence of modern science during the eighteenth century, being studied and mythologized by numerous scientists, explorers, and travelers (Giucci 2014; Silveira 2009). For example, while Charles Darwin often disparaged Patagonia (and its people) as a site of "death and decay" and a "useless" territory "without habitation, without water, without trees," he fully recognized the impact this place had on him, "owing to the free scope given to the imagination" (Darwin 1889, p. 503). With the Argentine annexation of most of Patagonia in the mid-to-late nineteenth century, national knowledges reconfigured those visions to contest homogenizing colonial perspectives regarding Argentina's geography. Simultaneously, Patagonia became a key figure of "otherness" to construct a modern, postcolonial Argentine identity that put the urban and European descendants of the Buenos Aires capital at the center. Contrasting modern urban cities and citizens with the rest of the country, the North became a site of traditional rural peoples (Chamosa 2016), while Patagonia continued to be imagined as extensive uncultivated lands, represented as a "desert" ready to be filled, occupied, and optimized (Dicenta 2021; Navarro Floria 2002). Multiple forces and scales have continued to interact in the region, whereby elements of the "desert" for national colonization can today be recognized by its designation as one of the world's last remaining "pristine wilderness areas" for international conservation (Briones and Delrio 2007; Mittermeier et al. 2003). There is also an increasing interest in understanding the effects of global drivers of social-ecological change, such as transportation and telecommunications that "telecouple" Patagonia and the rest of the world, particularly via tourism (see Raya-Rey et al. 2017).

In this chapter, we examine how changing notions and values of nature and tourism have shaped the creation of PAs in Argentine Patagonia. Previous work has analyzed how parks and tourism have helped to affirm national values, especially during the first half of the twentieth century (e.g., Carreras Doallo 2012, 2016; Núñez 2014). Here, we expand that debate to explore how those values have shaped the natures of PAs and how those values have changed over time and in response to multi-scalar forces.

National PAs are considered one of the most successful environmental policies in the Americas, touted as a sustainable use of nature that provides multiple benefits for humans (IPBES 2018). Hence, while we recognize that other human-nature activities also occur in the region—forestry, livestock ranching, mining, fisheries, or urbanization—we were keen to explore changing social imaginaries of both nature and tourism in national PAs given their prominence as both a conservation and development strategy. Furthermore, it is well established that national PAs arose as a management approach in neo-European contexts (e.g., North America, Australia, western Europe) at the end of the nineteenth century, often seeking to exclude local communities or humans entirely from particular sites (Dicenta 2021; Fisk et al. 2021; West et al. 2006; Worster 1994). Yet, mainstream conservation is under contestation and now attempts to incorporate diverse peoples and multi-dimensional sustainability as a way to reconcile biodiversity conservation with human well-being and justice via the inclusion of people (IPBES 2019, 2022; Mace 2014). This study responds to these shifts and aims to analyze the local to global production of discourses, practices, and institutions around nature and tourism that configure national PAs in Argentine Patagonia.

4.2 Methodology

We studied the creation of a variety of PAs, including national reserves (NRs), national parks (NPs), natural monuments (NMs), natural wilderness areas (NWRs), marine-coastal inter-jurisdictional parks (MCIPs), marine inter-jurisdictional parks (MIPs), natural military reserves (NMRs), and marine protected areas (MPAs). First, we created a database of legal instruments available in the Argentine *Sistema de Información de Biodiversidad* (https://sib.gob.ar) that pertained to the establishment of PAs, including transfers of lands to establish or add to PAs, changes in PA status, or PA management plans. For the purpose of delimitation, PA expansions or those with multiple legal instruments were only considered as "creation" when there was also a change in legal status (e.g., NR or NM to NP). These official documents were used to determine (a) reasons PAs were created, (b) visions of nature, (c) visions of tourism in relation to PA objectives or management, and (d) stakeholders and institutional contexts.

Then, we searched archives for historical documents from local (i.e., Ushuaia: *Museo del Fin del Mundo, Biblioteca Sarmiento, Biblioteca CADIC*; Punta Arenas: *Archivo del Instituto de la Patagonia*) and national (i.e., Argentina: *Biblioteca Nacional, Archivo del Museo de Ciencias Naturales Bernardino Rivadavia, Archivo Histórico Naval, Archivo Intermedio de la Nación*) repositories to investigate social-political contexts more broadly. Specifically, archives were examined for the themes of "national territories," "national colonization plans," "agriculture," "military reports on Tierra del Fuego," and "national parks." Similarly, academic literature searches were conducted in Google Scholar to incorporate secondary information sources, using combinations of keywords "Argentina," "Patagonia," "protected areas," "national parks," "conservation," and "tourism," and their Spanish translations, as well as the names of the PAs themselves. Other relevant references were gathered using a snowballing technique via reviewing citation lists from the originally identified literature.

Information from these sources was assessed using the social imaginary concept (Castoriadis 1993). Social imaginaries are defined as shared discourses and practices that reproduce a social group's values, norms, and beliefs (Taylor 2004) and have been used previously in Patagonia to study biological invasions (Archibald et al. 2020) and land use (Laterra et al. 2021). While they are subject to change over time or between social groups, social imaginaries are powerful systems that organize ideas and behaviors in particular times and spaces and orient individuals toward the collectively defined "normal," "reasonable," or "correct" behaviors, including political decisions.

As an analytical tool, we applied the social imaginary framework to understand how the "behavior" of creating PAs was based on particular "discourses" and "institutions" associated with nature and tourism, and how those discourses were also

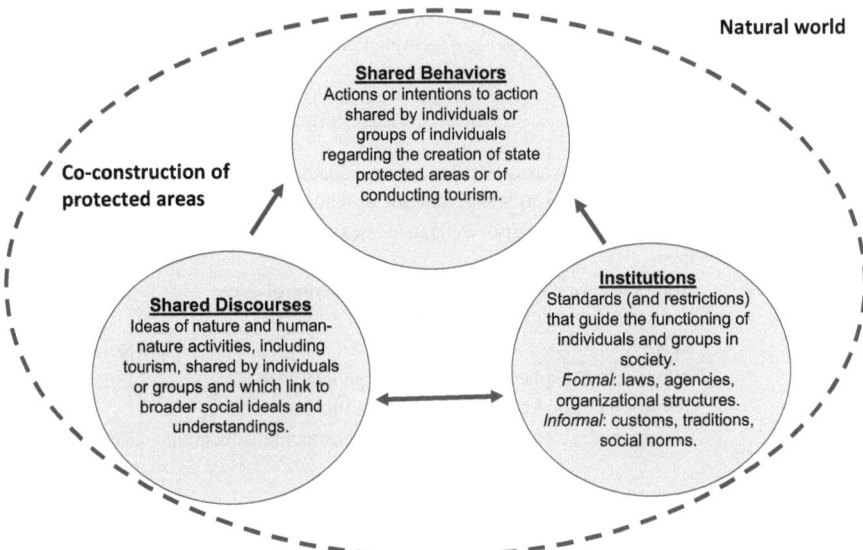

Fig. 4.1 Social imaginary conceptual framework

affected by local, national, and international contexts in different time periods. In particular, we examined: (a) the collectively held discourses that gave sense to and ordered ideas about nature and tourism; (b) the shared practices associated with managing and ordering nature conservation and tourism in particular ways; and (c) the formal and informal institutions (including stakeholders) that legislated and/or influenced conservation and tourism within Argentine Patagonian PAs (Fig. 4.1).

Our analysis looked for both ruptures and continuities in discourses, behaviors, and institutions. We compared and complemented each level of assessment with iterative clusters and themes detected in documents (i.e., grounded analysis, Strauss and Corbin 1990). We then created meta-categories around imaginaries of nature and tourism that helped to construct a chronological diagram to visually comprehend the dynamics influencing PAs. In each document, we identified and categorized the ideals and values that shape understandings of nature and tourism. Our analysis of nature's values drew upon Kellert's typology (1996) to create intermediate categories, which were then expanded to include other use values and later organized into instrumental, relational, and intrinsic values (Gale and Ednie 2020; Table 4.1). Regarding discourses in tourism, we focused on analyzing its conceptualization as mostly an elite privilege, a commercial enterprise, or a public/social endeavor. Finally, each category of nature's values was associated with a specific value justification: instrumental, relational, and intrinsic. Tourism was classified based on elite, commercial, and social values.

As part of the institutional analysis, we determined the organizational structures developed to administer PAs and tourism (e.g., creation of secretaries or directions and their placement into the ministerial organization). Finally, these data were related to a timeline of key ideas, people, events, and organizations. Attending to the ruptures in observed discourses, institutions, and behaviors, we categorized and

Table 4.1 Reasons to protect nature and promote tourism, as identified in the legal instruments that created national protected areas (PAs) in Argentine Patagonia and other supporting literature

Meta-categories	Description	Value justification
Nature ideals		
Utilitarian	Rational use, overexploitation, sustainable development, natural resources, forestry, tourism, sustainable development	Instrumental
Dominionistic	Delimitation (property), mitigate overexploitation, control fire	Instrumental
Geopolitical	Border security, military purposes, lands control over Indigenous peoples and foreigners	Instrumental
Aesthetic	Beauty	Relational
Moral	Social order, displacement of certain people, activities, definition of good behaviors and moralities	Relational
Education-scientific	Educational, scientific, awareness, environmental education	Relational
Cultural	Identity, practices, traditions	Relational
Historical	Archeological, paleontological, modern history	Relational
Spiritual	Connections to place that transcend physical experience	Relational
Ecological	Flora, fauna, species, biodiversity, ecosystem representation, habitats	Intrinsic
Tourism ideals		
Elite	High end, foreign models, exclusive hotels	N.A.
Commercial	Economic sector, mass infrastructure, promotion	N.A.
Social	Workers, vacation camps, unions, education	N.A.

periodized the social-ecological history of nature conservation and tourism in Argentine Patagonia, delimiting three social imaginaries related to efforts: (a) to incorporate the territory into a Euro-nationalist enterprise (Euro-nationalism); (b) to expand the notion of nationalism to explicitly include workers and some previously marginalized sectors (New Argentina); and (c) efforts to mitigate diverse environmental crises, which themselves change based on perspectives more aligned with orientations that exclude or include people from nature (Environmental Crisis).

4.3 Results

4.3.1 PA Creation in Argentine Patagonia

Table 4.2 summarizes our analysis of how ideas of nature and tourism manifested within PAs (i.e., NP, NR, NM, NMR, MCIP, MIP, NWR, and MPA) created in Argentine Patagonia from 1922 to 2021, as well as structural policies that shaped the management of these PAs. For each PA, primary and secondary sources were used to determine date of establishment, ideas associated with nature and tourism, stakeholders responsible for the implementation of ideas, and references [national decrees and national laws]. The first section of the table presents the results for the national system general legislation and the second section presents results for specific PAs.

From 1922 to 2021, 25 national PAs were created in Argentine Patagonia (Table 4.2). During this time, two periods clustered most PA creation in Patagonia: 1930s–40s ($n = 7$) and 2000s–2010s ($n = 10$). The "gaps" between these periods, however, should be understood in the broader national context. For example, while in the 1980s–1990s there were no PAs established in Patagonia, 13 PAs were created in other parts of the country.

Table 4.2 Analysis of how ideas of nature and tourism manifested and structural policies that shaped the management within national protected areas (PAs) created in Argentine Patagonia from 1922 to 2021

PA actions	Created (modified)	Ideas of nature	Ideas of tourism	Institutional actors	Ref.
General legislation for the national system of PAs					
National Parks Direction	1934	Control hunting, fishing, timber, mining & fire, importance of beauty & scientific value	Promote & regulate conservation, research & tourism, designating land for tourism infrastructure	Min. Agriculture	2
General Administration of National Parks & Tourism	1945	Importance of beauty & natural values	Promote tourism, teach the nation, including workers; NPs as tourist attraction	Min. Public Works	3
National Parks Administration	1980	Prohibit extractive uses (except hunting & fishing exotic species & tourism), protect flora & fauna, importance of aesthetic, historical & ecological values	Promotes tourism as economic activity & provides appropriate infrastructure for visitors	Min. Economy, Sec. Agriculture & Livestock	14
Argentine natural reserve categories	1990, 1994	Conserve strict, wilderness & educational & scientific values for ecosystem representation	Visiting for educational & scientific purposes	NPA	16, 17

(continued)

Table 4.2 (continued)

PA actions	Created (modified)	Ideas of nature	Ideas of tourism	Institutional actors	Ref.
Argentine National System of Marine PAs	2014, 2019	Sustainable use, represent & protect ecosystems, biodiversity & geological elements, use for education & enjoyment in the present & future, promote research & education	No explicit mention	NPA, Sec. Environment, Sec. Science, Min. Exterior, Min. Defense, Min. Security, Min. Agroindustry, scientific & university organizations, NGOs	24, 25
Specific PA creation					
del Sud NP	1922	Control overexploitation, protect forests, conserve flora & fauna	Regulate tourism to not affect nature	Provisional director	1
Nahuel Huapi NP	1934	Define boundaries	No explicit mention	NPD	2
Los Glaciares NR (NP)	1937 (1945)	Define boundaries, mitigate overexploitation, importance of aesthetic, scientific & cultural values, conserve species	Promote tourism, public enjoyment & use, implement appropriate infrastructure (including workers & economic resource)	NPD, mentions "local population" support (GANPT, unions, companies & municipalities)	3, 4
Perito Moreno NR (NP)	1937 (1945)	Define boundaries, mitigate overexploitation, importance of aesthetic, scientific & cultural values, conserve species	Promote tourism, public enjoyment & use, implement appropriate infrastructure (including workers & economic resource)	NPD, mentions "local population" support (GANPT, unions, companies & municipalities in 1945)	3, 4
Lanín NR (NP)	1937 (1945)	Define boundaries, mitigate overexploitation, importance of aesthetic, scientific & cultural values, conserve species	Promote tourism, public enjoyment & use, implement appropriate infrastructure (including workers & economic resource)	NPD, mentions "local population" support (GANPT, unions, companies & municipalities in 1945)	3, 4

(continued)

Table 4.2 (continued)

PA actions	Created (modified)	Ideas of nature	Ideas of tourism	Institutional actors	Ref.
Los Alerces NR (NP)	1937 (1945)	Define boundaries, mitigate overexploitation, importance of aesthetic, scientific & cultural values, conserve species	Promote tourism, public enjoyment & use, implement appropriate infrastructure (including workers & economic resource)	NPD, mentions "local population" support (GANPT, unions, companies & municipalities in 1945)	4
Laguna Blanca NR (NP)	1940 (1945)	Protect from disappearance, conserve flora & fauna	Promote tourism, public enjoyment & use, implement infrastructure (including workers & economic resource)	NPD (GANPT in 1945)	3, 5, 6
Petrified Forests NM (NP)	1954 (2012)	Define boundaries, importance of paleontological values, (ecosystem representation)	Public use	GANP, National Tourism Direction-Min. Transport & Social Tourism (NPA and Sta. Cruz Prov. in 2012)	7, 8
Tierra del Fuego NP	1946 & 1960	Define boundaries, allow grazing	No explicit mention	GANP, National Tourism Direction	9, 10
Lago Puelo NP	1971	Define boundaries	No explicit mention		11
Los Arrayanes NP	1971	Define boundaries	No explicit mention	NPS—Min. Agriculture & Livestock	11
Lihué Calel NP	1976, 2003	Importance of scientific-educational, archeological, cultural & ecological values	No explicit mention	NPS—Min. Agriculture & Livestock, La Pampa Prov.	12, 13
Right Whale NM	1984	Importance of aesthetic, historic, scientific & ecological values	Excludes all uses, except scientific & "visitors"	NPA	14, 15
Huemul NM	1996	Importance of aesthetic, historic, scientific & ecological values	Excludes all uses, except scientific & "visitors"	NPA, National Flora & Fauna Direction-Sec. Environment	14, 18

(continued)

Table 4.2 (continued)

PA actions	Created (modified)	Ideas of nature	Ideas of tourism	Institutional actors	Ref.
Monte León NP	2004	Importance for global biodiversity conservation	No explicit mention in legal documents	NPA, Sec. Tourism-Sta. Cruz Prov., CLT, GEF	19
Punta Buenos Aires NMR	2008	Importance of military, scientific, cultural & ecological values	No explicit mention	NPA, Min. Defense, Chubut Prov.	20
Southern Patagonia MCIP	2009	Right to a safe and balanced environment (article 41, *National Constitution*), manage natural resources, biological diversity, representation of ecoregions, conservation and rational use of species and habitats, protect landscape/natural/cultural heritage, environmental education, promote sustainable uses, public use for spiritual/mental well-being	No explicit mention	NPA, Chubut Prov., CENPAT, WCS, GEF	21
Makenke MIP	2012	Public use, sustainable development, rational use, environmental education, physical/cultural/spiritual health, scientific monitoring/research, aesthetic, healthy/clean environment, biodiversity, preservation, conservation, habitat	Develop provincial tourism	NPA, Santa Cruz Prov., Puerto San Julián Muni., rural property holders, local population, environmental NGOs, scientific-technical institutions, municipalities	22

(continued)

Table 4.2 (continued)

PA actions	Created (modified)	Ideas of nature	Ideas of tourism	Institutional actors	Ref.
Pingüino Island MIP	2012	Public use, sustainable development, rational use, environmental education, physical/cultural/spiritual health, scientific monitoring/research, aesthetic, healthy/clean environment, biodiversity, preservation, conservation, habitat	Develop provincial tourism	NPA, Sec. Tourism-Sta. Cruz Prov., rural property holders, local population, environmental NGOs, scientific-technical institutions, municipalities	23
Patagonia NP	2015	Define limits, protect flora and fauna (particularly hooded grebe), right to a clean and balanced environment, protect natural heritage, wilderness, unique environmental & landscape	No explicit mention	NPA, Min. Environment, Sta. Cruz Prov., FFF. text evokes art. 41, *National Constitution* and *Convention for Biological Diversity*	26, 27
Staten Island & New Year Archipelago NWR	2016	Sustainable development, scientific research, importance of historical, cultural & wilderness values (present & future), use & enjoyment, protect biological diversity	Tourism referenced in the TDF Prov. constitution & decree associated with the national PA	Min. Defense, TDF Prov. Sec. Environment, NPA, Min. Interior Min. Public Works, Min. Housing, Sec. Legal Affairs	28, 29

Abbreviations: National Parks Direction (NPD), General Administration of National Parks & Tourism (GANPT), National Parks Administration (NPA), General Administration of National Parks (GANP), National Parks Service (NPS), Conservation Land Trust (CLT), Global Environmental Facility (GEF), National Patagonia Center (CENPAT), Wildlife Conservation Society (WCS), non-governmental organizations (NGOs), Flora & Fauna Foundation (FFF), National Scientific & Technical Research Council (CONICET), Tierra del Fuego (TDF)
References (*NL* National Law, *ND* National Decree): 1. ND 8.4.1922, 2. NL12.103/34, 3. ND9.504/45, 4. ND105.433/37, 5. ND63.601/40, 6. NL13.895/49, 7. ND7.252/54, 8. NL26.825/12, 9. NL12.103 T.O., 10. NL15.554/60, 11. NL19.292/71, 12. ND609/76, 13. NL25.755/03, 14. NL22.351/80, 15. NL23.094/84, 16. ND2.148/90, 17. ND453/94, 18. NL24.702/96, 19. NL25.945/04, 20. Protocolo Adicional n°1 ANP-Min. Defense, 21. NL26.446/09, 22. NL26.817/12, 23. NL26.818/12, 24. NL27.037/14, 25. NL27.490/18, 26. NL27.081/15, 27. ND838/18, 28. ND929/16, 29. ND888/19

During the study period, there was also an increase in the types and structures of PAs. The first PA (i.e., del Sud NP) predates the formal system and legal structure; it was created by presidential decree. Early on, a series of seminal PAs were created first as NRs and subsequently NPs upon establishing the more formal legal structure and associated procedures (e.g., Nahuel Huapi NP, Glaciers NP). Only one park expansion was also associated with a change in status several decades after its initial creation, whereby Petrified Forests went from a NM (1954) to a NP. This process began with land purchases in the 1990s but was only changed legally to a NP in 2012.

Beginning in the 1990s, PA conservation objectives became more diverse; new administrative structures appeared like "strict," "wilderness," and "education" reserves. Plus, there was a focus on unrepresented ecosystems, such as coastal and marine areas (e.g., MCIPs); however, fully marine PAs were only created beginning in 2018. Given that the 1994 National Constitution devolved to provinces the rights over "natural resources," novel legal structures also were created to incorporate these environments into the national systems, when provinces were not willing to cede jurisdiction to the national government (e.g., inter-jurisdictional coastal PAs).

The overall role of tourism in creating these PAs has varied from being an explicit reason to being implicit. It is often absent from the legal instruments for the specific PAs evaluated in this study, except for more recent cases. However, it is frequently cited in other related documents, such as the legislation establishing the mission or functions of the structures that administer PAs (e.g., NL12.103/34, 3. ND9.504/45). During the 1940s, there was an explicit emphasis on "social tourism." However, the legal instruments themselves do not allow much interpretation regarding the earlier or later periods, and it was necessary to return to secondary sources (see Fig. 4.2). Nonetheless, it is notable that tourism first entered into the national government organization structure as a part of the General Administration of Protected Areas and Tourism. Subsequently, tourism evolved and gained importance, becoming a national ministry in 2000. For a period of time, the National Parks Administration was a subordinate of this ministry, although legally, it has had an "autonomous" status since the 1950s.

4.3.2 Local-Global Connections in Argentine Patagonia

4.3.2.1 Euro-nationalist Imaginary

As evidenced by the writings of F.P. Moreno (1942), the initial idea of national PAs in Patagonia arose in the late 1800s and early 1900s, emulating the United States NP movement that began under the presidency of Ulysses S. Grant in 1872, with the designation of Yellowstone NP and expanded during the presidency of Theodore Roosevelt (1901–1909). Roosevelt and Moreno reveled in one another's company, when the two traveled together during the former US president's visit to the area that would become Nahuel Huapi NP in 1913 (Scarzanella 2002). The idea of

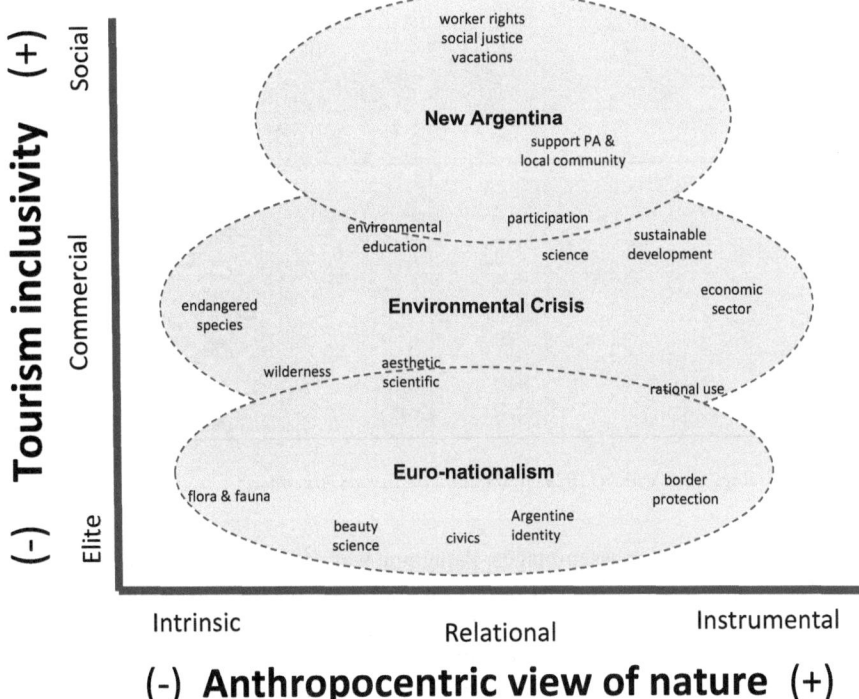

Fig. 4.2 Conceptual model of *Euro-nationalist*, *New Argentina*, and *Environmental Crisis* social imaginaries detected in official documents associated with Argentine national protected areas

conserving nature through parks was informed by international scientific understandings and initiatives (e.g., creating museums and collections to exhibit biological and geological features) and by notions related to education and social concerns (e.g., the ideas of progress, nation-building, national identity, civic character; Fortunato 2005; Wakild 2017). These two concepts exemplify a duality that is pervasive in Western thought, whereby nature is conceived both as a source of divinity and inspiration and as a wilderness to be mastered, civilized, and domesticated (Callicott 2008; Quijano 2000). This duality implies that nature is not just to be conserved by PAs but also "engineered" in many cases based on international (even global) standards of nature that promoted the introduction or prioritization of some (mainly European and North American) species or landscapes at the expense of others (Archibald et al. 2020; Dicenta 2021).

Valle's (1929) newsreel, *Por Tierras Argentinas*, illustrates early efforts to extend national sovereignty over this territory (Fig. 4.3a). Nature was portrayed as being national property, which extended from the tropical jungles to Antarctica. Nevertheless, the film highlighted that these areas, passed through a full range of European temperatures and that nature offered its various products, consistent with these temperatures. In keeping with the general tendency to create neo-European

Fig. 4.3 Frames from Valle's (1929) newsreel, *Por Tierras Argentinas*

landscapes and species assemblages, Patagonia was associated with forest resources (Archibald et al. 2020). Images were also contrasted with the concept of zoos, an urban element of conservation. Through this metaphor, a modern kind of tourism was promised to citizens that would expand from Buenos Aires to the peripheries thanks to the construction of new roads and railways. When the newsreel's depiction of this future tourism addressed Patagonia, images focused on the native richness of industrial forestry (Fig. 4.3b).

This conceptualization of nature underpinned land management decisions during Europe's colonization of the Americas. This period was characterized by deep social inequalities and the power of a dominant White elite, including oligarchic families. While defending the postcolonial autonomy of the republic, these elite founders of the nation emphasized Argentina's European origins, while silencing its non-White citizens (Geler and Rodríguez 2020). Their power, however, did not go unchallenged; a series of events show the state's repressive reaction against change and contestation, including the "Infamous Decade," military coups, Indigenous genocide, and worker massacres (Bayer and Lenton 2010). The desire to live in and engineer a national space made of White and European subjects, or what Gordillo (2016) has called "White Argentina," was also a desire to live among "White animals" and landscapes (Dicenta 2021). The desire was also reflected in the creation of the first PAs, whose legislation did not mention the Indigenous peoples or others living in the territories and included mechanisms to regulate who could occupy lands, circulate, and conduct activities in the parks, and who could be evicted, criminalized, and fined. As evidenced by NL12.103/34's mandate to "evict intruders," while at the same time European immigrants could receive revocable settlement and grazing permits on those same lands (Picone et al. 2020).

The first Patagonian PAs were created as this social imaginary began to wane in political influence. Therefore, it is not cohesively reflected in the formal legislation and legal instruments analyzed in the previous section. We can place the creation of del Sud NP (1922) and the initiation of Nahuel Huapi NP (1934) in this period, but subsequent formalization of these parks (1946) straddles the "New Argentina" imaginary. This difficulty of exact placement demonstrates the malleability of imaginaries and how their influences and affects are not discrete, but rather a continuum of dynamic, interacting forces.

4.3.2.2 New Argentina Imaginary

We also see this dynamism in an "elite" versus "inclusive" tension that played out in the conceptualization and operationalization of tourism. Beginning in the 1930s, tourism was seen as an emerging industry that would help the state's economy and geopolitical development interests, particularly in the area of Bariloche (Nuñez 2014). Overall, the initial tourism model in Patagonian PAs was imported from the Global North, with explicit reference and influence of the USA, Canada, Italy, and Switzerland (Torres 1954). Despite these "elite" origins, from its inception, tourism in PAs also had some recognition of its educational, civic, and patriotic functions. In his study of Patagonia, Sarobe (1935) built upon writings by North American engineer Bailey Willis, who had traveled around Patagonia to argue for social tourism, explicitly attacking the "elitist and colonialist" policies of the National Parks Direction. Under the leadership of Exequiel Bustillo, this agency promoted a vision for the Bariloche region as a European-style city for tourism, via elite hotels and summer homes (Navarro et al. 2012). In this elite tourism vision, nature was mostly conceived as an aesthetic resource for urban citizens in search of purity through contemplation and as a scientific resource for curiosity and exploration.

Simultaneously, this period's legislation permitted nature in PAs to be considered for other instrumental uses besides tourism, including hunting, fishing, logging, and mining (NL12.103/34). Overall, PAs were conceived as a management strategy for the national government to regulate and mitigate overexploitation of natural resources and to protect certain flora and fauna perceived as valuable due to rarity, autochthonous, or aesthetic values. In a mixed version of intrinsic and instrumental values, nature's aesthetic values and some native species were to be conserved, while simultaneously exploiting natural resources and promoting other introduced species (e.g., hunting and fishing exotic species, extracting timber). Plus, while there is little explicit mention of geopolitical interests in the legal instruments that created PAs in this period, these laws conceived of PAs as property, focusing a great deal of attention on their delimitation. Secondary sources also re-enforce this notion of PAs as "property" in a geopolitical sense to define borders with Chile, or in the case of Moreno's original vision, to create binational parks as a diplomatic strategy (Núñez 2014).

The 1920s–1950s were marked by multi-scalar social-political processes affecting Argentina. Following the global market failures of the late 1920s and the post-World War II reconstruction efforts, national planned economies began to shape countries across Europe, the Soviet Union, and North America (Dinoto 1994). In Argentina, this international-level restructuring prompted modifications of the

Argentine state after decades of failed governments and military coups. There was a significant expansion of government purview, which included not just economic influences but also social and political aspects. Argentina's Justice Party consolidated under the leadership of President J.D. Perón, and placed a novel spotlight on workers' rights, bringing about further social protections and paid vacation. Chapter IV of the reformed 1949 National Constitution conceived property, including nature, as a resource "for" social justice and national development through the rational planning of a strong state (see *Constitución de la Nación Argentina de 1949* 2014). In this way, nature was portrayed as a resource for a "new" Argentina, based on modernity and democracy. Previously disposable workers became the central subject of the nation with the Peronist government, a welfare vision that also reconfigured notions of tourism, which became a social benefit for workers. Simultaneously, tourism also became a vehicle for social justice by teaching the national richness to every citizen and by expanding public access to nature, rather than restricting it to the elites. The "New Argentina" period further incorporated aspects of the previous imaginary and—like Moreno—Perón considered nature not only as a right but also as a "duty of knowing the fatherland" via its geography and environment. Nature tourism became a responsibility of Argentines to know their territory and become a fully nationalized citizen (Carreras Doallo 2012).

Indeed, a 1950 government manual (Anonymous 1950) presented projects for the nation's future, contrasting past injustices, misery, and uncultivated lands with the prospect of a new Argentina that would protect workers, children, senior citizens, and women. Within the worker's rights section, a page included the slogan, vacations for all who work (Fig. 4.4a). This page portrayed an image representing a past in which only urban and White citizens could enjoy natural resources, and a series of other images, representing the present (1950s), in which the ocean, the mountains, the country, the sun, and the purest air were for everyone. It promised that no one was excluded in the exercise of a real democracy that grants equality with respect to both duties and rights. In the 1940s and 1950s, many such images and initiatives were promoted in Argentina. For example, summer camps emerged for children and propaganda portrayed images of happy workers, packing their cars, and traveling to the mountains or the coast. Also, a family board game, *Rutas Nacionales,* was developed (Fig. 4.4b, photo: M. Dicenta); to win, the object was to reach as many Argentine destinations as possible.

Furthermore, to optimize every region's contribution to Argentina's economy, the rational extraction of nature became central to PAs and Argentine environmental legislation. For example, the Law for the Defense of Forestry Wealth (*Ley de Defensa de la Riqueza Forestal*, NL13.273/48) prohibited the devastation of forested land and the irrational use of forest products (Carreras Doallo 2016). During this period, protectionism also shaped Patagonian PAs, which would become sites for promoting civic values, but unlike in the Euro-nationalist vision, this "new" way of thinking about PAs would be more inclusive of broader swaths of Argentines. Central to the New Argentina imaginary was redefining who was considered the "public" for use and enjoyment of these sites. Explicit emphasis was placed on families and workers' right to vacation and a duty to engage in tourism as a vehicle

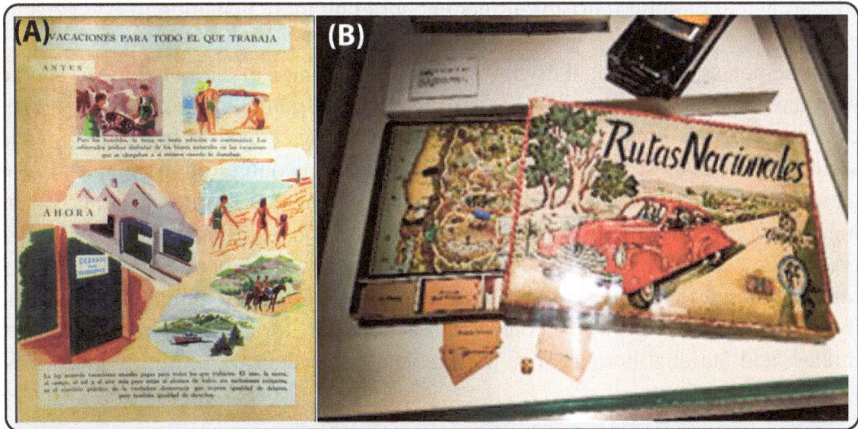

Fig. 4.4 Examples of materials used to socialize Patagonia in the 1940s and 1950s

for developing the PAs, local communities, and interior regions more broadly. The particularity of the model was its way of "synthesizing" the nation through its regions but also the intention to actively make it known to all social classes (Carreras Doallo 2012; Dicenta 2021). In 1941, the government institutionalized social tourism efforts with the National Tourism Direction. Additionally, paid worker vacations became law in 1945, and the encouragement to travel and explore the nation's territory led to more social tourism practices, opening tourism to the middle class (Carreras Doallo 2012; Picone et al. 2020).

As part of these social tourism practices, national PAs soon became a hot spot for Argentine working-class tourists (Picone et al. 2020; Rasmussen 2019). Indeed, in 1945, the National Parks Direction was fused with Tourism to become the General Administration of National Parks and Tourism under the Ministry of Public Works, whose Minister, Pistarini, had declared that:

> [w]e undertake tourism as a social goal. We want the masses to enjoy the beauties of the national parks so that in this way they will admire and love their fatherland more. The Ministry of Public Works can coordinate the transportation and soon will begin the construction of economical hotels. (Pistarini discourse, cited in Scarzanella 2002)

The incorporation of the NP system into Public Works further illustrated the emphasis on inclusive tourism for the working class, as well as the desire to utilize tourism as a national industry, especially through the development of hotels and transportation routes (particularly trains) to increase accessibility and use throughout the country (Carreras Doallo 2012). In 1951, the governmental tourism apparatus came to depend on the Eva Perón Foundation, while NPs transitioned back to the Ministry of Agriculture and Livestock (Carreras Doallo 2016). During this period, the system of national PAs was important to the second "Five Year Plan" as part of the country's broader planned national economic development (Carreras Doallo 2012). Additionally, tourism values were expanded upon in the 1960s, when new programs started to focus not only on recreational tourism but also on environmental education and cultural tourism, further encouraging more visitation to PAs (Picone et al. 2020).

Nonetheless, the national myth of White exceptionalism did not come to an end with the socialist turn. Previous ideas from the Euro-nationalist period (e.g., genocide) were updated with softer mechanisms, including education, moralization, and assimilation policies. PAs at the time reflected the reconfiguration of older Euro-nationalist visions in socialist ideals with policies oriented toward silencing, evicting, and controlling the peoples who did not assimilate sufficiently. In this sense, PAs became a tool for the government to control and censor populations, to control foreigners, and to register who became sedentary and formed a monogamous family. It also became a way to control Indigenous territories, immigrants, political threats, and national borders with mechanisms other than war, military narratives, and guards.

4.3.2.3   Environmental Crisis Imaginary

Patagonia has long been part of the Western social imaginary as a "remote" or "desolate" place (Moss 2008). However, more recently, it has been at the forefront of global conservation and tourism discourse as one of Conservation International's designated areas of last remaining "wilderness areas" (Mittermier et al. 2003). Their criteria for this designation included an area greater than 10,000 km^2, with greater than 70% intact native vegetation, and population density of less than 5 persons per km^2. Patagonia is 147,200 km^2, with 95% intact native vegetation, and population density of 0.14 persons per km^2. This global status as one of the last areas of remaining wilderness, justifies both Patagonia's conservation (e.g., "we must protect it because there are so few left"), and also its use (e.g., "we must visit it while we can!"). It also can be leveraged by academics to obtain funding and recognition (Rozzi et al. 2012). In this way, Patagonia is "singularized" at an international scale, which can erode attention to local and national forces, values, and interests.

By the 1950s–1970s, the notion of *environmental crises* began to affect the international understanding of nature, based on factors like contamination increase, climate change, and biodiversity loss (Reboratti 2000a, b; Estensorro Saavedra 2007). This movement was affected by multiple factors, but in particular beginning in the 1950s, the emerging discipline of ecology consolidated, especially in North America and Europe, and to gain legitimacy, it focused on the quantitative study and management of nature (Golley 1993; MacIntosh 1986). At this time, ecology also initially rejected normative or ethical positions associated with advocacy (Callicott 2008; Fiege 2011; Worster 1994). However, by the 1980s, the new field of conservation biology consolidated within this academic domain, explicitly having a mission-driven focus to staunch the crisis of "biodiversity," a newly coined term that expanded conservation's purview from charismatic species to the "diversity of life" in all of its expressions (Meine et al. 2006).

Changing ideas of nature at the international scale were complemented by new environmental management strategies. In particular, a series of intergovernmental meetings and programs arose that continue to this day. For example, the International Union for the Conservation of Nature (IUCN) was created in 1948 to help mitigate harmful human impacts on nature and has since played an influential role in standardizing and globalizing approaches to species conservation among its approximately 1,400 member states and organizations (Dudley et al. 2010). For example, in 1978, the IUCN created a typology of PAs that unified criteria, including ecological indicators like ecosystem representation. At the same time, thinkers "from the South" were also making local to global proposals with an explicit reference to humans in nature. For instance, in response to the "limits of growth" report (Meadows et al. 1972) that proposed reducing consumption at a global level based on purportedly "neutral" mathematical models to ameliorate the environmental crisis, the Bariloche Foundation in Argentina composed a multi-disciplinary team that recognized the need to incorporate normative and ethical dimensions of environmental decisions, including the ultimate goal of overcoming human misery in addition to nature conservation (Castro-Díaz et al. 2019). The local-to-global South-North connections are expressed by an early adoption of these integrated social-ecological ideas in PA management via the UNESCO Man and the Biosphere Program's model for biosphere reserves, developed in the 1970s, that explicitly integrated human well-being with nature conservation via zoning of core, buffer, and transitions zones (Araya-Rosas and Clusner-Godt 2007).

Similarly, in 1972, the seminal Stockholm Meeting on the Human Environment was the first world conference to make the environment a major issue, beginning a globalized debate about a rational use of nature that would ensure its continued existence for current and future generations, but without questioning the precept that economic expansion was part-and-parcel of human well-being (Brundtland 1987). These ideas have become mainstreamed in multilateral instruments like the Convention for Biological Diversity, which was signed in 1993 and, in addition to recognizing nature's multiple values, calls for local participation, equitable access, and benefits sharing *vis-à-vis* biodiversity. Increasingly, there is an enhanced integration of issues highlighted by scholars from the Global South, such as equity, justice, and health, as seen in the UN's 2030 Agenda for Sustainable Development and its associated Sustainable Development Goals (United Nations 2015; Castro-Díaz et al. 2019).

Environmental discourses from this period are based on ideas of "environmental crisis," where predominant academic disciplines and international approaches have held sway in the definition and valuing of nature largely from a "natural" perspective (e.g., ecological sciences). However, at the same time, these are not hegemonic influences, and the notion of "crisis" itself has responded to both local resistances and efforts to create more holistic and effective outcomes that conceive of the environment more broadly (Mace 2014). Indeed, today it is recognized that Indigenous peoples and local communities host the majority of the world's biodiversity, illustrating the importance of diverse human-nature interactions (IPBES 2019). Increasingly, the paradigm of "pristine" nature is disputed by alternative proposals that recognize humanity's role in niche construction, which refers to anthropogenic modification of environments (Ellis et al. 2010). For this reason, while this period is characterized here as a response to the "environmental crisis," there is also an active transformation of what is understood as "environment" at local, national, and international scales (IPBES 2022).

In Argentine Patagonia specifically, we find an emerging expression of PAs that explicitly incorporate not only interjurisdictional legal structures (national-provincial) but also inter-institutional management strategies (co-management committees), including greater and broader participation (explicit recognition of local communities and other stakeholders). The outcomes of these approaches are seen in the creation and implementation of several PAs in Argentine Patagonia. For example, the coastal areas, including Southern Patagonia MIP and Monte Leon NP, were promoted under the aegis of the Global Environment Facility (GEF) and with the active engagement of scientists. Newly created marine PAs were also an effort of the national government at the time to meet the Convention for Biological Diversity's Aichi Target 11, which entailed attaining protected status for 17% of terrestrial and 10% of marine national surface areas.

However, PAs in this period also express geopolitical objectives, as evidenced by the involvement of the Ministries of Defense, Security and Foreign Affairs in their co-management committees (Table 4.2). Furthermore, these efforts in the 1990s and 2000s were supported by academic and civil society organizations, such as the Forum of NGOs Dedicated to the Conservation of the Patagonian Sea and Areas of Influence, which brings together the efforts of more than 20 partner organizations from Argentina, Chile, and Uruguay (Foro para la Conservación del Mar Patagónico y Áreas de Influencia 2019). In this way, the more recent PAs display a broader understanding of nature and tourism but maintain many of the traditional values regarding instrumental use of these PAs for other purposes, such as military and economic goals.

4.4   Conclusions

Patagonia is not only affected by biophysical drivers but also social representations that are produced through the circulation of national and international ideas and their local encounters. This is especially evident when certain individuals or organizations obtain benefits by fetishizing the region both for its natural beauties or cultural singularities (e.g., aesthetic, ecological, or intrinsic values, Mittermeier et al. 2003; Rozzi et al. 2012), for its natural resources (e.g., livestock ranching, Caro et al. 2017; oil and gas exploitation, Hadad et al. 2021) and for tourism branding (Rodríguez et al. 2014). Today, the region faces unprecedented globalizing dynamics, but our analysis shows that these dynamics have been ongoing, albeit through different guises (imaginaries), for the past century (Fig. 4.5).

While in these PAs nature and tourism have been affected by international factors for over a century, not all drivers of social imaginaries are based on the Global North. Indeed, local and national efforts, and other international efforts emerging from the Global South, have also influenced particular management strategies. A

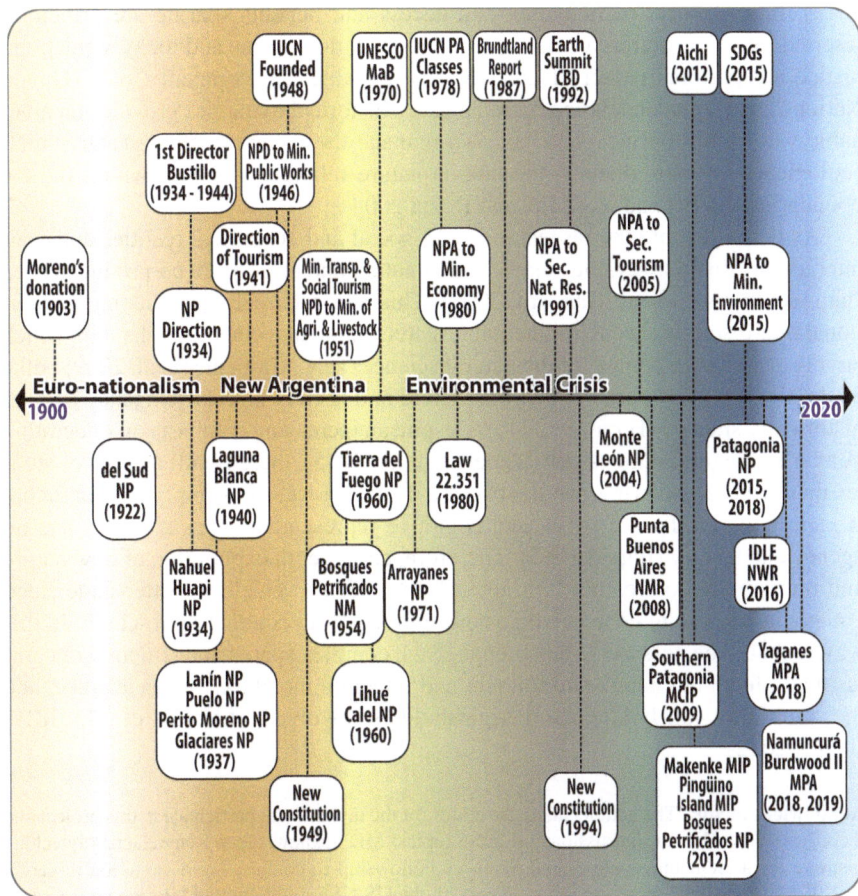

Fig. 4.5 A timeline of key events, legislation, and organizations help visualize the interacting dynamics that affect protected area creation in Argentine Patagonia

case in point is UNESCO's Man and the Biosphere Program, which arose in the 1970s (Araya-Rosas and Clüsener-Godt 2007). Since the 1990s, globalized conservation based on "Northern" paradigms has been critiqued (and resisted) for often reducing nature to market-based solutions, cost-benefit analyses, and for imposing Western scientific values and top-down agendas over peripheral regions (Gudynas 2003). These agendas can displace local actors and—despite increased resources—not necessarily reach promised outcomes for local communities (Rodríguez et al. 2007). In some parts of Latin America, local communities have even been persecuted for denouncing that conservation often justifies land-grabbing and control over Indigenous territories (West et al. 2006; Trentini 2017; Nuñez et al. 2017), which has also been a concern in Argentine Patagonia (Martín and Chehébar 2001).

At the same time, PA co-construction based on local to global perspectives leaves room for addressing nature and tourism challenges in an inclusive way (IPBES

2022). Indeed, local participation and access and benefits sharing are principles inserted into Convention for Biological Diversity documents and today, equity and justice are part-and-parcel of the UN's Sustainable Development Goals (United Nations 2015). Understanding national PAs' historical dynamics provides an amenable way to not only explore but also manage the complex, multi-scalar social-ecological system implied in the human-nature relationships both within the PA boundaries and beyond (e.g., Jax and Rozzi 2008).

PAs help illustrate the inseparability of social and ecological realities and their interactions. In this way, they teach us that nature and society are co-produced (i.e., "naturecultures," sensu; Haraway 2008). This approach complements more traditional interdisciplinary environmental research and management scholarship, which largely arose in the context of the natural sciences (e.g., Carpenter et al. 2009; Folke et al. 2011). It reinforces the fact that nature and society both have agency and are historically situated (Haraway 2008). Despite an increasing emphasis on integrating human and natural dimensions (Anderson et al. 2015), there is still a need for studies that avoid considering that the two are distinct domains or implying that nature is *merely* the product of social and cognitive representations, as if nature had no agency and history (Ingold 1993). Therefore, an in-depth exploration of how nature and tourism are co-creating PAs not only provides a vehicle to better understand Patagonia but also to conceptualize new approaches to conducting research. In this way, it should be possible to better engage the complex spatial and temporal dynamics that affect both nature and tourism and integrate the plural values at stake, and the multiple stakeholders and rightsholders involved (Mrotek et al. 2019; IPBES 2022).

Acknowledgments The authors thank the editor for the invitation to participate in this project and the Grupo SocioEco for discussions on these topics. MD acknowledges Rensselaer Polytechnic Institute for a Humanities and Social Sciences Fellowship to conduct her dissertation research. JLA, CBA, and AEJV were supported in part by the US NSF grant entitled "Patagonia Research Experiences for Students in Sustainability" (OISE 1261229) administered by Northern Arizona University. We also especially thank C. Chehébar and A. Sapoznikow for their time during interviews to detect information and sources associated with PAs in northern Patagonia.

References

C.B. Anderson, J.C. Pizarro, R. Estévez, A. Sapoznikow, A. Pauchard, O. Barbosa, A. Moreira-Múñoz, A.E.J. Valenzuela, ¿Estamos avanzando hacia una socio-ecología? Reflexiones sobre la integración de las dimensiones "humanas" en la ecología en el sur de América [Are we moving towards a socio-ecology? Reflections on the integration of "human" dimensions in ecology in the American South]. Ecol. Austral **25**(3), 263–272 (2015)

Anonymous, *La Nación Argentina Libre, Justa y Soberana [The Free, Just and Sovereign Argentine Nation]* (Peuser, 1950)

P. Araya-Rosas, M. Clüsener-Godt, *Reservas de la Biosfera: Un Espacio para la Integración de Conservación y Desarrollo: Experiencias Exitosas en Iberoamérica [Biosphere Reserves: A Space for the Integration of Conservation and Development: Successful Experiences in Iberoamerica]* (UNESCO, 2007)

J.L. Archibald, C.B. Anderson, M. Dicenta, C. Roulier, K. Slutz, E.A. Nielsen, The relevance of social imaginaries to understand and manage biological invasions in southern Patagonia. Biol. Invasions **22**, 3307–3323 (2020). https://doi.org/10.1007/s10530-020-02325-2

O. Bayer, D. Lenton, *Historia de la Crueldad Argentina: Julio A. Roca y el Genocidio de los Pueblos Originarios [History of Argentine Cruelty: Julio A. Roca and the Genocide of Native Peoples]* (Red de Investigadores en Genocidio y Política Indígena (RIGPI), 2010)

E.L. Bridges, *Uttermost Part of the Earth* (E.P. Dutton & Co, 1948)

C. Briones, W. Delrio, La "Conquista del Desierto" desde perspectivas hegemónicas y subalternas [The "Conquest of the Desert" from hegemonic and subaltern perspectives]. Runa **27**, 23–48 (2007). https://doi.org/10.34096/runa.v27i1.2639

G.H. Brundtland, *Our Common Future: Report of the World Commission on Environment and Development* (UN-Dokument A/42/427, 1987) Retrieved January 8, 2020, from https://sustainabledevelopment.un.org/content/documents/5987our-common-future.pdf

J.B. Callicott, ¿Qué *wilderness* en ecosistemas de frontera? [What wilderness in frontier ecosystems?] Environ. Ethics **30** , 235–249 (2008). https://doi.org/10.5840/enviroethics200830330

J. Caro, S.C. Zapata, J.I. Zanón, A. Rodríguez, A. Travaini, Ganadería ovina y usos alternativos del suelo en la Patagonia austral, Argentina [Sheep ranching and alternative land use in southern Patagonia, Argentina]. Multequina **26**, 33–50 (2017)

S.R. Carpenter, H.A. Mooney, J. Agard, D. Capistrano, R.S. DeFries, S. Díaz, A. Whyte, Science for managing ecosystem services: Beyond the Millennium Ecosystem Assessment. Proc. Natl. Acad. Sci. **106**(5), 1305–1312 (2009). https://doi.org/10.1073/pnas.0808772106

X.A. Carreras Doallo, Parques nacionales y peronismo histórico: La patria mediante la naturaleza [National parks and historical peronism: Homeland through nature]. Estud. Perspect. Tur. **21**(5), 1318–1335 (2012)

X.A. Carreras Doallo, Discurso y política forestal en el peronismo histórico. Entre la protección al ambiente y el productivismo [Discourse and forest policy in historical Peronism. Between environmental protection and productivism.]. Estudios Rurales **6**(11), 1946–1955 (2016). https://doi.org/10.48160/22504001er11.343

C. Castoriadis, *La Institución Imaginaria de la Sociedad [The Imaginary Institution of Society]* (Tusquets, 1993)

R. Castro-Díaz, M. Perevochtchikova, C. Roulier, C.B. Anderson, Studying social-ecological systems from the perspective of social sciences in Latin America, in *Social-Ecological Systems of Latin America: Complexities and Challenges*, ed. by L.E. Delgado, V.H. Marín, (Springer, 2019), pp. 73–94

O. Chamosa, People as landscape: The representation of the *Criollo* interior in early tourist literature in Argentina 1920–1930, in *Rethinking Race in Modern Argentina*, ed. by P. Alberto, E. Elena, (Cambridge University Press, 2016), pp. 53–72

Constitución de la Nación Argentina de 1949 [Constitution of the Argentine Nation of 1949] (Ediciones Infojus, 2014) Retrieved January 11, 2023, from http://www.bibliotecadigital.gob.ar/items/show/1571

C. Darwin, *Journal of Researches into the Natural History and Geology of the Countries Visited During the Voyage of HMS Beagle Round the World, Under the Command of Capt. Fitz Roy* (Ward, Lock and Company, 1889)

M. Dicenta, White animals: Racializing sheep and beavers in Argentinian Tierra del Fuego. Latin American and Caribbean Ethnic Studies. Advanced online publication (2021). https://doi.org/10.1080/17442222.2021.2015140

M. Dicenta, G. Correa, Worlding the end: A story of colonial and scientific anxieties over beavers' vitalities in the Castorcene. Tapuya: Lat. Am. Sci. Technol. Soc. **4**(1), article 1973290 (2021). https://doi.org/10.1080/25729861.2021.1973290

M. Dinoto, Centrally planned economies. Am. J. Econ. Sociol. **53**(4), 415–432 (1994). https://doi.org/10.1111/j.1536-7150.1994.tb02614.x

N. Dudley, J. Parrish, K. Redford, S. Stolton, The revised IUCN protected area management categories: The debate and ways forward. Oryx **44**(4), 485–490 (2010). https://doi.org/10.1017/S0030605310000566

E.C. Ellis, K. Klein Goldewijk, S. Siebert, D. Lightman, N. Ramankutty, Anthropogenic transformation of the biomes, 1700 to 2000. Glob. Ecol. Biogeogr. **19**, 589–606 (2010). https://doi.org/10.1111/j.1466-8238.2010.00540.x

J.F. Estenssoro Saavedra, Antecedentes para una historia del debate político entorno al medio ambiente [Antecedents for a history of the political debate around the environment]. Revista Universum **22**, 92–111 (2007). https://doi.org/10.4067/S0718-23762007000200007

Foro para la Conservación del Mar Patagónico y Áreas de Influencia, *Taller técnico sobre implementación de áreas marinas protegidas nacionales en el Mar Argentino. Reporte final [Technical workshop on the implementation of national marine protected areas in the Argentinean Sea. Final report]*. (2019). Retrieved January 28, 2023, from https://marpatagonico.org/publica/informe-tecnico-sobre-implementacion-de-areas-marinas-protegidas-nacionales-en-el-mar-argentino/

M. Fiege, The nature of the west and the world. West. Hist. Q. **42**(3), 305–312 (2011). https://doi.org/10.2307/westhistquar.42.3.0305

J.J. Fisk, L.A. Jacobs, B.U.K. Russo, E. Meier, Nakachi, Alohi, K.K.P. Spencer, K. Kaulukukui-Narikawa, A.W. Datta, K. Quiocho, Cultivating sovereignty in parks and protected areas: Sowing the seeds of restorative and transformative justice through the #LANDBACK movement. Parks Stewardship Forum **37**(3), 517–526 (2021). https://doi.org/10.5070/P537354734

C. Folke, A. Jansson, J. Rockstrom, P. Olsson, S.R. Carpenter, F.S. Chapin III, et al., Reconnecting to the biosphere. Ambio **40**, 719–738 (2011). https://doi.org/10.1007/s13280-011-0184-y

N. Fortunato, El territorio y sus representaciones como fuente de recursos turísticos: Valores fundacionales del concepto "parque nacional" [The territory and its representations as a source of tourism resources: Foundational values of the "national park" concept]. Estudios y Perspectivas en Turismo **14**(4), 314–348 (2005)

T. Gale, A. Ednie, Can intrinsic, instrumental, and relational value assignments inform more integrative methods of protected area conflict resolution? Exploratory findings from Aysén, Chile. J. Tour. Cult. Chang. **18**(6), 690–710 (2020). https://doi.org/10.1080/14766825.2019.1633336

L. Geler, M.E. Rodríguez, Mixed race in Argentina: Concealing mixture in the 'White' nation, in *The Palgrave International Handbook of Mixed Racial and Ethnic Classification*, ed. by Z.L. Rocha, P.J. Aspinall, (Palgrave Macmillan, Springer Publishing, 2020), pp. 179–194

G. Giucci, *Tierra del Fuego: La Creación del Fin del Mundo [Tierra del Fuego: The Creation of the End of the World]* (Fondo de Cultura Económica, 2014)

F.B. Golley, *A History of the Ecosystem Concept in Ecology: More Than the Sum of Its Parts* (Yale University Press, 1993)

G. Gordillo, The savage outside of White Argentina, in *Rethinking Race in Modern Argentina*, ed. by P. Alberto, E. Elena, (Cambridge University Press, 2016), pp. 241–267

E. Gudynas. Ecología, *Económia y Ética del Desarrollo Sostenible [Ecology, Economy and Ethics of Sustainable Development]*. 4th edition, Ediciones Abya Yala, Quito, Ecuador, **132** pp. (2003)

M.G. Hadad, T. Palmisano, J. Wahren, Socio-territorial disputes and violence on fracking land in Vaca Muerta, Argentina. Lat. Am. Perspectives **48**(1), 63–83 (2021). https://doi.org/10.1177/2F0094582X20975009

D.J. Haraway, *When Species Meet* (University of Minnesota Press, 2008)

T. Ingold, The temporality of the landscape. World Archaeol. **25**(2), 152–174 (1993)

IPBES, *The IPBES Regional Assessment Report on Biodiversity and Ecosystem Services for the Americas*. (2018). Retrieved January 26, 2023, from https://doi.org/10.5281/zenodo.3236253

IPBES, *Summary for Policymakers of the Global Assessment on Biodiversity and Ecosystem Services of the Intergovernmental Science-Policy Platform on Biodiversity and Ecosystem Services*. (2019). Retrieved January 26, 2023, from https://doi.org/10.5281/zenodo.3553579

IPBES, *Summary for Policymakers of the Methodological Assessment of the Diverse Values and Valuation of Nature of the Intergovernmental Science-Policy Platform on Biodiversity and Ecosystem Services*. (2022). Retrieved January 26, 2023, from https://doi.org/10.5281/zenodo.6813144

K. Jax, R. Rozzi, Ecological theory and values in the determination of conservation goals: Examples from temperate regions of Germany, United States of America, and Chile. Rev. Chil. Hist. Nat. **77**, 349–366 (2008). https://doi.org/10.4067/S0716-078X2004000200012

S.R. Kellert, *The Value of Life: Biological Diversity and Human Society* (Island Press, 1996)

P. Laterra, L. Nahuelhual, M. Gluch, P.L. Peri, G. Martínez Pastur, Imaginaries, transformations, and resistances in Patagonian territories from a socio-ecological perspective, in *Ecosystem Services in Patagonia: A Multi-Criteria Approach for an Integrated Assessment*, ed. by P. Peri, G. Martínez Pastur, L. Nahuelhual, (Springer Publishing: Natural and Social Sciences of Patagonia, 2021), pp. 397–428

G.M. Mace, Whose conservation? Science **345**, 1558–1560 (2014). https://doi.org/10.1126/science.1254704

R.P. MacIntosh, *The Background of Ecology: Concept and Theory* (Cambridge University Press, 1986)

A. Malm, A. Hornborg, The geology of mankind? A critique of the Anthropocene narrative. Anthr. Rev. **1**, 62–69 (2014). https://doi.org/10.1177/2053019613516291

C.E. Martín, C. Chehébar, The national parks of Argentinian Patagonia—Management policies for conservation, public use, rural settlements, and indigenous communities. J. R. Soc. N. Z. **31**(4), 845–864 (2001). https://doi.org/10.1080/03014223.2001.9517680

C. McEwan, L.A. Borrero, A. Prieto, *Patagonia: Natural History, Prehistory, and Ethnography at the Uttermost Ends of the Earth* (Princeton University Press, 1997)

D.H. Meadows, D.L. Meadows, J. Randers, W.W. Behrens III, *The Limits of Growth: A Report for the Club of Rome's Project on the Predicament of Mankind* (Universe Books, 1972)

C. Meine, M. Soulé, R. Noss, A mission-driven discipline: The growth of conservation biology. Conserv. Biol. **20**(3), 631–651 (2006). https://doi.org/10.1111/j.1523-1739.2006.00449.x

R.A. Mittermeier, C.G. Mittermeier, T.M. Brooks, J.D. Pilgrim, W.R. Konstant, Wilderness and conservation. Proc. Nat. Acad. Sci. **100**(18), 10309–10313 (2003). https://doi.org/10.1073/pnas.1732458100

F. Morello, L. Borrero, M. Massone, C. Stern, A. Garcia-Herbst, R. McCulloch, et al., Hunter-gatherers, biogeographic barriers and the development of human settlement in Tierra del Fuego. Antiquity **86**, 71–87 (2012). https://doi.org/10.1017/S0003598X00062463

C. Moss, *Patagonia: A Cultural History* (Oxford University Press, 2008)

A. Mrotek, C.B. Anderson, A.E.J. Valenzuela, L. Manak, A. Weber, P. Van Aert, et al., An evaluation of local, national and international perceptions of benefits and threats to nature in Tierra del Fuego National Park (Patagonia, Argentina). Environ. Conserv. **46**(4), 326–333 (2019). https://doi.org/10.1017/S0376892919000250

P. Navarro-Floria, El desierto y la cuestión del territorio en el discurso político argentino sobre la frontera sur [The desert and the question of territory in the Argentine political discourse on the southern border]. Revista Complutense de Historia de América **28**, 139–168 (2002). https://revistas.ucm.es/index.php/RCHA/article/view/RCHA0202110139A/28668

A. Núñez, E. Aliste Almuna, Á. Bello, M. Osorio, *Imaginarios geográficos, prácticas Y Discursos de Frontera: Aisén-Patagonia Desde el Texto de la nación [Geographical Imaginaries, Practices and Border Discourses: Aisén-Patagonia from the Text of the Nation]* (Pontificia Universidad Católica de Chile, Instituto de Geografía, 2017)

P.G. Núñez, La región del Nahuel Huapi en el último siglo. Tensiones en un espacio de frontera. Pilquen-Sección Ciencias Sociales **17**(1), 1–13 (2014)

G. Oliva, E. dos Santos, O. Sofía, F. Umaña, V. Massara, G. García Martínez, et al., The MARAS dataset, vegetation and soil characteristics of dryland rangelands across Patagonia. Sci. Data **7**(1), Article 327 (2020). https://doi.org/10.1038/s41597-020-00658-0

S.E. Picone, I.J. Liscovsky, A.F. Schweitzer, Territories for conservation? Capitalist strategies for appropriating nature in Los Glaciares National Park in the Argentinean Patagonia, in *Socio-Environmental Regimes and Local Visions: Transdisciplinary Experiences in Latin America*, ed. by M. Arce Ibarra, M.R. Parra Vázquez, E. Bello Baltazar, L. Gomes de Araujo, (Springer, 2020), pp. 241–252

A. Quijano, Coloniality of power, eurocentrism, and Latin America. Nepantia: Views from South 1(3), 533–580 (2000). https://doi.org/10.1177/0268580900015002005

A. Raya-Rey, J.C. Pizarro, C.B. Anderson, F. Huettmann, Even at the uttermost ends of the Earth: How seabirds telecouple the Beagle channel with regional and global processes that affect environmental conservation and socio-ecological sustainability. Ecol. Soc. 22(4), 31 (2017). https://doi.org/10.5751/ES-09771-220431

M.B. Rasmussen, Institutionalizing precarity: Settler identities, national parks and the containment of political spaces in Patagonia. Geoforum 119, 289–297 (2019). https://doi.org/10.1016/j.geoforum.2019.06.005

C. Reboratti, Los profetas de la catástrofe y los optimistas [The prophets of doom and optimists], in Ambiente Y Sociedad [Environment and Society], ed. by C. Reboratti, (Planeta/Ariel, 2000a), pp. 172–194

C. Reboratti, Sociedad y ambiente: Las miradas clásicas [Environment and society: Classic views], in Ambiente y sociedad [Environment and Society], ed. by C. Reboratti, (Planeta/Ariel, 2000b), pp. 149–171

J.P. Rodríguez, A.B. Taber, P. Daszak, R. Sukumar, C. Valladares-Padua, S. Padua, et al., Globalization of conservation: A view from the south. Science 317(5839), 755–756 (2007). https://doi.org/10.1126/science.1145560

J.C. Rodríguez T., P. Medina H., S.E. Reyes H. Territorio, paisaje y marketing global. Imaginarios en la construcción de la Patagonia como marca [Territory, landscape and global marketing. Imaginaries in the construction of Patagonia as a brand. Magallania 42(2): 109–123 (2014)

R. Rozzi, J.J. Armesto, R. Gutiérrez, F. Massardo, G.K. Likens, C.B. Anderson, et al., Integrating ecology and environmental ethics: Earth Stewardship in the southern end of the Americas. BioScience 62, 226–236 (2012). https://doi.org/10.1525/bio.2012.62.3.4

J.M. Sarobe, La Patagonia y sus problemas. Estudio geográfico, económico, político y social de los territorios nacionales del sur [Patagonia and its problems. Geographical, economic, political and social study of the southern national territories] (Aniceto López, 1935)

E. Scarzanella, Las bellezas naturales y la nación: Los parques nacionales en Argentina en la primera mitad del siglo XX [Natural beauties and the nation: National parks in Argentina in the first half of the 20th century]. European Review of Latin American and Caribbean Studies 73, 5–21 (2002)

M.J. Silveira, Lady Florence Dixie en la Patagonia Austral (1879), in XII Jornadas Interescuelas/ Departamentos de Historia, (Universidad Nacional del Comahue, San Carlos de Bariloche, 2009)

W. Steffan, P.J. Crutzen, J.R. McNeill, The Anthropocene: Are humans now overwhelming the great forces of nature? Ambio 36, 614–621 (2007). https://doi.org/10.1579/0044-7447(2007)3 6[614:TAAHNO]2.0.CO;2

A. Strauss, J.M. Corbin, Basics of Qualitative Research: Grounded Theory Procedures and Techniques (Sage, 1990)

C. Taylor, Modern Social Imaginaries (Duke University Press, 2004)

C.R. Torres, Ciudades de Invierno y Verano [Winter and Summer Cities] (Biblioteca del Congreso de la Nación, 1954)

F. Trentini, Ecología política y conservación: El caso del "co-manejo" del parque nacional Nahuel Huapi y el pueblo Mapuche [Political ecology and conservation: The case of the "co-management" of the Nahuel Huapi national park and the Mapuche people]. Rev. Pilquen Secc. Cienc. Sociales 15(1), 84–94 (2017). https://revele.uncoma.edu.ar/index.php/Sociales/article/view/1568

United Nations, Resolution 70/1. 2030 Agenda for Sustainable Development (2015). Retrieved January 13, 2023, from https://www.un.org/en/development/desa/population/migration/gener-alassembly/docs/globalcompact/A_RES_70_1_E.pdf

F. Valle, Por tierras argentinas [Through Argentine Lands]. Archivo General de la Nación, Departamento de Cine, Audio y Video (360.C16.1.A) (1929)

E. Wakild, Protecting Patagonia: Science, conservation and the pre-history of the nature state on a South American frontier, 1903–1934, in *The Nature State*, ed. by W. Hardenberg, M. Kelly, C. Leal, E. Walkid, (Routledge, 2017), pp. 37–54

P. West, J. Igoe, D. Brockington, Parks and peoples: The social impact of protected areas. Annu. Rev. Anthropol. **35**, 251–277 (2006). https://doi.org/10.1146/annurev.anthro.35.081705.123308

D. Worster, *Nature's Economy: A History of Ecological Ideas* (Cambridge University Press, 1994)

Chapter 5
Western Patagonia: From Frontiers of Exploration to the Commodification of Nature

Alejandro Salazar-Burrows, Claudia Matus, and Jorge Olea-Peñaloza

Abstract Western Patagonia is a territory whose historical trajectory has been described in terms of the harshness of its colonization processes and the constant struggle to domesticate nature. These ideas are still in use today. This chapter reflects on two major transformations: first, the historical and environmental configuration of western Patagonia, including the colonization processes and their continuity up to the present day, and the continuities and ruptures of economic activities, mainly tourism. Second, it describes the current occupation strategies, mainly focused on the touristification of western Patagonia, understood as a process of commodification of nature. As a result, we discuss the impact of normative notions of nature and culture in the production of narratives about western Patagonia, which constitute the basis of the practices of the contemporary tourism industry. At the same time, we identify and consider that the previous production and use of data

A. Salazar-Burrows (✉)
Estación Patagonia de Investigaciones Interdisciplinarias, Pontificia Universidad Católica de Chile, Santiago, Chile

Observatoire Homme-Milieu International Patagonia-Bahía Exploradores, LabEx DRIIHM (Programme Investissements d'avenir: ANR-11-LABX-0010), CNRS-INEE, Paris, France

Institute of Geographie, Pontificia Universidad Católica de Chile, Santiago, Chile
e-mail: asalazab@uc.cl

C. Matus
Estación Patagonia de Investigaciones Interdisciplinarias, Pontificia Universidad Católica de Chile, Santiago, Chile

Center for Educational Justice, Pontificia Universidad Católica de Chile, Santiago, Chile
e-mail: cmatusc@uc.cl

J. Olea-Peñaloza
Estación Patagonia de Investigaciones Interdisciplinarias, Pontificia Universidad Católica de Chile, Santiago, Chile

Environmental Science Department, Universidad Católica de Temuco, Temuco, Chile

Observatoire Homme-Milieu International Patagonia-Bahía Exploradores, LabEx DRIIHM (Programme Investissements d'avenir: ANR-11-LABX-0010), CNRS-INEE, Paris, France
e-mail: jolea@uct.cl

T. Gale-Detrich et al. (eds.), *Tourism and Conservation-based Development in the Periphery*, Natural and Social Sciences of Patagonia,
https://doi.org/10.1007/978-3-031-38048-8_5

about the territory provide a particular reality in which the definition of both social and natural aspects of the territory must be considered.

Keywords Patagonia · Colonization · Touristification: commodification of nature · Data production

5.1 Introduction

In recent decades, contemporary social science perspectives (critical, post-structural, and posthumanist) have highlighted the importance of how social, cultural, and environmental dynamics are conceptualized (Alaimo 2016; Barad 2007; Bennett 2010; Guillion 2018; Haraway 2016; Houser 2014; Nixon 2011), and how the rise of tourism associated with nature is interpreted. These perspectives help us to consider the effects of economic globalization, and associated processes of economic deregulation as dynamics dependent on the commodification of nature for their production, circulation, and consumption processes (Ávila-García 2016; Bustos and Prieto 2019; Núñez et al. 2014). These dynamics have led to increased national and international tourism investments, including in developing countries (Zoomers 2010). Under these logics, new processes of resignification and territorialization emerge, promoted by the State-Market, (Alessandri 2008), that affect tourism-nature relationships and the communities that host them.

Thus, while studies of land privatization have traditionally focused on large-scale land grabbing for consumption and resource exploitation, in recent decades, land grabbing has increasingly been associated with tourism and conservation practices, transforming non-capitalist spaces, and resources into commodities. In this context, sites of environmental relevance and global scientific interest have been reconceptualized and mutated as the new raw material of capitalist production linked to tourism and ecosystem services (Kelly 2011). For example, the process of green grabbing, where ecosystems are for sale, for current, future, and speculative uses, appears to be experiencing significant growth around the world (Fairhead et al. 2012). To a certain degree, the development of protected areas (PAs), nature reserves, and ecotourism can also be understood as forms of land grabbing and expropriation of natural spaces. Under the tutelage of international and national organizations, private individuals are buying large tracts of land for conservation and the development of "special interest tourism," provoking the creation of full-service private nature reserves with the potential to generate high income (Rivera and Vallejos-Romero 2015; Holmes 2014).

Since the 1990s, there has been a constant buying and selling of land from former settlers to private individuals who are interested in this territory for its high natural value and biodiversity in Western Chilean Patagonia (Núñez et al. 2014). Properties located on the margins of protected ecosystems, natural attractions, or on the boundaries of State PAs are especially sought out. These changes in land use and occupation cause increasing anthropic pressures on sensitive natural spaces (Jorquera et al. 2017; Núñez et al. 2020). This trend can be linked to capitalist practices and green development discourse, which we propose represent a new form of conservation-oriented colonization, aimed at the protection of natural environments. Conservation colonization, such as this, involves not only physical land grabbing, but also the privatization of the rights of nature (Núñez et al. 2020; Corson and MacDonald 2012).

In this chapter, we describe these processes of land grabbing in Patagonia through two main foci: (1) the tensions between the environmental-tourist history of the place including the initial and recent colonization processes; and (2) the unequal ways in which tourism development has been understood as a process of nature commodification.

5.2 Study Area: Western Patagonia and the Exploradores Valley

Western Patagonia, and in particular the Aysén Region, is characterized by its extreme and isolated physical and cultural geographic conditions. The Aysén region has been exposed to different socio-cultural, economic-productive, environmental, and planetary interferences that have redefined recent processes of socio-territorial re-signification and transformation (Martinic 2005; Urbina 2013; Olea-Peñaloza et al. 2021). The region is a political, natural, touristic, and ecological frontier zone, constructed by diverse actors including pioneers, neo-settlers, and neo-ecologists (Bourlon 2020). These groups have different agendas that help to produce the social heterogeneity that we currently find in the area.

One catalyst of the current boom in tourism occurred during the 1970s when geopolitical strategies were developed to achieve decentralized development in the country. This led to policies and initiatives to integrate isolated and marginalized territories, promote occupation and connectivity, and to leverage the comparative advantages of the territories (Bustos 2014). Toward the end of the 1990s, processes of re-territorialization and resignification of nature began in the region, which became seen as a great reserve of potential use; an orientation that aligns with the neoliberal political-economic logics prevailing in the country. Thus, the strengthening of the image of nature as the wealth of the country promoted the protection and conservation of environmentally relevant spaces (Brigand et al. 2011; Nuñez et al. 2014, 2019).

It is important to consider that western Patagonia, according to official data, concentrates the largest amount of Natural Protected Areas (83% at the country level), with the Aysén Region being the most preserved in Chile (Chilean National Forestry Corporation 2020; Chilean Ministry of the Environment 2018). Since the 1990s, the Aysén Region, along with the association of the image-objective of life

reserve, has experienced an increase in anthropic pressure, not only with a population growth of 292% between 1952 and 2017 (26,262 to 103,159 inhabitants), but
also with the increase of floating population (9959 to 580,046 visitors between 1990
and 2017). The region stands out for having the largest number of companies in the
tourism industry, and for generating the highest concentration of tourism sales
(17.8%) at the regional level, with 5.8% of the region's economic activity in accommodation and tourism services, exceeding the national percentage of 4.4% (Chilean
National Tourism Service 2017, 2018, 2019).

In the last decade, the growing tourism activity linked to the nature experience
and natural attractions can be understood as an important economic sector, which
has been established in various parts of the region (Nuñez et al. 2019). Patagonia,
therefore, is experiencing tourism sector growth (Flores and Martínez 2020), especially in the area of special interests. Among the popular destinations are Lake
General Carrera, the Northern Ice Field, Caleta Tortel, and the national parks
Queulat, San Rafael, and the recently inaugurated Patagonia Park.

However, nature tourism activity actually began in the early decades of the twentieth century, as evidenced by tourism magazines and postcards from the 1930s
(Flores and Martinez 2020) (see Figs. 5.1 and 5.2). Since the beginning of the twentieth century, Western Patagonia was intended to be part of the most exclusive natural tourist destinations. This tourism vision centered around the desire to be able to
contemplate and feel elements of nature, which were aesthetically and convention-
ally defined as beautiful and inspiring. Further evidence of this vision were the
intentions of the State to build a hotel and airstrip on the shores of San Rafael

1871 Chile.—Ofqui, Ventisquero San Rafael   Prop. 5785
                                                                               Foto Mora

Fig. 5.1 1939 tourist postcard of San Rafael Lagoon and Glacier. (Mora Ferraz 1889–1958a)

Fig. 5.2 Photo of an antique postcard of the Isla de Mármol from the personal collection of Daniel Buck. (Published by Roberto Rosauer, Buenos Aires (ca 1905), photo by Clemente Onelli)

Fig. 5.3 1930 tourist postcard of Glacier and San Rafael Lagoon. (Mora Ferraz 1889–1958b)

Lagoon (Fig. 5.3), along with other State-supported actions from the private sector such as the introduction of salmonid species and development of sport fishing as a tourist activity (Camus and Jaksic 2009). The latter today remains one of the main factors of change and impact on freshwater and terrestrial ecosystems within the region (Reid et al. 2021). The contradiction is that the elite activity of sport fishing is one of the main tourism resources, promoting an image associated with the pristine nature of Patagonia.

The Exploradores Valley, with its proximity to the San Rafael Lagoon Biosphere Reserve and National Park, has acquired national and international relevance. The valley's accessibility, together with unique landscape features such as the Northern Ice Field, and the presence of large extensions of temperate rainforest have been essential for transforming it into a hub of tourism development. These conditions have consequently brought increasing anthropic pressures. The latter is evidenced by the increase of tourist visits to San Rafael Lagoon National Park in the last decade, with 187 visitors in 2010, compared to 8222 visitors in 2018 (Chilean National Forestry Corporation 2018).

To further explore the impact of normative notions of nature and culture in the production of narratives about western Patagonia, we (re)analyzed various data from official secondary sources at the regional level. We aimed to identify the trajectory and development of tourism, and forms of territorial transformation over time, with an emphasis on infrastructure for tourism development. We sought to understand both the tourism phenomenon itself and the construction of narratives associated with the activity. Our process included a case study designed to understand tourist behavior in the Exploradores Valley. Our analysis was supported by geographic information systems (GIS) and included spatial data related to land ownership, as well as urban and rural infrastructure linked to tourism and the transformation. Our process allowed visualization of the changes that have occurred in recent years within the study area.

Finally, our mixed comparative analysis helped us understand the construction of the tourism process, including how tourism information is constructed and what elements stand out. In this way, we seek to stress the practices of information production, as well as how tourism is studied, and we consider how the approach could be improved in the Exploradores Valley.

5.3 Tourism in the Exploradores Valley: Local History of a Global Trajectory

Patagonia has become a global tourist destination (Blair et al. 2019). In this sense, we are faced with a development that is part of a global industry and whose effect in each territory is accompanied by local processes. In addition, tourism in Patagonia falls into the category of special interests, because its main attractions are its landscapes and activities associated with magnificent natural scenery (Rovira and Quintana-Becerra 2019).

The trajectory of western Patagonian tourism, therefore, is associated with it being a remote and pristine territory, full of adventures, and offering the possibility of seeing something exclusive. The last decades of the century, pushed by a shift toward postmaterial values together with the emergence of a greater sensitivity toward the ecological crisis, have promoted natural destinations, which despite the difficulty of access, attracted the attention of specific groups of society (Rojek and Urry 1997). The turn to this type of tourism is a reaction to mass tourism growth throughout the twentieth century. This is due in part to advances in transportation and communication, along with real estate development. These allowed for tourism development in areas of modest income and upended traditional mass tourism that was concentrated mainly in beach destinations (Zuelow 2016; Cañada and Murray 2019). This summarized history has its own version in the Exploradores Valley. The result of what we see today is part of different stages of historical occupation, beginning with modern colonization in the first half of the twentieth century. After the failure of the exploitation companies in the Aysén region (Harambour 2017), the territory was temporarily left out of public expansion policies, initiating a process based on individual land occupation initiatives. In this way, valley-by-valley settlers who initially dedicated themselves to cattle ranching began to occupy the region (Millar 2017). Various means were used to clear the land, with fires being a prominent approach whose mark is seen to this day. The Exploradores Valley had a late occupation in relation even to Patagonia itself.

In parallel, exploration trips were undertaken in these territories in order to assess the possibilities of an effective expansion of the Chilean State, beginning in the mid-nineteenth century (Bello 2017). One of the most relevant has been that of Hans Steffen, who in the late nineteenth and early twentieth centuries, established a route through the confines of Patagonia, visiting places that did not have a first-sourced record. Likewise, along the coast, there was a greater flow of trips since the passage through Cape Horn had been used for quite some time and the knowledge of the steep coast was key both for navigation and for the marine fauna that existed at the time. In the case of the Exploradores Valley, the mission was entrusted to Augusto Grosse, an official of the Ministry of Public Works, who in the first attempt entered by sea, and second starting from Lake General Carrera. In 1946 he succeeded in opening a route to make the entry of new inhabitants more efficient (Grosse 1955, 1986).

These two processes occurred simultaneously and are the foundations of modern occupation in the valley. Both are the basis of a definitive settlement in the area, which allowed, among other things, recognition of the attractions of the area, as well as the establishment of possible routes to be followed by visitors. Gradually, the valley became better known and more visited by people other than residents or those who wanted to claim a site. Colonization and the search for a road were key initial forces for tourism development as a modern phenomenon, which requires both elements: attractions and known routes.

However, the initial stage of tourism development was through the coastal zone. The marine approach to the region was the most recognized and understood. In an attempt to search for Chilean Switzerland, an effort to develop tourism similar to that in the Alps began in the area of Lake Villarrica (Flores and Martínez, 2020). The first attraction was the San Rafael Lagoon (see Fig. 5.3), which offers an imposing natural setting with a glacier named the same as part of the Northern Ice Field. Here, two simultaneous ideas converged for the Chilean State, which sought to exercise sovereignty in these territories: the opening of a canal in the Ofqui isthmus (which falls into the San Rafael lagoon), and the installation of a hotel and airstrip in 1940. The canal sought to be an outlet for the products of the inhabitants of Lake General Carrera (Lake Buenos Aires) by connecting the coast of Aysén through the fjords (Martinic 2013), the Exploradores Valley, and the interior of Aysén with the coast. For its part, the hotel sought to attract tourists to stay in the area (see Fig. 5.4). The hotel was developed in relation to a maritime route that would be the extension of the railroad since land access was very difficult at the time. The only way that existed until then was a fleet of steamships leaving from Puerto Montt, which were the "maritime fleet" of the Empresa de Ferrocarriles del Estado (Rosetti 2018). However, this hotel began to be designed and built in 1939 but was short-lived, as the project was abandoned after the canal work was stopped.

The construction of the Carretera Austral (Southern Highway) opened the possibility of traveling through Patagonia by land. The Carretera Austral connects Aysén in its interior but also puts the region in contact with other areas of the country (Urrutia 2020). The road redesigns the old routes used by the population of the area. The arrival of the highway meant a rearticulation of the routes in the Aysén area. It also coincided with other processes, such as the opening to international markets and the consolidation of an economic model that sought to diversify the region's economic activities. The stage was set for tourist activity that valued

Otro aspecto del hotel en construcción

Fig. 5.4 Hotel under construction in San Rafael Lagoon. (Durand 1941, p. 60)

natural attractions such as Lake General Carrera, the various National Parks, Ice Fields, and communities such as Caleta Tortel or Chile Chico.

The road connected the regional capital and airport with Lake General Carrera, which had been the region's gateway since Grosse sought to connect the region via the sea. The town of Puerto Río Tranquilo gradually became an obligatory point of interest along the Aysén route, as it was a stop for those who wanted to continue south, mainly to reach Cochrane and Caleta Tortel. This early stage of tourism development attracted mostly adventurous people, who sought access to those spaces little known until then.

Puerto Río Tranquilo became a key gateway community for visits to the Marble Cathedral and Chapel (Bachmann-Vargas and van Koppen 2020). The presence of these marble structures became well known, due to the scenic beauty they represented. The old mining activity in the area had developed a transport circuit of people and goods across Lake General Carrera, connecting the different ports. The commune of Puerto Río Tranquilo had been one of the food producers in the area, from the first settlers' livestock and agriculture on the shores of the lake.

In this way, a re-signification of a previous productive activity (transportation and commerce linked to mining and agricultural activities), emerged in the service of tourist transportation. The Carretera Austral quickly became a very attractive opportunity for tourists, as did the lake ports, for tourists visiting the Chapels and Cathedrals in the area.

With this reorganization, the Exploradores Valley entered the regional tourist circuit. The Exploradores Valley began to experience its own process of land expansion, but much later. Although the Carretera Austral began to be built in the 1980s, the section that goes into the valley began to be developed in the mid-1990s (1996), reaching Bayo Lake at the turn of the millennium (see Fig. 5.5). The old path that

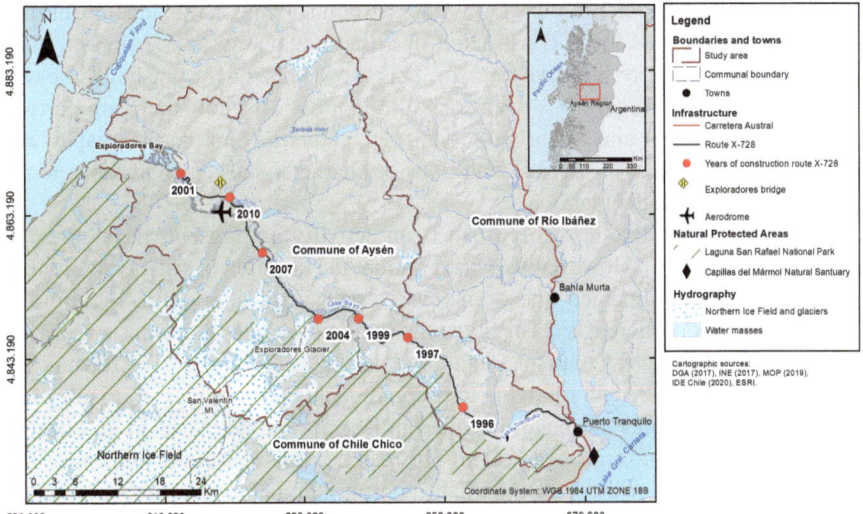

Fig. 5.5 Timeline of road and bridge development with the Exploradores Valley

used to take the settlers through hills, rivers, and ravines gave way to a land road that connected the deepest part of the valley with Puerto Río Tranquilo. By 2010, a large part of the road had already been built, but it was not until 2017 that the bridge over the Teresa River was completed, connecting the entire valley up to Exploradores Bay.

Route X-728 allowed the possibility of visiting the Exploradores Glacier and reaching the San Rafael Lagoon on a land-water trip directly from Coyhaique. The Exploradores Glacier provides an entrance to the Northern Ice Fields and represents the northern limit of the great continental ice masses of the southern hemisphere. It is not only an attraction for the sublime landscape and visitors wanting to observe it from viewpoints or take a walk on the ice, but also attracts scientists to learn what this space can tell us about climate change.

5.3.1   In Search of Patagonian Tourism: Pristine Scenery and Exclusivity

This brief history of tourism in Valle Exploradores shows us two central things. In the first place, tourism in the area has been built from the resignification of historical economic activities. The processes of occupation by the first western settlers and their families, together with their agricultural and livestock exploitations, ultimately created a gateway for tourism development. The tourism boom is based on the appeal of nature, sometimes romantic, sometimes as an economic support, and sometimes as a system of capitalist expansion (Núñez et al. 2019). Under the apparent uniformity of the activity, a series of territorial tensions are hidden, where the conversion to tourism was practically an imposition. Second, how tourism activity is constituted in the territory, which currently presents a resignification of nature that is always in dispute in relation to other activities. How is this idea of tourism constructed? It seems to be constructed in a traditional way and aspires to mass tourism through a more selective tourism. This shows that, on the one hand, the way in which nature is *enjoyed* has been transforming, but also how there are parts of these attractions that aim at mass tourism (e.g., cathedral and marble chapel) and others that are subject to special conditions or characteristics, such as visits to the glacier or the San Rafael lagoon.

Thus, in quantitative terms, tourism is an activity that presents an increase in visitors (see Fig. 5.6). However, this brings about the question of what type of infrastructure has accompanied these increases, and what this infrastructure has been designed for. There is a clash and at the same time a complementarity between traditional forms of mass tourism and alternative forms of tourism. The town is filling up with cabins for visitor lodging to accommodate the more mass tourism, while lodges and nature sites are being installed in less accessible places in an effort to attract the alternative forms of tourism. This can be observed as the process of late rural (neo)colonization associated with the appropriation of nature. In this sense, we

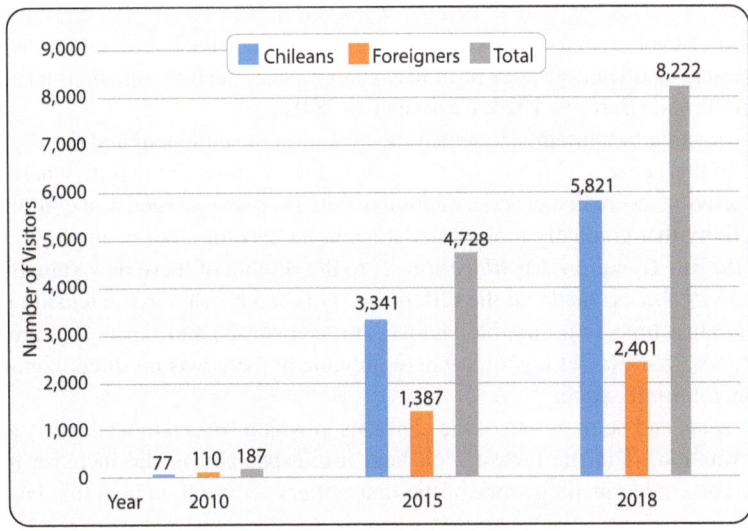

Fig. 5.6 Evolution of domestic and international visitors to San Rafael Lagoon National Park 2010–2018. Based on visitor statistics from the SNASPE Unit, Chilean National Forestry Corporation, 2010–2018

believe that it is necessary to continue the study of land ownership in combination with other studies of tourism development since the processes of urbanization and ruralization of the territory contain a tourism component. Doing so may show that the growth of the built and urbanized environment is not a product of the growth of the local population, but more a result of the arrival of new permanent and seasonal inhabitants whose main motivation is the tourism development of the area.

Among the main consequences of this process, we find transformations in the identity and daily narrative of the area. As has been seen in other places, tourist activity needs to be evaluated from different perspectives, since its impact reaches spheres of everyday life (Ojeda 2019; Zamirato and Tomazzoni 2015). Whether by the development activities that were not previously carried out in the place, or by resignifications to what was previously done, local inhabitants have had to get used to the constant arrival of people, as well as to the seasonality of visitor flows.

The Exploradores Valley is the last of the valleys in western Patagonia to be opened to land accessibility, making it possible to reach the international natural attraction, San Rafael Glacier and Lagoon on the same day by a combination of land and sea transport. This recent accessibility marks a point from which the processes of socio-territorial transformation and change accelerated. It is in this sense that the natural aspects become relevant again. The valley has gained recognition with respect to scientific interest, focusing on the monitoring of socio-environmental factors starting from a basal state (zero): an isolated valley whose conditions could be close to a pristine state of the environment. This interest illuminates one of the main discourses and narratives from which Patagonian tourism is understood, the search for spaces unaltered by human beings. However, human action was present within the region's evolution, as observed by evident fires since the beginning of the twentieth century and later decades or in less visible traces such as aquatic microorganisms in the soil (Barrows 1923; Castree et al. 2009).

The opening to tourism also meant the evolution of settlement and the use of the valley. In this sense, a similar process occurred throughout the region, where traditional activities were in a process of adjustment. Tourism emerged as a complementary activity that gradually took center stage in the territory. For example, the town of Puerto Río Tranquilo was transforming to the rhythm of these new valuations of the territory. Access made all the difference. This can be seen in the tension placed on local attractions (e.g., marble chapels and cathedrals) and others that are more distant (e.g., San Rafael lagoon), where previously there was no direct connection with the maritime space.

Likewise, and perhaps one of the elements in which we can most directly see the transformation, is in the pressure on land use, especially in the increase in land prices. The cattle ranching work of the first settlers occurred on land that had to be *cleared* for their use. Each group that came to settle requested from the State an extension of several square kilometers that would allow the installation of livestock. All these properties had a very low economic value, both because of poor connectivity and because of their challenging conditions for agricultural activity, which was the main mode of use. However, the opening of transportation routes, as well as a change in the discourse on nature and its consequent commodification, caused a considerable increase in the value of the land, as there was constant pressure to obtain a portion of this paradise.

If we observe the number of land transfers between 1959 and 2019 (see Fig. 5.7), we find that there is a first stage of the initial configuration of land ownership that goes from 1959 to 1978, involving few transactions of large acreage. In the second stage, from 1979 to 2008, the number of land purchases and sales increased

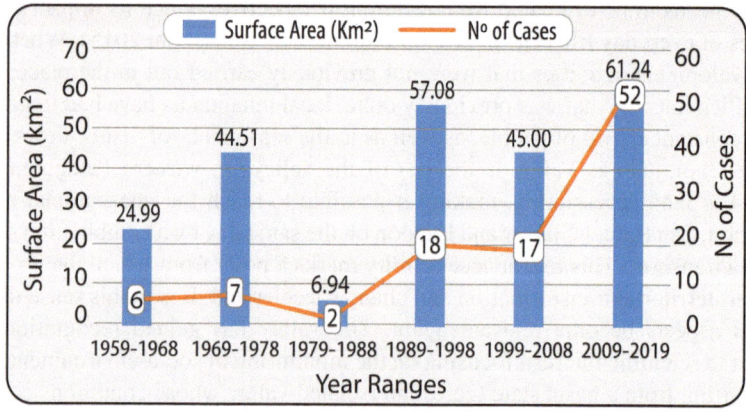

Fig. 5.7 Evolution of land ownership transfers 1959–2019 in the Exploradores Valley of the Aysén Region, Chile. Based on Aysén and Coyhaique Real Estate Conservators

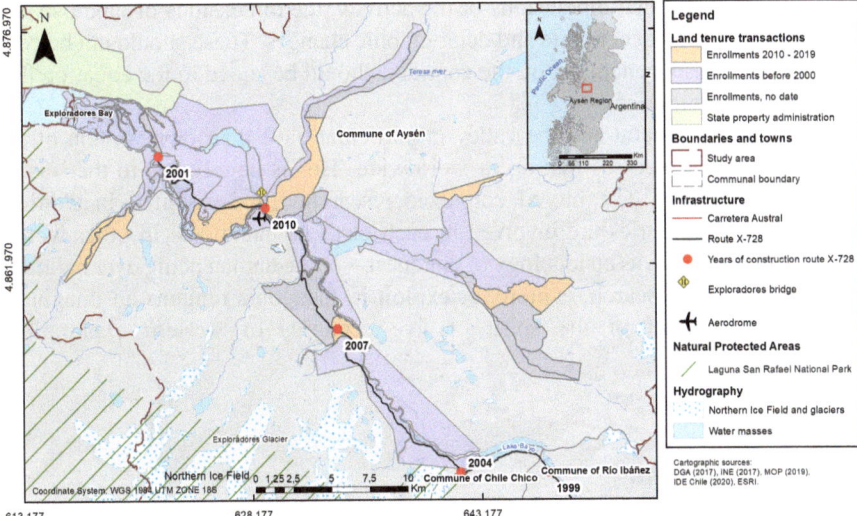

Fig. 5.8 Land tenure transactions in the Exploradores Valley, Bahía Exploradores-Lago Bayo section

considerably, with less acreage per parcel than in the first stage. In the third stage, from 2009 to 2019, land purchases continued to increase while the average acreage of each sale decreased more drastically. This trend demonstrates how, as time goes by, properties are becoming smaller and smaller.

In Fig. 5.8, it is possible to interpret the impact of connectivity created by the opening of the road in this isolated valley, from 2009 onward. Land ownership transfers shifted to new regional and extra-regional landowners with dissimilar transactions in terms of surface areas. This same process has been recorded in other sectors of western Patagonia, where the term "neo colonists" has been associated with these new landowners (Núñez et al. 2020). However, the phenomenon of subdivision and relative fragmentation of the properties has been observed in the study area since the 1990s. In this area, the sale of properties of the old and first settlers to corporations and individuals that did not necessarily occupy the properties, makes it difficult to label them as new settlers in the area (see Fig. 5.8).

These processes have also had an impact on residential settlement composition in the area. In direct relation to the tourist activity focused on lodging and the amenities of the surrounding landscape, Puerto Rio Tranquilo has spatially concentrated the installation of tourist infrastructure such as cabins for lodging, initiating a process of socio-spatial segregation. Considering tourism development as a socio-territorial process broadens our view of tourism in the Exploradores Valley. Its trajectory shows us that its tourist attractions have been in the spotlight for several decades and therefore are not a recent phenomenon. What is recent is the possibility of accessing them through a much more accessible network of transportation routes. Thus, we propose that efforts to understand tourism development in the area must consider the traditional dimensions of the activity, the dimensions of land use and ownership, and socioeconomic and demographic changes. These should not be considered as a consequence, but on the contrary, should be added to the equation that is at the basis of tourism.

Patagonia in general and the valley in particular have their own version of the socio-territorial processes that occur worldwide. This is largely due to the way in which the area has been explored, colonized, researched, and exploited. In addition, the western imaginaries are involved in each of these intentions. In other words, western imaginaries, as an ideology of conquest—of the human being over nature—affect those who research, explore, or exploit it. Patagonia remains an imaginary territory, a product of the spirit of five centuries of western explorations (Bourlon 2020).

5.4   Conclusions

The ways Patagonia and its inhabitants (locals and visitors) have been referred to and are referred to today portray a highly desirable region, which in many ways creates a status for the person who has had the experience of being in Patagonia. The economic activity of tourism in Patagonia involves a range of physical activities (backpacker lodges, cruises that visit the glacier, hiking itineraries, nature reserves, fishing in untamed rivers, etc.), that can be understood as a construction of the irresistibility of Patagonia as a tourist destination for travelers who enjoy the most daring activities. Thus, the construction of Patagonia, as a wild and untamed territory, invites a tourism oriented toward the isolated and remote and, in turn, attracts a certain type of tourist.

The Aysén region of Patagonia, in western Patagonia, is unlike other sectors of Patagonia that are mostly described in terms of relationships between Indigenous groups and their territories (such as Tierra del Fuego and the chronicles that document the relationships between Selknam, Yaghan, and missionary groups). The western part of Patagonia—referred to in this chapter—is characterized by a more corporate way of being, and this understanding is narrated by the available data. Wilderness, desolation, adverse climate, and exuberant nature, far from the idea of western civilization, provide a sense of extreme territory that is easily marketed. However, as Ogden (2021) points out, all places in the world are real and imagined at the same time, which in many ways suggests that we should ask ourselves how many Patagonias are possible based on the available information.

Considering the activities of different kinds that have occurred over time in this part of Patagonia, it seems important to offer some reflections on two questions. First, we consider how the available data shape the way Patagonia is developed; and second; how ideas of nature and culture, understood as two distinct spheres, underlie the ways in which research on Patagonia is produced.

A variety of types of secondary data are available for reviewing information about Patagonia: housing and population censuses, data on the number of visitors to the Region from the Chilean National Tourism Service (SERNATUR), data on protected areas and nature reserves, and other data produced by the Central Bank, Chilean National Forestry Corporation (CONAF), the Ministry of the Environment, among others. We propose that these data should be seen as empirical material with the purpose of understanding how, through the data, certain ideas about Patagonia and processes of tourism development, are perpetuated. It seems appropriate to inquire about the secondary data used to describe Patagonia about nature and culture; human/non-human; male/female; pristine/intervened; wild/civilized; local/tourist. Integrating this analysis of how information about Patagonia is produced, allows us to ask about other issues that are fundamental to understanding processes of tourism development.

We propose that the data used for both research and policy design should be interrogated in order to understand how dominant concepts and assumptions about nature, culture, gender, territory, and progress, underlie the production of both quantitative and qualitative information. We suggest that our research and practitioner communities consider how these concepts and assumptions might contribute to the persistence of systems that produce territorial inequality.

For example, we could explore how the binary gender norm affects the production of secondary data that is then used, in an unproblematized way, to talk about Patagonia. If the available data are shown using a distribution of work associated with the normative male/female dichotomy, we would have to assume that the distribution of work in Patagonia is *gendered*, that is, the activities and roles assigned to men and women from the gender norm are not questioned. This focus might make it difficult to account for and access other forms of work organization that could help to think about Patagonia in other ways. Just as many disciplines are reevaluating the reciprocal relationships between human and nature, we find challenging proposals that "identify nature as a category of social analysis as important as (and articulated with) class, race, and gender" (Cruikshank 2005, p. 4).

Likewise, perspectives from the social sciences and natural sciences—framed within more contemporary perspectives such as posthumanist theoretical lines and new materialisms—have highlighted the importance of the conceptualizations of nature and culture in the ways in which the environment is investigated (Alaimo 2016; Bennett 2010; Cruikshank 2005; Haraway 2016; Houser 2014; Nixon 2011; Tsing 2015). Primarily, these positions suggest that by separating nature from culture, we perpetuate a fictitious division of the world that has served to justify the use of nature as an exploitable resource, and in turn, culture is reduced to what groups of people do. This separation perpetuates an ontological, epistemological, and ethical order that defines, among other things, how these two spheres are studied (social sciences study the human, and natural sciences study the natural world) (Matus et al. 2021). This nonrelationship between human and non-human in how information is produced about regions such as Patagonia, is problematic. The relationship between the human and non-human worlds deserves consideration, in our opinion, in how the future of territories such as Patagonia is imagined.

Acknowledgments This work was funded by the Chilean National Agency for Research and Development (ANID) through: Regular Fondecyt N°1191865, Fondecyt N° 1160732, under Grants PIA SOC 180023, Center for Educational Justice under Grant CIE 160007; the OHM-I Patagonia-Bahía Exploradores, Labex DRIIHM ("Programme Investissements d'avenir": ANR-11-LABX-0010), INEE-CNRS, Paris, Francia; and the Centro de Investigación en Ecosistemas de la Patagonia (CIEP), through the PATSER ANID Project RF0F0002, Coyhaique. We thank Felipe Jorquera for his fieldwork and initial management of this document and Francisca Flores for her editing support.

References

S. Alaimo, *Exposed: Environmental Politics and Pleasures in Posthuman Times* (University of Minnesota Press, 2016)

A. Alessandri, De la "geografía de la acumulación" a la "geografía de la reproducción": Un diálogo con Harvey [From the "geography of accumulation" to the "geography of reproduction": A dialogue with Harvey]. Scripta Nova **12**(270) (2008)

P. Ávila-García, Hacia una ecología política del agua en Latinoamérica [Towards a political ecology of water in Latin America]. Rev. Estud. Soc **55**, 18–31 (2016). https://doi.org/10.7440/res55.2016.01

P. Bachmann-Vargas, C.S.A. van Koppen, Disentangling environmental and development discourses in a peripheral spatial context: The case of the Aysén Region, Patagonia, Chile. J. Environ. Dev. **29**(3), 366–390 (2020). https://doi.org/10.1177/1070496520937041

K. Barad, *Meeting the Universe Halfway. Quantum Physics and the Entanglement of Matter and Meaning* (Duke University Press, 2007)

H. Barrows, Geography as human ecology. Ann. Assoc. Am. Geogr. **13**(1), 1–13 (1923). https://doi.org/10.1080/00045602309356882

A. Bello, Exploración, conocimiento geográfico y nación: La "creación" de la Patagonia occidental y Aysén a fines del siglo XIX [Exploration, geographical knowledge and nationhood: The "creation" of western Patagonia and Aysén at the end of the 19th century], in *Imaginarios geográficos, prácticas y discursos de frontera. Aisén-Patagonia desde el texto de la nación [Geographical Imaginaries, Border Practices and Discourses. Aisén-Patagonia from the Text of the Nation]*, ed. by A. Núñez, E. Aliste, A. Bello, M. Osorio, (GEOlibros N°25, Instituto de Geografía, Pontificia Universidad Católica de Chile, 2017), pp. 61–86

J. Bennett, *Vibrant Matter: A Political Ecology of Things* (Duke University Press, 2010)

H. Blair, K. Bosak, T. Gale, Protected areas, tourism, and rural transition in Aysén, Chile. Sustainability **11**(24), 1–22 (2019). https://doi.org/10.3390/su11247087

F. Bourlon, *Voyager en Patagonie: Les usages touristiques de la nature [Travelling in Patagonia: The Tourist Uses of Nature]* (L'Harmattan, 2020)

L. Brigand, I. Peuziat, F. Arenas, A. Salazar, A. Núñez, I. Escobar, Aislamiento geográfico en las islas costeras francesas y en la Región de Aysén en Patagonia Chilena: Primeros elementos comparativos [Geographic Isolation in the French Coastal Islands and in the Aysén Region of Chilean Patagonia: First Comparative Elements], in *Aislamiento geográfico: ¿Problema u oportunidad? Experiencias, interpretaciones y políticas públicas [Geographic Isolation: Problem or Opportunity? Experiences, Interpretations and Public Policies]*, ed. by F. Arenas, A. Salazar, A. Núñez, (GEOlibros N°15, Instituto de Geografía, Pontificia Universidad Católica de Chile, 2011), pp. 23–33

B. Bustos, Territorialidad de la intervención estatal en contextos de crisis. El caso del virus ISA, la industria salmonera y la Región de Los Lagos, Chile [Territoriality of state intervention in crisis contexts. The case of the ISA virus, the salmon industry and the Los Lagos Region, Chile]. Revista Geográfica del Sur **5**(7), 77–94 (2014)

B. Bustos, M. Prieto, Ecología política en (desde y por) Chile: Posibilidades, desafíos y contribuciones [Political ecology in (from and by) Chile: Possibilities, challenges and contributions], in *(Las) otras geografías en Chile: Perspectivas sociales y enfoques críticos [(The) Other Geographies in Chile: Social Perspectives and Critical Approaches]*, ed. by A. Núñez, E. Aliste, R. Molina, (Lom, 2019), pp. 85–104

P. Camus, F. Jaksic, *Piscicultura en Chile: Entre la productividad y el deterioro ambiental 1856–2008 [Fish Farming in Chile: Between Productivity and Environmental Deterioration 1856–2008]* (GEOlibros N° 13, Instituto de Geografía, Pontificia Universidad Católica de Chile, 2009)

E. Cañada, I. Murray, *Turistificación global. Perspectivas críticas en turismo [Global Touristification. Critical Perspectives in Tourism]* (Icaria, 2019)

N. Castree, D. Demeritt, D. Liberman, B. Rhoads, *A Companion to Environmental Geography* (Wiley-Blackwell, 2009)

Chilean Ministry of the Environment, *Cuarto reporte del estado del medio ambiente [Fourth State of the Environment Report]*. (2018). Retrieved January 12, 2023, from https://sinia.mma.gob.cl/wp-content/uploads/2019/01/Cuarto-reporte-del-medio-ambiente-compressed.pdf

Chilean National Forestry Corporation, *Estadística visitantes unidad SNASPE [Visitor Statistics SNASPE Unit]*. (2018). Retrieved January 14, 2023, from https://www.conaf.cl/wp-content/files_mf/1561061927EstadisticaTot_a%C3%B1o_2018.pdf

Chilean National Forestry Corporation, *Catastro de los recursos vegetacionales nativos de Chile [Survey of Native Vegetation Resources of Chile]*. (2020). Retrieved January 14, 2023, from https://sit.conaf.cl/varios/Catastros_Recursos_Vegetacionales_Nativos_de_Chile_Nov2021.pdf

Chilean National Tourism Service, *Anuario de turismo 2017 [Tourism Yearbook 2017]* (2017). Retrieved January 15, 2023, from http://www.subturismo.gob.cl/wp-content/uploads/2015/10/ANUARIO-TURISMO-2017.pdf

Chilean National Tourism Service, *Anuario de turismo 2018 [Tourism Yearbook 2018]*. (2018). Retrieved January 15, 2023, from http://www.subturismo.gob.cl/wp-content/uploads/2015/10/Anuario-de-Turismo-2018.pdf

Chilean National Tourism Service, *Anuario de turismo 2019 [Tourism Yearbook 2019]*. (2019). Retrieved January 15, 2023, from http://www.subturismo.gob.cl/wp-content/uploads/2015/10/ANUARIO-TURISMO-2019_29092020.pdf

C. Corson, K.I. MacDonald, Enclosing the global commons: The convention on biological diversity and green grabbing. J. Peasant Stud. **39**(2), 263–283 (2012). https://doi.org/10.1080/03066150.2012.664138

J. Cruikshank, *Do Glaciers Listen? Local Knowledge, Colonial Encounters & Social Imagination* (The University of British Columbia, 2005)

G. Durand, En el Istmo de Ofqui y la Laguna San Rafael [On the Ofqui Isthmus and San Rafael Lagoon], in *En Viaje: Revista Mensual de Los Ferrocarriles Del Estado – Chile [En Viaje: Monthly Magazine of the State Railroads – Chile]*, (1941), pp. 60–61. https://www.memoriachilena.gob.cl/602/w3-article-72203.html

J. Fairhead, M. Leach, I. Scoones, Green grabbing: A new appropriation of nature? J. Peasant Stud. **39**(2), 237–261 (2012). https://doi.org/10.1080/03066150.2012.671770

J. Flores, P. Martínez, The touristification of the territory: Travelers and tourism magazines in southern Chile, 1853–1950. Rev. Geogr. Espacios **10**(20), 32–51 (2020). https://doi.org/10.25074/07197209.20.1876

A. Grosse, *Visión de Aisén [Vision of Aisén]* (Self-Published, 1955)

A. Grosse, *Visión histórica y colonización de la Patagonia occidental [Historical Vision and Colonization of Western Patagonia]* (Ministerio de Obras Públicas, 1986)

J.S. Gullion, *Diffractive ethnography: Social sciences and the ontological turn* (Routledge, 2018)

A. Harambour, Soberanía y corrupción. La construcción del Estado y la propiedad en Patagonia austral (Argentina y Chile, 1840–1920) [Sovereignty and corruption. The construction of state and property in southern Patagonia (Argentina and Chile, 1840–1920)]. Historia **50**(2), 555–596 (2017). https://doi.org/10.4067/s0717-71942017000200555

D. Haraway, *Staying With the Trouble. Making Kin in the Chthulucene* (Duke University Press, 2016)

G. Holmes, What is a land grab? Exploring green grabs, conservation, and private protected areas in southern Chile. J. Peasant Stud. **41**(4), 547–567 (2014). https://doi.org/10.1080/0306615 0.2014.919266

H. Houser, *Ecosickness in Contemporary U.S. Fiction: Environment and Affect* (Columbia University Press, 2014)

F. Jorquera, A. Salazar-Burrows, C. Montoya-Tangarife, Nexos espaciotemporales entre la expansión de la urbanización y las áreas naturales protegidas. Un caso de estudio en la Región de Valparaíso, Chile [Spatial-temporal links between the expansion of urbanization and natural protected areas. A case study in the Valparaíso Region, Chile]. Investig. Geogr. **54**, 41–60 (2017). https://doi.org/10.5354/0719-5370.2017.48041

A. Kelly, Conservation practice as primitive accumulation. J. Peasant Stud. **38**(4), 683–701 (2011). https://doi.org/10.1080/03066150.2011.607695

M. Martinic, *De la Trapananda al Aysén. Una mirada reflexiva sobre el acontecer de la Región de Aysén desde la prehistoria hasta nuestros días [From Trapananda to Aysén. A Reflective Look at the Events of the Aysén Region from Prehistoric Times to the Present Day]* (Pehuén Editores, 2005)

M. Martinic, Apertura del istmo de Ofqui: Historia de una quimera consideraciones sobre la vigencia de sus razones [Opening of the Ofqui Isthmus: History of a Dream, Considerations on the Validity of Its Reasons]. Magallania **41**(2), 5–50 (2013). https://doi.org/10.4067/s0718-22442013000200001

C. Matus, P. Bussenius, P. Herraz, V. Riberi, M. Prieto, Nature is for trees, culture is for humans: A critical reading of the IPCC report. Sustainability **13**(21), 1–9 (2021). https://doi.org/10.3390/su132111903

S. Millar, *La conquista de Aysén. Memorias y cartas de colonización de Aysén [The Conquest of Aysén. Memoirs and Letters of Colonization of Aisen]* (Ñirre Negro, 2017)

E. Mora Ferraz, 1871 Chile. – Ofqui, ventisquero San Rafael [photography] Photo Mora. Photographic archive. Available at Digital National Library of Chile (1889–1958a). Retrieved January 19, 2023, from http://www.bibliotecanacionaldigital.gob.cl/bnd/629/w3-article-613647.html

E. Mora Ferraz, Chile. – Ofqui, Laguna San Rafael [photography]. Photographic archive. Available at Digital National Library of Chile (1889–1958b). Retrieved January 19, 2023, from http://www.bibliotecanacionaldigital.gob.cl/bnd/629/w3-article-613646.html

R. Nixon, *Slow Violence and the Environmentalism of the Poor* (Harvard University Press, 2011)

A. Núñez, E. Aliste, Á. Bello, El discurso del desarrollo en Patagonia-Aysén: La conservación y la protección de la naturaleza como dispositivos de una renovada colonización. Chile, siglos XX–XXI [The development discourse in Patagonia-Aysén: Conservation and nature protection as devices of a renewed colonization. Chile, XX–XXI centuries]. Scripta Nova **18**(493), 1–13 (2014) https://revistes.ub.edu/index.php/ScriptaNova/article/view/15035/18388

A. Núñez, M. Benwell, E. Aliste, Interrogating green discourses in Patagonia-Aysén (Chile): Green grabbing and eco-extractivism as a new strategy of capitalism? Geogr. Rev. **112**(5), 688–706 (2020). https://doi.org/10.1080/00167428.2020.1798764

A. Núñez, F. Miranda, E. Aliste, S. Urrutia, Conservacionismo y desarrollo sustentable en la geografía del capitalismo: Negocio ambiental y nuevas formas de colonialidad en Patagonia-Aysén [Conservationism and sustainable development in the geography of capitalism: Environmental business and new forms of coloniality in Patagonia-Aysén], in *(Las) otras geografías en Chile: Perspectivas sociales y enfoques críticos [(The) Other Geographies in Chile: Social Perspectives and Critical Approaches]*, ed. by A. Núñez, E. Aliste, R. Molina, (Lom, 2019), pp. 23–46

J. Olea-Penaloza, A. Salazar-Burrows, & F. Jorquera-Guajardo, La patagonia como frontera científica: Exploraciones contemporáneas desde una ciencia global. Diálogo andino, (66), 95–105 (2021). https://dx.doi.org/10.4067/S0719-26812021000300095

L.A. Ogden, *Loss and Wonder at the World's End* (Duke University Press, 2021)

D. Ojeda, La playa vacía, el bosque exuberante y el otro exótico: Herramientas para el análisis crítico del turismo de naturaleza [The empty beach, the lush forest and the exotic other: Tools for the critical analysis of nature tourism], in *Turistificación global. Perspectivas críticas en turismo [Global Touristification. Critical Perspectives in Tourism]*, ed. by E. Cañada, I. Murray, (Icaria, 2019), pp. 463–474

B. Reid, A. Astorga, I. Madriz, C. Correa, Estado del conocimiento y conservación de los ecosistemas dulceacuícolas de la Patagonia occidental austral [State of knowledge and conservation of freshwater ecosystems of western southern Patagonia], in *Conservación en la Patagonia chilena: Evaluación del conocimiento, oportunidades y desafíos [Conservation in Chilean Patagonia: Assessing Knowledge, Opportunities and Challenges]*, ed. by J.C. Castilla, J.J. Armesto, M.J. Martínez-Harms, (Ediciones Universidad Católica de Chile, 2021), pp. 429–471

C. Rivera, A. Vallejos-Romero, La privatización de la conservación en Chile: Repensando la gobernanza ambiental [The privatization of conservation in Chile: Rethinking environmental governance]. Bosque (Valdivia) **36**(1), 15–25 (2015). https://doi.org/10.4067/S0717-92002015000100003

C. Rojek, J. Urry, Transformations of travel and theory, in *Touring Cultures. Transformations of Travel and Theory*, ed. by C. Rojek, J. Urry, (Routledge, 1997), pp. 2–22

F. Rosetti, From infrastructure to landscape. The southern highway as an engine of resignification. ARQ **99**, 86–95 (2018). https://doi.org/10.4067/S0717-69962018000200086

A. Rovira, D. Quintana-Becerra, Conocimiento de base para el desarrollo del turismo científico en la Patagonia chilena [Knowledge base for the development of scientific tourism in Chilean Patagonia]. Cuad. Tur. **1**(44), 327–349 (2019). https://doi.org/10.6018/turismo.44.404871

A. Tsing, *The Mushroom at the End of the World: On the Possibility of Life in Capitalist Ruins* (Princeton University Press, 2015)

X. Urbina, Expediciones a las costas de la Patagonia occidental en el período colonial [Expeditions to the coast of western Patagonia in the colonial period]. Magallania **41**(2), 51–84 (2013). https://doi.org/10.4067/S0718-22442013000200002

S. Urrutia, Hacer de Chile una gran nación. La Carretera Austral y Patagonia Aysén durante la dictadura cívico militar (1973–1990) [Making Chile a great nation. The Southern Highway and Patagonia Aysén during the civil-military dictatorship (1973–1990)]. Rev. Geogr. Norte Gd **60**(75), 35–60 (2020). https://doi.org/10.4067/s0718-34022020000100035

S. Zamirato, E. Tomazzoni, Patrimonio, turismo y transfiguraciones en las relaciones identidarias. El Pelourinho (Salvador – Bahia) y Porto Rico (Paraná), Brasil [Heritage, tourism and transfigurations in identity relations. El Pelourinho (Salvador – Bahia) and Porto Rico (Paraná), Brazil]. Estud. Perspect. Tur. **24**(2), 222–243 (2015)

A. Zoomers, Globalisation and the foreignisation of space: Seven processes driving the current global land grab. J. Peasant Stud. **37**(2), 429–447 (2010). https://doi.org/10.1080/03066150003595325

E.G. Zuelow, Beginnings: The grand tour, in *A History of Modern Tourism*, ed. by E.G. Zuelow, (Palgrave, 2016), pp. 14–29

Chapter 6
Geographic Imaginaries in Dispute in Northern Patagonia: Tourism, Environmental Conservation, and Indigenous Territorial Rights in Quinquén, Chile

Bastien Sepúlveda and Nelson Martínez-Berríos

Abstract In this chapter, we pose that tourism participates in a process of reterritorialization, in dialogue with past and present dynamics of deterritorialization, and arises from the actions of diverse powers that constrain Indigenous agency. We discuss tourism both as a window that reveals the tensions between environmental conservation and Indigenous territorial rights and as a socio-political process that could resolve them. Methodologically, this work reflects extensive bibliographical review, document analysis, and fieldwork that has been conducted through numerous periodic stays in Quinquén and other Pewenche communities. First, we explore a theoretical perspective of the production of geographical imaginaries and its applications in Northern Patagonia. Next, we analyze the dual processes of exploitation/protection of the Araucaria, contextualizing them within the framework of the territorial dispossession that affected the Pewenche in the upper basin of the Bío-Bío River, an area of Chile located in the mountainous communes of Lonquimay, in the Araucanía Region, and Alto Bío-Bío, in the Bío-Bío Region, where 54% and 83% of the population identify themselves as Pewenche, respectively. This chapter continues with consideration of the Quinquén territory of northern Chilean Patagonia, where the Pewenche struggle for the Araucaria tree has resulted in a tourism development project.

Keywords Patagonia · Indigenous agency · Geographic imaginaries · Tourism · Reterritorialization

B. Sepúlveda (✉)
Research affiliate at UMR SENS, Montpellier, France

N. Martínez-Berríos
Corporación Municipal de La Florida, Education Area, Santiago, Metropolitan Region, Chile

© The Author(s) 2023
T. Gale-Detrich et al. (eds.), *Tourism and Conservation-based Development in the Periphery*, Natural and Social Sciences of Patagonia,
https://doi.org/10.1007/978-3-031-38048-8_6

6.1 Introduction

More Indigenous communities throughout Latin America are incorporating tourism into their economies and ways of life. In the 1990s, an incipient development of Indigenous tourism was evident in southern Belize (Steinberg 1994), in the Ecuadorian Amazon (Wesche 1993), and in south-central Chile (Volle 1999). Not only has Indigenous tourism increased and expanded geographically over recent years, but it has also diversified to include activities that link aspects of rural tourism, agro-tourism, ecotourism, and ethno-tourism. This range of activities has been labeled *Indigenous tourism* (Pereiro 2015), a variant of community-based tourism that places the community (rural/peasant/Indigenous) as the articulating axis of a business venture that focuses on local livelihoods and claims to be economically, socially, and environmentally sustainable (Skewes et al. 2012).

In Chile, Indigenous tourism has undergone significant development, even becoming a specific focus of attention in national tourism policy, where it is defined as "that which is offered by Indigenous individuals, families, or organizations, and [whose] offer incorporates part of the Indigenous culture" (De la Maza and Huisca 2020, p. 105). Part of this development has been concentrated in the territory referred to as the Alto Bío-Bío (ABB), or upper basin of the Bío-Bío River. The ABB includes the mountainous communes of Lonquimay (La Araucanía Region) and Alto Bío-Bío (Bío-Bío Region), where, respectively, 54% and 83% of the population identify themselves as Pewenche, according to the 2017 Census (Chilean National Statistics Institute 2017). The Pewenche constitute a branch of the Mapuche people, whose way of life is closely linked to the Araucaria forests (*pinalerías*) found in their territory. In fact, the symbiotic relationship that the Pewenche maintain with these forests has led them to mobilize against the logging of Araucaria that was practiced without restraint by forestry companies in the mid-twentieth century.

As a result of these mobilizations, the Araucaria tree was declared a Natural Monument, thus strictly prohibiting its exploitation and preserving it within several Chilean protected areas (PAs). It is interesting to note that beyond the Pewenche mobilizations, these declarations responded to disputes that originated during colonial times over the establishment of divergent conceptions of space and geographical imaginaries in the region. For example, the Pacification of the Araucanía (1861–1883), which led to the annexation of Mapuche territory by the Chilean government, spatially inscribed its authority through the demarcation and creation of land designations, including *land concessions*, *forest reserves*, *land auctions*, *colonization territories*, and *Indigenous reductions*, among others. The main outcome of Pacification of the Araucanía was the deterritorialization of the Pewenche, who lost control of, and access to, their lands, forests, and waters. In some cases, Pewenche communities were even forcibly removed and relocated.

In this chapter, we are interested in exploring the capacity of Indigenous tourism to help communities reverse historic processes of deterritorialization by facilitating a symbolic and effective re-appropriation of certain environmental resources through processes of formal protection. We present the emblematic case of Quinquén's Pewenche community, which resisted the appropriation of their lands by the logging conglomerate, the Galletué Society. The Galletué Society sought to evict Pewenche families from their lands and had planned the exploitation of their Araucaria groves. Indigenous tourism development was proposed as a potential mechanism resource for protecting the Araucaria (Reyes 2006). We examine how the longstanding Pewenche struggle for the Araucaria tree became the catalyst for tourist development activity, that was centered around a broader rural identity (e.g., rural tourism, agro-tourism, ecotourism, and ethno-tourism), but fundamentally related to the Pewenche connection with the *ñuke mapu*, which refers to Mother Earth, or the Earth, in a deeper sense of the term.

In continuity with other works that discuss Indigenous tourism in the ABB (Palomino-Schalscha 2015; Krell 2020) and other sectors of Chilean (Pilquimán 2016, 2017) and Argentine Norpatagonia (Impemba and Maragliano 2019), we propose that in an Indigenous context, tourism development is not solely limited to the economic dimension. Rather, we posit that Indigenous tourism contributes to a process of reterritorialization that interacts with past and present dynamics of deterritorialization by diverse powers that seek to constrain Indigenous agency. We discuss the spatially intertwined and temporally simultaneous processes of Indigenous de/re/territorialization by interpreting tourism both as a process that reveals the tensions between environmental conservation and Indigenous territorial rights, and as a socio-political process that could resolve them (Haesbaert 2011).

In the first section, we delve into the understanding of the production of geographical imaginaries from a theoretical perspective (Zusman 2013), along with exploring its applications in northern Patagonia. We then analyze the dual process of exploitation/protection of the Araucaria, contextualizing it within the framework of the territorial dispossession that affected the Pewenche in the ABB. In the third section, we focus on the case of Quinquén, where the Pewenche struggle for the Araucaria tree resulted in a tourism development project. Finally, we outline some conclusions on the articulations between tourism, environmental conservation, and Indigenous territorial rights in the northern Patagonia region of Chile.

This chapter synthesizes data obtained by both authors over the last 15 years. Extensive fieldwork was conducted through numerous stays in Quinquén and other Pewenche communities, where both formal (community assemblies, meetings, and encounters with local authorities, etc.) and informal (conversations in family environments, participation in cultural activities, etc.) exchanges with various local actors took place. Furthermore, we complement these data with bibliographical revision of the archives of administrative-juridical-legal documentation from different institutions.

6.2 The Production of Norpatagonia as a Geographic Imaginary

The military conquest and incorporation of the Mapuche territory into the jurisdictions of Chile and Argentina at the end of the nineteenth century subjected the northern reaches of Patagonia to a competitive process of territorial appropriation and expansion between two nation-states that were still under formation. The most remote and isolated regions of northern Patagonia saw the creation of the first PAs, especially in the mountainous sectors of the Andes where the international border between the two countries had been established in 1882. Along with reserving spaces for timber production, Chile and Argentina sought to strengthen state sovereignty in these lands and restrict the lifestyles of Indigenous communities whose cultural practices and livelihood activities were soon limited through restrictions on use and access (Sepúlveda and Guyot 2016).

A global process of rural transformation has been underway since the final decades of the twentieth century. Directly associated with growing environmental awareness, rural transformation has imposed a new landscape paradigm on the rural world. Once considered as productive spaces for a range of extractive activities (e.g., ranching, mining, forestry), rural peripheries and the protected areas they shelter are now being valued as recreational spaces by and for conservation, under the wing of the sustainable development paradigm and global environmental governance (Blair et al. 2019; Olea 2020). This shift in global thinking allows regions that were historically constructed as remote and peripheral to become the new focus of rural development policies that promote ecotourism and other variants of community-based tourism. Thus, the mountainous areas of the former Mapuche territory, which was historically referred to as the border territories, are being reconceptualized as part of an interconnected Norpatagonia ecological unit. The environmental conservation paradigm has defined Norpatagonia as an interconnected ecological unit with diffuse and porous limits. Thus, environmental conservation acts as a transboundary process that manifests through the creation of PAs along and across different political borders (Sepúlveda and Guyot 2016).

This restructuring of space occurs in conjunction with transformations that occur in geographic imaginaries. A geographic imaginary is a set of beliefs or ideas held about a place that, when materialized in social practices, shape the way in which those places are perceived (Scott 1999; Zusman 2013). Geographic imaginaries result largely from the environmental, cultural, economic, and geopolitical narratives circulated by powerful actors and manifest through territorialities that are superimposed on one another over time, revealing competing ideological and discursive formations (Peet and Watts 1996).

The concept of Norpatagonia illustrates the sedimentary character of these imagined spaces, which have been constantly reimagined, according to the state's changing needs. During the ninetieth century, Norpatagonia was conceptualized as *desert*, which justified the eviction and genocide carried out against the Indigenous communities in the so-called Desert Campaign in Argentina (Navarro-Floria 1999). Afterward, on both sides of the Andes, sectors of Norpatagonia were (re)imagined as territories for colonization, enabling the concession of small landholdings to national and foreign settlers, and the dedication of large tracts of land for environmental conservation as forest reserves, which would later become national parks, geoparks, biosphere reserves, and more recently, Indigenous conservation territories.

With the 2019 designation of Lonquimay as a Zone of Tourist Interest (*Zona de Interés Turístico* [ZOIT]), tourism development has added another layer to perpetuate this geographic imaginary. Chilean ZOIT designations are accompanied by standards for how associated lands ought to be managed. For example, the Lonquimay ZOIT declaration decree specified how Indigenous communities would be inserted within this new geographic order, by designating their lands as *biodiversity conservation territories*. It described notable characteristics of the Lonquimay ZOIT associated with Indigenous lands:

> Quinquén's Pehuenche Park, which is the first biodiversity conservation territory managed by Indigenous communities, and the Los Arenales snow park, which is also managed by an Indigenous community, stand out. In addition, the commune of Lonquimay belongs to the UNESCO Araucarias Biosphere Reserve and the Kütralcura Geopark. (Chilean Ministry of Economy, Development and Tourism and Undersecretariat of Tourism 2019, p. 2).

Environmental and conservation-centered framing of Norpatagonia's geographical imaginary perpetuates narratives that simultaneously enable and justify powers of control over these territories and their inhabitants through a process of *environmentalization* (Agrawal 2005). For example, overlapping territorial claims and designations that are implemented in the name of conservation and conservation-based development (e.g., national protected areas, ZOITs, UNESCO Biosphere Reserve, Geoparks) are facilitated through state and international policy, law, treaty, or decree. Under these designations, agencies impose rules and norms regarding how lands can (and cannot) be used within Norpatagonia. These *environmental governmentalities* have serious consequences for Indigenous communities that depend on these lands for their survival and way of life (Agnew and Oslender 2010; Li 2007), requiring them to either adapt, negotiate, or resist. But, implementing these strategies normally requires the specialized capacities (e.g., technical, financial, and legal) to navigate and interact with an environmental bureaucracy of institutions that span multiple scales, from communities to supranational entities. Moreover, this bureaucracy is made up of institutions and actors who also compete for access, ownership, and rights to these environmental resources, among themselves. Predictably, disparate skill levels and social capital among local groups result in repeated cycles of domination by larger entities and resistance from local communities, with little advance (Peet and Watts 1996; Peluso 1992; Robbins 2019).

We understand Indigenous agency as the capacity of Indigenous individuals, communities, and organizations (whether or not they are formally recognized as such) to exercise control within the space they operate and counteract powers at various scales that limit this capacity. Moreover, Indigenous agency includes the capacity to contest, resist and/or negotiate the terms for defining their own development priorities, using a wide repertoire of actions and resources such as the invocation of ancestors, the alliance with nature, and performance, thereby preserving their way of life.

Over the last three decades, tourism has emerged as a powerful tool for northern Patagonia Indigenous communities. Through Indigenous tourism, communities have retained control and power over the re-territorialization process and resisted the geographical imaginaries that seek to constrain their livelihoods. Quinquén, and other communities, have intensified their participation in community-based tourism initiatives that are supported by environmental approaches to global governance, like the Indigenous Conservation Territory designation promoted by the International Union for Conservation of Nature (IUCN). Nevertheless, as discussed previously, these land designations still limit Indigenous agency. While theoretically these initiatives originate from within the communities themselves, in practice they respond to highly authoritative, schematic, and prescriptive visions of the territories as defined by international environmental agencies operating at larger scales (Fig. 6.1). Even when all, or part, of these Indigenous communities, participate in the implementation of these models, they often do so to challenge and/or counter-territorialize the space, motivated by ancestral (Martínez 2015), ontological (Palomino-Schalscha 2015), heritage (Andrade and Pilquimán 2020), and cultural survival (Pilquimán 2016).

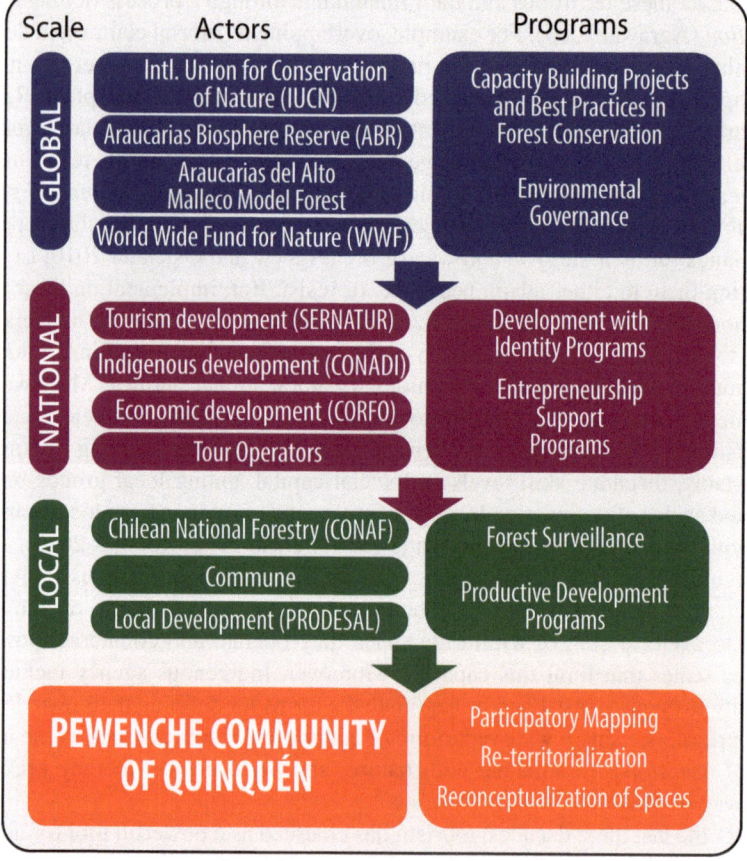

Fig. 6.1 Model for the development of an Indigenous Conservation Territory, adaptation of a model developed by the World Wide Fund for Nature Chile. (Molina and Pávez 2012, p. 15)

6.3   Land, Forests, and Conflicts in the Alto Bio-Bio Region

The Pewenche define themselves as people (*che*) of the Araucaria (*pewen*), primarily because of the intimate relationship they have with this tree. The Araucaria's fruit, called *piñón* or *nguilliu*, is harvested at the end of the summer season between March and April, at which time their livestock are taken to graze on the high plains of the mountain range. The tree's pine nuts also form the basis of their diet and occupy a central place in the Pewenche economic system, as they are sold to traders who supply the regional markets. Furthermore, the Pewenche use the Araucaria as an after (*rewe*) in the traditional celebration (*nguillatun*), a ceremony that brings together members of various communities to thank God (*Ngenechen*) for the harvests and goodness granted and pray for the coming year. These examples highlight the importance of the Araucaria tree for the Pewenche identity. This insight is fundamental to understanding the conflict that has arisen around the tree's use and appropriation in the ABB. The following section dissects this conflict by exploring the dynamics of territorial dispossession that continue to affect Pewenche communities.

6.3.1   Pewenche Territorial Dispossession

When the Mapuche territory was annexed to the State of Chile at the end of the ninetieth century, the mechanism that would establish Indigenous property rights had already been defined. The 1866 Titles of Merced (TM) Law would establish Indigenous property rights for reduced areas within the territories they had traditionally occupied; the remaining lands would be available for auctions, cessions, concessions, and settlement of Chilean and European settlers. Article 6 of this law specified that, "[...] all lands for which effective and continuous possession of at least one year has not been proven shall be considered uncultivated lands and, consequently, government property" (Chilean National Congress Library 1866, para. 12). This statute meant that the Pewenche's summer pasture lands in the AAB, where their prized *pinalerías* were located, were often excluded from the reductions. The Pewenche only occupied these areas for part of the year during which time they constructed mobile *outposts*; therefore, they could not prove *effective and continuous possession* (Azócar et al. 2005).

In addition, the resettlement of Indigenous people was sometimes carried out after some of the auctions and cessions had been made to settlers. For example, in Lonquimay, prior to the arrival of the Indian Registration Committee (*Comisión Radicadora de Indígenas* [CRI]) in 1896, the installation of a large number of settlers limited the possibilities of establishing Pewenche settlements. In other cases, including some sectors of the current commune of ABB, local landowners managed to prevent the arrival of the CRI, so the resettlement never took place (Molina and Correa 1996). As a result of these irregularities, many Pewenche were left without land and/or without documentation to prove their land claim, while others experienced legalized theft via land grants that did not encompass all the land they actually occupied. By 1930, when resettlement practices were terminated with the repeal of the Law of December 4, 1866, 12 land reductions had occurred in the ABB, representing a total of 370 km² (Fig. 6.2 and Table 6.1).

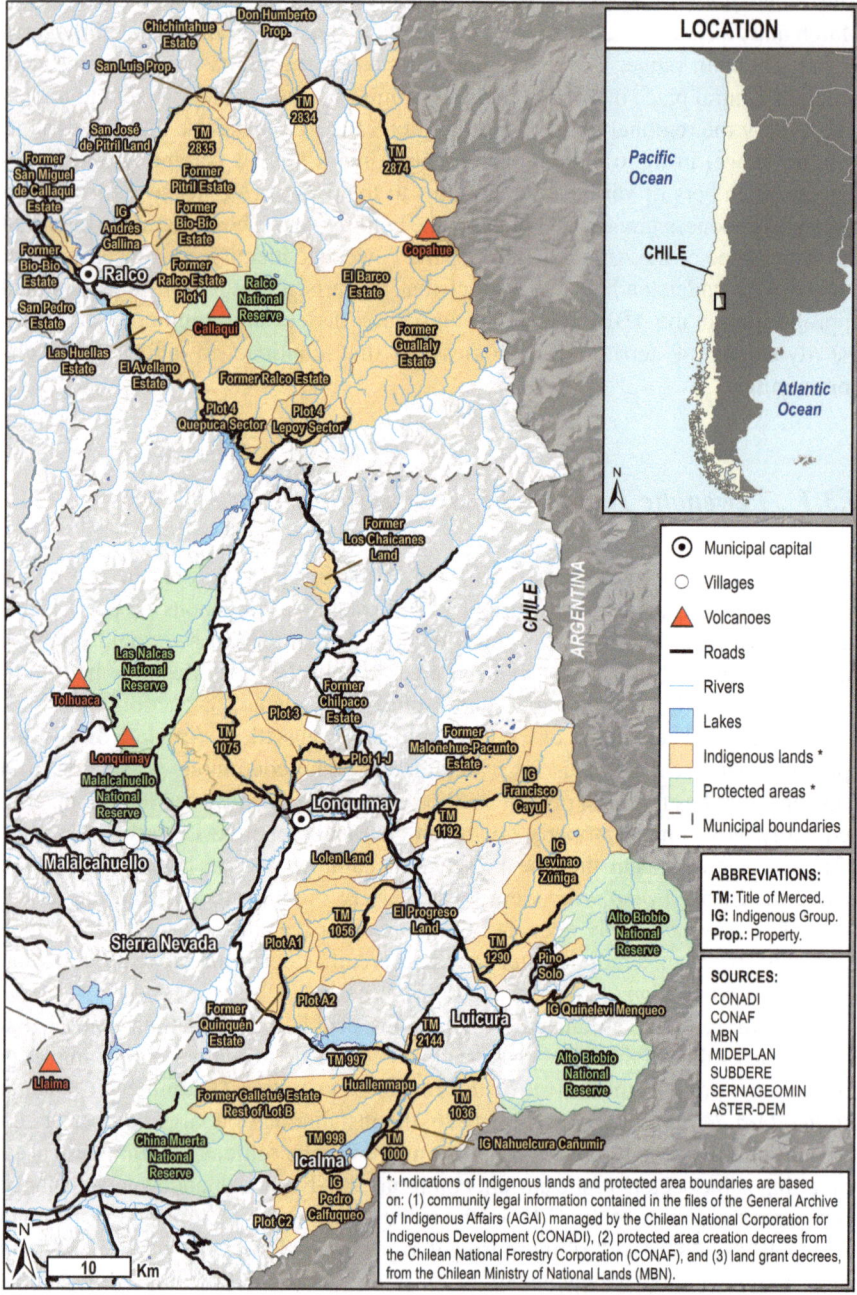

Fig. 6.2 Pewenche landholdings and protected areas in the Alto Bio-Bio region

Table 6.1 Reductions formed in the Alto Bío-Bío region, Chilean General Archive of Indigenous Affairs, Chilean National Indigenous development corporation (CONADI)

TM	Holder	Sector	Km^2	Date
997	Huenucal Ivante	Icalma	7.5	1905
998	Pedro Calfuqueo	Icalma	10	
1000	Benancio Cumillán	Cruzaco	9	
1036	Manuel & Samuel Queupu	Mariepumenuco	42	
1056	Paulino Huaiquillan	Pedregoso	28	1906
1075	Bernardo Ñanco	Naranjo	65	
1192	Francisco Cayul	Collipulli	22	1907
1290	P. Curilem, L. Zúñiga, and G. Ñehuen	Pehuenco	21	1908
2144	José Quintriqueo	Cerro Redondo	5	1911
2834	Antonio Marihuan	Malla Malla	34.44	1919
2835	José Anselmo Pavián	Cahuenucu	41.34	
2874	Antonio Canío	Trapa Trapa	84.3	1920

Half a century later, the military regime (1973–1990) resurveyed the properties in order to adapt their boundaries to the true occupation of their inhabitants, increasing the size of the Pewenche landholdings in the areas that had previously been reduced. In addition, Decree Law No. 2568, enacted in 1979, systemized the process of land division and sale within Indigenous communities, and also established the possibility for Indigenous families to petition the ceding of fiscal lands that for various reasons had not been settled. Thus, the Institute for Agricultural Development (INDAP) received 14 plots of land in the ABB to assign to Pewenche families who had been occupying them without land titles. Many Pewenche were thus able to recover their *veranadas*. When added to the lands remapped from the reductions, these transfers increased the area held by the Pewenche by a little more than 1580 km^2 (Table 6.2 and Fig. 6.2).

In 1993, Chile passed Law No. 19,253 (referred to as "the Indigenous Law") which established the Fund for Indigenous Lands and Waters. This fund helps subsidize or finance the re-purchase of land for Indigenous individuals and communities. In the ABB, slightly more than 160 km^2 were retransferred to Pewenche families thanks to this statute. Lands were also reacquired in other ways. For example, the El Barco estate was purchased at the end of 1994 by the *Empresa Nacional de Electricidad Sociedad Anónima* (ENDESA) to relocate families whose lands were going to be (and were) flooded by the filling of the Ralco reservoir in 2004 (Hakenholz 2004). Despite these legal reacquisitions, only 50 properties (just over 2000 km^2) have been recognized as the property of the Pewenche since the establishment of the first TMs in 1905. This total represents just 36% of the traditional Pewenche territory in the ABB (Table 6.3).

Table 6.2 Pewenche lands (re)measured by the Institute for Agricultural Development (INDAP) in the Alto Bio-Bío region, Chilean General Archive of Indigenous Affairs, Chilean National Indigenous development corporation (CONADI)

| Pewenche lands in the Alto Bio-Bío region | (Re)-Surveying | |
	Date	Km²
San Miguel de Callaqui Estate	1985	3.97
Ralco Estate: Lot 1		37.1
Ralco Estate: Lot 3–4		176.003
Guallaly Estate		183.83
Bío-Bío Estate: Lot A		99.69
Bío-Bío Estate: Property 48	1986	5.49
Pitril Estate		106.62
Gallina Indigenous Group		24.79
Title of Merced 2834		38.40
Title of Merced 2835		92.95
Title of Merced 2874		137.59
Sub-total Alto Bío-Bío		**816.49**
Title of Merced 997	1983	13.78
Title of Merced 1036		67.64
Huenucal Ivante Indigenous Group		2.94
Nahuelcura Cañumir Indigenous Group	1984	16.86
Pedro Calfuqueo Indigenous Group		56.47
Francisco Cayul Indigenous Group		83.498
Levinao Zúñiga Indigenous Group		134.63
Quiñelevi Meliqueo Indigenous Group		10.211
Title of Merced 1000		13.81
Title of Merced 1056		99.14
Title of Merced 1075		146.94
Title of Merced 1192		25.20
Title of Merced 1290		37.15
Title of Merced 2144		5
Title of Merced 998	1987	51.12
Sub-total Lonquimay		**764.42**
Total		**1580.91**

Table 6.3 Pewenche landholdings in the Alto Bio-Bio, Chilean General Archive of Indigenous Affairs, National Indigenous development corporation (CONADI)

Acquisition method	Property	Km²
Commission for the Settlement of Indians (1905–1920)	12	728.78
Institute for Agricultural Development (1987–1991)	14	852.14
Fund for Indigenous Lands and Waters (since 1993)	12	160.95
Other modalities	12	445.63
Total	**50**	**2187.51**

6.3.2 Araucaria Harvesting and Protection

The delayed and partial recognition of the Pewenche lands directly affected the conflicts that arose when forestry companies entered the ABB with the specific intention of exploiting the *pinalerías*. Luis Otero (2006) recalled that the construction of the Las Raíces tunnel, in 1938, "opened to exploitation the entire araucaria zone of the Alto Bío-Bío" (p. 119). This tunnel, which extended 4.5 km, facilitated rail connectivity between the area and the coast; much later, in 2005, this route was converted to facilitate automobile traffic. He recounted that nearly 300 km^2 of Araucaria forests were harvested for lumber within a period of 60 years. As a result of this pressure, Araucaria logging was regulated for the first time in 1969, based on the argument made in the Supreme Decree N°94 of the Ministry of Agriculture, "[...] that there is a risk of extinction of the Araucaria Araucana species due to the exploitation to which it has been subjected."

By this time, logging companies had already intruded on Pewenche lands. For communities that possessed formal recognition of their territorial rights, these conflicts could be resolved through just compensation to the Pewenche, generated by revenue from the timber that had previously been harvested on their lands. In Pedregoso, for example, two lumber companies that exploited the neighboring lands, subsequently encroached within the boundaries of TM 1056; as the Pewenche had a clearly established TM, they had recourse and ended up signing several purchase and sale contracts with the lumber company between 1964 and 1974.

Cases where communities did not yet have land titles were much more complicated to resolve. In Ralco, for example, the Pewenche were still considered illegal occupants of their lands when Araucaria exploitation began. The Ralco estate had been registered in the land registry in 1881 by Rafael Anguita, then ex-mayor of Los Angeles, and through successive transfers, in 1949, it transferred to Dionisio González, who contributed it to the working capital of the Ralco Lumber Holding Company (*Maderas Ralco Sociedad Anónima*) in 1962. The cutting of Araucaria trees began 4 years later. The Pewenche quickly petitioned the government to stop this logging, and in 1972, Chile declared "the Araucaria forests of the Ralco Estate, in the commune of Santa Bárbara" a national park (Chilean Ministry of Lands and Colonization 1972). While the decree of 1969 had only established harvesting regulations, this new regulation conferred absolute protection to the Araucaria within the limits of the Ralco estate (Molina and Correa 1996).

The Chilean government's response to the logging of Araucarias in Ralco has not been favorable for the Pewenche to date, as the new PA deprived the Pewenche of possession and access to their summer pastures (Aagesen 1998). Nevertheless, under Chile's National Reserve PA mandate, the possibility of integrating the needs of local communities into the management of the PA exists so, over time, this situation may continue to evolve (Sepúlveda and Guyot 2016). It is interesting to note that, far from being unique to the Ralco case, the creation of Chilean PAs for conservation purposes on lands claimed by Indigenous communities is widespread and has perpetuated the process of territorial alienation initiated during the conquest of Mapuche territory (Sepúlveda and Guyot 2016). The same logic was at work in Quinquén where, in addition to opposing the logging of their forests, the Pewenche families had to deal with the Galletué Society's attempts to evict them from their lands. This conflict gained notoriety at the national level (Bengoa 1992), but interestingly, in this case, the desire to create a National Reserve on the lands recovered from the Galletué Society was opposed by an alternative project that proposed the development of tourism as a primary mechanism for the protection of the Araucaria forests (Reyes 2006).

6.4 Tourism as an Alternative Protection Resource: The Case of Quinquén

The Quinquén community is made up of approximately 250 people, distributed among 52 families who live at the headwaters of the Biobío River in the commune of Lonquimay. They depend heavily on natural resources for both food security and cultural subsistence. Quinquén is also a community recognized internationally for its historical struggle against landowners and logging companies in the late 1980s. This prolonged struggle resulted in the declaration of the Araucaria as a Chilean Natural Monument in 1990, and much later—in 2007—the granting of formal land ownership to the Quinquén community (Molina 2015). This section describes the slow but progressive process of touristification (i.e., a complex process of territorial transformation through tourism) of Pewenche community spaces and the environmental and regulatory conditions that drove the families of Quinquén to participate in community-based tourism initiatives.

6.4.1 From the Creation of a Regulatory Landscape...

The Pewenche of Quinquén have coexisted in a close symbiotic relationship with the Araucaria forests. They have protected Araucaria habitat from anthropogenic threats, denouncing illegal logging and even confronting the logging companies that endangered it through overexploitation (e.g., in *Forestal Casagrande, Forestal Malleco,* and *Sociedad Galletué*; Bengoa 1992). As a result of these mobilizations and the support of national and international environmental organizations, Supreme Decree N°43 was promulgated in 1990, which established the protection of the Araucaria from any form of exploitation (see Table 6.4). This decree was of crucial importance for the inhabitants of Quinquén and all the Pewenche communities, since it recognized that the Araucaria "is intimately linked to values and principles that make up the historical, social, and cultural heritage of the Mapuche people and the nation as a whole."

Table 6.4 Chilean regulatory frameworks with impact on Quinquén

Legislation	Content	Validity
D.S. N°94/1969	Approves regulations for the exploitation of Araucaria Araucana timber.	Repealed (05/12/1970)
S.D. N°157/1969	Amends D.S. N°94/1969.	Repealed (05/12/1970)
D.S. N°439/1970	Approves new regulations for the exploitation of Araucaria Araucana timber and repeals the supreme decrees mentioned therein.	Repealed (04/26/1976)
Agreement N°2065/1971	Expropriation of Fundo Quinquén by the Corporation for Agrarian Reform (CORA).	Revoked (19/07/1974)
D.S. N°29/1976	Declares the Araucaria Araucana a Natural Monument.	Repealed (26/12/1987)
D.S. N°259 /1980	Regulation of D.L. No. 701/ 1974 on Forestry Development.	Amended (12/06/1998)
D.S. N°141/1987	Declares the Araucaria Araucana a Natural Monument in the places indicated and regulates its use in sectors located outside such places.	Repealed (03/04/1990)
D.S. N°43/1990	Declares the Araucaria Araucana a Natural Monument.	Current
D.S. N°56/1991	Creates the Lago Galletué National Reserve in the commune of Lonquimay.	
Law No. 19253/1993	Establishes norms for the protection, promotion, and development of Indigenous people, and creates the National Corporation for Indigenous Development.	Current
D.S. N°27/1997	It removes Lago Galletué National Reserve from its status as such.	
D.E. 525/2003	Authorizes the cutting of forest specimens indicated.	Repealed (12/23/2004)
Law No. 20283/2008	Law on native forest recovery and forestry promotion.	Current
D.E. 654/2009	Complements D.S. N°490/1976, N°43/1990 and N°13/1995, exempted, which declared natural monuments to different forest species.	Repealed (12/08/2011)
Decree N°146/2019	Declares Lonquimay a Zone of Tourist Interest.	Current

ᵃ*DE* Exempt Decree, *DL* Decree Law, *DS* Supreme Decree

This legal recognition was pivotal for the subsequent heritage protection of the Araucaria species. And, to give even greater protection to the ecosystems where the Araucaria grows, Supreme Decree No. 56 was issued in 1991, creating the Lago Galletué National Reserve (Fig. 6.3). This regulation established, among other considerations, that the habitat of the Araucaria:

> […] is very fragile, and its alterations are irreversible, which is why this species – despite having been declared a Natural Monument – is vulnerable to extinction, making it necessary to act with the utmost urgency in order to stop the process of deterioration that affects it. (D.S. N°56/1991)

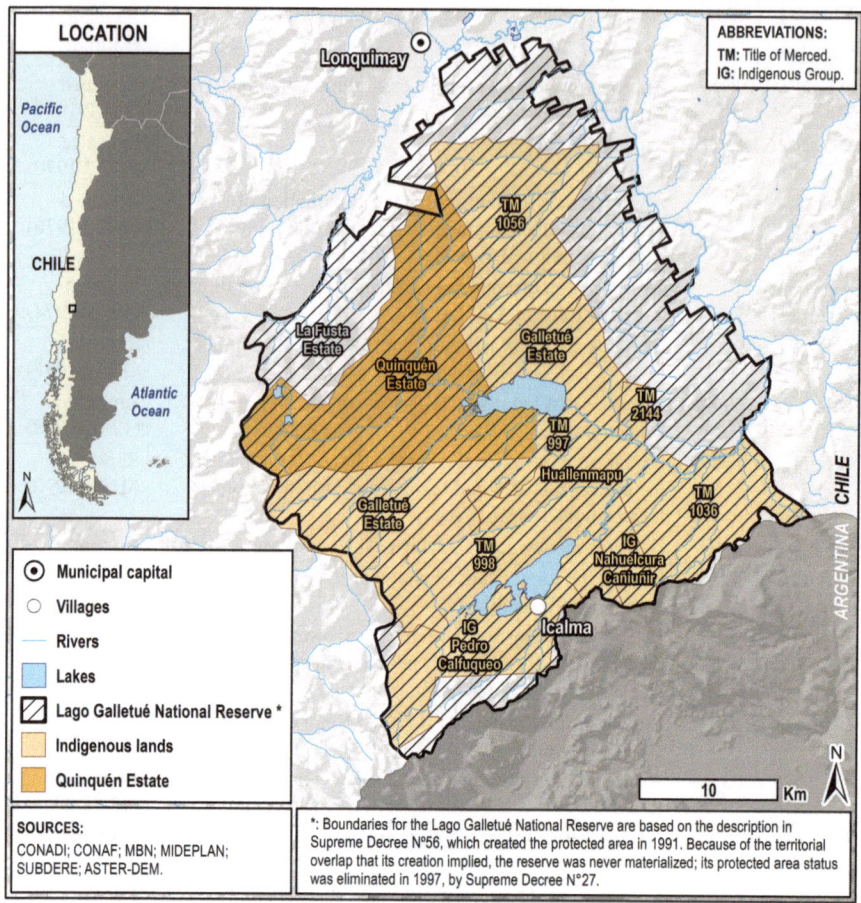

Fig. 6.3 Lago Galletué National Reserve, Lonquimay, according to D.S. N°56/1991

Through this act, the government placed several Indigenous communities, including Quinquén, within the limits of a 290.3 km² PA; thus, its implementation had important consequences in terms of both territorial configuration and access to the available natural resources. Actually, the Lago Galletué National Reserve never became a reality. In order to at least partially solve the problems caused by the territorial overlap that its creation implied, Supreme Decree N°27 was promulgated in 1997. This decree removed the protected area status to facilitate the transfer of land ownership to the National Indigenous Development Corporation (CONADI), which would subsequently grant land titles to the members of the Indigenous communities claiming territorial rights in the Reserve's area.

In 1995, the community of Quinquén was formally declared an Indigenous Community, under the protection of the Indigenous Law, which recognizes the importance of the land for the cultural existence of the Indigenous Peoples, ensuring its "adequate use, ecological conservation, and development." By virtue of this law, multiple mechanisms of subordination were activated, both of the Indigenous people through neo-developmentalist ventures, and of the geographic space through the environmentalization of the territory. This law also incorporated the creation of Indigenous Development Areas (*Áreas de Desarrollo Indígena* [ADI]), which are territorial planning instruments designed to facilitate and focus the transfer of resources to sectors considered a priority in terms of development. When the Indigenous Law was enacted, the Galletué-Icalma sector was identified as a key area in which the creation of an ADI could be integrated with the Lago Galletué National Reserve, a project that was opposed by the inhabitants of Quinquén who did not want to see an external development model imposed on them (Dodge and Reyes 1995).

Quinquen community agency has had to adapt to this succession of contradictory laws (e.g., environmental, forestry, Indigenous), which together make up the region's "regulatory landscape." This concept refers to how legal provisions manifest in geographical space, including through the creation of PAs, forest reserves, or Indigenous territories. The effects of these provisions have manifested in their capacity to produce different types of subjectivity; that is, the way in which people are defined collectively and individually in relation to the space they occupy (e.g., the inhabitants of Indigenous communities are *comuneros*). Indeed, it is through this exercise of government sovereignty that spaces are conceptualized, defined, and shaped based on the government's interests and vision for the land, its resources, and its inhabitants. Locations deemed appropriate for tourism development are constructed in this manner.

6.4.2   ...To the Touristification of Quinquén

In Quinquén, the coexistence of millenary *pinalerías* and Pewenche families constitutes an exemplary case to explain the geographic dynamics resulting from the wide repertoire of practices deployed by: (a) the regulatory action of the State, such as land designations, forestry laws, territorial ordinances, formal recognition of ancestral rights, development of entrepreneurship programs, and zoning; (b) the action of governmental and non-governmental, national, and international agencies, by way of forest monitoring, programs to support conservation, climate change mitigation, and ecosystem protection; and (c) Indigenous agency action to negotiate both intra- and extra-community agreements to gain access to natural resources in their territories, including a slow but steady process of counter-territorialization.

This interaction between government regulation, conservation interests and initiatives, and Indigenous rights advocacy in the region brought about the gradual touristification of Quinquén. The regulatory frameworks that were imposed to safeguard the region's resources from exploitation also restricted access to these resources to Indigenous communities, thereby impacting Indigenous livelihoods and leaving tourism as the only viable economic alternative. Tourism in Quinquén was seen as being a local management strategy for conservation, management, and environmental protection; also it represented the result (although not necessarily intentional) of strong essentialist pressures, since they installed subjectivation practices that placed the inhabitants of a territory as naturally gifted for the development of a single economic activity. In this sense, the aforementioned regulatory conditions have been decisive for the touristification of the territory where these families live or, in other words, for the emergence of what is known as the "tourist vocation" of the territory.

Tourism development in Quinquén can be summarized in three key phases: (a) Informal tourism development, (b) Initiatives of the Quimque Wentru Indigenous Association, and (c) Creation of the Quinquén Pewenche Park.

Informal Tourism Development Ever since the legal settlement in Quinquén, some of its inhabitants began to receive travelers on their way to other destinations that were attracted both by the scenic beauty of the surroundings and by the tourist promotion of the *Araucanía Andina* destination. Some of these travelers stayed for the entire summer season on the shores of the Galletué Lake. Along with these visitors came some undesired consequences. They did not bring resources to the community, or when they did, it was limited to minor forms of compensation such as exchanges for food. They damaged the bush flora to make firewood and left garbage and solid waste for the community to manage. Free transit to the lake area deteriorated the road and caused problems for community members who had their houses near the road or the lake area. These economic and environmental impacts motivated the community to seek external support in order to confront and organize against the emerging tourism activity.

Initiatives of the Quimque Wentru Indigenous Association In 1996, community leaders, supported by external institutions, created the Quimque Wentru Indigenous Association, a legal entity protected under the Indigenous Law that would allow them to advance their own development agenda (Arce et al. 2016). This association included members of the Huenucal Ivante, Kmkeñ (vernacular name of Quinquén), Marimenuco, Pedregoso, and Pedro Calfuqueo communities and aimed to recover land for families that had not benefited from the previous laws of Indigenous settlement and territorial recognition. As a result of the association's efforts, international aid funds were obtained for the development of a series of initiatives that sought to ensure the economic sustainability of the lands they had received through its productive use. For example, a building was purchased in the town of Lonquimay. Initially, this building allowed association members from distant areas to stay overnight in the town to carry out their business. Later it was converted into the Follil Pewenche Inn (Fig. 6.4).

To add property value and complement the services that would be offered at the Inn, the association created four trails (mainly for horseback riding) and a camping area at the Galletué Lake (Fig. 6.5). These trails were constructed in traditional use areas and, in summer, were offered to occasional visitors for hiking or flora and fauna observation. Along the same lines, in 2001 the community implemented the

Fig. 6.4 Follil Pewenche Inn in Lonquimay

Fig. 6.5 Campground in Quinquén

project known as *Ecotourism for the conservation of the Araucaria: a challenge for the Pewenche Quimque Wentru association of the Lonquimay commune*, in order to give greater impetus to tourism development. The objectives of this project were to promote both the conservation of the area's existing Araucaria forests and the dissemination of the Pewenche culture (Reyes 2006). To support the transition to tourism, the association procured resources to hire a full-time consultant to provide programming support, such as gastronomy, guide training, and administration. In developing these initiatives, the community established alliances with governmental and non-governmental forestry agencies such as the Chilean Committee for the Defense of Flora and Fauna (CODEFF), the Chilean National Forestry Corporation (CONAF), and the Model Forest program.

Creation of the Quinquén Pewenche Park Between 2009 and 2011, the community participated in the project: *Model of Areas Conserved by Indigenous Communities for the Development of Ecotourism and Biodiversity: proposal based on a pilot experience of a Pehuenche community park in the community of Quinquén, in the Andean Araucanía.* This project's primary goal was to transform the community open space into the first Indigenous Conservation Territory in Chile by creating Quinquén Pewenche Park. The initiative was implemented by the World Wildlife Fund, one of the largest international conservation organizations, with support from the Regional Government of La Araucanía, the Municipality of Lonquimay, and the Chilean National Tourism Service (SERNATUR). The community maintains a close collaborative relationship with WFF to this day (Aylwin and Cuadra 2011).

The project sought to make conservation-based tourism the main focus of the community. The Quinquén Pewenche Cooperative was created to promote intra-community connections and linkages by coordinating activities such as horseback riding, gastronomy, hiking service providers, as well as producers of leather, wood, and textile handicrafts. A welcome center (*Kukañwe Ruka*) was set up to provide information to tourists, along with a network of trails and signage reflecting the varying degrees of difficulty, with the Araucaria forests as the main attraction (Fig. 6.6). A participatory land-use planning process was also carried out to demarcate the areas that would be used for tourism and to protect the areas considered sacred and/or of high cultural value by the community.

Quinquén's location in the midst of millenary Araucaria forests has framed an imaginary of harmonious coexistence between its inhabitants and nature, which has been essential for the implementation of conservationist agendas and associated tourism development. This imaginary has not arisen exclusively as a result of the action of powers outside the community. The inhabitants of Quinquén have been instrumental to the protection of the Araucaria ecosystem and saw the touristification of their territory as a means to achieve this end. In the face of significant external pressures from both conservation and development interests, the Quinquén community saw tourism as a viable means of preserving its agency/sovereignty over its territories and taking control of the reterritorialization process.

Fig. 6.6 Welcome signs to the Quinquén Indigenous conservation territory

However, the promises of conservation tourism have not yielded the expected results. This is partly because of the strict safety standards and health codes that must be met in order to obtain permits and financing, and also because the income generated is neither consistent nor adequate to reverse the ongoing marginalization of the Quinquén community. In sum, although tourism development is not the definitive solution, it has provided the inhabitants of Quinquén with alternative opportunities for empowerment and control over unfolding reterritorialization processes that are also compatible with the emerging global conservation agenda.

6.5   Conclusion

As this chapter has demonstrated, the territories of the northern Patagonian Andes have been the object of an evolving geographical imaginary, which has aspired to control the region's development since the mid-nineteenth century. This imaginary was constructed to ensure the incorporation of these territories within the Chilean and Argentine nations and has inspired a series of concrete political strategies to strengthen private property rights, facilitate the introduction of agriculture, and promote the dispossession and resettlement of Indigenous lands. In the twentieth century, this same imaginary evolved, creating national areas for conservation and other areas that were deemed appropriate for timber harvesting. A lack of controls opened these areas to indiscriminate logging. The development of the railroad and later inclusion within bioceanic corridors, converted these logging areas into integration territories, facilitating tourism and their conception as biodiversity conservation areas of binational scope. As these strategies have evolved, the territory has been further integrated through environmental governmentalities that have been layered through their inclusion within the global network of UNESCO biosphere reserves, their rediscovery as sites of geological interest, or their designation as Indigenous conservation territories, and ZOITs. This constant transformation of *territorial orders* has resulted in the production of successive geographical imaginaries.

The participation of the Pewenche in community-based tourism initiatives resulted from a combination of external structuring forces and endogenous decisions. It allowed the Pewenche to take control of different elements of tourism in order to advance their own reterritorialization agenda, helping them retain sovereignty over their lands and erase the marks of colonialism. For example, for the Pewenche, the term *lof* refers to the community as a politically organized social group that is based on kinship. The reappropriation of this term *lof* in Quinquén, in connection with its community-based and centered development as a tourist destination, contrasts sharply with the term *Indigenous reduction* that has traditionally been employed by the state to describe Indigenous living spaces. This reappropriation illustrates how tourism can be an opportunity for identity revitalization.

In spite of the implementation and repercussions we have outlined, tourism has facilitated a shift in Indigenous agency. Strategically employed as a tool, tourism has allowed Indigenous communities to influence the transformation of geographical imaginaries, empowering communities to exercise control and counteract powers that have successively denied their existence and/or relegated them to the margins of the capitalist resource exploitation process, thus delegitimizing their territorial demands. Nevertheless, as in the case of the Pewenche communities and the Araucaria forests, associating Indigenous identity with environmental conservation through tourism development required they assume a central role in the production of a new imaginary based on the recognition of their ontologies and ways of life, and consequently of their territorial rights. In turn, this imaginary assumes a kind of strategic essentialism, calibrated by current movements (e.g., bioculturalism, multiculturalism, development with identity) that seek to make Indigenous subjectivities compatible with the demands of the neoliberal paradigm (Krell 2020). What is certain is that the struggle for the production and recomposition of geographical imaginaries in Norpatagonia is—and will continue to be—characterized by the growing participation and mobilization of Indigenous communities in the environmentalization of their territories.

Acknowledgments We would like to thank Andrés Adiego, of the Centro de Investigación en Ecosistemas de la Patagonia (CIEP), and the Department of Geography and Territorial Planning/ Management of the Universidad de Zaragoza, for his assistance with the development of the maps for our study area (Figs. 6.2 and 6.3).

References

D.L. Aagesen, Indigenous resource rights and conservation of the Monkey-puzzle tree (*Araucaria Araucana*, Araucariaceae): A case study from southern Chile. Econ. Bot. **52**(2), 146–160 (1998). https://doi.org/10.1007/BF02861203

J. Agnew, U. Oslender, Territorialidades superpuestas, soberanía en disputa: Lecciones empíricas desde América Latina [Overlapping territorialities, disputed sovereignty: Empirical lessons from Latin America]. Tabula Rasa **13**, 191–213 (2010)

A. Agrawal, *Environmentality: Technologies of Government and the Making of Subjects* (Duke University Press, 2005)

L. Andrade, M. Pilquimán, Percepciones sobre patrimonio inmaterial y turismo: Actores y experiencias en zonas precordilleranas, Panguipulli, Región de Los Ríos-Chile [Perceptions on intangible heritage and tourism: Actors and experiences in pre-mountain range areas, Panguipulli, Los Ríos Region-Chile]. Rev. Interam. Ambient. Tur. **16**(2), 164–174 (2020). https://doi.org/10.4067/S0718-235X2020000200164

L. Arce, F. Guerra, J. Aylwin, *Cuestionando los enfoques clásicos de conservación en Chile: El aporte de los pueblos Indígenas y las comunidades locales a la protección de la biodiversidad [Challenging Classical Approaches to Conservation in Chile: The Contribution of Indigenous Peoples and Local Communities to Biodiversity Protection]*. Observatorio Ciudadano. (2016). Retrieved January 7, 2023, from https://www.iwgia.org/images/publications/0754_Cuestionando-los-enfoques-clasicos-de-conservacion-en-Chile_1.pdf

J. Aylwin, X. Cuadra, *Los desafíos de la conservación en los territorios Indígenas en Chile [The Challenges of Conservation in Indigenous Territories in Chile]* (Observatorio Ciudadano, 2011) Retrieved January 7, 2023, from https://observatorio.cl/1287-2/

G. Azócar, R. Sanhueza, M. Aguayo, H. Romero, M. Muñoz, Conflicts for control of Mapuche-Pehuenche land and natural resources in the Biobio highlands, Chile. J. Lat. Am. Geogr. **4**(2), 57–76 (2005)

J. Bengoa, *Quinquen, 100 años de Historia Pehuenche [Quinquén, 100 Years of Pehuenche History]* (Ediciones ChileAmérica, 1992)

H. Blair, K. Bosak, T. Gale, Protected areas, tourism, and rural transition in Aysén, Chile. Sustainability **11**(24), Article 7087 (2019). https://doi.org/10.3390/su11247087

Chilean Ministry of Agriculture, *Aprueba reglamento para la explotación de maderas de Araucaria Araucana [Approves Regulations for the Exploitation of Araucaria Araucana Timber]*. Decree N°94, 26 February 1969 (1969)

Chilean Ministry of Economy, Development and Tourism & Undersecretariat of Tourism, *Declaración Zona de Interés Turístico Lonquimay [Lonquimay declared a Zone of Tourist Interest]*. Decree N°146, 19 August 2019 (2019)

Chilean Ministry of Lands and Colonization, *Declárense parques nacionales el Valle del Encanto, en la comuna de Punitaqui; el bosque petrificado de Pichasca, en la comuna de Samo Alto; los sectores de belleza autóctona llamados "Las Palmas de Cocalán", en la comuna de Las Cabras y "El Bollenar de las Nieves", en la comuna de Rengo, y los bosques de pino araucaria del fundo "Ralco", en la comuna de Santa Bárbara [Declare as national parks the Valle del Encanto, in the commune of Punitaqui; the petrified forest of Pichasca, in the commune of Samo Alto; the areas of native beauty called "Las Palmas de Cocalán", in the commune of Las Cabras and "El Bollenar de las Nieves", in the commune of Rengo, and the araucaria pine forests of the "Ralco" estate, in the commune of Santa Bárbara]*. Law N°17.699, Article 53, 14 August 1972 (1972)

Chilean National Congress Library, *Fundación de poblaciones en el territorio de los Indígenas [Foundation of Settlements on Indigenous Territory]*. Law without number, December 4, 1866 (1866)

Chilean National Statistics Institute, *Censo 2017 [Census 2017]* (2017). Retrieved November 15, 2022, from http://www.censo2017.cl/

F. De la Maza, E. Huisca, El relato turístico Mapuche: Identidad, territorio y políticas públicas [The Mapuche tourism story: Identity, territory and public policies]. CUHSO **30**(2), 98–118 (2020). https://doi.org/10.7770/2452-610x.2020.cuhso.05.a02

R. Dodge, C. Reyes, La implementación del área de desarrollo Pehuenche, sector Galletué-Icalma: Un proceso de contradicciones [The implementation of the Pehuenche development area, Galletué-Icalma sector: A process of contradictions], in *Tierra, territorio y desarrollo Indígena [Land, Territory and Indigenous Development]*, ed. by J. Aylwin, H. Carrasco, C. Martínez, (Universidad de la Frontera, 1995), pp. 205–213

R. Haesbaert, El mito de la desterritorialización [The myth of deterritorialization]. Siglo **XXI** (2011)

T. Hakenholz, Un peuple autochtone face à la « modernité » : la communauté mapuche-pewenche et le barrage Ralco (Alto Bío-Bío, Chili) [An Indgenous people facing «modernity»: The Mapuche-Pewenche community and the Ralco dam (Alto Bío-Bío, Chile)]. Les Cahiers d'Outre-Mer **228**, 347–366 (2004)

M. Impemba, G. Maragliano, Turismo y territorios en transformación en comunidades Mapuche del sur de la Provincia del Neuquén, Argentina [Tourism and territories in transformation in Mapuche communities in the south of the Province of Neuquén, Argentina]. Revista Antropologías del Sur **12**, 225–240 (2019). https://doi.org/10.25074/rantros.v6i12.1156

I. Krell, Turismo invasivo y turismo Mapuche: Territorio Indígena y emprendimiento con identidad en Laguna Icalma, Alto Biobío [Invasive tourism and Mapuche tourism: Indigenous territory and entrepreneurship with identity in Laguna Icalma, Alto Biobío]. CUHSO **30**(2), 119–148 (2020)

T. Li, *The Will to Improve. Governmentality, Development, and the Practice of Politics* (Duke University Press, 2007)

N. Martínez, Prácticas cotidianas de ancestralización de un territorio Indígena: El caso de la comunidad Pewenche de Quinquén [Everyday practices of ancestralization of an Indigenous territory: The case of the Pewenche community of Quinquén]. Rev. Geogr Norte Gd **62**, 85–107 (2015). https://doi.org/10.4067/S0718-34022015000300006

R. Molina, Quinquén y la tierra prometida: Política Indígena en una comunidad Mapuche-Pehuenche, Chile [Quinquén and the promised land: Indigenous politics in a Mapuche-Pehuenche community, Chile]. Rev. Austral de Cienc. Soc **29**, 89–105 (2015). https://doi.org/10.4206/rev.austral.cienc.soc.2015.n29-05

R. Molina, M. Correa, *Territorio y comunidades Pehuenches del Alto Bío-Bío [Territory and Pehuenche Communities of Alto Bio-Bio]* (CONADI, 1996)

J. Molina, C. Pávez, *Territorios Indígenas de conservación: Aprendizajes desde la práctica en el sur de Chile [Indigenous Conservation Territories: Learning from Practice in Southern Chile]* (WWF-Chile, 2012) Retrieved January 10, 2023, from https://wwflac.awsassets.panda.org/downloads/tic.pdf

P. Navarro-Floria, Un país sin Indios. La imagen de la Pampa y la Patagonia en la geografía del naciente estado argentino [A country without Indians. The image of the Pampas and Patagonia in the geography of the nascent Argentine state]. Scripta Nova **3**(32–54) (1999)

J. Olea, Gobernanza como contradicción: Reflexiones sobre el territorio en la configuración de la gobernanza ambiental [Governance as contradiction: Reflections on territory in the configuration of environmental governance]. Investig. Geogr **60**, 4–17 (2020). https://doi.org/10.5354/0719-5370.2020.57251

L. Otero, *La huella del fuego. Historia de los bosques nativos, poblamiento y cambios en el paisaje del sur de Chile [The Footprint of Fire. History of Native Forests, Settlement and Changes in the Landscape of Southern Chile]* (Pehuen Editores, 2006)

M. Palomino-Schalscha, Descolonizar la economía: Espacios de economías diversas y ontologías Mapuche en Alto Biobío, Chile [Decolonizing the economy: Spaces of diverse economies and Mapuche ontologies in Alto Biobío, Chile]. Rev. Geogr. Norte Gd **62**, 67–83 (2015). https://doi.org/10.4067/S0718-34022015000300005

R. Peet, M. Watts, *Liberation Ecologies. Environment, Development and Social Movements* (Routledge, 1996)

N. Peluso, *Rich Forests, Poor People. Resource Control and Resistance in Java* (University of California Press, 1992)

X. Pereiro, Reflexión antropológica sobre el turismo Indígena [Anthropological reflection on Indgenous tourism]. Desacatos **47**, 18–35 (2015). https://doi.org/10.29340/47.1419

M. Pilquimán, El turismo comunitario como una estrategia de supervivencia. Resistencia y reivindicación cultural Indígena de comunidades Mapuche en la Región de Los Ríos (Chile) [Community-based tourism as a survival strategy. Resistance and Indigenous cultural vindication of Mapuche communities in the Los Ríos Region (Chile)]. Estud. Perspect. Tur **25**(4), 439–459 (2016)

M. Pilquimán, Turismo comunitario en territorios conflictivos. El caso de las comunidades Indígenas Mapuche en la Región de Los Ríos en Chile [Community-based tourism in conflictive territories. The case of the Mapuche Indigenous communities in the Los Ríos Region in Chile]. Geopolítica(s) **8**(1), 11–28 (2017). https://doi.org/10.5209/GEOP.49479

C. Reyes, Ecoturismo para la protección de la araucaria: Un desafío para la asociación Pehuenche Quimque Wentru de Lonquimay [Ecotourism for the protection of the araucaria: A challenge for the Pehuenche association Quimque Wentru of Lonquimay], in *Bosques Y Comunidades del Sur de Chile [Forests and Communities in Southern Chile]*, ed. by R. Catalán, P. Wilken, A. Kandzior, D. Tecklin, H. Burschel, (Editorial Universitaria, 2006), pp. 300–307

P. Robbins, *Political Ecology: A Critical Introduction* (Wiley-Blackwell, 2019)

J. Scott, *Seeing Like a State. How Certain Schemes to Improve the Human Condition Have Failed* (Yale University Press, 1999)

B. Sepúlveda, S. Guyot, Escaping the border, debordering the nature: Protected areas, participatory management and environmental security in northern Patagonia (i.e., Chile and Argentina). Globalizations **13**(6), 767–786 (2016). https://doi.org/10.1080/14747731.2015.1133045

J. Skewes, C. Henríquez, M. Pilquimán, Turismo comunitario o de base comunitaria: Una experiencia alternativa de hospitalidad vivida en el mundo Mapuche. Tralcao sur de Chile [Community-based tourism: An alternative experience of hospitality in the Mapuche world. Tralcao, southern Chile]. Cultur **6**(2), 73–85 (2012)

M. Steinberg, Tourism development and Indigenous people: The Maya experience in southern Belize. Focus. Geogr. **44**(2), 17–20 (1994). https://doi.org/10.1111/j.1949-8535.1994.tb00080.x

A. Volle, Le tourisme communautaire mapuche contre l'écotourisme chilien ? [Mapuche community tourism versus Chilean ecotourism?]. L'Ordinaire Latinoaméricain **177**, 73–78 (1999)

R. Wesche, Ecotourism and Indigenous peoples in the resource frontier of the Ecuadorian Amazon. Yearb. Conf. Lat. Am. Geogr. **19**, 35–45 (1993)

P. Zusman, La geografía histórica, la imaginación y los imaginarios geográficos [Historical geography, imagination and geographical imaginaries]. Rev. Geogr. Norte Gd **54**, 51–66 (2013). https://doi.org/10.4067/S0718-34022013000100004

Part II
Contemporary Conservation-Based Development: Challenges for Green Integration

Part II examines the potential, approaches, and challenges associated with contemporary conservation-based development in Patagonia.

Chapter 7
The Production of Space in the Frontiers of Tourism: Critical Analysis of the Huella de Glaciares Circuit Between El Chaltén, Argentina, and Villa O'Higgins, Chile

Alejandro F. Schweitzer, Brenda Sofía Ponzi, and Sabrina Elizabeth Picone

Abstract Nature-based tourism (NBT) has been positioned as being aligned with conservation objectives and capable of contributing to social and ecological sustainability in contemporary contexts of socio-ecological crisis and global change. Nevertheless, some have characterized this strategy as a capitalist accumulation circuit, a form of commodification and production of nature. In southern Patagonia, NBT exploits roads, trails, and protected areas as tourism resources. This chapter focuses on the production of frontier spaces in south Patagonia through an analysis of cross-border NBT strategies. We present an intrinsic case study of the *Huella de Glaciares* (Trail of the Glaciers) NBT product, situated in the Santa Cruz Province of Argentina, and the Aysén Region of Chile. Through interviews, participant observation, field visits, and documentary analysis, we identify that the Huella de Glaciares, along with other tourism routes that connect with the larger Patagonia territory, respond to renewed cross-border processes. We observed that the *Huella de Glaciares* NBT circuit was being advanced and promoted by the governments of both countries and by private tourism agents between Villa O'Higgins and El Chaltén, without adequate consideration of the project's sustainability in this area of Patagonia.

Keywords Patagonia · Nature-based tourism · Circuit of capital accumulation · Commodification of nature · Cross-border processes

A. F. Schweitzer (✉) · B. S. Ponzi · S. E. Picone
Consejo Nacional de Investigaciones Científicas y Técnicas (CONICET), Santa Cruz
Research and Transfer Center, Rio Gallegos, Santa Cruz, Argentina
e-mail: alejandro.schweitzer@conicet.gov.ar; brendaponzi@conicet.gov.ar;
sabrina.picone@conicet.gov.ar

© The Author(s) 2023
T. Gale-Detrich et al. (eds.), *Tourism and Conservation-based Development in the Periphery*, Natural and Social Sciences of Patagonia,
https://doi.org/10.1007/978-3-031-38048-8_7

7.1 Introduction

Contemporary spatial research has extended earlier considerations of peripheraliza-tion to focus on process, in addition to place. In this way, peripheralization entails processes that lead to structural marginalization, leading to social marginality and poverty, political dependency or exclusion, and economic polarization, among other outcomes (Bachmann-Vargas and van Koppen 2020; Kühn 2015; Kühn et al. 2017). Southern Patagonia can be considered a *peripheral space* in relation to the central, capital areas of the countries in which it is situated: Chile (CL) and Argentina (AR). In turn, CL and AR are peripheries with respect to the global North. Moreover, in a dual sense, southern Patagonia is a space of frontiers. First, there are political bor-ders. These borders were *interjurisdictional* during the colonial period and *inter-state* since formal agreements were established between AR and CL at the beginning of the nineteenth century (Porto and Schweitzer 2018). Second, since the mid-eighteenth century, the interfaces between capitalism and nature have created new commodity chains and evolved others, which has reconfigured and adapted the natural domains, shifting the lines between mountains and plateaus, forests and steppe, icefields, and oceans. Through social processes of the production of space, a commodification of nature has unfolded in this sector of Patagonia, giving way to *commodity frontiers* that have expanded (Moore 2020), as capitalism has spread across the interior of Patagonia, from the coasts. Both of these cases represent polit-ical frontiers because they alter pre-existing power relations and generate unequal geographic development (Smith 2020).

French Marxist theorist Henri Lefebvre (1901–1991) argued that humans are inherently social, and as such, they produce their lives and their world through social processes (Lefebvre 2013). Lefebvre (2013) viewed space as a product of this social production, involving the interrelations between living beings, activities, con-cepts, undertakings, and discourses; an interrelation of all that is produced by nature or society. Humans produce their spaces through a rational mobilization of resources and tools, and a series of actions that are organized around specific objectives and logics. Their decisions and processes are based on social relations and use values; thus, rather than being a thing or container, space is *relational*, meaning that it is socially made and remade over time (Lefebvre 2013; Massey 1999). As such, it has both physical and conceptual frontiers, or boundaries, which are established, overlap, and superimposed with other social spaces (Fuchs 2019).

The delimitation of borders is a technique used in the social production of space; thus, it is necessary to study both the characteristics of borders as containers and also the configurations of the spaces they contain (Foucher 1991). Processes of *borderization* and *cross-borderization* alternate in the construction of borders, both in political-administrative terms and in terms of the expansion of commodification. In these processes, and in the context of border demarcation and definitions, govern-ment scales (e.g., municipal, provincial, regional, and national) foster the configura-tion and deployment of circuits, which respond to an economic and productive logic. These processes of *borderization* and *cross-borderization* affect the frontiers of nature commodification. For example, the accumulation circuits that historically developed in the southern Patagonia area around ranching and wool exports depended on fluid interactions between agents operating on both sides of the international boundary, including a movement of animals from one coast to the other. These cross-borderization processes facilitated rapid expansion of commodification frontiers (Pereira Carneiro 2016; Schweitzer 1998). Later, during the 1980s, when AR and CL underwent a period of geopolitical conflict, strict borderization processes were employed, closing what had historically been termed a *sheep frontier* and constraining related capitalist activity. When the conflict ended and integration processes regained momentum, processes of borderization relaxed somewhat (Foucher 1991).

In the production of space, a number of agents employ different strategies to compete for the control of space, each proposing territorial alternatives that are built upon distinct values and meanings. Territorial configurations end up being the result and reflection of these processes and power disputes. The advance of the commodification frontier is achieved through the provision of *natural production conditions* (Castree 1995; O'Connor 2001): landscapes and access to water or land, general conditions and infrastructure related to access (e.g., roads, trails, and ferries), energy, communications, equipment, and services (usually public). Despite fundamental differences in the values and meanings associated with agent preferences for these production circuits, the production conditions required for these competing activities often overlap, causing tensions to rise.

In many cases, to achieve the production conditions required for particular social productions of space, fundamental changes are provoked within the territory that can affect the conditions required for *the reproduction of life* (Torres et al. 2014), defined here as, "the process by which a society [or organism] reproduces itself from one generation to another and also within generations" (Burton 2014, p.1). When the social production of space impinges on the social, physical, and/or biological conditions required for the reproduction of life, disputes can arise. Common examples include disputes among productive sector agents and disputes between the productive sector and social sector agents (Barbosa e Souza and Chaveiro 2019; Valenzuela et al. 2021).

The study of processes of the social production of space can be carried out from a range of approaches including conventional economic, geographic, and environmental science approaches, and political ecology and critical geography approaches. These disciplines provide a range of perspectives and interpretations along a theoretical-ideological gradient of environmental issues and the conception of nature. In geographical terms, this research was developed under the assumption that space is *relational*, meaning that it is socially made and remade over time (Lefebvre 2013; Massey 1999). From a broader perspective, political ecology theories suggest the importance of extending analysis of territorial agents and conditions beyond humans and social reproduction to include the conditions needed for the reproduction of all life (human and non-human beings) (Moore 2020). Thus, evaluating the prospects for different territorial development alternatives and strategies requires analysis of the various agents, conditions required for productive activities, and the conditions required for the reproduction of life (Gumuchian and Pecqueur 2007; Schweitzer 2004, 2008). Lajarge and Roux (2007) defined territorial building projects as being the strategies and actions of state actors, undertaken to produce a desired territorial configuration. In our research, we applied these concepts more broadly, positing that territorial building strategies are not uniquely driven by the state; rather, they can also be undertaken by a variety of actors including large corporations and landholders, or even as expressions of alternative visions proposed by social organizations (Schweitzer 2008).

In recent decades, social frontiers have advanced in southern Patagonia through nature-based tourism (NBT) processes in and around protected areas (PAs) based on commoditization processes and capitalist logic. These PAs, which are often publicly owned, facilitate NBT by providing several of the primary production conditions necessary for success (e.g., access, infrastructure, unique natural landscapes, and established support service systems like trail maintenance or search and rescue). The case study presented within this chapter sought to better understand the recent advances in nature commodification frontiers in southern Patagonia through a transboundary case study that examined the production of space and nature on both sides of the Andean Mountain Range. Our study area focused on the *Huella de Glaciares* (Trail of the Glaciers) NBT product, which is accessed in AR via El Chaltén and Provincial Route 41. In CL, the *Huella de Glaciares* is accessed via Villa O'Higgins and the southernmost regional extension from the Carretera Austral (Route 7). The Carretera Austral is included within the *Route of the Parks of Patagonia* (rutadelosparques.org), a conceptual circuit connecting 17 national parks and 60 communities between Puerto Montt and Cape Horn in Chilean Patagonia. Our intrinsic case study approach identified the main agents involved in these processes and explored the similarities and differences in their strategies of appropriation and capitalization of space. Our work employed critical approaches to understanding the social production of space. We understand NBT, real estate development, private protected areas, and other forms of conservation-based development (CBD) as operating through a capitalist logic. Thus, our work assumes these logics require constant expansion and advance of capital to maintain their rates of profit and empower their reproduction.

7.2 Literature Review

During the twenty-first century, a number of studies have explored the dynamics surrounding nature exploitation, tourism, and conservation in this area of Argentine and Chilean Patagonia (Blair et al. 2019; Borrie et al. 2020; Gale et al. 2013; Grenier 2003; Núñez et al. 2014, 2018; Pérez 2019; Picone 2020; Picone et al. 2020; Ponzi 2022; Schweitzer 2020). French author Grenier (2003) offered early advances in the social production of space research, observing neoliberal globalization dynamics in Chilean Patagonian over the course of 30 years that were advancing hand in hand with the exploitation of untouched, peripheral, isolated nature. His book featured chapters dedicated to tourism, real estate speculation, and land grabbing for conservation purposes, providing an important background to the issues addressed in our chapter. A number of more contemporary works continue to enrich our understanding of social production of space in these areas in CL. Gale et al. (2013) examined NBT visitor perceptions of authenticity and change in several of the rural southern Patagonia spaces within our study area, including Villa O'Higgins (CL) and El Chaltén (AR). Their research with international travelers indicated that communities who maintained their traditional identity instead of developing a tourist-oriented culture were perceived as being less "welcoming." Núñez et al. (2014) analyzed discourses of development linked to conservation, finding that CBD discourse has resulted in a new form of territorial colonization. Núñez et al. (2018) examined touristification in the Aysén region of Chilean Patagonia, focusing on the Carretera Austral and green businesses. Blair et al. (2019) examined nature-based tourism and local perceptions of changes in rural spaces around CL's Cerro Castillo National Park, identifying deep-rooted skepticism, resentment, and distrust toward central policy shifts that promoted conservation-based development. Borrie et al. (2020) examined the dynamics and tensions associated with the advance of private PAs in Chilean Patagonia, calling for integrated governance between PAs and local communities based on "transparency, fair and legal adjudication, establishment of collaborative platforms, cross-jurisdictional coordination, and peer enforcement" (p.13).

Important contributions have also been made from an Argentine perspective within the study area, including Pérez's (2019) research on transboundary border inhabitation within Villa O'Higgins (CL) and El Chaltén (AR), which identified renewed locally driven practices of exchange, circulation, and economic transformation centered on tourism despite ongoing national level geopolitical conflicts in the area. Schweitzer's (2020) study of Argentine conservation strategies and practices within the Santa Cruz Province identified tensions around the production of space that threatened local well-being. Picone et al. (2020) identified a shifting social construction of Los Glaciares National Park that has resulted from the gradual incorporation of green economic strategies. Picone et al. (2022) further analyzed this shift, identifying land access/scarcity and land appropriation processes around the northern limits of Los Glaciares National Park as critical dimensions affecting the quality of life for residents of El Chaltén (AR). Lastly, Ponzi's (2022) article on convergent ideas of nature in two Argentine national parks (Los Glaciares and Perito Moreno) identified a convergence of imaginaries and an intensification of commercialization processes of nature through tourism promotion.

The originality of our approach lies in consideration of the processes occurring on both sides of the mountain range, including the agents involved, the ways in which they secure and deploy production conditions in their promotion of cross-border NBT circuits, and the relationship between these agents, their actions, and processes of expansion of the frontiers of commodification associated with the production of nature (Castree 1995; O'Connor 2001). Thus, drawing on the existing theory and research presented in the previous sections, we aim to contribute to the collective understanding of the dynamics of frontierization, in its many forms, and the ways in which the commodification of nature can advance over interstate borders through NBT.

7.3 Methods

7.3.1 The Study Area

Southern Patagonia is marked by a cold climate, glacier masses, and, especially in Chile, a rugged geography that is difficult to access. The Chilean government considers this portion of its territory as an extreme zone, a designation based on low levels of connectivity and population which lead to underdevelopment and relative isolation as compared with the urban centers of the nation's central zone (Gale et al. 2013). For more than two and a half centuries, the lower links of the first global textile accumulation circuit have been deployed in the southern Patagonia region, originating with the commercialization of oil, fats, and skins obtained from the hunting of whales and other marine mammals. This circuit became more complex in 1875, when it evolved to focus its accumulation process on the production and export of wool. It secured the necessary production conditions through the introduction of sheep from the Falkland Islands and the labor of migrant workers and Indigenous peoples who were proletarianized through forced acculturation (Barbería 1995; Bascopé 2018; Coronato 2017). With the implementation of the meat packing plants at the beginning of the twentieth century, this circuit evolved again to accommodate a new international export of lamb meat. Similar circuits evolved on the Chilean side of the frontier during this period. Other accumulation circuits in southern Patagonia included Argentine energy-related circuits that began to evolve in 1907 with the discovery of hydrocarbons in Comodoro Rivadavia and—later—coal exploitation in Río Turbio beginning in the 1940s.

More recent accumulation circuits in Patagonia (AR, CL) have focused on metalliferous mining and aquaculture. Similarly, during the last two decades of the twentieth century, another type of accumulation circuit has begun to emerge that involves the exploitation of nature through CBD. Many CBD circuits have been implemented within the same spaces used for older production circuits, particularly in the mountain areas of the Santa Cruz Province (AR) and two Chilean regions: Aysén del General Carlos Ibáñez del Campo (Aysén) and Magallanes and Chilean Antarctica Region (Magallanes). NBT accumulation circuits are one of the main ways in which CBD manifests. The attractiveness of this area is based on spectacular mountain, lake, glacier, and forest environments, like Mount Fitz Roy, Viedma Lake, Glacier, O'Higgins/San Martin Bi-national Lake, and the eastern reaches of the Southern Patagonia Icefields. Moreover, these landscapes show little evidence of the visual transformations caused by the deployment of previous circuits, so the NBT sector has been able to frame marketing narratives around an imaginary of pristine nature. As a result of improved transportation and communication access, southern Patagonia began to be incorporated within global CBD circuits during the latter decades of the twentieth century and has gained strength during the first few decades of the twenty-first century.

Our study focused on the mountainous areas of Capitán Prat Province of Chile's Aysén Region and the department of Lago Argentino in the Santa Cruz Province of

AR. Even within the context of low population density and isolation that has characterized southern Patagonia, this sector was considered extreme. Population centers were few and far between. There are four within the study area: El Calafate (AR), El Chaltén (AR), Tres Lagos (AR), and Villa O'Higgins (CL). At the time of this study, the Lago Argentino department was estimated to have a population density of around 0.5 persons per square kilometer, with the majority concentrated in the town of El Calafate. The population density of the Capitan Prat Province (CL) was estimated at 0.3 persons per square kilometer. Thus, NBT infrastructure and services (e.g., lodging, food services, transportation services, and fuel) were limited and geographically dispersed. Visitors who arrived by plane normally accessed the area through airports in either El Calafate (AR; population 23,065) or Balmaceda (CL; population 550) near the regional capital of Coyhaique (CL; population 57,818). After spending a few nights within these cities, visitors normally circulated the territory via overland tours, visiting natural attractions and PAs, and lodging in private lodges or smaller towns.

On the Argentine side, the Huella de Glaciares NBT product involved two natural PAs: Los Glaciares National Park and Lago del Desierto Provincial Reserve. Los Glaciares National Park encompasses approximately 7269.27 km² (Argentine Administration of National Parks 2019). Its natural characteristics include a representative sample of the Patagonian Forest ecoregion composed mainly of the Magellanic Forest subunit, and to a lesser extent, the Patagonian Steppe subunit (Argentine Administration of National Parks 1997). To the north, following the De las Vueltas River valley, lies another natural PA, Lago del Desierto Provincial Reserve, which is under provincial jurisdiction (Honorable Chamber of Deputies of the Province of Santa Cruz of Argentina 2005). In addition to conservation, the Lago del Desierto Provincial Reserve permits cattle ranching, tourism, and vacation home development projects (Picone 2020). Almost all of the Argentine surface area of the Southern Patagonian Ice Field is contained within these two PAs.

These two PAs form a biological corridor for several animal species including the endangered huemul (*Hippocamelus bisulcus*) and torrent duck (*Merganetta armata*). The huemul is particularly threatened and has been classified as *in danger of extinction* by the Red Book of Threatened Mammals of Argentina (Argentine Ministry of Environment and Sustainable Development of the Nation and the Argentine Society for the Study of Mammals 2019), and the Red Book of Terrestrial Vertebrates of CL (Glade Carreño 1993). The huemul was also included in the Convention on International Trade in Endangered Species of Wild Fauna and Flora (CITES) (2015) and in the Convention on the Conservation of Migratory Species of Wild Animals (CMS) (2015), as it is a species shared between two countries and regularly moves between their boundaries.

On the Chilean side, the Huella de Glaciares NBT traversed fiscal lands along the buffer area of Bernardo O'Higgins National Park, which are considered to be part of the Chilean Border Zone. The Park has an estimated surface area of 35,259.01 km² distributed between the Última Esperanza Province in the Magallanes Region and the Capitán Prat Province in the Aysén Region (Rosenfeld and Sekulovic Ltda 2000). This PA conserves forest environments, part of the steppe, a diverse hydrological system (i.e., canals, fjords, islands, lakes, rivers), and is home to most of the Southern Patagonian Ice Field (Fig. 7.1). In addition to being the world's third largest reserve of fresh water in solid form, the glaciers of the Southern Patagonian Ice Field hold valuable information regarding climate and past glacial and interglacial periods. CL and AR are the South American countries with 90% of the continent's glaciers, so their study, monitoring, and conservation represent a strategic issue in terms of climate change for the continent (Argentine Institute of Nivology, Glaciology and Environmental Sciences 2010).

Historically, this area has been the subject of border disputes between AR and CL. Patagonia was incorporated into the Chilean and Argentine national spaces after the colonial era. Until the beginning of the nineteenth century, the border was drawn as an east-west strip, inhabited and crossed by colonial armies, Creole settlers, and Indigenous peoples. In the 1856 *Tratado de Paz, Amistad, Comercio y Navegación entre la República de Chile y la Confederación Argentina* (Treaty of Peace, Friendship, Commerce and Navigation between the Republic of CL and the Argentine Confederation), both countries agreed to promote free movement in this zone until the delimitation of the borders inherited from the colonial administration was resolved (Porto and Schweitzer 2018).

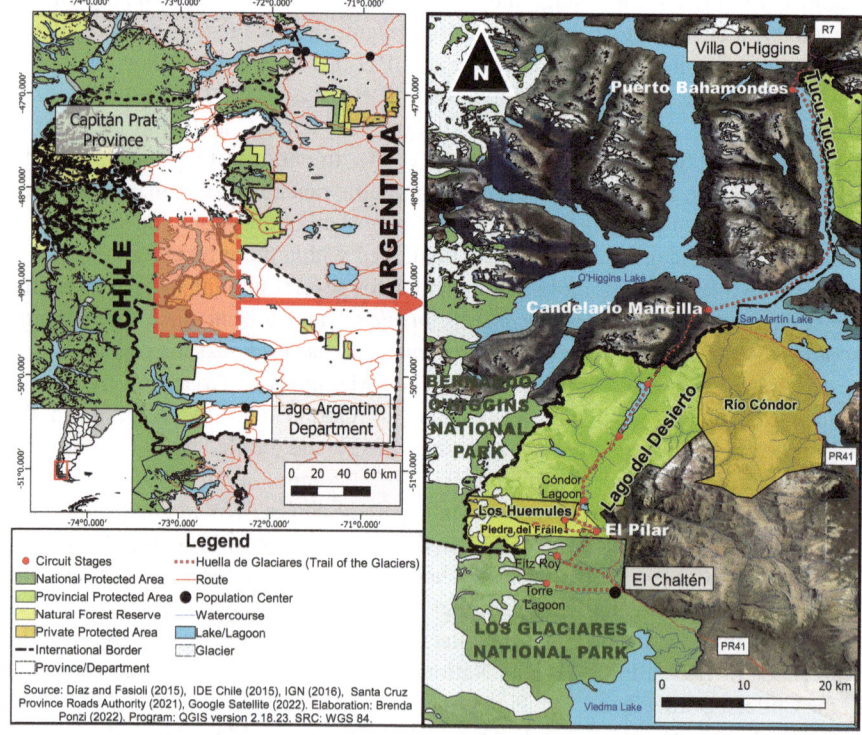

Fig. 7.1 Binational Study Area, including the natural protected areas, population centers, and Huella de Glaciares nature-based tourism circuit

This treaty enacted border limits, affecting the flow of people and goods throughout Patagonia. It was implemented in a concrete manner in most passes along the border, with the exception of the study area, in the areas south of Buenos Aires/ General Carrera Binational Lake and north of Punta Arenas – Puerto Natales, which were very peripheral, isolated, and unknown, at the time. The Llanquihue Province (CL) created in 1861 and the Santa Cruz National Territory (AR) in 1884 were the first national territorial administrations in the study area. Thus, for the following decades, this zone was barely affected by the arrival and presence of the State, including the new border limits and procedures. For example, unlike other Patagonian areas to the north and south, it was not the scene of the genocidal campaigns and territorial dispossession undertaken by the Chilean and Argentine states against the Mapuche and Tehuelche peoples. The religious missions that had been installed in other areas of Patagonia were also absent.

In the 1881 Limits Treaty and those that followed (i.e., Protocol of 1893, Act of 1898), until final limits were established in 1902, both AR and CL affirmed their sovereignty over Patagonia and agreed on the peaceful definition of the border: both considered the Andes Mountains as the western boundary but disputed sovereignty over the Strait of Magellan and the Beagle Channel Islands. Since the establishment of formal limits in 1902, different legal and cartographic interpretations have led to the continued dispute, marked by periodic protocols and addendums (e.g., 1941, 1991, 1996; De la Rosa 1998).

Conservation has played an important role in this nation-building process on both sides of the international border; often used as a geopolitical tactic to consolidate national sovereignty (Picone 2020). For example, in the Argentine portion of southern Patagonia, the Perito Moreno y Los Glaciares national parks were established in 1937. At this time, they grandfathered in existing estancias using a special permit, which allowed the parks to assume ownership and the estancias to manage and graze the lands, albeit with restrictions. Over time, these restrictions have increased as Argentina has moved to a stricter conservation mandate. Beginning in the 1990s, estancias were encouraged to convert into tourism providers as a tactic to align use and conservation (Picone 2020). In Chile, the Jeinimeni and Lago Cochrane (Tamango) national reserves were established in southern Patagonia in 1967, followed closely by the Bernardo O'Higgins National Park in 1969. The use of the national reserve figure in Chile allowed for greater use diversity within Jeinimeni and Cochrane, including timber harvesting and even some mining endeavors, while Bernardo O'Higgins was established under the stricter norms of national park.

The establishment of national parks in the borderlands of southern Patagonia ensured territorial sovereignty over partially unexplored spaces, as is the case of the area under study. For example, the mountain range area between the Fitz Roy and Daudet peaks was poorly defined in 1898 by the Boundary Commission. Throughout the decades, both Argentina and Chile produced maps with profound differences in their delimitation; thus, the conflict over this area increased. For example, towards the end of the 1970s, border tensions with CL flared after a presentation by the Argentine government to UNESCO. In 1979, UNESCO invited the Argentine embassy to submit natural or cultural properties of outstanding universal value for inclusion in the *World Heritage of Humanity* list. For these purposes, UNESCO requested information about grazing and tourism controls within the national reserve that would ensure the area would always maintain its natural state (United Nations Educational, Scientific and Cultural Organization 1980). Chile formally protested UNESCO's consideration of the inclusion of Los Glaciares National Park (AR) within the *World Heritage of Humanity* list, complaining that the area of the park remained under international dispute. Nevertheless, it was included in 1981.

In 1984, CL and AR signed the Treaty of Peace and Friendship, and in 1985, they established the Binational Commission for Cooperation and Physical Integration. In 1991, Economic Cooperation Agreement N° 16 was established, and later that same year the two countries signed an Environmental Treaty accompanied by a Specific Additional Protocol on Shared Water Resources. This laid the foundations for the creation of an Argentine-Chilean Water Resources Technical Group, formally constituted in Buenos Aires in December 1996 (Schweitzer 2000). In 1997, the two countries took another step forward in the integration process, signing the Treaty on Mining Integration and Cooperation, and the Cooperation Protocol. This treaty determined a strip along the binational border for mining activity, suspending restrictions in both countries regarding property access, mining rights, and the exercise of the activity. Although this treaty remains in force, it has not had repercussions to date within the study area (Schweitzer 2019).

In the context of the successive agreements that had been reached between President Carlos Menem (AR) and presidents Patricio Aylwin and Eduardo Frei (CL) in the 1990s, a solution to the border controversy was sought. In 1991, a 19-point polygon was proposed but not accepted by their respective Congresses (De la Rosa 1998). Consequently, in 1994, the matter was taken up by an international arbitration tribunal whose decision defined the current boundaries in place between the two countries. The tribunal ruled in favor of AR, granting it more than 532 km^2 in and around Lago del Desierto Lake. In 1996, an additional protocol established that, even though the area remained in dispute, the existence of adjacent PAs required an integrated and coordinated management between the two countries (De la Rosa 1998). In 1998, an agreement was signed establishing the current limits in the area of the Southern Patagonia Ice Field by largely respecting the topographic line determined at the end of the nineteenth century. However, the preparation of the agreed upon 1:50,000 map, which was to be drawn up by a Joint Commission for the Demarcation of Limits, was never completed; thus, tensions remain. For example, after the presentation of the 2018 Argentine National Glacier Inventory, CL again denounced that AR had projected its mapping of the AR portion of the Continental Ice Zone between the Santa Cruz Province (AR) and the Magallanes Region (CL), over territory that CL considered its own.

Today, both countries have specific legislation and restrictions within their border areas. For example, the Border Zones in CL, including zones within the Capitán Prat province, are controlled by the Chilean National Directorate of Borders and Boundaries (DIFROL) under the Ministry of Foreign Affairs, which must authorize the foreign land purchase and any type of assignment, concession, or contract related to land access (Chilean National Directorate of Borders and Boundaries 2022). The Argentine Directorate of Limits and Borders, within the Ministry of Foreign Affairs and Traditions, functions in a similar manner. Additionally, these two directorates coordinate the operations of a binational integration committee for the southern Patagonia region, which is responsible for the binational management of border areas and crossings. This committee is based on the model created within the framework of the Treaty of the La Plata Basin that was later adopted as a model organization for the integration processes of South America.

Conservation initiatives in this sector of southern Patagonia continue to grow and evolve. For example, in 2014, Patagonia National Park (AR) was established to the north of Perito Moreno National Park in AR, at the same latitude as Jeinimeni National Reserve in CL (Ponzi 2020). A few years later (2018), the Jeinimeni and Lago Cochrane National Reserves (CL) were joined with the donated lands of the Patagonia Private Park in the Chacabuco valley to form Patagonia National Park (CL). This expansive (and growing) network of State PAs within CL and AR is complemented by provincial PAs in AR, Chilean conservation zones, and PPAs on both sides of the border (Argentine Information System for Biodiversity and Argentine Administration of National Parks 2022; Chilean National Forestry Corporation 2019).

The Huella de Glaciares NBT product involves several border crossings, both as part of the core product and more generally in terms of its access routes. From north to south these include: a) the Río Mayer crossing which provides access between Villa O'Higgins (CL), Perito Moreno National Park, and Tucu Tucu Provincial Reserve (AR); b) the Río Mosco crossing which connects Villa O'Higgins (CL) with El Chaltén (AR) via the Cocoví Pass; c) the frontier area between Lago del Desierto (AR) and Candelario Mancilla (CL); and d) San Martín (AR) – O'Higgins (CL) binational lake and the Marconi control, located south of El Chaltén between San Martín Lake (AR) and Bernardo O'Higgins National Park (CL) (Fig. 7.2). Circulation through these crossings is constrained by the limits of the geography, infrastructure, and operations, and limited to pedestrian, bicycle, or maritime passage via lake crossings. Travel is further constrained by sporadic operations, which must be arranged in advance with the appropriate authorities, local agents, and attributions of the committee and the directions of each country.

7.3.2 The Huella de Glaciares Project

The Huella de Glaciares is a cross-border NBT product that has been created by tourism actors and government agents within CL and AR over the past decade. For most of its route, the Huella de Glaciares product uses existing trails and navigations (Fig. 7.1). In AR, it extends from the town of El Chaltén, which is located within the northern zone of Los Glaciares National Park, through privately protected lands, the Lago Del Desierto Provincial Reserve, and Los Glaciares National Park. In CL, the trail begins in Villa O'Higgins, crossing Chilean waters within the

Fig. 7.2 Banner displayed at the bus terminal in El Chaltén

binational lake (O'Higgins Lake, CL/San Martin Lake, AR) to Candelario Mancilla, where the trail winds along the fiscal lands that border the edge of the Southern Patagonia Ice Field and Bernardo O'Higgins National Park (Fig. 7.1). The Huella de Glaciares initiative proposal seeks to strengthen the cross-border NBT circuit between Chile and Argentina, in the southern reaches of the Andes Mountains, as projected in successive sustainable tourism plans in both Argentina (Argentine National Touristry Ministry 2005, 2011, 2015) and Chile (Chilean Subsecretary of Tourism 2017). In AR, it has mainly been driven by the Santa Cruz Province through the Provincial Secretary of Tourism, with the support of the Ministry of Tourism of the Nation and backing of the El Chaltén Chamber of Commerce and the Tourism

Secretary of the El Chaltén Municipality. In Chile, the project has been driven by the Villa O'Higgins Chamber of Tourism and representatives of the Villa O'Higgins Municipality, along with the Corporation of Los Glaciares Tourism Interest Zone (ZOIT), which is supported by the regional branch of the National Tourism Service (Sernatur), and the National Ministry of Economy, Development, and Tourism. Since 2019, the Huella de Glaciares cross-border NBT product has been promoted at international tourism fairs, which has further advanced the tourism frontier.

The Huella de Glaciares circuit extends more than 120 km through forested mountain landscapes where participants can see and enjoy more than 10 of the area's glaciers. Developers propose that the circuit be hiked in seven stages with an estimated time of seven nights and 8 days. If hiking the route from AR, the first two stages employ two trails within Los Glaciares National Park that are extremely busy during the summer high season: Laguna Torre and Fitz Roy. The third stage follows the Fitz Roy trail, beginning from the Poincenot camp (National Park) and ending at the Piedra del Fraile refuge (Natural Wildlife Reserve), which is accessed via the Piedras Blancas trail adjacent to the El Pilar lodge (private). The fourth stage connects the Piedra del Fraile refuge to the Cóndor Lagoon (private in Provincial Reserve), passing through Route 41 (Provincial Reserve). The fifth stage occurs in the Provincial reserve, following the Huemul trail from the Cóndor Lagoon to the southern end of Lago del Desierto and the Huemul trail (private in the Provincial Reserve). The sixth stage crosses Lago del Desierto by boat ending on the northern shore of the lake, which is also in the Lago del Desierto Provincial Reserve. The seventh stage begins with the border crossing into Chile, near the northern shore of Lago del Desierto in AR passing through milestone 62 to the Candelario Mancilla border post on the shores of O'Higgins Lake (CL). Finally, the eighth stage involves a boat trip across O'Higgins Lake from Candelario Mancilla to the pier at Puerto Bahamondes, and a seven-kilometer hike (or transfer) from the port to Villa O'Higgins. The hike can also be experienced beginning in Villa O'Higgins (CL) and doing the stages in reverse order, ending in El Chaltén (AR). Either way, the daily hikes range between nine and 15 km, with overnight stays in wild campsites or private inns (Figs. 7.1 and 7.2).

7.3.3 Data Collection and Analysis

This chapter presents an intrinsic cross-border case study, which has employed a qualitative approach. We focus on the recent advances in the frontiers of nature commodification in southern Patagonia through a transboundary analysis of the production of space and nature on both sides of the Andean Mountain Range. We did not choose this case because it was representative of a larger phenomenon, nor because it illustrates a particular issue; rather, it was chosen because we were particularly interested in this case and wanted to understand its "particularities and ordinariness" (Stake 2003, p.136). Data production techniques included field visits, participant observation, documentary analysis, and semi-structured interviews.

With respect to field visits, one of the researchers observed and kept field diaries as part of a participatory action research process in and around El Chaltén (AR) and conducted a field trip of Lago del Desierto from south to north (AR). Participant observation occurred during the public hearing for the PR41 paving project (AR), the presentation of Huella de Glaciares NBT project at the Hotel Chaltén Suites in September 2019 (AR), and the public hearing on the Environmental Impact Study of the El Chaltén-Lago del Desierto section of Scenic Route No. 41 (AR). Documentary analysis expanded the dataset to include consideration of dynamics occurring in AR and CL. Analysis included regional tourism agency websites, local newspapers, official organizations, land-use plans, management plans of the PAs involved in the case, and environmental impact studies of recent development projects in both CL and AR. Finally, semistructured interviews were conducted with personnel of the Argentine Administration of National Parks and the Chilean National Forestry Corporation between 2019 and 2022.

7.4 Results and Discussion

This case sought to understand evidence of an advance of the frontiers of commodification and capitalization in southern Patagonia, based on the different strategies and multiple articulations expressed between agents. The complexity of these advances increased due to the binational character of the study area and the relatively recent occupation and territorial organization of these spaces, which required contact and interaction with different territorial administration systems across several scales.

Results and discussion are organized in three sections. First, we present an analysis of the similarities and differences that arose with respect to territorial planning, political-administrative functions, and PA management on each side of the international border. Second, we describe the Huella de Glaciares project, identifying the agents involved, their relationships, and their strategies and/or interests. Finally, we describe transit-related advances taking place in the study area to enable and promote one of the core production conditions necessary for the Huella de Glaciares NBT circuit and other forms of CBD advancement. Discussion focuses on our interpretation of the production conditions that facilitate the advance of the commodification frontier and the tensions this generates with local inhabitant conditions for the reproduction of life.

7.4.1 Territorial Organization and Natural Protected Areas

At the binational political level, we observed there to be profound differences between CL and AR, with respect to territorial organization and planning. As a unitary republic, CL maintained a centralized system of government with little regional autonomy. AR, on the other hand, has a federal republic system, in which the provinces are autonomous states with broad powers over land management and natural resources. National government decisions in AR are limited to the powers that have not been ceded to the provinces, like defense, economic policy, and some education and health agreements. Specific to this case, Santa Cruz acquired provincial autonomy in 1957 (Fig. 7.3).

In the Chilean territorial system, the political-administrative functions of regional, provincial, and municipal governments were extended within combined urban/rural territories. In contrast, with Santa Cruz (AR), municipal political-administrative institutions' representation was limited to specific urban areas (e.g., the urban limits of El Chaltén). Thus, more than 90% of the Argentine territory was not covered by any local-scale authority, and therefore dependent on decisions made within the provincial capital (Fig. 7.3). This dynamic manifested within our data in several ways, contributing to interference and crossover between national, provincial, and municipal governments on the Argentine side of the case study. Although the Chilean provinces had less autonomy, their local area land-use plans and PA management plans seemed to carry greater weight.

For example, in CL, the Huella de Glaciares NBT product has been deployed within areas with defined guidance from political-administrative institutions within the Capitán Prat Province and the Villa O'Higgins Municipality. In AR, the Lago Argentino Department did not have any political-administrative functions and the El Chaltén Municipality's political-administrative functions were limited to the 1.35 km^2 urban area of El Chaltén, which represents a very small proportion of the departmental territory. These asymmetries hinder possible advances in the definition of local policies.

Another difference between the two countries relates specifically to issues of ownership and jurisdiction of natural PAs. Chilean State PAs are owned as public lands, under the control of the Ministry of Public Lands. Currently, the administration of these areas is entrusted to the National Forestry Corporation (CONAF) which reports to the Chilean Ministry of Agriculture. Two other ministries are directly involved in policy and decision making: the Ministry of the Environment and the Ministry of the Economy, Development, and Tourism. Yet, while the Ministry of the Environment is in charge of environmental policy, plans, and programs, they do not manage all aspects of the environment: the Environmental Evaluation Service administers environmental impact assessments for Chile, the Directorate General of Waters controls all freshwater related management issues, and the Superintendency of the Environment ensures policy and legislative compliance (Borrie et al. 2020). This complex situation was designed and implemented within the neoliberal framework instilled by the military government (1973–1990), with marginal change since return to democracy. In theory, successful environmental oversight is possible in Chile if effective coordination and collaboration are achieved within this bureaucracy (Latta and Aguayo 2012). In practice however, the complexity of Chile's public/private environmental governance system has proven weak, and a number of accumulation circuits (e.g., NBT, salmon aquaculture,

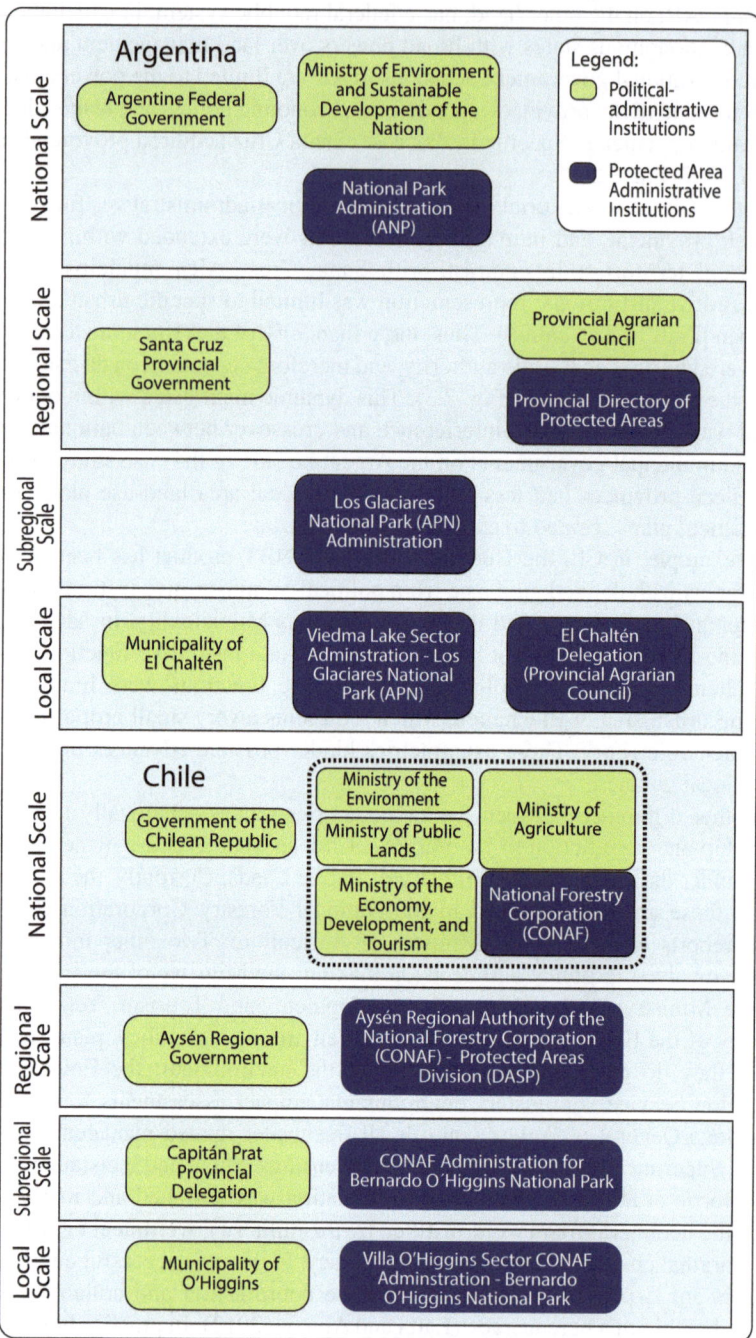

Fig. 7.3 Territorial organization within the study area

sphagnum moss extraction, mining) have successfully installed themselves within and around Chilean PAs (Segura 2022).

Since all public lands in Chile belong to the republic, they can be reconfigured as national PAs through national-level legislative approval rather than regional agreement. In contrast, Argentina's government structure requires agreement to be reached between the federal and provincial governments before changes can be made to the constitution of national parks and/or reserves, requiring one or the other entities to cede jurisdiction and/or dominion. Furthermore, on the Argentine side, national PAs are managed by the Argentine Administration of National Parks (APN), which depends on the Argentine National Ministry of Environment and Sustainable Development. There is also a board of directors and a provincial directorate or intendancy that is in charge of the local administration of the PAs, and some Argentine provinces even have their own agencies in charge of PA management. For example, Santa Cruz has a provincial system administered by the Provincial Directorate of Protected Areas, under the Provincial Agrarian Council (Fig. 7.3). Both CL and AR permit private entities to develop private PPAs and associate in order to coordinate actions and channel common proposals through their respective political-administrative and PA administrative institutions.

Some of the jurisdictional and domain combinations we have described in the previous paragraphs can be found in the study area. For example, on the Chilean side, Bernardo O'Higgins National Park is divided between the Magallanes and Aysén Regions. In Aysén, the park lies within the southern part of the Capitán Prat Province, divided between the O'Higgins and Tortel municipalities. Other Chilean lands in the study area are protected as fiscal areas, Border Zones, or municipal parks. In AR, the northern sector of Los Glaciares National Park falls under federal authority, managed by the Lago Viedma Section of the Argentine Administration of National Parks. In contrast, the Piedra del Fraile Natural Wildlife Reserve falls under provincial jurisdiction, along with the Lago del Desierto and Tucu-Tucu Provincial Reserves, while three other PAs (Los Huemules, Laguna Cóndor, and Río Cóndor) are considered private nature reserves.

To date, a coordinated binational system of biodiversity governance has not been achieved for this complicated web of PAs and administrative institutions. In fact, only one coordinated binational governance initiative was detected in Patagonia, lying to the north of our study area. In the framework of the 15th Meeting of the CL-AR Environment sub-commission in 2017, the CL-AR Joint National Parks Specialized Committee was created. Its first meeting took place in 2019 to coordinate the management of Nahuel Huapi and Lanín National Parks in AR, and Puyehue and Vicente Pérez Rosales National Parks in CL. AR and CL have also defined some agreements for economic cooperation and cultural integration. For example, in 2009, the Maipú Treaty established regional working committees to resolve conflicts and establish agreements (Pérez 2019). In 2014, CL and AR signed a memorandum for tourism cooperation related to the Huella de Glaciares NBT project (Argentine National Touristry Ministry 2015). And, in 2018, a feasibility study was conducted south of Parallel 42° to evaluate the potential for CL-AR integration within the framework of a strategic plan for the coordinated development of the Patagonia territory (Argentine Ministry of the Interior 2018).

7.4.2 Huellas de Glaciares Nature-Based Tourism Project

Although the trail crosses two Argentine PAs, the public institutions in charge of their administration were included in the project in late August 2020. At that time, the Argentine National Park Administration and Provincial Agrarian Council authorities signed a cooperation agreement, ignoring the concerns of local representatives about the effects that the increased flow of visitors would have on the management of PAs in the area. As the Chilean side of the Huella de Glaciares crosses fiscal lands in the area of Candelaria Mancilla and does not officially enter Bernardo O'Higgins National Park, coordination has focused on the O'Higgins Municipality, tourism actors, and the Los Glaciares Tourism Interest Zone Corporation rather CONAF who is in charge of the park administration.

Until 2021, Argentine investments for this project were allocated under the *50 Destinations Program* with funds from the Country Tax. Four construction projects have been planned for the AR portion of the Huella de Glaciares to be paid for through AR financing: trail signage, control posts, a waiting room with restrooms at the southern end of Lago del Desierto, and a cable crossing over the Eléctrico River. All of these represent infrastructure improvements that will facilitate production conditions for the intensification and advancement of the tourist frontier. On the Chilean side of the trail, project plans, developed within the Los Glaciares Zone of Tourism Interest action plan, concentrate on infrastructure improvements to accommodate tourism growth, including construction of a new airstrip, amplification of the port in Bahía Bahamonde, development of trails to connect the Huella de Glaciares with the O'Higgins Glacier in the Chico Lake sector, and the development of local tourism capacity within Bernardo O'Higgins National Park through the Marconi Pass route (Chilean Subsecretary of Tourism 2017). These four projects would provide tourists with additional activities and options on the Chilean side of the route while simultaneously advancing the tourism frontier by extending the Huella de Glaciares through new routes and usage within the national park.

Our analysis of the Argentine plans suggests that only the first two initiatives (trail signage and control posts) would contribute to improved visitor registration and management within the Lago del Desierto Provincial Reserve. The Reserve does not currently have any type of public infrastructure in its approximately 600 km² of extension. Thus, signage might help to encourage visitors to remain along designated trails, thereby reducing off-trail impacts. The control posts would facilitate the addition of a park ranger or fauna ranger, which has not existed within the PA to date (there were only four people dedicated to the PA in 2022, working from offices in El Chaltén). While these two initiatives also represent infrastructure improvements to facilitate tourism frontier advancement, they could also be considered to advance conditions needed for the reproduction of life, as they could help ensure the control, supervision, and conservation of the PA. Nevertheless, we posit that the other two initiatives, a waiting room with restrooms at the southern end of Lago del Desierto, and a cable crossing over the Eléctrico River, would employ public tax funding to the benefit private tourism operators, essentially enabling tourism capital production conditions at the expense of public and/or conservation good.

There are other restroom services nearby, however, because they are currently concessioned to another tourism service provider under a grandfathered agreement related to a land donation expanding the Los Glaciares National Park, separate restroom facilities were petitioned (Picone 2020). Given the environmental impact of restroom facilities in sensitive environmental areas, it seems important to resolve the ongoing issues between tourism operators, concessionaires, and administrative institutions, rather than invest more public funds in duplicative infrastructure. Similarly, the cable crossing project would likely contribute to, and perhaps escalate, existing tensions between Huella de Glaciares actors, by proposing to connect the Los Huemules Private Reserve with the Piedra del Fraile Refuge that is located inside this area of contention.

Furthermore, our research suggested that these situations were linked with NBT accumulation circuits and the expansion of NBT frontiers through the Huella de Glaciares. In June 2019, a key informant from Los Glaciares National Park explained,

> Those are the big problems I have: charging for access or not? If Parks do not charge for access, how is a private company able to charge for access within Park territory? It seems illogical to me, but well, that's the approach they are taking, in exactly those terms. They do not make any kind of contribution to the Park. I cannot control anything they do because, in fact, I do not have jurisdiction so I cannot control a third party. In other words, I have someone inside my house, doing whatever he wants to do, according to the terms he once agreed to with someone who is no longer the owner. They gave me a little gift with strings attached (Key informant Los Glaciares National Park 16, 01/06/2019).

And seemingly, the expansion of NBT frontiers was intimately linked with other types of conservation-based accumulation circuits. For example, agents that were linked to the gift referred to in the prior quote (i.e., the donation of Ricanor Estancia to Los Glaciares National Park), were also connected with other forms of CBD, including NBT, real-estate development positioned around conservation, private PA development, and a foundation dedicated to bond issues for the purchase of forest lands in AR. Finally, the December 2020 inauguration of the Chilean-backed Explora Hotel Chain suggested a transfrontierization of these practices in the Los Huemules sector near the Piedra del Fraile Refuge on the north bank of the Eléctrico River.

The land ownership irregularities that were observed on the AR side of the circuit were not observed on the CL side of the circuit. Rather, popular press articles and press releases from the O'Higgins Municipality suggest that the Municipality and the Chilean Ministry of Housing and Urbanism are advancing a formal project to develop the Candelario Mancilla sector as a small town adjacent to the Dos Lagunas border control, to provide services to the more than 2000 tourists that make the crossing along the Huella de Glaciares each year (Municipality of O'Higgins 2016;

Tehuelche Noticias 2022). These sources report that the funding to develop the project in Candelario Mancilla has been secured through Chile's national *Lugares que Crecen* (Places that Grow) program, administered through the Chilean Ministry of Housing and Urbanism. According to the Chilean Subsecretary of Tourism (2017), the rest of the projects identified in the Los Glaciares Tourism Interest Zone (ZOIT) action plan (e.g., new airstrip, port expansion, trail development, augmented tourism capacity within Bernardo O'Higgins National Park), represent specific initiatives that have been approved through the Chilean Committee of Ministers of Tourism to be defined in the officially designated ZOIT area. Each project within the plan establishes responsibilities, deadlines, and proposed sources of funding; however, funding is not guaranteed and must be secured through the actions and petitions of the ZOIT's public-private governance system. If funding is not secured and projects do not advance, the validity of the ZOIT comes into question and may be revoked (Chilean Subsecretary of Tourism 2017, 2020).

Several Chilean scholars have proposed that these types of initiatives represent a new form of *green* colonization within the territory involving land redistribution in Aysén (Aliste et al. 2018; Miranda Cabaña 2016; Núñez et al. 2014, 2018, 2020). These scholars contend that this reterritorialization process has been framed and enabled by the State by positioning the Aysén region as a *Reserve of Life*, a new social narrative that aligns with the neoliberal environmental imaginary. *According to these authors,* State policy, strategies, and funding have supported the *Aysén Reserve of Life* narrative and CBD for several decades, influencing social understanding, discourse, and action within the region and beyond (Aliste et al. 2018; Miranda Cabaña 2016; Núñez et al. 2014, 2018, 2020). While green colonization operates through a neoliberal environmental imaginary and discourse that frames an urgent global need for the protection of sensitive natural environments, these authors have argued that its practice combines land grabbing with a privatization of nature based on progressive environmental values and capitalist speculation (Corson and MacDonald 2012; Goodman and Stroup 1991; Núñez et al. 2020). Thus, as these authors assert, it is not surprising that green colonization has coincided with recent tourism growth in Aysén (Aliste et al. 2018; Blair et al. 2019).

7.4.3 Access, as a Primary Condition of Production

In many respects, the binational Huellas de Glaciares circuit replicates the production models employed by other long-distance trails and routes in CL and AR, including the *Huella Andina* (lahuellaandina.com.ar) model of Northern Patagonia (AR) and the Chilean projects: *Sendero de Chile,* (fundacionsenderodechile.org) and *Route of Patagonia Parks* (https://www.rutadelosparques.org). Each of these projects employs a strategy that joins a series of existing tourist attractions and trails via existing roads and maritime infrastructure to achieve essential production conditions for packaging and positioning large-scale NBT products on the global market. In this final section of our results, we outline development considerations that arose within our data with respect to the Huella de Glaciares strategy for leveraging AR Route 41 in El Chaltén and CL Route 7 in Villa O'Higgins.

The development of Provincial Scenic Route No. 41 (AR) represents an ongoing project for the Santa Cruz Provincial government that, when completed, will provide one of the main structuring axes for tourism and CBD in western Santa Cruz. The project involves a set of road sections that are in varying stages of completion. Some portions of the route have not yet been defined, others are gravel, and others are completely paved. Although the collection of routes has been designed within a range of road standards, the route is gradually emerging in its entirety, with the objective of facilitating access and connectivity for tourism destinations to the west of National Route 40 (Argentine General Provincial Roads Administration, ESUCO Sociedad Anónima, and National University of Southern Patagonia 2022).

The design of this section of the route and the components of the associated territorial project have been in development since the 2010s but have yet to be fully defined. Within the framework of the Provincial Program for the Productive Development of the Santa Cruz Valleys (PROVASA), an international loan was granted to finance consulting work which was carried out between 2010 and 2011 on the route and its tourism attractions. Some road segments were renamed as Provincial Scenic Route No. 41, connecting protected areas of different jurisdictions and domains with gateway towns. Some of these gateway communities already existed, while others have advanced projects for their creation, as in the case of Tucu-Tucu, a future town near Villa O'Higgins. The route connecting the town of El Chaltén with Lago del Desierto, previously known as Provincial Route 23, was renamed *De la Soberanía Nacional en la Senda del Huemul* (Of National Sovereignty along the Huemul Trail) as part of Scenic Route N° 41 in 2017 (Legislative Branch of the Province of Santa Cruz, Argentina 2017). The choice to include this section in the route was made by the Provincial Roads Authority, and planned upgrades and improvements (e.g., paving) will be financed by the Argentine National Roads Authority.

The Environmental Impact Study (EIS) for the new layout and paving of this section of Provincial Scenic Route No. 41 (AR) includes three stages, two of which deal specifically with the route, and a third involving the construction of jetties and the improvement of the Lago del Desierto-Chilean border section of the Huellas de Glaciares trail. The construction of jetties at Lago del Desierto complements the proposed checkpoint and restrooms to be built at the southern end of the lake, financed through the *50 Destinations Program*. This demonstrates the close linkages between the provincial and national tourism development strategies and the ways in which they have furthered NBT and CBD in areas that were not previously open for this type of use.

Construction of the first two stages, which were addressed by the EIS, will be financed by the Argentine National Roads Authority and carried out by the ESUCO S.A. Two of the activities planned are of great concern in terms of biological and ecological aspects: deforestation and the impacts affecting the huemul (*Hippocamelus bisulcus*). On the one hand, the deforestation planned for the new route and the widening of the current road fall within categories I and II of the Law of Minimum

Budgets for the Development of Native Forest Land No. 26,331. Article 14 of this regulation prohibits the native forest removal in these areas. Hence, the removal implied by the widening of the road would not be allowed (Honorable Congress of the Argentine Nation 2017). Furthermore, the huemul would be threatened by this project, both during the construction stage and in later phases of operation. The project is moving forward despite the facts that (1) the huemul is categorized as *in danger of extinction*, (2) the huemul is protected in AR as a Natural Monument, and (3) the huemul who inhabit the De las Vueltas River valley represents one of the few populations with reproductive success, with a proposal for the construction of wildlife crossings as a mitigation measure.

Route No. 7, also known as Carretera Austral, was developed by Chile's central government in the 1970s with the purpose of establishing an overland route that would connect the Aysén Region (CL) with Puerto Montt (CL) and points north (Adiego et al. 2018). Technically, the road to Villa O'Higgins (Route X-91) extends 89.5 km south from where it joins with the Carretera Austral, to reach Bahamondes Port, along the shores of O'Higgins Lake. Nevertheless, most recognize the end of Route X-91 as the southernmost point of the Carretera Austral.

In 2017, an agreement was reached between the Chilean Government and Rewilding Chile (formerly Tompkins Conservation Chile), for the donation of 45,000 km² of private PA lands to the central government (CL) to be managed as national parks. One of the terms of this agreement involved the creation of a joint project between the Chilean government and Rewilding Chile to develop and promote a new NBT circuit, the Route of Patagonia Parks, which involves a 2800 km route, joining 17 national parks and 60 gateway communities, from Puerto Montt to Cape Horn (CL). The entire length of the Carretera Austral is included in the Route of Patagonia Parks, as are many other overland and maritime routes (e.g., Route 9; Y-85, the Austral Broom ferry route between Puerto Yungay and Puerto Natales).

Rewilding Chile had been working in CL since the 1990s, under a number of different NGO names and figures (e.g., Tompkins Conservation; Conservación Patagónico, Fundación Yendegaia, Foundation for Deep Ecology). One of their main objectives was to purchase large tracts of lands in areas of conservation interest, rewild these areas, and donate them back to the state for the expansion and/or declaration of numerous National Parks, including Corcovado (established 2005, reclassified 2018), Cerro Castillo (established 1970, reclassified 2018), Magdalena Island (established 1983, reclassified 2018), Melimoyu (established 2018), Pumalín Douglas Tompkins (established 2018), Yendegaia (established 2013, reclassified 2016; Chilean National Register of Protected Areas 2020). In 2017, various ministries, CONAF and other foundations linked to Tompkins signed a Protocol of Agreement for the development of the *Chilean Patagonia National Parks Network* project. This protocol included provisions for the creation, expansion, or recategorization of PAs in areas of high biodiversity value and potential for NBT in the regions of Los Lagos, Aysén, and Magallanes (Chilean National Forestry Corporation 2019). To support the 17 national parks and development of NBT in the 60 gateway communities, the *Route of the Parks Fund, Protecting Patagonia Forever* was established in 2018 by the Chilean government, Rewilding Chile, and the Pew Charitable

Trusts. The fund's design guarantees private capital as a complement to the progressive increase in public investment in PA management. Recently, the Route of Patagonia Parks designed a *passport* for the route, offering visitors the opportunity to obtain stamps in each of the PAs, as a mechanism to promote responsible visitation and consumption. For example, the official website states, "When you get your passport, you will accept the zero-footprint commitment as a responsible tourist, assuming your role as guardian of the Route of Parks of Patagonia" (Rewilding Chile Foundation 2018).

The Huella de Glaciares (AR-CL), the Scenic Route 41 in Santa Cruz (AR), and the Route of Patagonia Parks (CL) represent some of the many examples of CBD that are evolving on both sides of the border in Patagonia. These projects are revaluing Patagonian landscapes and renaming roads, trails, and even towns, in order to build a unique imaginary around a product that can be positioned and marketed for NBT. Our evaluation of this project and the many side projects proposed and underway to facilitate production conditions, suggested that these projects carry negative impacts for endemic animal and plant species such as the Huemul, due to habitat degradation and fragmentation. As well, we have identified an increased risk of visitor-induced accidents and impacts due to crowding, changing weather conditions, and difficult access in case of emergencies or fire. These social and environmental impacts increase the risk for the conditions for the reproduction of life within this zone, particularly if these areas become more crowded and congested as a result of NBT configurations.

The Huella de Glaciares product and the proposed improvements of former Provincial Route 23, between El Chaltén and Lago del Desierto, also stress the conditions for the reproduction of life in El Chaltén and the De las Vueltas river valley. For example, although the Lago del Desierto Provincial Reserve was officially established in 2005 (Provincial Law No. 2,820/2005), it does not have sufficient staff or the necessary infrastructure to carry out conservation monitoring and control, meaning increased tourism will carry commensurate increases in risk and impact. Moreover, the PA had just recently approved their management plan by Provincial decree No. 1546, in December of 2022. As one of our participants from the Argentine National Park administration described the staffing and planning deficiencies that exist with respect to the Lago de Desierto Provincial Reserve create risk for the entire PA system in the area, as well as the community and the De las Vueltas Valley:

> There [in the Lago del Desierto Provincial Reserve] the shortcoming is with the [Provincial] Agrarian Council, which also does not have sufficient human resources to enforce the regulations. Or there is no one available to control the area and make sure that cows are not grazing within. But really, that is also due to a lack of human resources in the Agrarian Council. I think there are only two or three people working in the Council now. Many seasons there was only one person working. So, for us, that increases the risks in terms of fire or an accident. And, in the winter sometimes nobody takes responsibility when an accident occurs in the mountains. And we cannot take charge, in principle, because it is not our jurisdiction. We could do a rescue, because we know how to do it, but we cannot do it, in terms of regulations. Somehow, we have to coordinate and work with the Agrarian Council and with the private property [owners] or whoever it is. But well, you can't expect someone to come forward and say 'ah, this is my responsibility" (Key informant section Lago Viedma 1, 10/05/2019).

Similar human resource pressures affect the northern zone of Los Glaciares National Park. The number of personnel is higher than in other sections of the PA but is still not sufficient to support new trail development (Picone et al. 2020). According to key informant, Lago Viedma 2 section, 12/09/2019, the current levels of infrastructure and funding are also insufficient to support monitoring and conservation tasks in the new Piedra del Fraile Wildlife Nature Reserve, located more than 15 km from the Lago Viedma section at the entrance to El Chaltén.

These developments also impact local communities. The El Chaltén community has faced a housing emergency for several years due to the scarcity of land that arises from their location within a national park (Picone 2020). During the tourism season, many residents are displaced, and it is common for tent cities to arise to accommodate seasonal workers (Picone 2020). The inhabitants of El Chaltén have demanded the right to access decent housing and land; yet real estate projects continue to multiply in the valley of the De las Vueltas River associated with tourism. And the proposed road improvements will likely increase the value of land in the area by improving access to areas of great scenic value; thus, encouraging private urban development and real estate speculation, which will likely aggravate social problems. In sum, while the improvements proposed for the route favor the production conditions for the development of the Huella de Glaciares, NBT, and amenity-based real estate projects, they do not benefit (and may heighten) the concerns of local inhabitants of El Chaltén and the De las Vueltas River valley, the agencies involved in the management of provincial and national PAs, and the nature that brings these actors together.

Although we did not identify similar dynamics occurring in CL during the time of our study, the intentions of the O'Higgins Municipality and the Chilean Ministry of Housing and Urbanism to develop the Candelario Mansilla sector as a small town oriented around tourism could produce similar conflict and speculation in the future. The Aysén Region has also prioritized the pavement of the Carretera Austral for the coming years and the projects prioritized within the ZOIT are all designed to facilitate greater visitation and tourism flows. Moreover, projects like the Route of Patagonia Parks are contributing to tourism growth and interest in the region, as well as processes of natural amenity-based migration.

Thus, it seems critical that we continue to research and grow our understanding of the peripheralization processes that are occurring in these remote areas of southern Patagonia. Complexity increases within peripheral frontier spaces, where contemporary occupation and territorial organization are relatively new, and the actors, processes, laws, and systems differ along multiple scales (Schweitzer 2020). Yet, an integrated approach adds valuable perspective, permitting communities and decision-makers to learn from each other, avoid pitfalls, and improve the planning and controls required on both sides of the border to protect the conditions required for the reproduction of life.

7.5 Conclusions

The results of the case study suggest that the production of the space through NBT and other forms of CBD is the result of a combination of dynamics that are typical of interstate borders. These dynamics include policies or actions that create barriers or filters (i.e., *frontierization*), and policies or actions that promote greater interactions between local inhabitants and/or the advance of the commodification of nature over interstate borders (i.e., *transfrontierization*). We would like to close this chapter with a series of conclusions, or lessons learned, that may help researchers and communities as they face similar processes.

In our study of the Huellas de Glaciares NBT circuit in southern Patagonia, we observed large investments that have been made to facilitate and promote circulation, scenic enjoyment, and access to marketable natural attractions on both sides of the mountain range (AR and CL). Overland, lacustrine, and maritime access infrastructure, along with the presence of PAs and their infrastructure that have been developed by government investments in CL and AR provide important production conditions that enable the development of binational NBT strategies like the Huella de Glaciares (AR – CL), Scenic Route 41 (AR), and the Route of Patagonia Parks (CL).

We observed a change in the (CL – AR) border logic with the advance of NBT. During the twentieth century, border disputes and borderization processes reinforced the frontier between AR and CL through barriers and boundary functions. Yet, in recent decades, cross-border processes have (re)configured this space as transboundary, as the creation of the Huella de Glaciares NBT circuit and associated private strategies demonstrate. Although these strategies were met with tension and conflict from other social productions of space, they were largely reinforced through public policies. In both countries, the public sector provided support to enable the Huella de Glaciares NBT at a variety of scales, including a legal and regulatory framework, financial and economic contributions, and a wide range of initiatives and projects to support the promotion of the area. Public processes were advancing hand in hand with the business strategies of large and medium-sized real estate investors and actors within the NBT sector. Thus, our evidence suggests that the area has changed from being an area configured to establish sovereignty in the dispute over Lago del Desierto, to a space of accumulation "without borders" based on integration and private interests and supported through public resources.

Though traditionally the production of space in southern Patagonia has been marked by multiple visions between agents that act under similar capitalist logics (e.g., ranching, mega-mining, and salmon aquaculture), the social production of space also involves social logics that operate around the defense of conditions of social reproduction, or in recognition of socio-ecological interdependence with other natures. While at times capitalist logics may conflict, they can also be compatible, or even, complementary by promoting and defending one another. For example, in CL, mining baron Andrónico Luksic Craig owns some 350 km^2 of land around Villa O'Higgins which he manages as private conservation and ranching areas (Segura 2022). We observed that contradictions and resistance to capitalist logics created greater tension when they involved opposing logics, and a fight for the conditions necessary for the reproduction of life (Moore 2020; Navarro Trujillo and Linsalata 2021). This is evidenced by the dispute over urban land in El Chaltén and the complaints of inhabitants about the ecological and social impacts of real estate projects located along former Provincial Route 23 between El Chaltén and Lago del Desierto (AR).

The dynamics and trajectories we exposed within the case study of the Huella de Glaciares NBT suggest a likelihood of continued *transfrontierization* practices and related advances for the frontiers of CBD in Patagonia. The proposed road improvements for Provincial Route 23 between El Chaltén and Lago del Desierto (AR) continued southward advances of the pavement of the Carretera Austral (CL), and the infrastructure projects projected for Bahamondes port and Candelario Mancilla (CL) all evidence a likely convergence for NBT within the study area that is likely to attract luxury tourism capitals in alliance with other CBD capitals.

Left unchecked, this dynamic may eventually extend toward the north and south, incorporating other PAs along the border and giving rise to similar reconfigurations of Patagonia's largely intact natural territories in proximity to the southern reaches of the Andes. Evidence of this trend is already visible in the southwest sectors of the Río Negro Province and the western reaches of Chubut, around Esquel and Trevelin (AR). Based on the literature, theory, and dynamics that surfaced from our study, we posit that the implementation of this model will require further financialization of nature and fragmentation of ecosystems, negatively impacting the natural conditions of these spaces and their attractiveness for NBT. We foresee an increase in disputes over how transit networks get deployed, both between productive sectors and with social logics that operate around the defense of conditions of social reproduction, or in recognition of socio-ecological interdependence with other natures. And we believe similar conflicts are likely to arise around other scarce resources, like water. Thus, we urge caution and additional research about CBD and NBT in cross-border environments that can inform policy and decision-making based on principles of social justice and sustainability.

References

A. Adiego, A. Antas, T. Gale, M. Grossman, C. Mardones, M.J. May, R. Merino, G. Orellana, M.J. Oyarzo, *Carretera austral: Catalizador de turismo en Aysén Patagonia [Carretera Austral: Catalyst for Tourism in Aysén Patagonia]* (2018). Retrieved 21 Jan 2023, from https://drive.google.com/file/d/0ByuKTEJGEnPWTWptVWNlSk9ubVpuNnJsWURTZ2FBVGZhb 0Nj/view?usp=sharing

E. Aliste, A. Núñez, M. Galarce, Geografías de lo sublime y el proceso de turistificación en Aysén-Patagonia. Turismo, territorio y poder [Geographies of the sublime and the process of turistification in Aysén-Patagonia. Tourism, territory and power], in *Araucanía-Norpatagonia II: La fluidez, lo disruptivo y el sentido de la frontera [Araucanía-Northpatagonia II: Fluidity, the*

Disruptive and the Sense of Frontier], ed. by P. Núñez, A. Núñez, M. Tamagnini, (Universidad Nacional de Río Negro, 2018), pp. 249–269

Argentine Administration of National Parks, *Plan preliminar de manejo Parque Nacional Los Glaciares [Preliminary Management Plan for Los Glaciares National Park]* (Administración de Parques Nacionales de Argentina, 1997)

Argentine Administration of National Parks, *Plan de Gestión del Parque Nacional Los Glaciares 2019–2029 [Los Glaciares National Park Management Plan 2019–2029]* (Administración de Parques Nacionales de Argentina, 2019). Retrieved 13 Jan 2023, from https://sib.gob.ar/archivos/PN_LosGlaciares.pdf

Argentine General Provincial Roads Administration, ESUCO Sociedad Anónima & National University of Southern Patagonia, *Estudio de impacto ambiental pavimentación Ruta Escénica Provincial Número 41 El Chaltén- Lago Del Desierto, Provincia de Santa Cruz [Environmental Impact Study for the Paving of Provincial Scenic Route Number 41 El Chaltén-Lago Del Desierto, Province of Santa Cruz]* (Ministerio de Obras Públicas, 2022)

Argentine Information System for Biodiversity & Argentine Administration of National Parks, *Estadísticas de visitantes de las áreas protegidas nacionales de Argentina considerando los registros desde 2003 a 2020 [Visitor Statistics for Argentina's National Protected Areas Considering Records from 2003 to 2020]* (2022). Retrieved 15 Jan 2023, from https://sib.gob.ar/institucional/visitantes-apn

Argentine Institute of Nivology, Glaciology and Environmental Sciences, *Inventario nacional de glaciares y ambiente periglaciar: Fundamentos y cronograma de ejecución [National Inventory of Glaciers and Periglacial Environment: Rationale and Implementation Schedule]* (2010). Retrieved 21 Jan 2023, from https://www.glaciaresargentinos.gob.ar/wp-content/uploads/legales/fundamentos_cronograma_ejecucion.pdf

Argentine Ministry of Environment and Sustainable Development of the Nation & Argentine Society for the Study of Mammals, *Categorización 2019 de los mamíferos de Argentina según su riesgo de extinción. Lista Roja de los mamíferos de Argentina [2019 Categorization of the Mammals of Argentina According to Their Risk of Extinction. Red List of Argentina's Mammals]* (2019). Retrieved 1 Feb 2023, from http://cma.sarem.org.ar

Argentine Ministry of the Interior, *Plan estratégico territorial: Avance 2018 (1 edn. ampliada) [Territorial Strategic Plan: Advance 2018 (1 expanded edn.)]* (Ministerio del Interior, Obras Públicas y Vivienda, 2018). Retrieved 18 Jan 2023, from https://www.argentina.gob.ar/sites/default/files/plan_estrategico_territorial_2018_baja.pdf

Argentine National Touristry Ministry, *Plan federal estratégico de turismo sustentable: Turismo 2016 [Federal Strategic Plan for Sustainable Tourism: Tourism 2016]* (Ministerio de Turismo y Deportes, 2005). Retrieved 18 Jan 2023, from http://biblioteca.cfi.org.ar/wp-content/uploads/sites/2/2005/01/44793.zip

Argentine National Touristry Ministry, *Plan federal estratégico de turismo sustentable: Turismo 2020 [Federal Strategic Plan for Sustainable Tourism: Tourism 2020]* (Ministerio de Turismo y Deportes, 2011). Retrieved 18 Jan 2023, from https://issuu.com/asap1/docs/planfederal_estrategico_deturismosustentable#:~:text=Las%20metas%20deben%20orientarse%20tanto,hacia%20un%20desarrollo%20tur%C3%ADstico%20sustentable

Argentine National Touristry Ministry, *Plan federal estratégico de turismo sustentable: Turismo 2025 [Federal Strategic Plan for Sustainable Tourism: Tourism 2025]* (Ministerio de Turismo y Deportes, 2015). Retrieved 18 Jan 2023, from https://www.mininterior.gov.ar/planificacion/pdf/Plan-Federal-Estrategico-Turismo-Sustentable-2025.pdf

P. Bachmann-Vargas, C.S.A. van Koppen, Disentangling environmental and development discourses in a peripheral spatial context: The case of the Aysén Region, Patagonia, Chile. J. Environ. Dev. **29**(3), 366–390 (2020). https://doi.org/10.1177/1070496520937041

E.M. Barberia, *Los dueños de la tierra en la Patagonia Austral [Landowners in Southern Patagonia]* (UNPA, 1995)

L. Barbosa e Souza, E.F. Chaveiro, Território, ambiente e modos de vida: Conflitos entre o agronegócio e a Comunidade Quilombola de Morro de São João, Tocantins [Territory, environment and ways of life: Conflicts between agribusiness and the Quilombola Community of Morro de São João, Tocantins]. Soc. Nat. **31**, 10.14393/sn-v31n1-2019-42482 (2019)

J. Bascopé, *En un área de tránsito polar: Desde el establecimiento de líneas regulares de vapores por el estrecho de Magallanes (1872) hasta la apertura del canal de Panamá (1914) [In an Area of Polar Transit: From the Establishment of Regular Steamship Lines Through the Strait of Magellan (1872) to the Opening of the Panama Canal (1914)]* (CoLibris, 2018)

H. Blair, K. Bosak, T. Gale, Protected areas, tourism, and rural transition in Aysén, Chile. Sustainability (Switzerland) **11**(24), 1–22 (2019). https://doi.org/10.3390/su11247087

B. Borrie, T. Gale, K. Bosak, Privately protected areas in increasingly turbulent social contexts: Strategic roles, extent, and governance. J. Sustain. Tour. **30**(11), 2631–2648 (2020). https://doi.org/10.1080/09669582.2020.1845709

M. Burton, Social reproduction, in *Encyclopedia of Critical Psychology*, ed. by T. Teo, (Springer, New York, 2014), pp. 1802–1804

N. Castree, The nature of produced nature: Materiality and knowledge construction in Marxism. Antipode **27**(1), 12–48 (1995). https://doi.org/10.1111/j.1467-8330.1995.tb00260.x

Chilean National Directorate of Borders and Boundaries, *Integración fronteriza [Border Integration]* (2022). Retrieved 13 Oct 2022, from https://difrol.gob.cl/

Chilean National Forestry Corporation, *Resolución N°450 [Resolution N°450]* (Corporación Nacional Forestal, 2019)

Chilean National Register of Protected Areas, *Búsqueda en áreas protegidas [Search for Protected Areas]* (2020). Retrieved 1 Nov 2021, from http://bdrnap.mma.gob.cl/buscador-rnap/#/busqueda?p=14

Chilean Subsecretary of Tourism, *Plan de acción para la gestión participativa de las Zonas de Interés Turístico (ZOIT), Zona de Interés Turístico Provincia de los Glaciares [Action Plan for the Participatory Management of the Zones of Tourist Interest (ZOIT), Glacier Province Zone of Tourism]* (2017). Retrieved 22 Jan 2023, from http://www.subturismo.gob.cl/wp-content/uploads/2015/10/PLAN-ACCI%C3%93N-PROVINCIA-DE-LOS-GLACIARES.pdf

Chilean Subsecretary of Tourism, *¿Qué es una ZOIT? [What Is a ZOIT?]* (2020). Retrieved 1 Feb 2023, from www.subturismo.gob.cl/zoit/

Convention on International Trade in Endangered Species of Wild Fauna and Flora, *Appendices I, II, and III of the Convention on International Trade in Endangered Species of Wild Fauna and Flora (CITES)* (2015). Retrieved 25 Jan 2023, from https://cites.org/sites/default/files/eng/app/2023/E-Appendices-2023-01-11.pdf

Convention on the Conservation of Migratory Species of Wild Animals, *Appendices I and II of the Convention on the Conservation of Migratory Species of Wild Animals (CMS)* (2015). Retrieved 26 Jan 2023, from http://www.cms.int/en/species

F. Coronato, *Ovejas y ovejeros en la Patagonia [Sheep and Sheep Farmers in Patagonia]* (Prometeo, 2017)

C. Corson, K.I. MacDonald, Enclosing the global commons: The convention on biological diversity and green grabbing. J. Peasant Stud. **39**(2), 263–283 (2012). https://doi.org/10.1080/03066150.2012.664138

C.L. De la Rosa, *Acuerdo sobre los hielos continentales [Continental Ice Agreement]* (Ediciones Jurídicas Cuyo, 1998)

M. Foucher, *Fronts et frontières. Un tour du monde géopolitique [Frontiers and Borders. A Geopolitical World Tour]* (Fayard, 1991)

C. Fuchs, Henri Lefebvre's theory of the production of space and the critical theory of communication. Commun. Theory **29**(2), 129–150 (2019). https://doi.org/10.1093/ct/qty025

T. Gale, K. Bosak, L. Caplins, Moving beyond tourists' concepts of authenticity: Place-based tourism differentiation within rural zones of Chilean Patagonia. J. Tour. Cult. Chang. **11**(4), 264–286 (2013). https://doi.org/10.1080/14766825.2013.851201

A. Glade Carreño, *Libro rojo de los vertebrados terrestres de Chile, 2nd edn. [Red Book of Terrestrial Vertebrates of Chile, 2nd edn.].* Proceedings of the Symposium on the Conservation Status of the Terrestrial Vertebrate Fauna of Chile. Corporación Nacional Forestal, April 21–24, 1987, Santiago, Chile (1993)

J.C. Goodman, R.L. Stroup, *Progressive Environmentalism: A Pro-human, Pro-science, Pro-free Enterprise Agenda for Change* (The Mackinac Center for Public Policy, 1991). Retrieved 18 Jan 2023, from https://www.mackinac.org/S1991-03

P. Grenier, *Des tyrannosaures dans le paradis. La ruée des transnationales sur la Patagonie chilienne [Tyrannosaurs in Paradise. The Rush of Transnationals on Chilean Patagonia]* (L'Atalante, 2003)

H. Gumuchian, B. Pecqueur, *La ressource territoriale [The Territorial Resource]* (Economica, 2007)

Honorable Chamber of Deputies of the Province of Santa Cruz of Argentina, *Crea Reserva Provincial "Lago del Desierto" [Creation of "Lago del Desierto" Provincial Reserve].* Provincial Law No. 2820. Río Gallegos, Argentina (2005). Retrieved 1 Feb 2023, from http://www.saij.gob.ar/LPZ0002820

Honorable Congress of the Argentine Nation, *Presupuestos mínimos de protección ambiental de los bosques nativos. Buenos Aires, Argentina [Minimum Budgets for the Environmental Protection of Native Forests. Buenos Aires, Argentina].* Law N°26.331 (2017). Retrieved 1 Nov 2021, from http://servicios.infoleg.gob.ar/infolegInternet/anexos/135000-139999/136125/norma.htm

M. Kühn, Peripheralization: Theoretical concepts explaining socio-spatial inequalities. Eur. Plan. Stud. **23**(2), 367–378 (2015). https://doi.org/10.1080/09654313.2013.862518

M. Kühn, M. Bernt, L. Colini, Power, politics and peripheralization: Two eastern German cities. Eur. Urban Reg. Stud. **24**(3), 258–273 (2017). https://doi.org/10.1177/0969776416637207

R. Lajarge, E. Roux, Ressource, projet, territoire: Le travail continu des intentionnalités [Resource, project, territory: The continuous work of intentions], in *La ressource territoriale [The Territorial Resource]*, ed. by H. Gumuchian, B. Pecqueur, (Economica, 2007), pp. 133–146

A. Latta, B.E.C. Aguayo, Testing the limits neoliberal ecologies from Pinochet to Bachelet. Lat. Am. Perspect. **39**(185), 163–180 (2012). https://doi.org/10.1177/0094582X12439050

H. Lefebvre, *La producción del espacio [The Production of Space]* (E. Martínez Trans.). (Capitán, Madrid, 2013). Original work published 1974

Legislative Branch of the Province of Santa Cruz, Argentina, *Ruta escénica "De la soberanía nacional en la senda del huemul" a la ruta Provincial N° 41 [Scenic route "Of the national sovereignty on the huemul trail" to Provincial Route N° 41].* Law No. 3541. Río Gallegos, Argentina (2017)

D. Massey, Philosophy and politics of spatiality: Some considerations. The hettner-lecture in human geography. Geogr. Z. **87**(1), 1–12 (1999)

F. Miranda Cabaña, Políticas del estado y la incorporación de espacios en la geografía del capitalismo: El caso de Patagonia Aysén [State policies and the incorporation of spaces in the geography of capitalism: The case of Patagonia Aysén]. Bol. Electrón. de Geogr. (BeGEO) **4**, 50–70 (2016)

J. Moore, *El capitalismo en la trama de la vida. Ecología y acumulación de capital [Capitalism in the Web of Life. Ecology and Capital Accumulation]* (Traficantes de sueños, 2020)

Municipality of O'Higgins, Municipio de O'Higgins y Minvu trabajan para levantar poblado en sector Candelario Mancilla [Municipality of O'Higgins and Minvu work to raise population in Candelario Mancilla sector]. *Ilustre Municipalidad de O'Higgins.* http://www.municipalidadohiggins.cl/web/2016/03/06/municipio-de-ohiggins-y-minvu-trabajan-para-levantar-poblado-en-sector-candelario-mancilla/#:~:text=%E2%80%9CEstuvimos%20con%20funcionarios%20del%20Minvu,comercio%20y%20centro%20de%20salud (2016, March 6)

M.L. Navarro Trujillo, L. Linsalata, Capitaloceno, luchas por lo común y disputas por otros términos de interdependencia en el tejido de la vida. Reflexiones desde América Latina [Capitalocene, struggles for the commons and disputes for other terms of interdependence in the fabric of life. Reflections from Latin America]. Relac. Int. **46**, 81–98 (2021). https://doi. org/10.15366/relacionesinternacionales2021.46.005

A. Núñez, E. Aliste, Á. Bello, El discurso del desarrollo en Patagonia Aysén: La conservación y la protección de la naturaleza como dispositivos de una renovada colonización. Chile. Siglos XX–XXI [The development discourse in Patagonia Aysén: Conservation and nature protection as devices of a renewed colonization. Chile. Siglos XX–XXI]. Scr. Nova **18**(493), 1–12 (2014)

A. Núñez, M. Galarce, E. Aliste, Geografías de lo sublime y el proceso de turistificación de Aysén-Patagonia. Turismo, territorio y poder [Geographies of the sublime and the process of touris-tification of Aysén-Patagonia. Tourism, territory and power], in *Araucanía-Norpatagonia II: La fluidez, lo disruptivo y el sentido de la frontera [Araucanía-Northpatagonia II: Fluidity, the Disruptive and the Sense of Frontier]*, ed. by P. Núñez, A. Núñez, M. Tamagnini, (UNRN, 2018), pp. 249–269

A. Núñez, M.C. Benwell, E. Aliste, Interrogating green discourses in Patagonia-Aysén (Chile): Green grabbing and eco-extractivism as a new strategy of capitalism? Geogr. Rev. **112**(5), 688–706 (2020). https://doi.org/10.1080/00167428.2020.1798764

J. O'Connor, *Causas naturales. Ensayos de marxismo ecológico [Natural Causes. Essays on Ecological Marxism]* (Siglo Veintiuno Editores, 2001)

C. Pereira Carneiro, *Fronteiras irmãs. Transfronteirizações na bacia do Prata [Sister Borders. Crossing Borders in the La Plata Basin]* (UFRGS, 2016)

S. Pérez, Geopolíticas del cotidiano en la frontera patagónica: Las dinámicas del habitar en torno a los campos de hielo patagónicos [Geopolitics of the everyday in the Patagonian frontier: The dynamics of inhabiting around the Patagonian ice fields]. Rev. Austral de Cienc. Soc. **37**, 185–205 (2019). https://doi.org/10.4206/rev.austral.cienc.soc.2019.n37-11

S. Picone, Áreas protegidas e acesso à terra: O caso de El Chaltén no Parque Nacional Los Glaciares (Patagônia Argentina) [Protected areas and land access: The case of El Chaltén in Los Glaciares National Park (Patagonia Argentina)]. Confins **47**, 1–16 (2020). https://doi. org/10.4000/confins.32276

S.E. Picone, I.J. Liscovsky, A.F. Schweitzer, Territories for conservation? Capitalist strategies for appropriating nature in Los Glaciares National Park in the Argentinean Patagonia, in *Socio-environmental Regimes and Local Visions: Transdisciplinary Experiences in Latin America*, ed. by M. Arce Ibarra, M.R. Parra Vázquez, E. Bello Baltazar, L. Gomes de Araujo, (Springer International Publishing, Cham, 2020), pp. 241–252

S.E. Picone, I.J. Liscovsky, A.F. Schweitzer, Entre las fronteras del turismo sustentable: Desigualdades, tensiones y continuidades en la producción espacial de El Chaltén (Santa Cruz, Argentina) [Between the frontiers of sustainable tourism: Inequalities, tensions and continu-ities in the spatial production of El Chaltén (Santa Cruz, Argentina)]. Rev. Pueblos y Front. Digit. **17**, 28 (2022). https://doi.org/10.22201/cimsur.18704115e.2022.v17.566

B. Ponzi, Han tomado la parte del fondo: La territorialización del Parque Nacional Patagonia, Cruz (Argentina) [They have taken over the back part: The territorialization of the Patagonia National Park, Santa Cruz (Argentina)]. Ambientes: Rev. de Geogr. e Ecol. Polít. **2**(1), 228–269 (2020). https://doi.org/10.48075/amb.v2i1.24284

B. Ponzi, ¿Dónde los guanacos se pasean crudos? Las ideas de naturaleza en parques naciona-les [Where do guanacos roam wild? Ideas of nature in national parks]. Geogr. Bull. **44**(1), 103–127 (2022)

J. Porto, A. Schweitzer, *Estrategias territoriales para la ocupación del continente sudameri-cano: Inserción de la periferia e institucionalización espacial [Territorial Strategies for the Occupation of the South American Continent: Insertion of the Periphery and Spatial Institutionalization]* (UNIFAP, 2018)

Rewilding Chile Foundation, *La Ruta de los Parques. Quiénes somos [The Parks Route. Who We Are]* (2018). Retrieved 9 Sept 2022, from https://www.rutadelosparques.org/quienes-somos/

Rosenfeld and Sekulovic Ltda, *Guía de manejo del sector norte del Parque Nacional Bernardo O'Higgins. Versión final [Management Guide for the Northern Sector of Bernardo O'Higgins National Park. Final version]* (2000). CORFO-CONAF XII Región. Retrieved 30 Jan 2023, from https://bibliotecadigital.ciren.cl/bitstream/handle/20.500.13082/28969/1382465907PNB ernardoOhiggins.pdf?sequence=1&isAllowed=y

A. Schweitzer, Viejas fronteras y nuevas regiones al interior del MERCOSUR [Old borders and new regions within MERCOSUR], in *El espacio en la cultura latinoamericana [Space in Latin American Culture]*, ed. by B. Lisocka-Jaegermann, (Center for Latin American Studies (CESLA), University of Warsaw, 1998), pp. 131–147

A. Schweitzer, *Aménagement du territoire et intégration régionale dans le MERCOSUR: Frontières, réseaux et dynamiques transfrontalières [Regional Planning and Integration in MERCOSUR: Borders, Networks and Cross-Border Dynamics]* (ANRT, 2000)

A. Schweitzer, Territorios en proyecto en la Patagonia Austral [Projected territories in Southern Patagonia], in *Dinámicas mundiales, integración regional y patrimonio en espacios periféricos. Hacia un plan de desarrollo para la Patagonia Austral [Global Dynamics, Regional Integration and Heritage in Peripheral Spaces. Towards a Development Plan for Southern Patagonia]*, ed. by R. Zarate, L. Artesi, (UNPA, 2004), pp. 48–61

A. Schweitzer, Desarrollo territorial y ajuste espacial. La difícil relación entre políticas públicas y planificación "privada" en el norte de Santa Cruz [Territorial development and spatial adjustment. The difficult relationship between public policies and 'private' planning in northern Santa Cruz]. Rev. de Estud. Reg. y Merc. de Trab. **4**, 201–216 (2008)

A. Schweitzer, Fronteras de la minería en la Patagonia sur Argentina [Mining frontiers in southern Patagonia Argentina]. Rev. Pós Ciênc. Soc. **16**(32), 145–166 (2019). https://doi.org/10.18764/2236-9473.v16n32p145-166

A. Schweitzer, Territórios cercados, territórios esvaziados e conservação da natureza no oeste da província de Santa Cruz, Patagônia Sul [Fenced territories, emptied territories and nature conservation in western Santa Cruz province, southern Patagonia]. Confins **47** (2020). https://doi.org/10.4000/confins.32551

P. Segura, Jugando al Monopoly verde en la Patagonia: Conservación y democracia ¿una posible combinación? (3° parte) [Playing green Monopoly in Patagonia: Conservation and democracy: A possible combination (Part 3)]. Patagon J. https://www.patagonjournal.com/index.php?option=com_content&view=article&id=4393%3Atercera-parte-jugando-al-monopoly-verde-en-la-patagonia&catid=187%3Aguest-blog&Itemid=340&lang=es (2022, March 15)

N. Smith, *El desarrollo desigual: Naturaleza, capital y producción del espacio* (L. F. Téllez, Trans.) [*Uneven Development: Nature, Capital, and the Production of Space* (L.F. Téllez, Trans.)]. (Traficantes de sueños, Madrid, 2020). Original work published 2008

R.E. Stake, Case studies, in *Strategies of Qualitative Inquiry*, ed. by N.K. Denzin, Y.S. Lincoln, 2nd edn., (SAGE Publications Inc, Thousand Oaks, 2003), pp. 134–164

Tehuelche Noticias, Trabajo conjunto entre Minvu y el municipio busca concretar nuevo poblado en Candelario Mansilla [Joint work between Minvu and the municipality seeks to build a new town in Candelario Mansilla]. Tehuelche Noticias (2022, December 21). Retrieved 2 Feb 2023, from https://tehuelchenoticias.cl/nuevo_sitio/2020/12/21/trabajo-conjunto-entre-minvu-y-el-municipio-busca-concretar-nuevo-poblado-en-candelario-mansilla/

L. Torres, S. Moreno, D. Pessolano, La reproducción social como eje analítico a lo largo del tiempo [Social reproduction as an analytical axis over time], in *Ventanas sobre el territorio. Herramientas teóricas para comprender las tierras secas [Windows on the Territory. Theoretical Tools for Understanding Drylands]*, ed. by L. Torres, E. Abraham, G. Pastor, (EDIUNC, 2014), pp. 31–115

United Nations Educational, Scientific and Cultural Organization, *Letter to the Argentine Permanent Delegate of UNESCO*. Argentine National Park Administration File 1888/84 (1980)

H.C. Valenzuela, J.B. Sáez, C.C. Becker, Neoliberal economic, social, and spatial restructuring: Valparaíso and its agricultural hinterland. Urban Plan. **6**(3), 69–89 (2021). https://doi.org/10.17645/up.v6i3.4242

Chapter 8
Beyond the Border: Understanding Freshwater Resources, Shared Identity, and Transboundary Cooperation in Southern Patagonia

Sanober R. Mirza, Fiorella Repetto Giavelli, and Jennifer M. Thomsen

Abstract Rapidly increasing environmental threats that transcend political borders have highlighted the need for collaborative approaches to conservation that extend beyond protected areas. Transboundary conservation operates across political and spatial scales by involving two or more countries cooperating to conserve a border resource or ecosystem. Though the recognition of transboundary conservation is growing, there is limited understanding of the key factors that can support these initiatives or impede them from achieving their goals. This study focused on the Southern Patagonian Ice Field, shared between Chile and Argentina. To gain a greater understanding of transboundary conservation within this landscape, we conducted a case study using semi-structured interviews to explore stakeholder perspectives on key factors influencing transboundary collaboration. The findings from this project have underscored the role of freshwater resources in disputed, transboundary landscapes. Second, local community collaboration, rooted in shared identity, was the basis of existing transboundary collaboration in Southern Patagonia, demonstrating the need to emphasize the local scale in transboundary initiatives. Also, a need was identified for more meaningful engagement and inclusion of both local and Indigenous communities in this transboundary landscape.

Keywords Patagonia · Transboundary conservation · Freshwater resources · Shared identity · Transboundary collaboration

---

S. R. Mirza (✉) · J. M. Thomsen
College of Forestry and Conservation, Department of Society and Conservation, University of Montana, Missoula, MT, USA
e-mail: srmirza@uwalumni.com; jennifer.thomsen@umontana.edu

F. Repetto Giavelli
Torres del Paine Legacy Fund,
Puerto Natales, Magallanes & Chilean Antarctica Region, Chile
e-mail: fiorella@supporttdp.org

© The Author(s) 2023
T. Gale-Detrich et al. (eds.), *Tourism and Conservation-based Development in the Periphery*, Natural and Social Sciences of Patagonia,
https://doi.org/10.1007/978-3-031-38048-8_8

8.1 Introduction

Transboundary conservation is an effective way to preserve shared ecosystems that transcend borders to bolster conservation outcomes in the face of drastic environmental change (Thomsen & Caplow, 2017; Vasilijević et al. 2015). The International Union for the Conservation of Nature (IUCN) defines transboundary conservation as the "process of cooperation to achieve conservation goals across one or more international boundaries" (Vasilijević et al. 2015, p. xi). Transboundary conservation initiatives have grown in popularity in the last two decades with increasing global opportunities (Mason et al. 2020). However, the planning and implementation of both formal and informal transboundary conservation have their challenges and weaknesses (Vasilijević et al. 2015), due to the complex nature of shared governance with varying institutions, jurisdictions, and local communities.

Transboundary landscape conservation is often complex and involves broadening the scale of traditional boundaries of conservation, such as protected areas, by integrating human and social aspects into the conservation area (Sayer 2009). In an effort to support transboundary initiatives, the IUCN's World Commission on Protected Areas (WCPA) developed guidelines for translating transboundary principles into practice (Vasilijević et al. 2015). However, our understanding of the diverse factors and actors that influence transboundary border reconciliation remains largely underdeveloped (Ide 2018). To address these needs, this study investigates the following research question: What factors influence transboundary conservation in Patagonia?

8.1.1 Transboundary Conservation in Southern Patagonia

South America has high potential for transboundary cooperation due to its vast biodiversity and endemic species, particularly in the Andes and along the Chilean-Argentinian border (Mason et al. 2020; Thornton et al. 2020). However, environmental resource conflicts are prevalent within both countries due to economic pressures (Larocque 2020) and Chile and Argentina share a history of disputed territorial claims, particularly in the Patagonia region (Perry 1980). This background still influences modern relations between the countries and the potential for transboundary cooperation (van Aert 2016).

The Patagonia region encompasses one of the most remote areas shared between Chile and Argentina, which is home to the Southern Patagonian Ice Field. This study focuses on the Torres del Paine, Bernardo O'Higgins, and Los Glaciares protected areas in southern Patagonia. These three national parks provide an essential connectivity corridor for conservation and include the Southern Patagonian Ice Field, a crucial water resource for both countries with a still unmarked border (Fig. 8.1). While there is no formalized transboundary conservation area, a study has evaluated adaptation strategies in the area's internationally adjoining protected areas (Solórzano 2016).

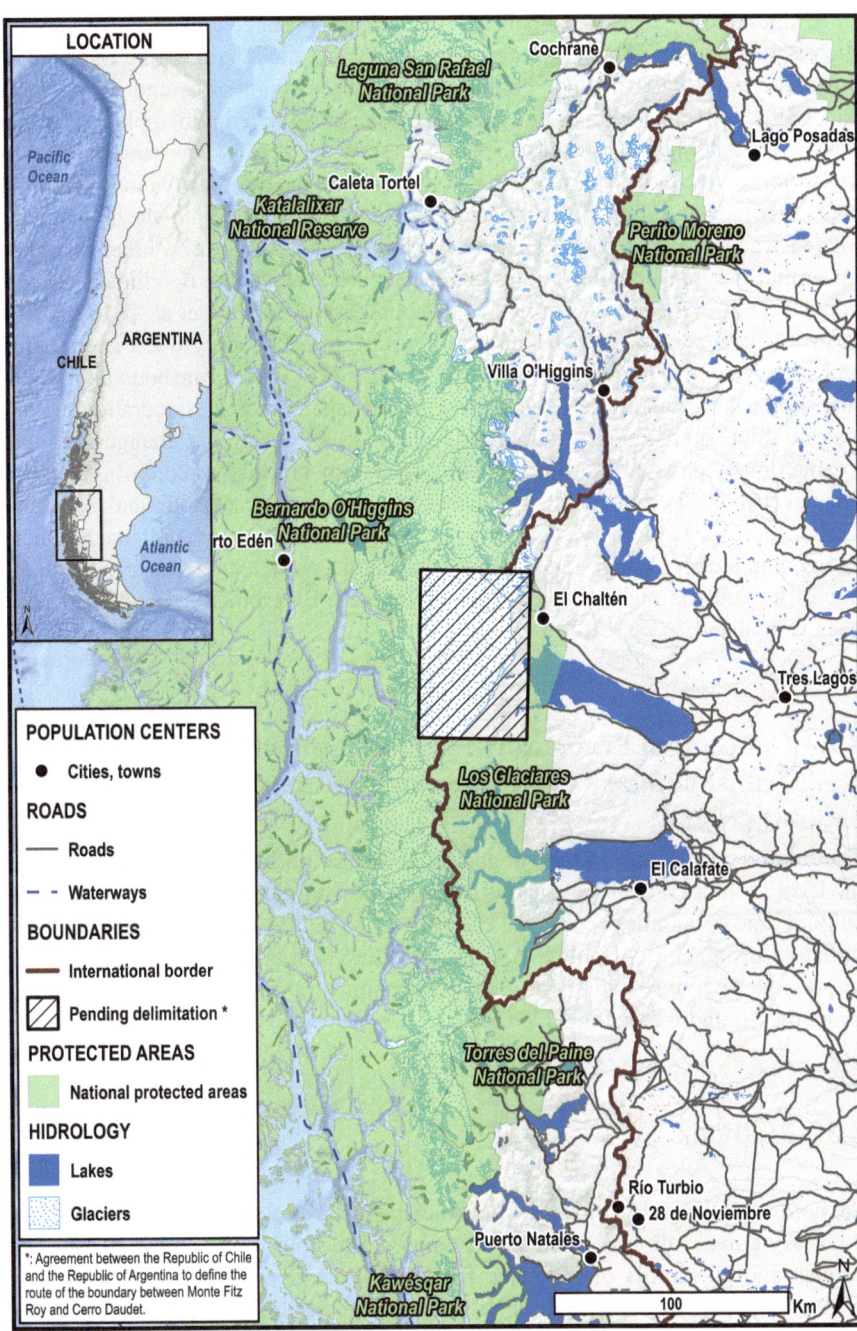

Fig. 8.1 Study area, including the Southern Patagonian Ice Field and Torres del Paine, Bernardo O'Higgins, and Los Glaciares National Parks

A key challenge for transboundary landscapes is shared freshwater resources, such as those of the Southern Patagonian Ice Field. Challenges facing these transboundary waterscapes include competition among resource uses and users, direct and indirect effects of climate change, and the tension of overlaying political boundaries with ecological ones (Zeitoun et al. 2013). Water scarcity has become increasingly more acute both on regional and global scales in the twentieth century (Kummu et al. 2016), positioning opportunities for escalated conflict around water resources. Shared water resources between states makes this challenge even more complex, given that the transboundary water challenges are exacerbated by climate change, population growth, and the overlap of independent states (Guo et al. 2016). A lack of trust between countries can significantly affect transboundary water cooperation and further contribute to social issues (Biswas 2011). Even transboundary waterscapes that have institutions for facilitating management and cooperation experience conflict (Battistello Espíndola and Ribeiro 2020). As climate change continues to affect many parts of the world, South America specifically is increasingly at risk (Hagen et al. 2022), with direct implications for its natural ecosystems and resources. These challenges of transboundary freshwater resources are evident in the Southern Patagonian Ice Field shared by Chile and Argentina, especially because the delineation of the international border within the Southern Patagonian Ice Field has never been formalized.

8.2   Conceptual Framework: Factors Influencing Transboundary Collaboration and Conservation

The multi-scalar nature of transboundary conservation includes interactions from the local to international scale and is influenced by political, environmental, social-cultural, and economic factors that can support or inhibit successful outcomes and initiative sustainability. Table 8.1 summarizes key factors that have been identified in previous literature and serve as the conceptual framework for this case study about the Southern Patagonian Ice Field.

8.3   Methods

We used a qualitative methodology to understand stakeholder perspectives on transboundary conservation in Southern Patagonia. The field work took place between August and November of 2019 and included 40 semi-structured interviews with managers, conservation practitioners, and local community members associated with Torres del Paine and Bernardo O'Higgins National Parks in Chile and Los Glaciares National Park in Argentina. A chain referral method (Noy 2008) was applied to identify participants beyond a core group identified by the researchers

Table 8.1 Supporting and inhibiting factors of transboundary collaboration and conservation

	Supporting factor	Inhibiting factor
Political	Formal agreements and processes (Chettri et al. 2007; Vasilijević et al. 2015) State-level support (Mackelworth 2012)	Historical and ongoing tensions between states (Portman and Teff-Seker 2017; Taggart-Hodge and Schoon 2016) Differences in political systems, laws, and agendas (Taggart-Hodge and Schoon 2016; Vasilijević et al. 2015) Security and border issues (Portman and Teff-Seker 2017; Vasilijević et al. 2015)
Environmental	Shared, charismatic species (Knight et al. 2011; Lambertucci et al. 2014; Mason et al. 2020) Shared goals and vision (Portman and Teff-Seker 2017)	Invasive species (Jaksic et al. 2002) Climate change and resource challenges (Zeitoun et al. 2013) Industry threats (Mackelworth et al. 2013)
Economic	Long-term, sustainable funding (Beever et al. 2014; Mackelworth 2012) Community-oriented governance of tourism (Chiutsi and Saarinen 2017) Stakeholder economic profitability (Portman and Teff-Seker 2017)	Imbalanced financial resources between countries (Portman and Teff-Seker 2017) Competition and national interests in tourism (Ioannides et al. 2006) Corruption and lack of transparency (Chiutsi and Saarinen 2017)

and their contacts. Interviews were conducted until thematic saturation was reached (Hennink et al. 2017). Table 8.2 provides background on the stakeholders interviewed with 24 interviews conducted in Chile and 16 interviews conducted in Argentina. The presence of two protected areas of interest in Chile influenced the higher number of Chilean participants. Interviews were mainly conducted in Spanish, with only four interviews conducted in English.

This project included a few participants who are members of some of the Kawesqar communities within the region. The findings are not a comprehensive account of Indigenous perspectives, or Kawesqar perspectives. However, perspectives from members of these communities were sought in particular because historical and continued oppression of Indigenous peoples is central to many landscapes, including Patagonia (Gasteyer and Flora 2000). Before the conquest of southern Patagonia led by European immigrants, there were at least four traditional Indigenous communities in the area: the Selknam, Kawesqar, Yaganes, and the Aonikenk. While it is uncertain how many descendants of these communities remain, there is still an Indigenous presence within the landscape. Indigenous involvement and management, allowing for agency over culture, land, and livelihoods, are possible and important for transboundary conservation in Patagonia (Sepúlveda and Guyot 2016).

The interview questions were developed from key understandings of transboundary conservation identified in the 2015 IUCN Transboundary conservation guidelines (Vasilijević et al. 2015). Participants were asked to discuss (a) their own experience with (and knowledge of) transboundary conservation in the area; (b) challenges they believed were evident to achieving transboundary collaboration; (c) what they thought was necessary for establishing more transboundary conservation;

Table 8.2 Interviews conducted in Chile and Argentina with stakeholder groups

Stakeholder group	Reasoning for inclusion	Chile interviews	Argentina interviews
Tourism worker or guide	Immense amount of tourism in the region	5	5
Protected area/ National Parks employees	National parks central to landscape and border area	5	6
Government and ministry employees	To discuss any previous transboundary collaborations and government views on binational collaboration	3	0[a]
Community and citizen leaders	Insight into what transboundary means for local communities now and in the future	3	1
Conservation organization employees	Provide insight on local conservation goals and efforts	5	4
Indigenous community members	Understand Indigenous representation in current context	3	0[b]

[a]The researcher was unable to contact any government representatives in Buenos Aires, most likely due to the transition in the government that was taking place after a presidential election in October 2019 in Argentina
[b]With limited time, the researcher was not able to make adequate efforts to contact Indigenous communities and their descendants in Argentina

(d) whether they believed there was even potential for transboundary collaboration; (e) how they thought transboundary conservation would affect them and their work; (f) the role the government and communities should play in transboundary conservation; and (g) what they would view as successes in transboundary conservation. Three sets of interview questions were used, with language that varied slightly depending on which type of stakeholder the participant was. There was one set of questions for local community stakeholders, one set for government and ministry representatives, and one for APN and CONAF employees. The questions aimed to allow participants to identify and detail key factors that support or inhibit transboundary conservation in the local landscape, allowing for comparison with those identified in the literature and compiled by the study authors as a conceptual framework of transboundary collaboration and conservation (Table 8.1).

All interviews were recorded, translated, and transcribed. Interview transcripts were coded using NVivo qualitative data software. Open and axial coding was used to categorize themes from interviews (Böhm 2004). The first round of coding emphasized a comprehensive collection of themes categorized into broad nodes. To support intercoder reliability, initial coding was conducted by several members of the research team to discuss thematic nodes and collectively develop a coding scheme for the remaining part of the data analysis. During the second round of coding, existing thematic nodes were organized into second and third-level nodes that had more specificity in content.

8.4   Results

In interviews with local stakeholders, all factors listed in Table 8.1 were identified by participants as relevant to transboundary conservation in the Southern Patagonian landscape, with the exception of language barriers. Many of the supporting factors, including state-level support and participatory local community engagement and management, were noted as absent by participants. In addition to factors identified in Table 8.1, other key factors particular to the landscape emerged from this study. Participants identified knowledge exchange as a supporting factor for transboundary conservation, as well as existing local collaborative efforts. Several inhibiting factors were identified, including the presence of freshwater resources, a lack of regional conservation standards, over-tourism in Patagonia, political turnover, and the undefined border, which is unique to this transboundary landscape. Select results as well as interview quotes and excerpts are highlighted in the section below.

8.4.1   Environmental Factors Influencing Transboundary Collaboration

8.4.1.1   Ecological Factors in the Shared Patagonia Landscape

Participants emphasized how the Southern Patagonia Ice Field, a freshwater reserve, are at the heart of the shared Patagonian landscape. Many of the participants shared their beliefs that transboundary collaborations are needed to protect and conserve the region's ecology, and specifically the freshwater reserve. For example, a Chilean Nacional Forestry Corporation (CONAF) employee noted: "Yes, ecosystems are the same, they know no borders. So, we should establish more work [together]". A local Chilean citizen leader noted, "I believe first in the planning... it would make sense to me for protection of the water resource, which belongs to both countries and the world."

One Argentina conservation organization employee noted the connectedness of the shared landscape in contrast to the dissonance of political management,

> Ecologically, it [the landscape] is the same...we have situation A,B, and a line through the middle...You have the same flora, the same fauna, the same water, the same everything...the agenda that one [country] sets, marks the fate of an area that is much more encompassing than the little dot you have to govern.

However, the area is associated with a history of complicated water politics between the two countries that include controversial damming and extraction projects. A Chilean citizen leader discussed the region's past examples of water conflicts,

> Many years ago, they wanted to install dams, five mega-dams, and there were many years of local fighting. We were involved with almost all of the [Patagonia] regions in the fight so that they did not construct the mega-dams.

Several participants highlighted the need for connectivity in the landscape. One CONAF employee noted, "I actually believe personally that this [is] more [of a] global vision, independent of specific work, the view should be used in terms of preservation and conservation in the eco-regional level." One Chilean tour guide stated that the connectivity that could be established through cooperation would combat issues of fragmentation in management, "I think it would give more meaning to the ecosystem we have there in the sense that it would no longer be isolated islands, but that we would be a completely whole ecosystem." Overall, there was a strong awareness of the shared landscape and the importance of conserving it.

8.4.1.2 Combatting Transboundary Threats

Environmental threats can serve as a motivation for transboundary collaboration. Some of these threats include the spread of invasive species, fires, and pressures for extracting resources. Mitigating threats, particularly climate change, were emphasized by participants as a need for cooperation. For example, an Argentina National Parks (APN) employee stated, "no matter what, working as Argentina or Chile alone won't stop climate change." One major example is collaboration to combat fires and other emergency responses. One APN employee shared that he worked with people from Chile *only in emergency situations*. Another APN employee commented that, "…there has been collaboration when the fires are really big, the crew has traveled to combat fires from one side to the other, as many Chileans have come here as Argentinians have gone to Chile."

Conflict in Patagonia is also shaped by the role that extractive industries and foreign actors play in the economic development of the region. The threat of extractive industries was noted by participants on both sides of the border. A Chilean government employee discussed the role of salmon harvesting and how that affects the Patagonian landscape, "the salmon business in Chile, it is the worst of examples. It settled in Patagonia, in the south of Chile… and has a way of raising and cultivating, harvesting salmon [that is] very harmful for the environment." This extraction industry is extremely controversial with local businesses displaying signs stating that the salmon industry is not welcome in Patagonia (Fig. 8.2).

Another Chilean ministry worker discussed challenges in working with extractive industries,

> I have to sit down and talk to him and convince the mining industry that it cannot get and exploit everything it can think of, or that the forest industry cannot come and load three hectares of forest and then leave. You realize, with the economic logic that exists in Chile, it is difficult.

Participants also emphasized how information exchange and dialogue were integral to transboundary conservation. One APN employee shared what would be useful for improving their own work, "Collaboration and the exchange of experiences, of knowledge, personal exchanges." In addition, a conservation organization employee from Chile described this importance in terms of evaluating transboundary success,

Fig. 8.2 A flier opposing the salmon industry in a local restaurant window in Puerto Natales

"coming together to dialogue at a common table would be a success for me."
Knowledge exchange was desired by several participants, particularly those who
worked directly with CONAF and APN to manage the national parks.

The native fauna and flora were also at the center of discussion for any future
collaboration, reflecting shared species is an important motivator for collaboration
(Table 8.1). An APN employee emphasized shared knowledge of species' populations as a goal,

> To have a real knowledge of the animal populations that live on both sides of the bor-
> der…maybe the huemul, the more emblematic species, to know the dynamic of the huemul
> populations and to know the quantity of populations and real needs of the species and to be
> able to get specific conservation tools for it.

Conservation of the huemul, the endemic Southern Andean deer that is increasingly threatened due to environmental change, was prioritized by CONAF and APN participants. An APN employee summarizes this sentiment,

> Our challenges are continuing work on the huemul…the observations they have in Chile, the huemul are six kilometers from the border with Argentina and it is very important for us to work together to get those animals back to Argentina or to start to circulate between Argentina and Chile.

The huemul was not the only shared species that was mentioned in interviews. One Argentinian conservation organization employee gave a description of what transboundary communication can bring for the protection of the puma, another emblematic species of the region by noting economic and policy disparities,

> On that side [Chile] they have made a fortune watching the animal [puma] live and on this side, poverty paid for killing it so that it doesn't eat the sheep that is worth nothing and that is negative for nature. We are talking about the difference being a border. If that border were eliminated… because it is precisely two parks, this [Chilean] positive experience could be capitalized so that these people [Argentinians] have more money and when they have money they realize 'The puma serves me alive, not dead'.

While many participants focused on shared and charismatic species of the region, some pointed to the threat that invasive species have on the landscape. An Argentinian interview noted that there is limited collaboration around invasive species,

> …here [Los Glaciares National Park] we invited people from Torres del Paine…It was all our staff, people from the province of Santa Cruz and Torres del Paine, because we have some problems with the invasion of mink and in Chile [they have it] too.

Invasive species still provided an opportunity for transboundary collaboration in this landscape, rather than being an inhibiting factor (Table 8.1).

8.4.2 Socio-cultural Factors Influencing Transboundary Conservation

8.4.2.1 Shared Identity

Participants reflected on the shared social identity between Chileans and Argentinians in Patagonia. A Chilean community leader described the nature of shared backgrounds between Chileans and Argentinians in Patagonia,

> We are more [like] siblings…In the north, in Santiago for example, there is a lot of rivalry between Chilean and Argentinians, here there is a lot of family connections, also because many people live 30 km from Argentina…they have worked many years there and they came to live in Chile or the reverse. There is much more human connection.

One Argentinian guide further explained the importance of overcoming differences, "Border is something with history, but if we want to have a better world, we need to work all together. In that sense, transboundary is not only Chile and Argentina, but also worldwide. It is the only way."

Although a shared identity exists, there were still a few instances of bias against the other side coming through in interviews. Many participants lamented the lack of connection between the two sides. There have been challenges in establishing relationships across borders between protected area staff as explained by an APN employee,

> I think it [transboundary conservation] would be useful, but difficult at the moment because we do not have any connection. This was the first time I met some people that worked there and in fact I don't know the national parks on the other side. Nobody has gone, not for work or for personal travel.

A couple of participants who worked in protected areas inquired about the other country's protected areas to the researcher, demonstrating how little connection and collaboration there is between the two governments' national park services. A Chilean conservation organization employee explained why they believed there was this lack of connection,

> [Chile] in the south...has a better relationship with the people of Argentina than the rest of the country, but I feel that they [the governments] are not creating instances of working more collaboratively and something more concrete between countries... the intention is there, the will is there, but they have not created the space for it.

8.4.2.2 Engagement and Conflict with Local Communities

Local communities in Patagonia, which includes Indigenous communities, are engaged in ongoing efforts to preserve their land and cultures. On the Chilean side, locals who are members of the Kawesqar communities, located in and around the Bernardo O'Higgins National Park and Torres del Paine National Park, still remain fighting to preserve their cultures and people while pursuing recognition by the Chilean government. Given the history of and current oppression, many in the community do not trust the Chilean government. One member of a local Kawesqar community expressed how the thought of more involvement from national governments is nearly unfathomable given the ongoing fight for recognition of Indigenous territories,

> I insist the first thing is to recognize and respect Indigenous territories; the first big step. Chile has never done it.... Because today for example, within Indigenous territory, there is sovereignty of other countries...entrepreneurs from other countries have been granted perpetuity in the delivery of concessions and they are the owners of those spaces. That is very criminal, especially if these countries do not respect the legislation that is here today and do not respect Indigenous territories.

When asked if they would support transboundary collaboration, a member of a Kawesqar community stated, "I support conservation, not transboundary, but by Indigenous people." Social conflict and oppression of Indigenous communities, like the Kawesqar communities, are a critical factor for all actors to consider moving forward if there is to be change that can lead to equitable Indigenous participation in any collaboration. These participants highlighted that the current model of conservation by the Chilean government is not working in regard to their rights and recognition.

The vision for many stakeholders of transboundary collaboration involved ample community participation and representation. A Kawesqar community member emphasized the role of consultation and equal treatment of communities with those at the national level,

> I think that the first to be consulted are members of the peoples that are on their own territory. More than a consultation because the ones who decide alone are the governments and the governments must decide together with the peoples to choose the best form of conservation.

This lack of community participation goes beyond Indigenous communities to include the social legacies of past historical conflict within both countries. One local community leader in Chile described how the Chilean dictatorship under General Augusto Pinochet has affected how Chileans participate and perceive their own power,

> Chile has not had 50 years since the last dictatorship...Becoming an actively civic and empowered and informed community takes more than a generation. Now there is tremendous change, we are many, but those in power still do not have the concept of civic participation of the communities.

Local community members also expressed frustration at the loss of their identity in Patagonia. One Chilean tour guide noted that the decisions made in the capital city, especially in regard to authorizing private sector concessions in the park, do not always represent local interests or cultural contexts,

> An authority that lives in the center, in the capital, will never come to see how that is and when they decide, usually the place does not resonate with them....However, when things are decided by local communities, everyone feels part of it...we in Puerto Natales have been suffering for a long time that the park is a concession to private people coming from other places in Chile.

However, not all participants believed that there was no hope for locals. One local Chilean community leader briefly stated how communities in Patagonia have been reclaiming their power, "But you know that the citizens have the power, the municipality doesn't, and that is the difference."

8.4.3 Political Factors Influencing Transboundary Conservation

8.4.3.1 Border Claims for Southern Patagonian Ice Field Resources

Participants reflected on how the area's unique political geography affects the plausibility of effective transboundary conservation. The Southern Patagonian Ice Field border is at the heart of the Patagonia landscape and is a massive freshwater reserve. This area's full demarcation is still not formalized, as both countries claim a specific section of the border.

A Chilean ministry worker noted how this ongoing controversy impedes binational relations and collaboration,

> They [the Chilean government] avoid using or mentioning the word "binational" because with Argentina we have some places throughout the frontier that are still unresolved. One piece of the frontier is unresolved, especially near to the ice caps, the big ice caps. So, it [transboundary work] is a very controversial point.

This formalization and demarcation have been at a standstill since the border claims made by each country in 1998. Another Chilean government employee discussed some skepticism about the motivation for transboundary work,

> In what exists in the case of Chile and Argentina with the southern ice fields, I understand it is a controversy over demarcation...it would be very convenient for Chile and Argentina to declare all of this a protected area, because it is a reserve of water, etc., but that is a political problem.

There are complex political histories between Chile and Argentina. One local community leader in Chile discussed this in the context of freshwater resources,

> They are strategic places, and the state has to worry, not the citizens. It is the state of Chile that has to worry about the border situation... they are strategic places in the water issue, it has a lot of water, the situation of freshwater is going to be one of the biggest problems that we are going to have in the future.

However, not all participants agreed that the Southern Patagonia Ice Field is a challenge. One tour guide described the connection in the shared history of the Southern Patagonia Ice Field as more of an opportunity for collaboration,

> I think that there are more things that unite us than divide us. Including between Bernardo O'Higgins and Los Glaciares Park, just in the middle of the Southern Ice Fields, there is a political sector without a border...That already starts to say that the border does not divide us but binds us.

8.4.3.2 Political Turnover and Distrust

During the time of the study, there were socio-political changes within both countries. A Chilean CONAF employee described their perception on the socio-political situation in Argentina and how that affects both conservation and binational collaboration,

> Today the unemployment in Argentina, the inflation in Argentina are social problems that are concerns there...First the Argentines need to resolve their basic needs and employment issues..., they have many other problems that in the long run make them not concerned about conservation actions...That also hinders the work that you can do between countries.

Political turnover is another challenge to transboundary collaboration efforts as explained by a local community leader in Argentina, "we have just started another four-year process in which to talk about the bad stuff the previous one [government] did, then there are no plans of sociocultural models and socioeconomic models that go beyond more than 4 years." This turnover and change in governments significantly affect binational relations and the potential for transboundary collaboration. One APN employee described this effect in terms of political leadership,

> Many years ago, there was more rivalry between Argentina and Chile. Today, for many, that
> has disappeared and there is more [collaborative] work. There is closeness between the two
> countries. It depends a lot on each country's political alignment.

There is also significant distrust of national governments, given the political histories of the countries. More recently, this has been an issue in Chile that led to the citizen movement demanding a more fair and just government system. One member of a Kawesqar community explained some of their distrust towards the government,

> Until today, we have applied Chilean politics to be able to reclaim [territory]. The Chilean
> politics are so bad, but there are others that allow you to do things...I could not say if it
> [transboundary conservation] would be good or bad, but from my perspective, already hav-
> ing a government involved, it is bad, so having two [governments] will be much worse.

Similarly, several participants lamented the direction of the Chilean government. A local community leader from Chile described their perspective on recent changes in the government and its actions, "The state of Chile was weak, now with the government it is worse, because for this government the private side develops, takes over, puts up the money, and the state does not take responsibility."

In terms of binational political relations, there are concerns around how the two countries would negotiate based on concerns about sovereignty. One representative of a Kawesqar community stated, "Now thinking in...these three units, Torres del Paine, Bernardo O'Higgins, and Los Glaciares, as to the extent of the territory, it is a lot of land for two countries to agree to share, so the question is difficult." These sovereignty issues reflected the political histories of Chile and Argentina and concerns about future transboundary cooperation. One municipal government tourism employee from Chile described the potential for political backlash,

> I don't think it is easy, it isn't easy...you can see it today in Europe, you can see it with
> Brexit, they united around a great proposal, but they are already having serious difficulties
> and that has nothing to do with countries, it has to do with how we organize ourselves as
> communities.

Many stakeholders emphasized the need for an established political process. A local citizen leader in Chile stated, "So for it [transboundary conservation] to happen, it is political decision, nothing but political will must exist to make an agreement." The lack of political forums to work on transboundary conservation issues was evident in interviews with various stakeholder groups.

8.4.4   Economic Factors for Transboundary Conservation

8.4.4.1   Limited Resources and Management Capacity

Limited resources and funding for protected area management are major factors that can influence any future transboundary collaboration. One conservation organization employee discussed their role supporting CONAF in Chile, given its inadequate resources from the national government,

> I think a lot of people think of CONAF the wrong way and think they don't do a lot to protect the park and they [CONAF] really do and there are a lot of initiatives happening and for the very little funding they have...we are filling gaps because there is nobody there right now filling those gaps.

Overall, there was a perception from both sides of the border that CONAF does not have adequate resources to manage Torres del Paine, an extremely popular protected area. Same for the less-visited Bernardo O'Higgins National Park; there are very few CONAF staff working to manage the largest protected area in Chile.

In addition to the need for more sufficient funding, there is a stark need for better allocation of resources. One Chilean conservation organization employee highlighted this need, "I think the optimization of resources is something that the protected areas lack, both in Chile and Argentina." Some participants believed transboundary cooperation could provide a solution to the lack of resources that is faced by protected area staff on both sides of the border. A Chilean conservation organization employee described this vision for future cooperation,

> The resources that are there to manage the parks, as much as in Argentina as in Chile, are extremely scarce. [With transboundary collaboration], you can share park rangers, you can share funds to use for research, as long as the duties and rights of each one are fair, and the duties and rights are clear.

8.4.5 Tourism Factors in Transboundary Conservation

A focus on economic prosperity from tourism has made conservation more difficult as parks have shown signs of overcrowding and degradation due to visitors. Perceived over-tourism in the Patagonia region was a common sentiment among participants. One Chilean tourism employee in the town of Tortel on the edge of Bernardo O'Higgins National Park described this aversion to more tourism,

> ...we do not want massive tourism...With the arrival of the ferry, that comes once a week with 140 passengers, Tortel already collapses in January and February, we do not have sufficient capacity to receive all of the passengers, we still aim to keep the environment as it is for many years to come.

Similarly, another Chilean tourism employee described the concern about increased tourism with transboundary conservation,

> I imagine that if they made an agreement of this big park of Patagonia, the big park of glaciers in Patagonia, that is going to be widely announced, all of the world is going to want to come and what happens to conservation? In the end, I think that what is the most important is to conserve and to not transform [the parks] into a Disneyland.

Overall, high levels of tourism are a controversial point at the center of any discussions around binational collaboration for Torres del Paine and Los Glaciares National Parks. Patagonia is a major international destination for visitors from around the world. Many small towns have developed solely around the tourism market. This increase in tourism had led to the rapid development of towns like

Puerto Natales in Chile and El Calafate in Argentina. Both towns grew exponentially with the high growth of international visitors, and largely consist of lodging, tourism, and dining options.

Currently, there is shared Patagonia tourism between the two countries. Several participants noted that it is common for visitors to cross borders during their tourist visit. There has been some engagement for coordinated tourism between tour guides in the region. However, there are significant differences in tourism operations between the two countries. For example, Chile does not have requirements for tour guide training, while Argentina requires guides to be university-educated. In addition to studying in universities, guides in Argentina must also get specific certifications to work within national parks. While several Argentinian stakeholders described this difference as a detriment to collaborative work, some Chileans did not find it problematic.

8.5 Discussion

8.5.1 National and International Politics of Transboundary Water Resources

Participants strongly recognized a shared landscape and shared water resources that do not adhere to Patagonia's political borders. The role of water within the landscape was emphasized by participants, reflecting the importance and complexity of transboundary freshwater resources like the Southern Patagonia Ice Field. According to participants, the geopolitical border and sovereignty dispute is shaped around freshwater resources, mirroring historical classification of Chilean and Argentinian border conflicts (Child 1983). In addition to these conflicts, the shared value and concern for water as a resource reflects the risk of increasing water scarcity in South America (Hagen et al. 2022). One participant noted that the term "binational" in relation to this area was avoided by both governments, further underpinning the constraint that the Southern Patagonia Ice Field's freshwater resources can be in shaping any transboundary interaction on a formal level. This reflects the tension between managing transboundary landscapes and recognizing political pressures (Zeitoun et al. 2013).

In addition, a level of skepticism was expressed about the motivation behind either government participating in transboundary collaboration around the Southern Patagonia Ice Field, which reflected concerns about sovereignty and territorial claims in the name of resource conservation. This strong awareness of the political aspect of transboundary water resources was not just expressed by government representatives, but also by local actors. Local participants were sensitive to the long histories of border disputes in the region, different from previous claims of a lack of diplomatic awareness from lower-level actors in transboundary conservation (Büscher and Schoon 2009). As climate change threatens global water security, geopolitics may become increasingly tense as the Southern Patagonian Ice Field could influence transboundary cooperation and conflict.

International dynamics in water resources provide an extremely challenging context (Zeitoun et al. 2013), with overwhelming considerations of both sovereignty and resource scarcity. Our findings illuminate the challenges of political turnover and changes in administrative priorities that disrupt progress and increase the vulnerability of transboundary efforts. Transboundary water interactions need to be situated in a multidimensional political context (Zeitoun and Mirumachi 2008). Shared freshwater resources in this case study contributed to state actors' aversion to utilize the term "binational." The lack of state actors, processes, and structures in place also limited the support that stakeholders had in localized efforts of collaboration. On the other hand, the participation of state actors in the case of Southern Patagonia can fuel distrust given political histories of both countries that include US imperialism and military dictatorships.

8.5.2   Shared Identity and Local Collaboration

Despite the challenges discussed in previous sections, this study also found strong potential for transboundary conservation between Chile and Argentina around Torres del Paine, Bernardo O'Higgins, and Los Glaciares National Parks. Shared identity underpins many of the existing collaborative efforts at the local scale despite non-formalized partnerships and political uncertainty at the national scale. This shared Patagonian identity is rooted in being geographically and culturally unique and connected with the natural environment. Some participants stated that they feel more like neighbors with the other country's Patagonian inhabitants than their fellow countrymen who live in the major cities. A similar shared identity was found in another transboundary conservation case study along the Zimbabwe-Mozambique border, which highlighted historical linkages between communities as well as ethnic ties (Kachena and Spiegel 2019). Overall, our findings highlight the artificiality of borders, whether imposed by colonialism or modern nation-states, in not only delineating resources and ecosystems but also defining community.

The shared objectives of conserving landscapes and combating climate change that emerged in this study build on the already high potential for transboundary conservation along the Andean range (Mason et al. 2020). Many local stakeholders emphasized the importance of working together regardless of past histories and current challenges, suggesting that community connections are more resilient to changes than national relations given frequent political turnover and competing state interests. While national level support is integral to starting and formalizing transboundary conservation initiatives, integrating and empowering local communities are important to support and sustain local economic and conservation outcomes (Schoon 2013). Some local community stakeholders have regular, informal communication, while other examples are more formalized like the natural reserves network that hosts regular meetings for knowledge sharing and includes representatives throughout Patagonia. Similarly, local participants mentioned their collaboration on search and rescues on the Southern Patagonia Ice Field, fighting wildfires, and cross-border tourism. These examples mirror the emphasis for locally driven development initiatives in Patagonia (Blair et al. 2019).

Our findings also indicate local stakeholder distrust in the national government, given the participant responses that highlighted years of Indigenous oppression and exclusion of local voices in Patagonia. Some areas in both Chile and Argentina have experienced this tension with the government over its economic priorities with ranching, tourism, and extractive industries (Blair et al. 2019; Reboratti 2012). Locals are more willing to be supportive of conservation areas when involved in planning processes (Andrade and Rhodes 2012); yet there is a lack of community inclusion in the establishment of conservation and tourism in Patagonia (Blair et al. 2019), which was noted by participants in this study. The shared identity at the local scale provides an opportunity to continue to work beyond political borders and use local relations as the key foundation for conserving Patagonia landscapes despite disputes at the national and international scales. This finding is crucial for shaping transboundary conservation dialogues in the future, as political borders may seem divisive, but locally there is a shared ecological and cultural landscape that creates opportunities.

8.5.3   Bridging Local, National, and International Scales

There is evidence of power imbalances between stakeholders in Patagonia's water conflict (Reboratti 2012) and in transboundary water interactions around the world (Zeitoun et al. 2014). Local referendums have been suggested to resolve multi-stakeholder water conflicts within Argentina (Reboratti 2012). The Southern Patagonian Ice Field may present an opportunity for a multi-stakeholder, multi-national referendum for delineating and/or managing these shared resources. However, any action within the region of Patagonia must come with a critical awareness of the systems and processes that catalyze injustice within transboundary waterscapes (Zeitoun et al. 2014).

Conflict is further exacerbated within southern Patagonia around recognition, inclusion, respect, and rights of the region's Indigenous communities, specifically the various Kawesqar communities. While Indigenous perspectives were not sought as a central component of this study, the complicated relationship with Indigenous peoples in Patagonia emerged in interviews, highlighting the important need to continue to address local and Indigenous rights in Patagonia. This is particularly concerning considering that Indigenous communities and other marginalized groups will most severely feel the effects of climate change in South America (Hagen et al. 2022). Complicated relationships with local communities within conservation spaces are not unique to the Kawesqar communities in southern Patagonia but extends to other Indigenous groups and local peoples within Chile and Argentina (Sepúlveda and Guyot 2016) and around the world (Muboko 2017). Rights-based approaches can be incorporated into transboundary conservation, requiring

Indigenous self-determination and sovereignty (Hsiao and Le Billon 2021), which are noticeably inadequate in conservation efforts in the Southern Patagonian landscape despite the efforts that have been made by both government agencies and conservation organizations.

This study demonstrates the need to further operationalize transboundary water conservation and governance as a multi-stakeholder endeavor. Multi-scalar conceptualizations of both the social and environmental threads of transboundary landscapes is critical (Hsiao and Billon 2021). The results of this study highlight the fact that locals still are not adequately included or enabled in regard to both conservation initiatives and transboundary interactions in their landscapes. Transboundary conservation operates in complex geopolitical situations (Barquet 2015) and the people within the landscapes illustrate complex and changing relationships. Local community engagement is crucial to avoid further disenfranchising those who inhabit the threatened regions. Transboundary water governance must recognize and prioritize marginalized peoples' needs in its operation (Upadhyay 2020). State actors in transboundary collaboration and negotiation can inflame conflict within the landscape (Barquet 2015) and drown out the voices of those most affected by the initiative (Petursson et al. 2013). States are obligated to simultaneously ensure both ecological and social peace in transboundary conservation areas (Hsiao and Le Billon 2021); yet this study's results imply that both the Chilean and Argentinian states are not meeting this responsibility.

The 2015 IUCN Guidelines on Transboundary Conservation highlight the need to identify and consult stakeholders as well as determine who should convene the transboundary conservation effort (Vasilijević et al. 2015). This implies that the state(s) involved are not needed to be the lead convener. There are active and passionate local stakeholders throughout southern Patagonia who are convening and organizing transboundary conservation in the face of continued oppression and marginalization. Transboundary cooperation can be successful through localized efforts, even in the face of diplomatic and political tension at the international scale (Portman and Teff-Seker 2017). This research demonstrates that any transboundary cooperation must be inclusive and representative of communities who inhabit that space while simultaneously de-emphasizing efforts that solely focus on states and their narratives. In both academic and professional realms, discussion of resource conflicts in terms of only state actors is a disservice to border communities who have endured within the landscapes.

8.6   Conclusions

The aim of this research was to investigate transboundary conservation efforts, interactions, and relationships around the Southern Patagonian Ice Field in Chile and Argentina. We found that the presence of freshwater resources has significantly shaped this landscape and border situation in a way that differentiates it from other transboundary landscapes, presenting the need for a distinguished subfield of transboundary research that focuses on transboundary freshwater resources. Second, despite the lack of official government spaces, opportunities, and resources for transboundary dialogue, several locals in Patagonia have committed to transboundary collaboration and conservation, demonstrating the importance of community efforts in defining transboundary collaboration and conservation. Also, like many other situations in conservation, there is a significant divide between national and local engagement in the shared landscape.

There were study limitations that can inspire future research. First, the researcher's positionality as a foreign, Western researcher, may have influenced participant responses and limited the understanding of the region's historical and present context. Second, additional time would have allowed for more extensive stakeholder representation compared to the shorter interview period. During or after the time period of data collection, several political changes happened, and social movements occurred which may have influenced the application of findings. Also, limitations in translations and understanding arose from the non-native understanding of Chilean and Argentinian Spanish of the researcher. In terms of future research, there is the need for an analysis of the history of binational relations between Chile and Argentina in order to better understand what governance structures are currently in place for supporting or inhibiting transboundary conservation in other regions of Chile and Argentina. There is also the opportunity to conduct comparative studies with similar border contexts to evaluate how these themes translate to other geographies.

This study highlighted ongoing local-scale transboundary collaboration around Torres del Paine, Bernardo O'Higgins, and Los Glaciares National Parks in southern Patagonia despite a lack of formalized cooperation at the national scale. Historically and presently, there are significant challenges and conflicts for meaningful engagement with local communities, particularly with Indigenous groups, that must be resolved to ensure equitability and efficacy of any transboundary conservation efforts. This has been paralleled by many other transboundary studies across geographies and highlights how transboundary conservation must be transformed to prioritize and support local communities along borders.

Acknowledgements The authors would like to express the deepest of gratitude to Germaynee Vela Ruiz for providing guidance to Sanober in developing the case study. In addition, we would like to thank those in Chile and Argentina who allowed Sanober the opportunity to interview them. Thank you for being kind enough to share your time and perspectives. Finally, the authors would like to thank Andrés Adiego, of the Centro de Investigación en Ecosistemas de la Patagonia (CIEP), for developing the map of our study area (Fig. 8.1).

References

G.S.M. Andrade, J.R. Rhodes, Protected areas and local communities: An inevitable partnership toward successful conservation strategies? Ecol. Soc. **17**(4) (2012). https://doi.org/10.5751/ES-05216-170414

K. Barquet, "Yes to peace"? Environmental peacemaking and transboundary conservation in Central America. Geoforum **63**, 14–24 (2015). https://doi.org/10.1016/j.geoforum.2015.05.011

I. Battistello Espíndola, W.C. Ribeiro, Transboundary waters, conflicts and international cooperation – Examples of the La Plata basin. Water Int. **45**(4), 329–346 (2020). https://doi.org/10.108
0/02508060.2020.1734756

E.A. Beever, B.J. Mattsson, M.J. Germino, M.P. van Der Burg, J.B. Bradford, M.W. Brunson,
Successes and challenges from formation to implementation of eleven broad-extent conservation programs. Conserv. Biol. **28**(3), 302–314 (2014). https://doi.org/10.1111/cobi.12233

A.K. Biswas, Cooperation or conflict in transboundary water management: Case study of South
Asia. Hydrol. Sci. J. **56**(4), 662–670 (2011). https://doi.org/10.1080/02626667.2011.572886

H. Blair, K. Bosak, T. Gale, Protected areas, tourism, and rural transition in Aysén, Chile.
Sustainability (Switzerland) **11**(24), 1–22 (2019). https://doi.org/10.3390/su11247087

A. Böhm, Theoretical coding: Text analysis in grounded theory, in *A Companion to Qualitative
Research*, ed. by U. Flick, E. Kardorff, I. Steinke, (Sage Publications, 2004), pp. 270–275

B. Büscher, M. Schoon, Competition over conservation: Collective action and negotiating transfrontier conservation in southern Africa. J. Int. Wildl. Law Policy **12**(1–2), 33–59 (2009).
https://doi.org/10.1080/13880290902938138

N. Chettri, R. Thapa, B. Shakya, Participatory conservation planning in Kangchenjunga transboundary biodiversity conservation landscape. Trop. Ecol. **48**(2), 163–176 (2007)

J. Child, The American southern cone: Geopolitics and conflict, in *Proceedings of the Conference
of Latin Americanist Geographers*, vol. 9, (1983), pp. 200–213. https://doi.org/10.1097/
EDE.0b013e3181

S. Chiutsi, J. Saarinen, Local participation in transfrontier tourism: Case of Sengwe community in
Great Limpopo Transfrontier Conservation Area, we. Dev. South. Afr. **34**(3), 260–275 (2017).
https://doi.org/10.1080/0376835X.2016.1259987

S.P. Gasteyer, C.B. Flora, Modernizing the savage: Colonization and perceptions of landscape
and lifescape. Sociol. Rural. **40**(1), 128–149 (2000). https://doi.org/10.1111/1467-9523.00135

L. Guo, H. Zhou, Z. Xia, F. Huang, Evolution, opportunity and challenges of transboundary
water and energy problems in Central Asia. Springerplus **5**(1) (2016). https://doi.org/10.1186/
s40064-016-3616-0

I. Hagen, C. Huggel, L. Ramajo, N. Chacón, J.P. Ometto, J.C. Postigo, E.J. Castellanos, Climate
change-related risks and adaptation potential in Central and South America during the 21st century. Environ. Res. Lett. **17**(3), Article 033002 (2022). https://doi.org/10.1088/1748-9326/ac5271

M.M. Hennink, B.N. Kaiser, V.C. Marconi, Code saturation versus meaning saturation:
How many interviews are enough? Qual. Health Res. **27**(4), 591–608 (2017). https://doi.
org/10.1177/1049732316665344

E.L.Y. Hsiao, P. Le Billon, Connecting peaces: TBCAs and the integration of international, social,
and ecological peace. Int. J. World Peace **XXXVIII**(1), 7–41 (2021)

T. Ide, Does environmental peacemaking between states work? Insights on cooperative environmental agreements and reconciliation in international rivalries. J. Peace Res. **55**(3), 351–365
(2018). https://doi.org/10.1177/0022343317750216

D. Ioannides, P.Å. Nielsen, P. Billing, Transboundary collaboration in tourism: The case of the
Bothnian Arc. Tour. Geogr. **8**(2), 122–142 (2006). https://doi.org/10.1080/14616680600585380

F.M. Jaksic, J.A. Iriarte, J.E. Jiménez, D.R. Martínez, Invaders without frontiers: Cross-border
invasions of exotic mammals. Biol. Invasions **4**(1–2), 157–173 (2002). https://doi.org/10.102
3/A:1020576709964

L. Kachena, S.J. Spiegel, Borderland migration, mining and transfrontier conservation: Questions
of belonging along the Zimbabwe–Mozambique border. GeoJournal **84**(4), 1021–1034 (2019).
https://doi.org/10.1007/s10708-018-9905-0

M.H. Knight, P.J. Seddon, A.A. Midfa, Transboundary conservation initiatives and opportunities in the Arabian peninsula. Zool. Middle East **54**, 183–195 (2011). https://doi.org/10.108
0/09397140.2011.10648909

M. Kummu, J.H.A. Guillaume, H. De Moel, S. Eisner, M. Flörke, M. Porkka, S. Siebert,
T.I.E. Veldkamp, P.J. Ward, The world's road to water scarcity: Shortage and stress in the 20th
century and pathways towards sustainability. Sci. Rep. **6**(1), Article 38495 (2016). https://doi.
org/10.1038/srep38495

S.A. Lambertucci, P.A.E. Alarcón, F. Hiraldo, J.A. Sánchez-Zapata, G. Blanco, J.A. Donázar, Apex scavenger movements call for transboundary conservation policies. Biol. Conserv. **170**, 145–150 (2014). https://doi.org/10.1016/j.biocon.2013.12.041

F. Larocque, Who fought for water and what did they fight for? A comparative analysis of open water conflicts in four South American countries between 2000 and 2011. Idées d'Amériques **15**, 0–18 (2020). https://doi.org/10.4000/ideas.7724

P. Mackelworth, Peace parks and transboundary initiatives: Implications for marine conservation and spatial planning. Conserv. Lett. **5**, 90–98 (2012). https://doi.org/10.1111/j.1755-263X.2012.00223.x

P. Mackelworth, D. Holcer, B. Lazar, Using conservation as a tool to resolve conflict: Establishing the Piran – Savudrija international Marine Peace Park. Mar. Policy **39**, 112–119 (2013). https://doi.org/10.1016/j.marpol.2012.10.001

N. Mason, M. Ward, J.E.M. Watson, O. Venter, R.K. Runting, Global opportunities and challenges for transboundary conservation. Nat. Ecol. Evol. **4**(5), 694–701 (2020). https://doi.org/10.1038/s41559-020-1160-3

N. Muboko, The role of transfrontier conservation areas and their institutional framework in natural resource-based conflict management: A review. J. Sustain. For. **36**(6), 583–603 (2017). https://doi.org/10.1080/10549811.2017.1320224

C. Noy, Sampling knowledge: The hermeneutics of snowball sampling in qualitative research. Int. J. Soc. Res. Methodol. **11**(4), 327–344 (2008). https://doi.org/10.1080/13645570701401305

R.O. Perry, Argentina and Chile: The struggle for Patagonia 1843–1881. Americas **36**(3), 347–363 (1980)

J.G. Petursson, P. Vedeld, A. Vatn, Going transboundary? An institutional analysis of transboundary protected area management challenges at Mt Elgon, East Africa. Ecol. Soc. **18**(4) (2013). https://doi.org/10.5751/ES-05729-180428

M.E. Portman, Y. Teff-Seker, Factors of success and failure for transboundary environmental cooperation: Projects in the gulf of Aqaba. J. Environ. Policy Plan. **19**(6), 810–826 (2017). https://doi.org/10.1080/1523908X.2017.1292873

C. Reboratti, Socio-environmental conflict in Argentina. J. Lat. Am. Geogr. **11**(2), 3–20 (2012)

J. Sayer, Reconciling conservation and development: Are landscapes the answer? Biotropica **41**(6), 649–652 (2009)

M. Schoon, Governance in transboundary conservation: How institutional structure and path dependence matter. Conserv. Soc. **11**(4), 420 (2013). https://doi.org/10.4103/0972-4923.125758

B. Sepúlveda, S. Guyot, Escaping the border, debordering the nature: Protected areas, participatory management, and environmental security in northern Patagonia (i.e., Chile and Argentina). Globalizations **13**(6), 767–786 (2016). https://doi.org/10.1080/14747731.2015.1133045

C. Solórzano, Connecting climate social adaptation and land use change in internationally adjoining protected areas. Conserv. Soc. **14**(2), 125–133 (2016). https://doi.org/10.4103/0972-4923.186334

T.D. Taggart-Hodge, M. Schoon, The challenges and opportunities of transboundary cooperation through the lens of the East Carpathians Biosphere Reserve. Ecol. Soc. **21**(4) (2016). https://doi.org/10.5751/ES-08669-210429

J.M. Thomsen, S.C. Caplow, Defining success over time for large landscape conservation organizations. J. Environ. Plan. Manag. **60**(7), 1153–1172 (2017). https://doi.org/10.1080/09640568.2016.1202814

D. Thornton, L. Branch, D. Murray, Distribution and connectivity of protected areas in the Americas facilitates transboundary conservation. Ecol. Appl. **30**(2), Article e02027 (2020). https://doi.org/10.1002/eap.2027

M. Upadhyay, Pragmatic approaches to transboundary river governance: Exploring pathways for co-operation on shared water resources. World Water Policy **6**(1), 101–114 (2020). https://doi.org/10.1002/wwp2.12030

P. Van Aert, The beagle conflict. Island Stud. J. **11**(1), 307–314 (2016)

M. Vasilijević, K. Zunckel, M. McKinney, B. Erg, M. Schoon, T. Rosen Michel, *Transboundary conservation: A systematic and integrated approach*, Best practice protected area guidelines, series no. 23 (IUCN, Gland, 2015). Retrieved January 4, 2023, from https://doi.org/10.2305/IUCN.CH.2015.PAG.23.en

M. Zeitoun, N. Mirumachi, Transboundary water interaction I: Reconsidering conflict and cooperation. Int. Environ. Agreements Polit. Law Econ. **8**(4), 297–316 (2008). https://doi.org/10.1007/s10784-008-9083-5

M. Zeitoun, M. Goulden, D. Tickner, Current and future challenges facing transboundary river basin management. Wiley Interdiscip. Rev. Clim. Chang. **4**(5), 331–349 (2013). https://doi.org/10.1002/wcc.228

M. Zeitoun, J. Warner, N. Mirumachi, N. Matthews, K. McLaughlin, M. Woodhouse, A. Cascão, T.J.A. Allan, Transboundary water justice: A combined reading of literature on critical transboundary water interaction and "justice", for analysis and diplomacy. Water Policy **16**, 174–193 (2014). https://doi.org/10.2166/wp.2014.111

Chapter 9
Values, Conflicts, and Narratives of Private Protected Areas: The Case of Tompkins Conservation in Chilean Patagonia and Argentina

Christopher Serenari and Pamela Bachmann-Vargas

Abstract There are many ways to approach an understanding of nature conservation values and narratives. This chapter employs a case study to focus on one of the most important contemporary nature conservation initiatives, because of its global coverage and impact. We trace the formation of narratives developed by researchers, conservation entities, politicians, and other actors who have given meaning to Douglas and Kristine Tompkins effort to create privately protected areas (PAs) in Chilean Patagonia and Argentina. We reviewed academic and popular press texts to construct and interpret stories and their evolution throughout the processes of land acquisition and PA development, management, integration, and donation. Findings reveal connections, tensions, and contradictions produced by the broader Tompkins project and different approaches by the more critical academicians than the upbeat and colorful popular press. In this chapter, we demonstrate how narratives can be a useful tool to analyze value and ideological conflicts about private protected areas (PPAs) in Chile and Argentina. Narratives are also helpful to determine how humans do and will use stories to advance PPAs to achieve global biodiversity conservation goals, such as the 30 × 30 initiative, or problematize their existence.

Keywords Patagonia · Conservation narratives · Douglas Tompkins · Kristine Tompkins · Private protected areas

C. Serenari (✉)
Texas State University, Department of Biology, San Marcos, TX, USA
e-mail: c_s754@txstate.edu

P. Bachmann-Vargas
Environmental Policy Group, Wageningen University & Research, Wageningen, Netherlands

Department of Geography, Umeå University, Umeå, Sweden
e-mail: pamela.bachmann@umu.se

T. Gale-Detrich et al. (eds.), *Tourism and Conservation-based Development in the Periphery*, Natural and Social Sciences of Patagonia,
https://doi.org/10.1007/978-3-031-38048-8_9

9.1   Introduction

Recently, representatives from 188 governments have adopted the Kunming-Montreal Global Biodiversity Framework (GBF), with the aim "to address biodiversity loss, restore ecosystems, and protect Indigenous rights" (UNEP 2022, para. 2). One of the framework's main goals is to protect 30% of land and ocean environments by 2030, exceeding earlier agreements to protect 17% of terrestrial and 10% sea (Roberts et al. 2020; Sala et al. 2021). In many instances, few details remain about how to achieve these *bold* and *sweeping* targets. One of the core concepts associated with the 30 × 30 movement acknowledges that social-ecological challenges like wildfire, drought, and wildlife population declines do not respect fences and geopolitical borders. Thus, while 30 × 30 goals are likely to center on the creation and extension of protected areas (PAs), effectively protecting biodiversity will need to go further, crossing geopolitical boundaries and integrating private lands in order to create contiguous zones that can account for migratory routes and shifts in climate.

Private protected areas (PPAs) that achieve both ecological and human benefits will continue to be important policy instruments to strengthen PA networks needed to achieve global integration and adoption of 30 × 30 initiative goals. They often fly under the radar, however, outshined by their public counterparts. Nonetheless, efforts largely led by wealthy individuals, corporations, and non-governmental organizations are increasingly protecting millions of square kilometers of land and water across the globe and are recognized at the highest levels of biodiversity governance (UNEP-WCMC and IUCN 2021). PPAs have been lauded and received support from governments as a legitimate tool for protecting biodiversity and enhancing human well-being (e.g., providing jobs). As evidenced by prior research (Aastrup 2020), the alignment of the ideologies and political agendas underpinning PPAs will be critical to their broad-scale legitimation.

PAs have an associated discourses—language-driven mediums that stimulate action and exercises in power— that underpin and legitimize their origins and propagation (Serenari and Lute 2021). Evolutions in PA discourse have emphasized, for example, their diversity, shortcomings, and a need for deliberative approaches to management (Paterson 2014). As commanding as discourses can be, narratives—or stories—help give meaning to these discourses by stabilizing assumptions about PPAs, simplifying criterion by which to evaluate them, or qualifying their benefits, amongst others. Narratives originate from unique framings or worlds and are powerful tools for minimizing or enhancing the legitimacy of protectionist conservation efforts. Helping defines success and renders decision-making as common sense; narratives can become embedded in networks and structures of power and provide communicative blueprints for broader discourse (Woodhouse et al. 2021). Despite their track record of helping legitimize conservation projects, few studies have yet to systematically explore the narratives that coincide with the rising popularity of PPAs and give them global legitimacy.

To close this gap, we examined narratives developed by researchers, conservation entities, politicians, and others who have given meaning to globally renown PPA projects. Specifically, we examine narratives that helped legitimize Doug and Kristine Tompkins' master PPA projects within Chilean Patagonia and Argentina. The couple created and donated perhaps the most renowned and contested system of PPAs in the world, with repercussions for cultures, politics, economics, and biodiversity (Di Giminiani and Fonck 2018). At a finer scale, our analysis reveals connections, tensions, and contradictions produced by the broader Tompkins Conservation's family of projects and traces their evolution throughout the processes of land acquisition and PA development, management, integration, and donation. Broadly, this investigation reveals the entangled meanings ascribed to PPA projects, which might help us to better understand the efficacy, or the lack thereof, to help solve global biodiversity loss.

9.1.1 Background on Tompkins Conservation

Tompkins conservation is a non-profit environmental organization co-founded by American environmental philanthropists, Doug Tompkins (1943–2015), and his wife Kristine Tompkins. Over the past 30 years, Tompkins Conservation has carried out extensive work on nature conservation throughout southern Chile and Argentina, setting a precedent for private as well as public conservation, by creating PAs and implementing rewilding programs (Bachmann-Vargas et al. 2021; Busscher et al. 2018; Hora 2018). Pumalín Park was their first large-scale project in Chile. In the early 1990s, the Tompkins confronted strong political and local resistance while purchasing the land to create Pumalín. Their aim was to create a PPA, with high-quality infrastructure for visitors. In 2005, Pumalín was designated as a Nature Sanctuary (Hora 2018). Later on, Patagonia Park became their second flagship project. The creation of the Patagonia Park, a former ranching estate located in the Aysén region (Chile), was not exempt of controversies either (Borrie et al. 2020; Louder and Bosak 2019). Meanwhile in Argentina, the Tompkins have been involved in the creation of a number of PAs, namely the Iberá National Park (NP), donating 1500 km^2 (UNEP 2018), the El Impenetrable NP, El Piñalito Provincial Park, El Rincón (Perito Moreno NP), and Monte Leon NP (Tompkins Conservation 2019).

In 2017, as part of a broader conservation strategy, Tompkins Conservation signed a cooperation agreement with the Chilean state aiming to create new NPs in Patagonia, enlarge the surface of three other parks, reclassify national reserves as NPs, and create the Route of Parks of Patagonia (Chilean Ministry of the Environment 2017). Launched in 2018, the Route of Parks is "a vision of economic development based on conservation and ecologically minded tourism" (Tompkins Conservation 2021). Inspired by a *last of the wild* imaginary (Promis et al. 2019), the route envisions the Patagonian territory (from Puerto Montt to Cape Horn) as a network of 17 NPs encompassing 60 local communities, wherein tourism should be developed as a key economic sector, enabling local communities to build livelihoods in line with nature conservation objectives (Bachmann-Vargas et al. 2021; Borrie et al. 2020). In 2018, Pumalín Park was designated as Pumalín Douglas Tompkins National Park (Chilean Ministry of Public Lands 2008). To date, Tompkins Conservation continues its work by developing wildlife programs and promoting the Route of Parks of Patagonia, among other activities. On August 24th, 2021, it was announced that Tompkins Conservation was renamed Fundación Rewilding Chile. In their website, Kristine Tompkins stated, "the name Rewilding Chile promotes the idea that we will bring back species where they have gone missing. Where the gems of the country haven't been protected, we will help protect them in perpetuity. I will continue to work alongside the Chilean team, as I always have, under a new name that strongly identifies our future path" (Rewilding Chile 2021).

9.2   Theoretical Approach

Theoretically, stories help order and annunciate life experiences and human understandings (Polkinghorne 1995; Sharp et al. 2019). Narratives require researchers to analyze the composition of different worlds (Lieblich et al. 1998). The narrative paradigm is popular among sociology, psychology, anthropology, and other fields interested in making sense of human existence and affords researchers the option to use narrative as a focal point of research or means to an end (Lieblich et al. 1998). Our narrative analysis is grounded in social constructivism, a theory of knowledge that posits a cyclical process of deriving socially shared meaning. The utility of this theory is timely in light of recent theoretical and framework developments calling for the wholesale remaking of human relations to save our planet (e.g., Büscher and Fletcher 2020; Escobar 2018). A narrative constructionist approach is less interested in using narrative inquiry to understand internal or cognitive states and focused more on making sense of stories as social events or phenomena (Esin et al. 2014).

Moreover, analysis of the broader narrative context can reveal where we have been and where we are going (O'Neill et al. 2008). Narratives tell a story with a beginning and an end, wherein emotions, affects, and meanings make sense of a particular issue (Louder and Wyborn 2020). Narratives of nature conservation are often contested through counter-narratives. Narratives and counter-narratives can shape public opinion and policymaking, delineating what should be acceptable or not. Narratives of nature conservation may convey different conservation imaginaries, ranging from protecting *the last wild place in the world* to a combined conservation approach considering both natural and cultural dimensions (Adams 2020; Bachmann-Vargas et al. 2021). Moreover, narratives of nature conservation can make sense of public and private conservation initiatives, and their global and local effects. We accept that narratives are social phenomena because they are articulated through widely read academic and media outlets designed to shape knowledge, understanding, and experiences, and inform or influence societal developments and—therefore—values (e.g., Shlapentokh 1982).

Several studies have used narrative analysis to reveal the deeper meanings and spatial nature of human-nature relations, as we propose for our analysis of PPA creation. For instance, O'Neill et al. (2008) argued that a focus on spatial considerations, such as PAs, helps us unite human values about nature, identity, and sense of place. Drenthen (2018) explored conflict between rewilding and cultural heritage stating that divergent interpretations of the environment can serve as a horizon or framing for self-understanding. This study divulged how protectionist types fought for the removal of humans, even though socially constructed landscapes are technically meaningless without them, and that rewilders focused on temporal factors to promote new human-nature relations. More recently, Frei et al. (2020) revealed regional and local symbolic functions of three narratives about natural forest regrowth and demonstrated these divergences shaped unique and conflicting perceptions across four case studies.

9.3   Methods

9.3.1   Case Study Context

In Chile, Tompkins Conservation has donated approximately 4000 km², which along with state owned lands were aimed at creating four new NPs, namely Pumalín Douglas Tompkins NP, Melimoyu NP, Patagonia NP, and Yendegaia NP. Moreover, the land donation comprised the expansion of three already existing NPs: Hornopirén, Corcovado, and Isla Magdalena, and the reclassification from national reserve to NP (Fig. 9.1). Former national reserves Cerro Castillo and Alacalufes became Cerro Castillo NP and Kawésqar NP, respectively (Chilean Ministry of the Environment 2017). Meanwhile in Argentina, the main focus of Tompkins Conservation has been the rewilding of the Gran Chaco, along with the expansion of PAs (Quammen 2020), with Iberá NP being one of their most important conservation projects.

9.3.2   Narrative Analysis

Narrative analysis is a technique to analyze consistent stories about a phenomenon within texts such as newspapers, reports, and interviews. We situated our research within Lieblich et al.'s (1998) *holistic-content* analysis, where a combination of texts, including stories, commentary, and analysis, is broadly analyzed to reveal the plot or structure of the story, its foci, and trajectory (e.g., comedy or tragedy). Our approach followed a set of linear steps summarized by Sharp et al. (2019), including reading the text, extracting principal sentences and passages, defining categories, and evaluating and sorting material into the categories and subcategories (e.g., influential factors; how many times mentioned). Finally, we drew conclusions on this

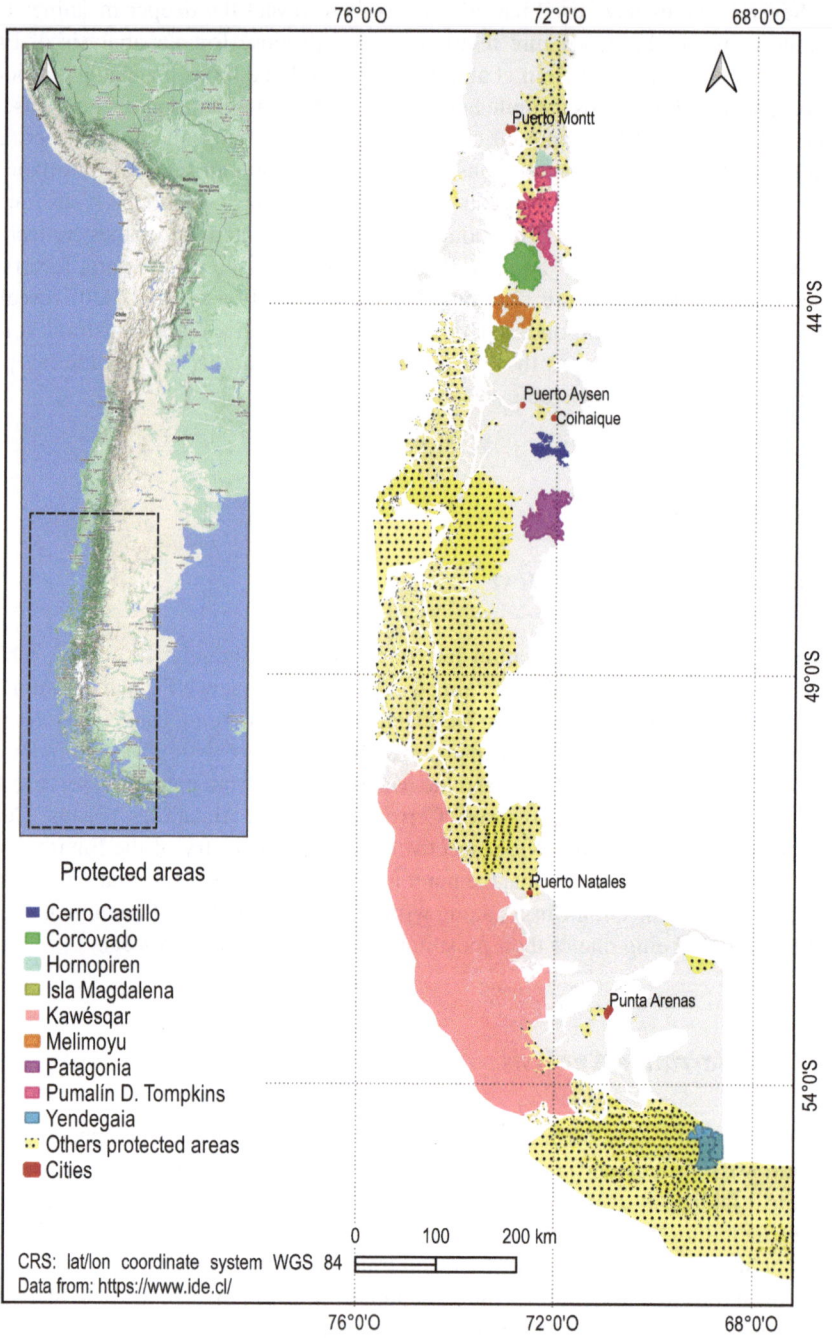

Fig. 9.1 Protected areas that were created and expanded based on the land donation made by Tompkins Conservation to the Chilean state, along with state owned lands

material to understand how narratives give meaning to the Tompkins case at a broader scale (e.g., interpretation, message). We employed an inductive rather than deductive approach to parse the meaning of PPAs from within academic, popular press, and political sources.

We cast a wide net and obtained peer-reviewed articles, popular press articles, and a few other documents such as undergraduate theses and conference proceedings. We searched Google Scholar (Doug + Tompkins + conservation + Chile), all databases—English and Spanish—within EBSCOhost and ProQuest (Tompkins + Patagonia; Doug + Tompkins + Chile) via Texas State University and Wageningen University libraries, and Google News ("Kris McDivitt"; "Kris Tompkins"), securing over 180 popular press articles. The search for Spanish popular press articles built upon previous bibliographic documents (e.g., Pinochet and Theroux 2016). We analyzed all text titles first, then scanned associated abstracts to determine if they would be further reviewed. After excluding articles for myriad reasons (Fig. 9.2), we settled on 32 academic articles—English and Spanish—between 1999 and 2020 focused on Chile, with one published in 2018 about Argentine projects. Select tangential documents helped capture the broader trend of foreign-owned land expansion in Argentina. In total, we analyzed 90 popular press articles focused on Doug and Kristine and their PPA projects, published between 1997 and 2021.

We did not bind ourselves to the assumption that narratives are mutually exclusive, given that narrative analysis relies on researcher discretion to build the interpretation of data and its meaning through systematic coding (Yanow 2000; Stelling et al. 2017). Soliva and Hunziker (2009), for example, submitted that the average individual stakeholder's views about landscape change are comprised of more than one narrative. Additionally, our analysis blended the approaches of a few narrative studies to create a framework for analysis. We started with older texts first and moved chronologically to newer texts to detect patterns in the narratives.

We coded texts with NVivo software version 12 (QRS International, Doncaster Australia), and with Atlas.ti software version 9 (Scientific Software Development GmbH). The coding procedure extracted relevant text elements and clustered them according to categories (Strauss and Corbin 1990). Deductive coding involved thoroughly reading sentences and paragraphs, organizing, and synthesizing data elements in a way that produced a coherent account. We started narrative development establishing codes related to problem definition/framing, aims, interests, values, assumptions, solutions, and strategies because they underpin the construction of stories (Frei et al. 2020) (Fig. 9.3).

We shifted to an indicative approach for axial coding to refine narratives and determine the relationships between PPAs and meaning. This involved using inductive analysis of the texts and the more common codes, enabling themes to emerge. For instance, preliminary themes such as *problem definition, author goals*, and *assumptions* gave way to refined themes such as *socio-political conflict, asymmetrical power relations, direct policy influence, ideological evolution/rural transition, neoliberal ideological incompatibility*, and *exclusionary decision-making*. The greatest number of files and references within each axial code determined which themes we prioritized for the final stage of analysis. In the final stage, we achieved

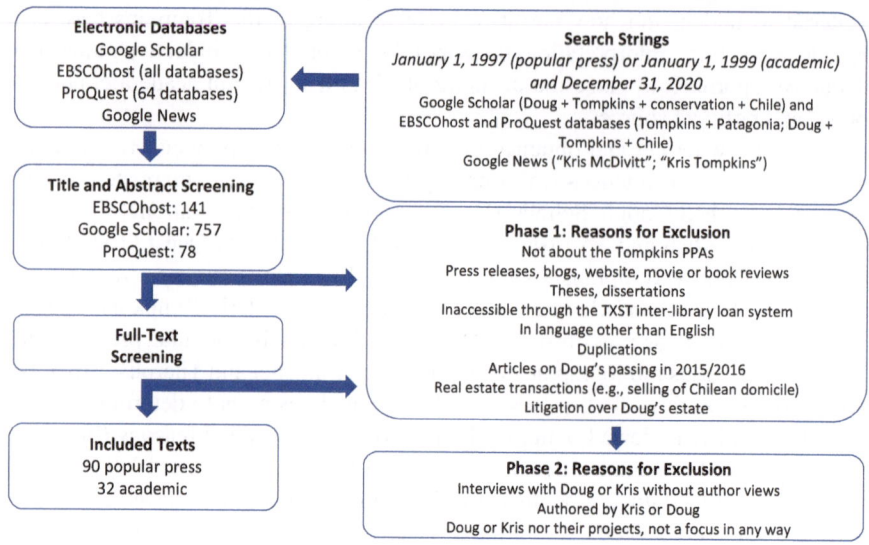

Fig. 9.2 Synopsis of narrative analysis sample selection process

a holistic interpretation of narratives by systematically selecting main codes and relating them to other codes through back-and-forth movement between the data elements and the emerging story (Sharp et al. 2019; Stelling et al. 2017). We followed Polkinghorne's (1995) story ending/outcome and criteria-focused approach, considering the story's ending or outcome once again to distinguish causal linkages and link disconnected data. This approach is particularly useful for academic articles framing their analysis within a particular theory. To improve the trustworthiness of our results, continual movement back and forth between the data required multiple rounds of triangulation. To reduce the likelihood of researcher views and biases being introduced into the analysis or interpretation of the findings, we reviewed articles independently, agreed on a coding scheme, and reconciled any discrepancies observed until agreement reached 100%. Additionally, we improved our approach through a rigorous peer review process.

9.4   Results

9.4.1   Academic Narratives

In our analysis, we identified six academic narratives (Table 9.1). These narratives are not mutually exclusive and overlap, comprised by analyzing the texts in aggregate.

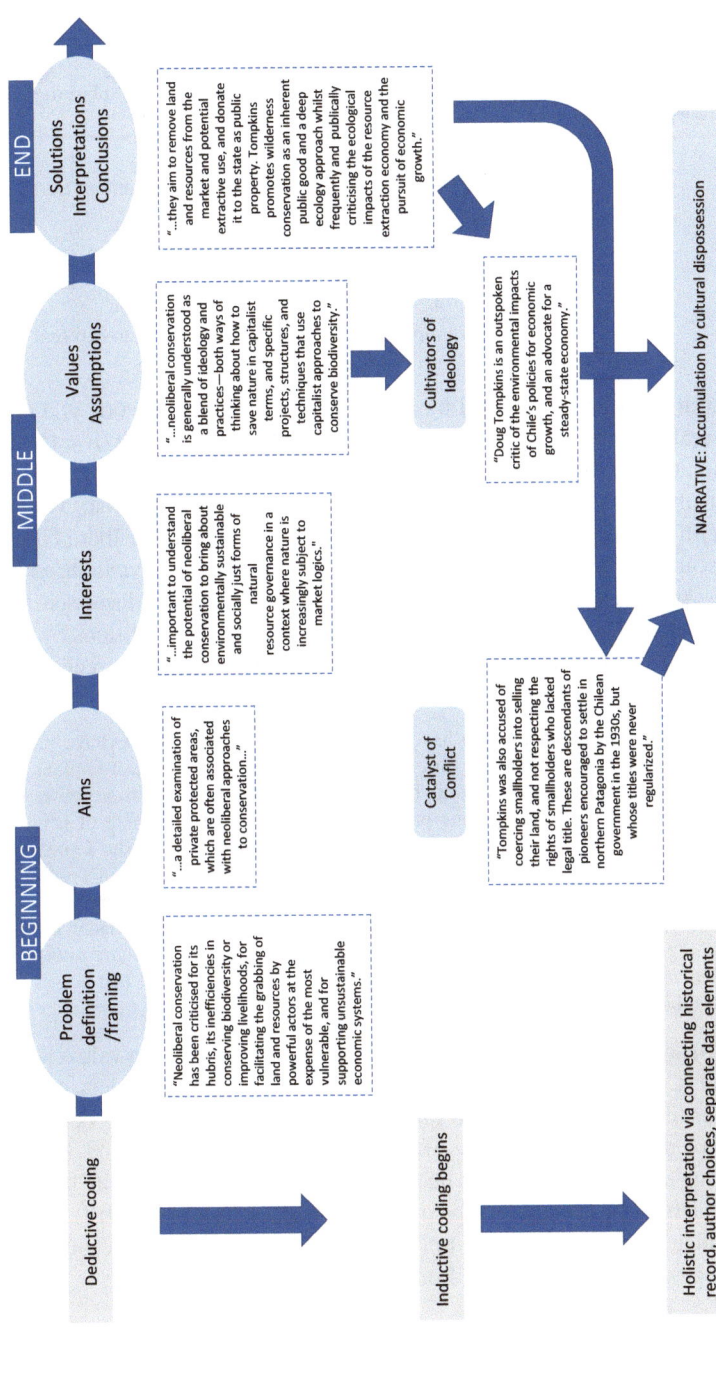

Fig. 9.3 Example of how the research team moved from deductive to holistic using narrative analysis. Sample text is from Holmes (2015)

Table 9.1 Academic and popular press narratives ascribing meaning to Doug and Kristine Tompkins private protected area projects within Chilean Patagonia and Argentina

Academic narratives	Critical	Accumulation by cultural dispossession
		Persistent imposition of an ideological force
		Neoliberal elite's contradictory conservation subterfuge
	Balanced/ Positive	Investment or green grab?
		PPAs advance sustainable development
		Paradigm shift in Chilean human-nature relations: the Tompkins effect
Popular press		Counter-culture eco-warriors
		The "gringo complot": ideological standoff at the end of the world
		The Triumphant Ecobaron/ess: from villain to hero(es)

A set of narratives were dominated by researchers framing their work as a critique of neoliberal conservation or situating PPAs as a neoliberal project ($n = 16$, "a capital accumulation strategy that commodifies nature" (Busscher et al. 2018, p. 579)). Most of the scientific inquiries have been conducted in English, with a focus on Pumalín NP and Patagonia NP both located in Chile, and the Iberá NP in Argentina. Researchers operating in this space often articulated an *accumulation by cultural dispossession* narrative. From this perspective, researchers discussed the discontent of local people with the disregard for culture and local traditions by the Tompkins archetype approach in the process of PPA and nature-based recreation development:

> For several years, some stakeholder groups have expressed feeling that their values have not been considered within the initiatives and processes undertaken by the Tompkins Conservation Foundation. Recent complaints have centered on the negotiation processes associated with the donation of the private parks, the subsequent establishment of new Chilean National Parks, and the creation and marketing of the Route of the Parks. Critics report that these negotiations were almost entirely contained in Santiago, the nation's capital, where members of the foundation interacted exclusively with high-ranking members of the government. Now that lands are transferring to the Chilean government under the administration of CONAF, these groups are calling for greater transparency, more intentional local involvement, and participatory practices that consider local values and priorities. (Gale and Ednie 2019, pp. 6–7)

Bourlon (2017b) articulated the replacement of, for example, utilitarian values with protectionist ones stating, "The rural world, entrepreneurs and defenders of the spirit of the pioneers, clearers of virgin lands, criticize his opposition to development and lack of respect for their way of life" (p. 1). Conservation Land Trust's focus on restoring *degraded* landscapes underpinned their efforts to re/write history in Latin America to create their ideal type of landscape (Busscher et al. 2018). Clear intent to erase identity and attachments to the land (Blair et al. 2019) to reconstruct culture in the region was chronicled by Louder and Bosak (2019):

> In one moment, when asked a question about Valle Chacabuco, (the park administrator) cut off the researcher mid-sentence, correcting, saying, "It's Patagonia Park [not Valle Chacabuco]," She said, "it's taken years to get people to stop calling it that." (p. 169)

Bantle (2010) stated that cultural dispossession encompassed disruption to daily activities and symbolic and material livelihoods, resulting in deficient support for Patagonia NP: "It seemed as though the more of the threat a respondent thought the park posed to the regional culture, the more negatively they perceived the park." (p. 6)

Academics sketched a narrative that a disregard for rural Chilean lived experiences was by design. Specifically, researchers dedicated much effort to unpacking an obvious ideological schism that was exacerbated by exclusionary tactics (Jones 2012). Over two decades, researchers outlined a persistent imposition of an ideological force, through which the local nature acquired a sort of museum status (Núñez et al. 2018, p. 149), that led to myriad problems and setbacks for the Tompkins, until they realized their ultimate goals of donation and, frankly, legacy-building. Borrie et al. (2020) inferred researcher rationale underpinning their analyses in that "PPAs that are seen as reflecting the values, priorities and political power of private interests" (p. 3). The Tompkins' worldview was rooted in an American or Western brand of protectionism through fortress conservation, elevating the importance of PAs in Latin America to safeguard landscapes imbued with symbolic meaning. As mentioned by most researchers, this schism was exacerbated by Doug's unapologetic Deep Ecology activism and devotion to Western environmental philosophy (e.g., Tecklin and Sepulveda 2014, p. 215). Ideological conflict was attributed to a "clash between two completely antagonistic paradigms or worldviews" (Ramírez and Folchi 1999, p. 2) or a "storm of imaginaries," as Bourlon (2017a, p. 88) colorfully put it. In the battle to define Patagonia (Louder and Bosak 2019), imposition of Doug's vision for a *Next Economy* to better or save humanity, and imaginary prioritizing the environment-recreation-culture nexus (Blair et al. 2019; Bourlon 2017b), was fought off for decades. It was first opposed by the central government in Santiago (Blair et al. 2019) and corporations "not interested in social movements, environmental or otherwise" because they were "an obstacle" in "development plans" (Wakild 2009, p. 116). Researchers began to chronicle local community experiences, with this narrative bolstered by findings that some locals were unsupportive of "Tompkins' idea of conservation" (Jones 2012, p. 253). Distrust of PPAs among locals was tightly linked to a history of exclusion and under-representation by government (Zorondo-Rodríguez et al. 2019) and the argument that the costs of PPAs to local people may in fact outweigh the benefits (Holmes 2018; Borrie et al. 2020).

A third academic narrative that we called the *neoliberal elite's contradictory conservation subterfuge* refers to that PPAs were a shrouded operation, cast as one thing to garner support but pursuing and achieving another aim. This narrative advances in at least two paths. The first narrative framing highlights that the Tompkins were a wealthy, well-connected elite class of citizens seeking, intentionally or not, to (re)colonize Patagonia for conservation (Bourlon 2017b). Holmes (2014) signaled that history may be repeating itself in Chile, writing that the

Tompkins had been "accused of coercing smallholders into selling their land, and not respecting the rights of smallholders who lacked legal title" (p. 558). The cultural dispossession narrative dovetails with this framing. Additionally, a few researchers pointed out the unspoken contradictions of PPA creation and operationalization. They made light of the fact that Doug and Kristine made their millions by taking advantage of the capitalist system (Bourlon 2017b; Wakild 2009). Jones (2012) asserted that the PPA phenomenon actually replicates and does not reject capitalist expansion by, for example, dispossessing local economies by eliminating the sheep grazing industry. Núñez et al. (2018) elaborated on the re-appropriation of Nature through conservation interests, underpinned by power and capitalist mechanisms in order to (re)create new businesses and support capitalist production, a phenomenon these authors labeled *eco-extractivism*. The other path characterized PPAs in an upbeat way. For instance, Wakild (2009) wrote,

> But some would say Parque Pumalín is a business, veiled as a gift. Tompkins is the biggest employer in the region and his investment of close to $50 million dollars makes up 10 percent of the total foreign investment in the region in the past ten years. His ecological footprint may be small, but his economic one is worth taking notice. (p. 117)

Bachmann-Vargas et al. (2021) referred to an optimism about tourism being a vehicle capable of financing nature conservation, and critiques that have ensued about the commodification of nature. The authors were chary in their discussions of tourism-driven initiatives, such as the Route of Parks of Patagonia, and their potential to materialize win-win scenarios for local communities.

We note that not all researchers were devoted to situating Tompkins' PPA projects and outlining their contradictions and controversy within a neoliberal conservation paradigm. A few researchers mentioned PPAs as fringe (Tecklin and Sepulveda 2014) or regional/global neoliberal conservation projects (Borrie et al. 2020), while others inserted a neoliberal model as a backdrop for alternative practices of territorialization (García and Mulrennan 2020). Additionally, we observed a notable split between researchers, exemplified by three additional narratives characterizing the Tompkins' achievements in a different, more positive light.

One divergence was brought forth by the *investment or green grab?* narrative which deconstructs PPAs as having a potentially dual meaning. Some researchers contrasted the capitalism-focused purchase, transaction, and green grabbing phrasing (Busscher et al. 2018; Holmes 2014) and framing with that of investment (e.g., Hora 2018; Wakild 2009). These researchers tended to offer a less critical, more balanced, and sometimes positive framing to the evaluation of Tompkins' projects but references to green grabbing carried a negative connotation.

In Argentina, local authors described the creation of the Iberá NP as a rapid process that transferred the Provincial estate and its jurisdiction to the National State (Mantegna et al. 2016). This process foregrounded the national debate on foreign-owned land and green grabbing. The authors indicate that conservationists—mostly foreign investors—possess most of the estates around the Iberá, thereby controlling its access. In 2011, Argentina passed a Bill called Protection of the National Territory and Rural Lands, which states the conditions for foreign citizens to acquire land (Pohl Schanake and Vallejos 2008).

Some authors employed a positive framing to the impacts of PPAs. Researchers not wholly embracing a critical lens informed a narrative that *PPAs advance sustainable development.* One way this was achieved was by chronicling positive effects of early PPAs on local economy. Research by Wakild (2009) found a positive ripple effect of Pumalín on the surrounding communities. Other researchers also noted efforts to "reappraise local culture and stimulate sustainable tourism" in Iberá, Argentina (Busscher et al. 2018, p. 579). Some researchers found noteworthy synergies between PPAs and environmental activism. For instance, Hora (2018) noted that the Tompkins' opposition to dam construction in the Aysén region placed PPA efforts in the good graces of Chileans, while the Tompkins' global influence helped bring global attention to the plight (Jones 2012). However, Bantle (2010) demonstrated that this endearment was not universal because the roots of infrastructure development or PA imposition are cut from the same colonialist cloth.

The final narrative articulates a *paradigm shift in Chilean human-nature relations,* which also might be termed as the *Tompkins effect.* This narrative is synergistic with the *ideological imposition* narrative, as well as with the Tompkins' influence on wealthy Chilean and foreign citizens who have been interested in environmental philanthropist initiatives ever since (Bourlon 2017a). Following this trend, the former President of Chile, Sebastian Piñera, among others, created their own PPAs (Bourlon 2017a). Hora (2018) suggests that researchers' conclusions about cultural disruption should not be viewed as a sweeping generalization. Findings determined that an evolutionary framing is more appropriate:

> Results of the investigation show that 27 years after the implementation, it is clear that Pumalín Park has changed this paradigm and other possibilities for development arise as opportunities for rural marginalized areas that are rich in nature and cultural landscapes such as the Patagonian fjord lands. (p. 17)

Quantitative data collected by Bantle (2010) also reinforced that some locals do take pride in the park, a change that can influence the manifestation of new values. Hora (2018) concluded that there is evidence that Pumalín Park positively shaped local values, identity, and economy since its foundation and, therefore, is well-integrated into the local community. Ramirez and Folchi (1999) foreshadowed several of the narratives we extracted, writing,

> Consequently, the Tompkins project represented a clear alteration of the society/nature exchanges that had been articulated culturally, socially and financially through a very long history. (p. 9)

To create a new space for conservation, Jones (2012) and Bourlon (2017b) concluded that the Tompkins' projects created new spatial, social, aesthetic, cultural, and political realities to design a green Utopia: "Doug Tompkins effected a change of paradigm in Patagonia" (p. 43). For instance, Bachmann-Vargas and van Koppen (2020) identified the *Patagonian wilderness* discourse, acknowledging the substantial influence Doug and Kristine had in the formation and reproduction of the ideas that helped reframe Chilean Patagonia as a pristine and untouched place, and one of the last wild places in the world. Bourlon (2017b) also downplayed criticism about value changes and local discontent about a major rural transition occurring in parts of Latin America, remarking because locals are complicit in making PPAs happen:

> The rural world, entrepreneurs and defenders of the spirit of the pioneers, clearers of virgin lands, criticize his opposition to development and lack of respect for their way of life. Nevertheless, they sell their land to rich Westerners and Chileans who want to possess their own private parks at world's end while hoping that tourism will ensure their future. (p. 1)

Additionally, Borrie et al. (2020) acknowledged the potential of a Tompkins-style approach, provided that other local and national actors incorporate its logic and strategies:

> The [Route of the Parks] promotes a collaborative public-private approach to what the Foundation describes as "conservation-based development." ...Still, the Route's success will ultimately depend on the willingness of local communities to incorporate its logic and strategies within their own planning, development, and priorities. For example, in early 2020, a publicly funded Territorial Integration Program (PTI), was created to focus on the integration of local interests with the Route of Parks Concept. This initiative has received funding for three years through the support of the government economic development agency (CORFO), the Lakes Regional Government, the National Tourism Agency (Sernatur), the National Forest Corporation (CONAF), and the Ministry of the Environment (MMA), as well as private participation from universities, six associated communities, and tourism operators (ptirutadelosparques.org). If initiatives like PTI prove successful, and are validated by local communities, they are likely to be replicated within other parts of the increasingly integrated public-private Chilean system. (Borrie et al. 2020, p. 12)

9.4.2 Popular Press Narratives

Analysis of popular press sources revealed three main narratives (Table 9.1). These narratives were enveloped within broader corpora, with few authors espousing just one narrative. Early texts focused only on Doug, characterizing him as having deep connections to the Patagonia landscape. Many authors situated their works within a wilderness ideal, referencing Doug's affinity for the "pristine, virgin, nameless, wild, dramatic, untouched, and untamed" Patagonian landscapes: "Pumalín is a revelation...the land that time forgot" (Graydon 2006). Against this backdrop, we termed the *counter-culture eco-warriors'* narrative. This narrative depicts a legendary mountain and whitewater adventurer who made his fortune as a fashion mogul but decided to follow his passion living "the wilderness dream in the real Patagonia" and uphold these landscapes in his image post-capitalist epiphany. All text connected the Tompkins' wealth to their capitalist pursuits and successes but pursued a "high school dropout-turned-billionaire" tenor to emphasize Doug's counter-culture lifestyle from his youth to his evolution from adventure recreationist, entrepreneur, and green philanthropist. He was often portrayed as a doer with a "well-practiced talent for making things happen" (Eckersley 1999). Most texts reference Doug's wealth due to ownership of Esprit and The North Face companies, highlighting the size of his land purchases and the amount of money he made or spent. Texts then progress to his "anti-capitalist epiphany" to "atone for his capitalist past," ultimately resetting his moral compass with Deep Ecology (Mark 2019).

Older articles rarely mentioned Kristine or her role because of their relatively new relationship. That tendency shifted in the mid/late 2000s when she took on the role of "peace broker" (Doherty 2020) to offset Doug's "blunt, outspoken, and critical ways" and commentary about, for example, logging, mining, and salmon farming (Bonnefoy 2018; Warburton-Lee 2001). In 2012, Kristine received her own headline calling her "Mrs. Patagonia" in an article published by *Revista Que Pasa* (López 2012). The article emphasized her leading role in the creation of the Patagonia NP, along with her opinions about the HidroAysén project. Most of the texts focused on Kristine, often via interviews with her, when she took the reins after Doug's death. A staple of most articles was a reference to her role as CEO of Patagonia, Inc. and her access to financial capital, to accentuate her business acumen as a complement to Doug's ambition. For example, Weir (2017) remarked, "she helped turn his company, Patagonia, into a global brand" (para. 2). A few authors mention her connections to outdoor recreation, hence preserving the broader explorer-capitalist turned ecowarrior narrative that remains popular in texts.

Several press articles elaborated on Doug's efforts and the controversies related to the nature conservation projects. Accordingly, we termed the second narrative that emerged from popular press articles *the "gringo complot": ideological standoff at the end of the world.* Texts often expressed the ideological divide caused by injecting Deep Ecology into a natural resource-dependent society. It is important to note that there are at least two thrusts to this narrative. The first is a pre-donation misfit. Green philanthropy donating lands back to the people of Chile and Argentina was a goal or dream of Doug and Kristine, but they were not yet such philanthropists until the official first donation, that of the Pumalín property, which received nature reserve status in 2005 and became a national park in 2018. It is appropriate to characterize early texts as portraying Doug and Kristine each as an "ecowarrior" (Frankel 2004) on a "mission" to "combat biodiversity loss by "rewilding" natural habitats and expanding protected lands" (Ruggiero 2017). Most texts referred to all goals of buying contiguous land, restoring or improving it from the ravages of livestock grazing and timber extraction, rewilding it by introducing native wildlife such as Huemules (*Hippocamelus bisulcus*) and jaguars (*Panthera onca*), then donating the land for national park creation while installing a nature-based tourism economy. The main political conflict was the act of purchasing of land. Chilean press dedicated a number of articles reporting the *gringo complot*, on the grounds of the extensive land acquisition in southern Chile and referring to how Chile was "being cut in two parts." The purchase of land raised a bevy of suspicions. Some local authorities argued that there was a *complot* to appropriate Patagonia, and authorities found it questionable that wealthy Americans would want to spend their money on environmental campaigns in Patagonia rather than energy projects (El Mostrador 2012). Meanwhile, Argentine news outlets were referring to Tompkins and his *natural empire* around the Iberá in 2005 (Zacarías 2005).

At least half of texts referenced the mismatch between protectionism versus economic development values, writing of their "strange purpose" to pursue conservation over land development (Rohter 2005). Eckersley (1999) wrote, "It is no small irony that had Doug Tompkins planned to develop the land, he would have been greeted with enthusiasm" (p. 32). Similar to academic texts, several journalists, particularly those writing in the 1990s and 2000s, described tensions that arose due to the lack of local community buy-in for wholesale changes to their livelihoods and, more rarely, mentioned "hostility" towards Doug and Kristine (Wieners 2014). Nearly one-third of articles referenced the manufacture of "suspicion and strong opposition by local politicians, loggers, power companies, and nationalists" (Cuevas and Luna 2018) that underpinned a coordinated "vitriolic smear campaign" (Eckersley 1999) to besmirch "wealthy outsiders" (Wieners 2014) Doug and Kristine and the formation of a "green conservation cartel aimed to stop development" (Vidal 2016). This sentiment was summarized by Langman (1997):

> In a May press conference, Tompkins and his attorney, accompanied by leaders of Chile's largest environmental groups, presented information to counter a "harassment campaign" they said had been waged by the government. Among the numerous false accusations, they said, were rumors that Tompkins was pressuring locals to sell their land, building a nuclear base, planning a Jewish colony, developing a secret gold mine and financing political opposition. (para. 4)

Consequently, the *gringo complot* narrative found fertile ground in conspiracy theories. For example, the idea that Doug and Kristine sought to establish a Jewish colony inside their purchases was inspired by the so-called *"Plan Andina"* (Bohoslavsky 2008) and also by the large number of Israeli tourists coming to Patagonia (e.g., Hamon 2015). Moreover, critics of Deep Ecology emerged from the Catholic Church, referring to the Church's opposition to abortion. According to an article published in 2006, the Church expressed its objection to the work of Doug and Kristine, asserting that "the Christian vision of the ecology stipulates the supremacy of the "human ecology," which oblige us to condemn the mistakes of the "deep ecology," when its vision tends to consider the human being as one more element of the universe of the living beings, and when it punishes the fertility of the men because their actions against nature" (Chile Sustentable 2006, para. 16).

Socio-political concerns surrounding sovereignty and national security highlighted authors' desire to portray political and legal ideological paradoxes within Argentina and Chile societies. For example, texts pitted nationalism against foreign investment: "what is more important, the private property of a few, or the sovereignty of everyone?," (Reel 2006, para. 8), quoting a government secretary who supported cutting fences bordering foreign land acquisitions. Texts also pitted sovereignty against cultural integrity: "…inspires further questions of legal regulations in Latin America to limit foreign purchases of real estate given painful colonial legacy" (Hayden 2012, para. 2). And texts pitted social justice against environmentalism: the latter are "often grouped together under the same 'progressive' label" (Reel 2006, para. 10). These tensions manifested in various ways. One example was an outright attempt to purchase coveted land before Doug and Kristine. The hydro power juggernaut formerly known as National Electricity Company, S.A. (ENDESA) Chile beat Doug and Kristine to the purchase of a 125-square-mile tract owned by the Catholic diocese of Valparaíso out of spite for their support for local anti-dam protesters (Earth Island Journal 2000; Salas 2015). The second thrust is the idea of donating land after its purchase to fulfill the green philanthropy role, which many authors—including academics—also found socially, economically, and politically problematic for Chile's free-market economy. Hayden (2012) highlighted a divergence in North and South American ideologies commenting, "While [Doug] believed he was going south to serve as a savior of sorts, in reality, he was perpetually referring to his northern and U.S.-driven worldview in his assumptions about Chilean and Argentine land and people" (para. 2). Byrnes (2009) summarized sentiment of projects occurring in both countries writing, "Here in South America, this type of charity is not common—and is even viewed with suspicion by some" (para. 7). Texts emphasized saving nature and sustainable agriculture as complementary appendages, which later were merged with mentions of an ecotourism economy. The latter was a main focus of texts chronicling the Iberá Wetlands case in Argentina, with these more recent texts highlighting the rewilding of jaguars (*Panthera onca*) after nearly a century of extinction and emphasis on ecotourism replacing "indiscriminate hunting or environmental changes" (Garzón 2017, para. 18).

After Doug's death in 2015, the narrative, as we termed *the Triumphant Ecobaron/ ess: from villain to hero(es)*, became more prominent. These narrative positions both Tompkins as overcoming the odds and using their capital to save landscapes from human degradation, portraying a shift in the media framing; from villain to hero(es) (Pinochet and Thenoux 2016). Further, this powerful narrative depicts the Tompkins as helping to align humans with sustainable behavior by helping them discover the intrinsic value of ecosystems. Within this narrative, the Tompkins emerge as having overcome criticism, attacks, and bureaucratic stonewalling to save the planet: "This is the story of the couple who purchased paradise, the neighbors who curse their name and the idea that might save life as we know it" (Weir 2017, para. 1).

They exercised patience and diplomacy—particularly Kristine—and seized an opportunity to finally see their dream become a reality. They won over their critics through a measured campaign (Warburton-Lee 2001) slowly winning the hearts and minds of Chileans (Langman 1997): "To some extent, the Tompkins have become more accepted, in part, because they did what they said they would do" (Wieners 2014). Fitzner (2018) wrote, "Over time relationships with locals have improved too, as the Tompkins worked to involve people in the project and hear out the concerns of locals" (para. 6). Their accomplishments served to build trust among their staunchest critics across local and national scales as well: "The Chilean government has warmed up to the conservation projects" (Daley 2017, para. 7).

One of Doug's milestones in this regard was his active involvement as an environmental activist, mobilizing his network and contributing with economic resources to campaigns such as *Patagonia Sin Represas* (Patagonia without dams). He launched this effort to protest the construction of five dams in the Aysén region (i.e., HidroAysén project). Tompkins' detractors seized the opportunity to promote the counter narrative of *Patagonia without Tompkins*. Sentiments of mistrust and the ideological standoff that emerged from the purchase of a former large ranch (i.e., Valle Chacabuco), which later on became the Patagonia NP, spurred the animosities and the local debate (Saverin 2015). Nonetheless, by 2015 the press detailed that the opposition to the hydro power project became a new "battle for the gringo" but one in which he won, as he finally defeated HidroAysén (Fernández 2015b).

In recent texts, the Tompkins were less a part of a problem and perspectives are skewed towards framing them with their system of value. Headlines read, *"The Fashion Executives Who Saved a Patagonian Paradise"* (Mark 2019) and *"Gave It all Away"* (Weir 2017) to become genuine conservation or *wildlands* philanthropists or eco barons (Franklin 2008). In the texts analyzed, the Tompkins were depicted, rather neutrally, as *gringos* spending Yankee dollars and appealing to good intentions (Quammen 2020), but, more recently, were regarded as heroes, conservation champions, vigilantes, fighters, rewilders, revivers, and eco-capitalists who created a society-altering legacy: "…in a generation the parks created by the Tompkins in Chile and Argentina will be cherished as national treasures embedded in their identity" (Wilkinson 2018). Garzón (2017) and Londoño (2020) extended this narrative to depict how citizens of Argentina were the ones experiencing an epiphany. Garzón (2017), quoting a source, wrote, "It's as if after his absence the locals have reevaluated their opinion, realizing that he didn't want to 'steal Iberá's water,' but instead to create a new model for conservation." In these more recent texts, the aim appears to be to leave the reader with a sense that the Tompkins now have more supporters than detractors, and that the critics were wrong all along. Fernández (2015a), quoting Andrés Azócar, author of Doug Tompkins biographic book, wrote,

> The passing of Tompkins is the most symbolic (event). He died much more legitimized and respected than his detractors are today. Chile has more respect for Tompkins than it has for his opponents. When he passed away, the conflictive and harsh man was gone, but you realize that the parks are there. His projects speak for himself. Everybody speaks about his parks, which no longer are his property, but are his parks, which will belong to the State. (Fernández 2015a, para. 13)

Moreover, there is a sense that Doug was, and Kristine is, a sagacious being, destined to overcome the odds and realize their PPA dreams because of the synergies between, in large part, their risk-taking adventure recreation feats, success in the corporate world, and strong sense of place. Some argued that their efforts did humanity *a favor*, "saving the environment and communities from an out-of-control economic model that each day is edging civilization 'closer to the abyss'" (Langman 2012, para. 1). This narrative has even been fashioned to portray Doug as a martyr of sorts (e.g., his battle with ENDESA). Mark (2019) quoted the superintendent of Pumalín to achieve this tone: "…he died fighting for what he wanted…When he died, people finally got it." He is also considered a role model for how to define and operationalize nature conservation and treat conservation philanthropy as a value. Fernández (2015b), quoting Patricio Rodrigo, one of Doug's friends, stated, "I would say that one of the values that [Doug] instilled in nature conservation in Chile, is the concept of philanthropy" (Fernández 2015b, para. 11).

9.5 Discussion

According to Bruner (1990), humans innately use stories to organize experiences that unfold chronically and through an unfolding plot. Additionally, it is important to consider how narratives influence the establishment of what experiences mean to human lives, the ways humans construct reality (Carter 2013; Sharp et al. 2019). To better understand what PPAs mean to the ways humans experience biodiversity conservation through park creation, we establish the point of the stories being told. As indicated by Riessman (2008), every story has a moral, lesson, or point. Hence, in the text that follows, we use this approach to make sense of narratives collectively and what they tell us about what PPAs mean to human-nature relations.

First, our results suggest that one meaning of PPAs is rooted in critical analyses intended to confront neoliberal conservation where competition and markets are elemental to problem definitions and solutions (Igoe and Brockington 2007). A notable result from our analysis indicates that, thus far, academia has demonstrated a growing preference for a critical lens. Deployment of a critical approach is at least aimed at linking "reason with transformation" (Lynch 2001, p. 352) to reveal how humans, for example, produce conflicting or unjust outcomes, degrade the environment, and alienate our species from important values while promoting unsustainable ones (Richardson and Fowers 1997). For these PPA researchers, a lesson of the story is that *history is repeating itself*. The Tompkins' PPAs in Chile and Argentina have an origin, replication, and resistance process that is similar to those of public PAs. Our analysis highlighted that PPAs are considered as an ominous reimagination, replication, or appendage of late-stage capitalism or neocolonialism (Serenari et al. 2017). Researchers found critical assessment useful to detail biases, injustices, and contradictions that underscore *neoprotectionist* values (e.g., people and nature cannot coexist, PAs are ideal, locals can protect Nature with monetary inducements; Brockington et al. 2012; Büscher and Dressler 2007). Social impacts of both public and PPAs include but are not limited to livelihood disruption, displacement, denunciation, and cultural death (Serenari et al. 2017; West et al. 2006). Narratives that express the needs, grievances, and requests of impacted communities, local people can be viewed as legitimizing subaltern human experiences within a top-down conservation-development paradigm. Yet, there is an awkwardness about trying to address the PPA-environmental justice nexus if for no other reason than trying to reconcile private property rights within the larger, problematic human project of enhancing environmental governance for the collective good.

A second lesson, specifically, of pro-PPA narratives, is that as long as property rights are embedded within societal values, landowner engagement in and the legitimization of private lands conservation (PLC) will be critical to achieving large-scale conservation goals. As evidenced by our results, narratives that communicate the successful bridging of politics, PLC, and positive social change may help cultivate shared values over time. Hence, we feel there is a need to consider the implications of engaging in an overindulgence in academic critique of PPAs. For instance, academic criticism about PPAs has been considered a bureaucratic challenge to notable conservation projects occurring outside of the public domain such as the Tompkins' PPA and rewilding efforts (Zamboni et al. 2017). While important for inducing positive change, critical narratives of PPAs may, in the interim, erode the support they need to, despite their capitalist underpinnings, be preferred over environmentally degrading projects such as megadams or mining. Therefore, we believe there is an opportunity to reframe PPAs in a constructive way that produces positive conservation outcomes and behavior change in the short and long terms (Nisbet and Scheufele 2009).

Third, our analysis revealed that both academics and journalists tell a story that suggests a lesson in that *rewilding Patagonia presents a new cultural and paradoxical landscape* (cf. Drenthen 2018). In our study area, and in many other parts of the world, PAs are Janus-faced, or understood as having two diametrically opposed sides. One argument for this depiction is that they have been situated within the archetype yet contradictory conservation-development paradigm that attempts to reconcile unbridled economic growth and nature conservation (Serenari et al. 2015). Representative of this clear discrepancy in values, the narratives we uncovered emphasize an ideological power struggle between conservation elites, the political class, and historically marginalized rural people. As these narratives unfolded or developed over time, highlighting an important temporal aspect of narrative development, the lesser-equipped rural combatants were worn down and displaced by the better-networked and resourced conservation elite, yet, keeping in mind that some locals were complicit in their own domination (i.e., conquest of their pre-Tompkins realities). Likewise, these narratives also demonstrated how a political class can be converted to adopting nontraditional neoprotectionist values when (a) ecotourism helps establish new symbolic and material values upon the landscape in support of a *new economy* and (b) the idea of building conservation legacies are embraced nationally. Additionally, strategies used by the Tompkins were inherently contradictory and obscured to avoid socio-political conflict during a grinding, elongated metamorphosis in which narratives helped shift the framing of or the conversation about PPAs to be more socially and politically friendly. Our findings indicate that the lengthy, multi-decade land accumulation and cultural shift processes highlighted by many texts is a temporal element that should not be overlooked. Texts often referenced the number of years it took to accumulate and, eventually, achieve national park status, as well as to change attitudes and influence a value shift. This particular temporal component plays a key role in a narrative that captures the persistent challenges experienced by the Tompkins over decades as they sought to propagate their ideology and become societal engineers.

The fourth lesson is that *ideological warfare reveals the limits and capabilities of seriously engaging in our most important nature conservation debates.* Our analysis demonstrated how narratives and their development and perpetuation can generate fear, uncertainty, doubt, conflict, and conspiracy, fueling ideological clashes and a battle for a supreme ideology. These sentiments underpin juxtapositions between cultural, economic, and political values produced by the persistent negative framing of the "structure of the ideological world" (Fowler 1991, p. 93). We consider the aforementioned outcomes as communicative emissions, transmissions discharged after exposure to and negotiating narratives that "create a world of conflict" (Carter 2013, p. 8). Our analysis suggests that this negativity reflects working through critical paradoxes and trade-offs that need to be made to make progress towards a more sustainable world. We also propose that negativity may be compounded by the popular press's fixation on wealth, adventure-seeking, and dreaming the impossible dream of PPA creation, particulars that deviate from and can directly or indirectly influence social and political life in Chile and Argentina.

One of the important tools used to craft narratives in our study was the use of powerful visual imagery, in this case, to remind us of our existential limits. Popular press articles we reviewed were often situated within symbolic values. *Loaded language* describing a humanless, wild, pristine, or primeval landscape orients, familiarizes, or indoctrinates the reader into an "alliance of shared meaning" (Lester 2005, p. 123). These stories are relatable to those living in societal enclaves where this type of language intensifies behaviors that can support the PPA model such as fund-raising or ecotourism. Moreover, such narratives might be effective at offsetting the dualist, contradictory aspects of PPAs as outlined by many academics. In the PPA context, these narratives are salient to a popular, global reality that asserts how we currently relate to and govern biodiversity conservation is sufficient. Wilderness, sustainable development, PAs, and like concepts have been problematized by academics and philosophers who believe humans are not separate from Nature (e.g., Cronon, 1996) and that we must reconfigure our relations with Nature to do right by our planet and each other (Büscher and Fletcher 2019; Escobar 2018; Leopold 1989).

A dominant neoprotectionist value system has underpinned global projects like the 30 × 30 initiative and aligns with PPA formation in Chile and Argentina. Therefore, while societies figure out how to reconcile divergent conservation-development values, there is a need to pay careful consideration to the stories told in the conservation arena. The narratives assigning meaning to PPAs expose and (re) produce positive and negative outcomes as well as foretell both our ability and inability to achieve societal changes that can produce net benefits for human and non-human species alike.

Indeed, setting aside large tracts of land for biodiversity often produces immediate net benefits for biological and ecological systems, and cultivating broad sociopolitical acceptance for these projects can happen though might take decades with the help of carefully constructed and reality-altering narratives. Problematically, given our current climate, moral, and biodiversity challenges, we may not have that long to wait for the next iteration of narratives to evolve that promote reconfigured human-nature relations.

9.6   Conclusion

PAs are important policy instruments and narratives about them are powerful tools for minimizing or enhancing the legitimacy of protectionist conservation, PA networks, and human development projects and their archetype discourses. In this chapter, we unveiled nine narratives and their lessons or points that have given meaning to one of the largest private conservation efforts in Latin America, and a growing conservation phenomenon with global implications. Narratives distilled from academic and popular press texts revealed the meaning of PPAs in society. In our analysis, we highlight how the Tompkins projects advanced the PPA movement in the face of social and political opposition, aided and stymied by the narratives about their efforts. Yet, our analysis of narratives produced more questions than it answers. For instance, considering the temporal and socio-political aspects of our case study findings, if change from unsustainable to sustainable living takes time and is hotly contested, how quickly can we expect to evolve beyond what we have already accomplished? And how will societies integrate private lands into our bigger species and planet saving projects? We note that researchers must also consider how environmental non-governmental organizations, such as the Tompkins, position themselves within the popular press, social media, and academic research, to build their narratives.

The Tompkins' projects both in Chile and in Argentina provide a window into what it takes for societies to confront different value systems, driven by the primary idea of nature conservation. Hence, we find the Tompkins case is somewhat of a microcosm of the world we reside in today. Alternative interpretations of reality and human involvement in creating, managing, or interfering in natural processes are hotly debated, and real cultural schisms require urgent attention. That said, achieving 30 × 30 or other goals will involve creating meaningful stories that unite academia, communities, politicians, the media, and other critical actors but also represent the various experiences linked to PPA projects. These stories, as shapers of reality, will help determine which conservation projects are feasible, beneficial, and logical, help cultivate a sense of proprietorship among stakeholders, and help societies evolve in a way that better balances competing meanings that prevent consensus about the role PAs play in safeguarding biodiversity.

Acknowledgements   Map elaborated by Marcela Torres-Gómez, based on data set publicly available at IDE Chile http://www.ide.cl.

References

M.L. Aastrup, Conservation narratives and conflicts over protected areas in post-socialist Romania. J. Polit. Ecol **27**(1), 84–104 (2020). https://doi.org/10.2458/v27i1.23481

W.M. Adams, Geographies of conservation III: Nature's spaces. Prog. Hum. Geogr. **44**(4), 789–801 (2020). https://doi.org/10.1177/0309132519837779

Anonymous, Around the world - a Chile reception. Earth Island J. San Francisco **14**(4), (Winter 1999/2000), 17 (2000) https://www.proquest.com/magazines/around-world/docview/213826747/se-2?accountid=27871

P. Bachmann-Vargas, C.S.A. van Koppen, Disentangling environmental and development discourses in a peripheral spatial context: The case of the Aysén region, Patagonia, Chile. J. Environ. Dev **29**(3), 366–390 (2020). https://doi.org/10.1177/1070496520937041

P. Bachmann-Vargas, C.S.A. van Koppen, L. Lamers, Protecting wilderness or cultural and natural heritage? An analysis of environmental discourses in northern Patagonia, Chile, in *Meanwhile in Aysén-Patagonia: Exploring Discursive Transformations on Environment and Development in a Remote Periphery*, ed. by I.P. Bachmann-Vargas, (Wageningen University, 2021), pp. 81–100. https://doi.org/10.18174/553259

E. Bantle, Creating Patagonia National Park: Understanding community response to national park creation by a private foreign non-profit organization. J. Undergrad. Res **XIII**, 1–13 (2010)

H. Blair, K. Bosak, T. Gale, Protected areas, tourism, and rural transition in Aysén, Chile. Sustainability **11**(24), Article 7087 (2019). https://doi.org/10.3390/su11247087

E. Bohoslavsky, Contra la Patagonia Judía. La familia Eichmann y los nacionalistas argentinos y chilenos frente al Plan Andina (de 1960 a nuestros días) [Against Jewish Patagonia. The Eichmann family and the Argentine and Chilean nationalists in the face of the Andina Plan (from 1960 to the present)]. Cuadernos Judaicos **25**, 223–247 (2008). https://doi.org/10.5354/0718-8749.2008.25631

P. Bonnefoy, With 10 million acres in Patagonia, a national park system is born. New York Times (Feb 19, 2018). https://www.nytimes.com/2018/02/19/world/americas/patagonia-national-park-chile.html

B. Borrie, T. Gale, K. Bosak, Privately protected areas in increasingly turbulent social contexts: Strategic roles, extent, and governance. J. Sustain. Tour. **13**(11), 2631–2648 (2020). https://doi.org/10.1080/09669582.2020.1845709

F. Bourlon, La bio-geografía de Douglas Tompkins, una mirada comprensiva de la conservación privada en la Patagonia chilena [The bio-geography of Douglas Tompkins, a comprehensive view of private conservation in Chilean Patagonia]. Revista de Aysenología (4), 86–98 (2017a)

F. Bourlon, La géographie esthétique de Douglas Tompkins, une utopie éco-philanthropique en Patagonie [The aesthetic geography of Douglas Tompkins, an eco-philanthropic utopia in Patagonia]. Études Caribéennes, 37–38 (2017b)

D. Brockington, R. Duffy, J. Igoe, *Nature Unbound: Conservation, Capitalism and the Future of Protected Areas* (Routledge, London, 2012)

J. Bruner, *Acts of Meaning* (Harvard University Press, Cambridge, MA, 1990)

B. Büscher, W. Dressler, Linking neoprotectionism and environmental governance: On the rapidly increasing tensions between actors in the environment-development nexus. Conserv. Soc. **5**(4), 586–611 (2007)

B. Büscher, R. Fletcher, Towards convivial conservation. Conserv. Soc. **17**(3), 283–296 (2019)

B. Büscher, R. Fletcher, *The Conservation Revolution: Radical Ideas for Saving Nature Beyond the Anthropocene* (Verso Books, London, 2020)

N. Busscher, C. Parra, F. Vanclay, Land grabbing within a protected area: The experience of local communities with conservation and forestry activities in Los Esteros del Iberá, Argentina. Land Use Policy **78**, 572–582 (2018). https://doi.org/10.1016/j.landusepol.2018.07.024

B. Byrnes, The philanthropists "paying rent" to planet Earth in Argentina. CNN (Oct 7, 2009). http://edition.cnn.com/2009/WORLD/americas/10/05/going.green.tompkins.biodiversity/index.html

M.J. Carter, The hermeneutics of frames and framing: An examination of the media's construction of reality. SAGE Open 3(2), 1–12 (2013). https://doi.org/10.1177/2158244013487915

Chile Sustentable, El polémico imperio de Tompkins [The controversial Tompkins' empire] (2006). Retrieved 22 Jan 2023, from http://www.chilesustentable.net/2006/07/el-polemico-imperio-de-tompkins/

Chilean Ministry of Public Lands, Crea el "Parque Nacional Pumalín Douglas Tompkins", en las comunas de Cochamó, de Hualaihué, de Chaitén y de Palena, Provincias de Llanquihue y de Palena, Región de los Lagos [Creation of the "Pumalín Douglas Tompkins National Park", in the communes of Cochamó, Hualaihué, Chaitén and Palena, Provinces of Llanquihue and Palena, Los Lagos Region]. Decreto 28, Biblioteca del Congreso Nacional (2008). https://www.bcn.cl/leychile/navegar?idNorma=1121563

Chilean Ministry of the Environment, Gobierno y Tompkins Conservation sellan acuerdo para donación de tierras y creación de Red de Parques Nacionales de 4.5 millones de hectáreas [Government and Tompkins Conservation seal agreement for land donation and creation of 4.5 million-hectare National Park Network]. Ministerio del Medio Ambiente (March 15, 2017). https://mma.gob.cl/gobierno-y-tompkins-conservation-sellan-acuerdo-para-donacion-de-tierras-y-creacion-de-red-de-parques-nacionales-de-45-millones-de-hectareas/

W. Cronon, The trouble with wilderness: Or getting back to the wrong nature. Environ. Hist. 1(1), 7–28 (1996)

M. Cuevas, P. Luna, US conservationist, Chile sign creation of national parks. Associated Press News (Jan 29, 2018). https://apnews.com/article/e800eb4eacda477ba3fa68b303d3d8b4

J. Daley, Chile designates 10 million acres of land as national parks. Smithsonian Magazine (March 20, 2017). https://www.smithsonianmag.com/smart-news/chile-adding-11-million-acres-national-parks-180962592/

P. Di Giminiani, M. Fonck, Emerging landscapes of private conservation: Enclosure and mediation in southern Chilean protected areas. Geoforum 97(August), 305–314 (2018). https://doi.org/10.1016/j.geoforum.2018.09.018

B. Doherty, Icon: Rewilding our parks, our hearts, and our wallets. Forbes (Dec 15, 2020). https://www.forbes.com/sites/bdoherty/2020/12/15/icon-rewilding-our-parks-our-hearts-and-our-wallets/?sh=1fb4838849bd

M. Drenthen, Rewilding in layered landscapes as a challenge to place identity. Environ. Values 27(4), 405–425 (2018). https://doi.org/10.3197/096327118X15251686827732

R. Eckersley, A paradise called Parque Pumalin: Philanthropy or colonialism? Arena Magaz. (Fitzroy, Vic) 40, 31–33 (1999)

El Mostrador, Nuevo Seremi de vivienda de Aysén asegura que existe complot Gringo para quedarse con la Patagonia [New housing Seremi of Aysén assures that there is a Gringo plot to take over Patagonia]. El Mostrador (Aug 30, 2012). https://www.elmostrador.cl/noticias/sin-editar/2012/02/07/nuevo-seremi-de-vivienda-de-aysen-asegura-que-existe-complot-gringo-para-quedarse-con-la-patagonia/

A. Escobar, Designs for the Pluriverse: Radical Interdependence, Autonomy, and the Making of Worlds (Duke University Press, Durham, 2018)

C. Esin, M. Fathi, C. Squire, Narrative analysis: The constructionist approach, in The SAGE Handbook of Qualitative Data Analysis, ed. by U. Flick, (SAGE Publications Ltd, London, 2014), pp. 203–216

B. Fernández, Andrés Azócar: "Belisario Velasco se transformó en el operador del gobierno de Frei para sacar a Tompkins del país" [Andrés Azócar: "Belisario Velasco became the Frei government's operator to get Tompkins out of the country"] El Mostrador (Dec 11, 2015a). https://www.elmostrador.cl/noticias/pais/2015/12/11/andres-azocar-belisario-velasco-se-transformo-en-el-operador-del-gobierno-de-frei-para-sacar-a-tompkins-del-pais/

O. Fernández, Las batallas del "Gringo" ambientalista que derrotó al proyecto HidroAysén [The battles of the environmentalist "Gringo" who defeated the HidroAysén project]. La Tercera (Dec 8, 2015b). https://www.latercera.com/noticia/las-batallas-del-gringo-ambientalista-que-derroto-al-proyecto-hidroaysen/

Z. Fitzner, The inspired conservationists behind Chile's new national parks. Earth (July 4, 2018). https://www.earth.com/news/chile-new-national-parks/

R. Fowler, Language in the News: Discourse and Ideology in the Press (Routledge, London, 1991)

A. Frankel, Doug Tompkins, Chile's ecowarrior. Outside (May 5, 2004). https://www.outsideonline.com/adventure-travel/doug-tompkins-chiles-ecowarrior/

J. Franklin, Can eco barons save the planet ? The Jerusalem Post (Feb 28, 2008). https://www.jpost.com/magazine/features/can-eco-barons-save-the-planet

T. Frei, J. Derks, C.R. Fernández-Blanco, G. Winkel, Narrating abandoned land: Perceptions of natural forest regrowth in Southwestern Europe. Land Use Policy **99**, Article 105034 (2020). https://doi.org/10.1016/j.landusepol.2020.105034

T. Gale, A. Ednie, Can intrinsic, instrumental, and relational value assignments inform more integrative methods of protected area conflict resolution? Exploratory findings from Aysén, Chile. J. Tour. Cult. Chang. **18**(6), 690–710 (2019). https://doi.org/10.1080/14766825.2019.1633336

M. García, M.E. Mulrennan, Tracking the history of protected areas in Chile: Territorialization strategies and shifting state rationalities. J. Lat. Am. Geogr. **19**(4), 199–234 (2020). https://doi.org/10.1353/lag.2020.0085

R. Garzón, How to create paradise: A tale of love, death and nature. El País (2017). https://english.elpais.com/elpais/2017/02/16/inenglish/1487238993_690246.html

N. Graydon, Buying sanctuary. Ecologist **36**(4), 28–35 (2006)

A. Hamon, Un nuevo Israel en la Patagonia: El mito de la conspiración judía [A new Israel in Patagonia: The myth of the Jewish conspiracy]. El Mostrador (2015). https://www.elmostrador.cl/noticias/pais/2015/06/21/un-nuevo-israel-en-la-patagonia-el-mito-de-la-conspiracion-judia/

K. Hayden, Soiled sovereignty: Land issues in the Southern Cone. Council on Hemispheric Affairs (Nov 19, 2012). https://www.coha.org/20349/

G. Holmes, What is a land grab? Exploring green grabs, conservation, and private protected areas in southern Chile. J. Peasant Stud. **41**(4), 547–567 (2014). https://doi.org/10.1080/03066150.2014.919266

G. Holmes, Markets, nature, neoliberalism, and conservation through private protected areas in southern Chile. Environ. Plan. A Econ. Space **47**(4), 850–866 (2015). https://doi.org/10.1068/a140194p

G. Holmes, Conservation Jujutsu, or how conservation NGOs use market forces to save nature from markets in southern Chile, in *The Anthropology of Conservation NGOs*, ed. by P.B. Larsen, D. Brockington, (Palgrave Macmillan, Cham, 2018), pp. 181–201

B. Hora, Private protection initiatives in mountain areas of southern Chile and their perceived impact on local development - the case of Pumalín Park. Sustainability **10**, Article 1584 (2018) https://doi.org/10.3390/su10051584

J. Igoe, D. Brockington, Neoliberal conservation: A brief introduction. Conserv. Soc. **5**(4), 432–449 (2007)

C. Jones, Ecophilanthropy, neoliberal conservation, and the transformation of Chilean Patagonia's Chacabuco valley. Oceania **82**, 250–263 (2012)

J. Langman, Thinking big: After founding Esprit and North Face, Doug Tompkins dresses up an 800,000-acre park. Environ. Magaz. **8**(5), 24 (1997)

J. Langman, Conversation: Doug Tompkins. Earth Island J (2012) https://www.earthisland.org/journal/index.php/magazine/entry/doug_tompkins/

A. Leopold, *A Sand County Almanac, and Sketches Here and There* (Oxford University Press, New York, 1989)

L. Lester, Wilderness and the loaded language of news. Media Int. Australia **115**(1), 123–134 (2005). https://doi.org/10.1177/1329878x0511500112

A. Lieblich, R. Tuval-Mashiach, T. Zilber, *Narrative Research: Reading, Analysis, and Interpretation* (SAGE Publications Inc, Thousand Oaks, 1998)

E. Londoño, Fixing the damage we've done: Rewilding jaguars in Argentina. The New York Times (Sept 1, 2020). https://www.nytimes.com/2020/09/01/world/americas/Jaguars-argentina-ibera.html

M.J. López, Mrs. Patagonia. La Tercera (Dec 6, 2012). https://www.latercera.com/revista-que-pasa/1-10703-9-mrs-patagonia/

E. Louder, K. Bosak, What the Gringos brought: Local perspectives on a private protected area in Chilean Patagonia. Conserv. Soc. **17**(2), 161–172 (2019). https://doi.org/10.4103/cs.cs_17_169

E. Louder, C. Wyborn, Biodiversity narratives: Stories of the evolving conservation landscape. Environ. Conserv. **47**(4), 351–359 (2020). https://doi.org/10.1017/S0376892920000387

B.K. Lynch, Rethinking assessment from a critical perspective. Lang. Test. **18**(4), 351–372 (2001). https://doi.org/10.1177/026553220101800403

S. Mantegna, S. Viña, V. Pohl Schanake, V.H. Vallejos, Parque nacional: Y tomaron tu nombre "Iberá" [National Park: And they took your name "Iberá"]. XVIII Jornadas de Geografía de la UNLP (2016) http://sedici.unlp.edu.ar/handle/10915/93949

J. Mark, The fashion executives who saved a Patagonian paradise. Sierra (Sept 4 2019). https://www.sierraclub.org/sierra/2019-5-september-october/feature/fashion-executives-who-saved-patagonian-paradise-doug-kris-tompkins

M.C. Nisbet, D.A. Scheufele, What's next for science communication? Promising directions and lingering distractions. Am. J. Bot. **96**(10), 1767–1778 (2009). https://doi.org/10.3732/ajb.0900041

A. Núñez, E. Aliste, A. Bello, J.P. Astaburuaga, Eco-extractivismo y los discursos de la naturaleza en Patagonia-Aysén: Nuevos imaginarios geográficos y renovados procesos de control territorial [Eco-extractivism and the discourses of nature in Patagonia-Aysén: New geographic imaginaries and renewed processes of territorial control]. Revista Austral de Ciencias Sociales (35), 133–153 (2018). https://doi.org/10.4206/rev.austral.cienc.soc.2018.n35-09

J. O'Neill, A. Holland, A. Light, *Environmental Values* (Routledge, London, 2008)

A. Paterson, Protected areas governance in a southern African transfrontier context, in *Transboundary Governance of Biodiversity*, ed. by I.L.J. Kotzé, T. Marauhn, (Brill Nijhoff, Boston, 2014), pp. 163–203

P. Pinochet, V. Thenoux, La evolución del relato biográfico de Douglas Tompkins en los medios de comunicación chilenos entre 1995 y 2015 [The evolution of the Douglas Tompkins biographical narrative in the Chilean media between 1995 and 2015]. Facultad de Comunicaciones, Universidad del Desarrollo (2016)

V. Pohl Schanake, V.H. Vallejos, Concentración y extranjerización de tierras en torno a los Esteros del Iberá, Provincia de Corrientes [Concentration and foreign ownership of land around the Iberá marshlands, Province of Corrientes]. V Congreso Nacional de Geografía de Universidades Públicas "Geografías por venir", 855–871 (2008)

D.E. Polkinghorne, Narrative configuration in qualitative analysis. Int. J. Qual. Stud. Educ. **8**(1), 5–23 (1995). https://doi.org/10.1080/0951839950080103

A. Promis, D. Cortés, I. Espinoza, Ruta de los Parques Nacionales de la Patagonia: Conservación de la última naturaleza al sur del mundo [Patagonia National Parks Route: Conservation of the last of nature in the southern part of the world]. Biodiversidata **8**, 95–108 (2019)

D. Quammen, How an unprecedented gift built a legacy of conservation in South America. National Geographic (April 23, 2020). https://www.nationalgeographic.com/magazine/2020/05/how-an-unprecedented-gift-built-a-legacy-of-conservation-in-patagonia-feature/

F. Ramírez, M. Folchi, La factibilidad histórica-ecológica de proteger la naturaleza. El caso del Parque Pumalín de Douglas Tompkins [The historical-ecological feasibility of protecting nature. The case of Douglas Tompkins' Pumalín Park.]. VI Encuentro Científico sobre el Medio Ambiente, 1–12 (1999) http://bdrnap.mma.gob.cl/recursos/SINIA/Biblio_AP/El caso del Parque.pdf

M. Reel, Argentine land fight divides environmentalists, rights advocates. Washington Post (Sept 24, 2006). https://www.washingtonpost.com/wp-dyn/content/article/2006/09/23/AR2006092301054.html

Rewilding Chile, Rewilding Chile is launched. Rewilding Chile, (2021). https://www.rewildingchile.org/en/news/rewilding-chile-is-launched/

F.C. Richardson, B.J. Fowers, Critical theory, postmodernism, and hermeneutics: Insights for critical psychology, in *Critical Psychology: An Introduction*, ed. by D. Fox, I. Prilleltensky, (SAGE Publications Inc, London, 1997), pp. 265–283

C.K. Riessman, *Narrative Methods for the Human Sciences* (SAGE Publications Inc, Thousand Oaks, 2008)

C.M. Roberts, B.C. O'Leary, J.P. Hawkins, Climate change mitigation and nature conservation both require higher protected area targets. Philos. Trans. R Soc. Biol. Sci **375**(1794) (2020). https://doi.org/10.1098/rstb.2019.0121

L. Rohter, An American in Chile finds conservation a hard slog. New York Times (Aug 7 2005). https://www.nytimes.com/2005/08/07/world/americas/an-american-in-chile-finds-conservation-a-hard-slog.html

A. Ruggiero, Historic donation creates 5 new national parks in Chile. Gear Junkie (March 17, 2017). https://gearjunkie.com/news/tompkins-conservation-chile-protects-11-million-acres

E. Sala, J. Mayorga. D. Bradley, R.B. Cabral, T.B. Atwood, A. Auber, et al., Protecting the global ocean for biodiversity, food and climate. Nature **592**, 397–402 (2021). https://doi.org/10.1038/s41586-021-03496-1

M.J. Salas, Pensando en el retiro [Thinking about retirement]. La Tercera (2015). https://www.latercera.com/paula/pensando-en-el-retiro/

D. Saverin, El empresario que quería salvar el paraíso [The entrepreneur who wanted to save paradise]. La Tercera (2015). https://www.latercera.com/noticia/el-empresario-que-queria-salvar-el-paraiso/

C. Serenari, M.L. Lute, Delegitimizing large carnivore conservation through discourse. Soc. Nat. Resour. **34**(1), 3–22 (2021). https://doi.org/10.1080/08941920.2020.1727593

C. Serenari, M.N. Peterson, Y.F. Leung, P. Stowhas, T. Wallace, E.O. Sills, Private development-based forest conservation in Patagonia: Comparing mental models and revealing cultural truths. Ecol. Soc. **20**(3) (2015). https://doi.org/10.5751/ES-07696-200304

C. Serenari, M.N. Peterson, T. Wallace, P. Stowhas, Private protected areas, ecotourism development and impacts on local people's well-being: A review from case studies in southern Chile. J. Sustain. Tour. **25**(12), 1792–1810 (2017). https://doi.org/10.1080/09669582.2016.1178755

N.L. Sharp, R.A. Bye, A. Cusick, Narrative analysis, in *Handbook of Research Methods in Health Social Sciences*, ed. by P. Liamputtong, (Springer, Singapore, 2019), pp. 861–880

V. Shlapentokh, The study of values as a social phenomenon: The Soviet case. Soc. Forces **61**(2), 403–417 (1982)

R. Soliva, M. Hunziker, Beyond the visual dimension: Using ideal type narratives to analyse people's assessments of landscape scenarios. Land Use Policy **26**(2), 284–294 (2009). https://doi.org/10.1016/j.landusepol.2008.03.007

F. Stelling, C. Allan, R. Thwaites, Nature strikes back or nature heals? Can perceptions of regrowth in a post-agricultural landscape in south-eastern Australia be used in management interventions for biodiversity outcomes? Landsc. Urban Plan. **158**, 202–210 (2017). https://doi.org/10.1016/j.landurbplan.2016.08.015

A. Strauss, J. Corbin, *Basics of Qualitative Research* (SAGE, Newbury Park, 1990)

D.R. Tecklin, C. Sepúlveda, The diverse properties of private land conservation in Chile: Growth and barriers to private protected areas in a market-friendly context. Conserv. Soc. **12**(2), 203–217 (2014). https://doi.org/10.4103/0972-4923.138422

Tompkins Conservation, All protected areas. Tompkins Conservation (2019). http://www.tompkinsconservation.com/all_protected_areas.htm

Tompkins Conservation, Ruta de los Parques de la Patagonia [Route of Parks of Patagonia]. Ruta de los Parques de la Patagonia (2021). https://www.rutadelosparques.org/en/#

UNEP, Kristine McDivitt Tompkins named UN environment patron of protected areas. UN Environment Programme (2018, May 18). https://www.unep.org/news-and-stories/press-release/kristine-mcdivitt-tompkins-named-un-environment-patron-protected

UNEP, COP15 ends with landmark biodiversity agreement. UN Environment Programme (Dec 20, 2022). https://www.unep.org/news-and-stories/story/cop15-ends-landmark-biodiversity-agreement#:~:text=The%20United%20Nations%20Biodiversity%20Conference,weeks%20for%20the%20important%20summit

UNEP-WCMC & IUCN, Protected planet report 2020 (2021). Retrieved 9 Jan 2023, from https://www.unep.org/resources/protected-planet-report-2020

J. Vidal, Love, death and rewilding – how two clothing tycoons saved Patagonia. The Guardian (Nov 28, 2016). https://www.theguardian.com/environment/2016/nov/28/how-two-clothing-tycoons-saved-patagonia-doug-tomkins-kris-mcdivitt-tomkins

E. Wakild, Purchasing Patagonia: The contradictions of conservation in free market Chile, in *Lost in the Long Transition: Struggles for Social Justice in Neoliberal Chile*, ed. by W.L. Alexander, (Lexington Books, Lanham, 2009)

J. Warburton-Lee, Conservation natural preserve. Nestled in the Chilean rainforest Pumalín Park stands as a beacon for privately funded conservation. Geographical, 65–72 (2001)

B. Weir, They purchased paradise... then gave it all away. CNN (Nov 13, 2017). https://edition.cnn.com/travel/article/wonder-list-bill-weir-patagonia/index.html

P. West, J. Igoe, D. Brockington, Parks and peoples: The social impact of protected areas. Annu. Rev. Anthropol. **35**, 251–277 (2006) https://doi.org/10.1146/annurev.anthro.35.081705.123308

B. Wieners, Patagonia dreaming: Kris Tompkins works to build the best national park. Bloomberg (April 23, 2014). http://www.businessweek.com/articles/2014-04-17/patagonia-dreaming-kris-tompkins-works-to-build-the-best-national-park

T. Wilkinson, The essential role of eco-capitalism in saving the best that remains. Mountain J (Jan 29, 2018). https://mountainjournal.org/the-essential-role-of-capitalism-in-protecting-nature

E. Woodhouse, C. Bedelian, P.R. Barnes, G.S. Cruz-Garcia, N. Dawson, N. Gross-Camp, et al., *Rethinking Entrenched Narratives About Protected Areas and Human Wellbeing in the Global South* (UCL Open: Environment Preprint, 2021)

D. Yanow, *Conducting Interpretive Policy Analysis. Qualitative Research Methods* (SAGE Publications Inc, Thousand Oaks, 2000)

A. Zacarías, Argentina, Corrientes: Tompkins y su imperio natural [Argentina, Corrientes: Tompkins and his natural empire]. Biodiversidad LA (Oct 18, 2005) https://www.biodiversidadla.org/Principal/Prensa/Argentina_Corrientes_Tompkins_y_su_imperio_natural

T. Zamboni, S. Di Martino, I. Jiménez-Pérez, A review of a multispecies reintroduction to restore a large ecosystem: The Iberá rewilding program (Argentina). Perspec. Ecol. Conserv **15**(4), 248–256 (2017) https://doi.org/10.1016/j.pecon.2017.10.001

F. Zorondo-Rodríguez, M. Díaz, G. Simonetti-Grez, J.A. Simonetti, Why would new protected areas be accepted or rejected by the public?: Lessons from an ex-ante evaluation of the new Patagonia Park network in Chile. Land Use Policy **89**, Article 104248 (2019) https://doi.org/10.1016/j.landusepol.2019.104248

Chapter 10
Exploring Social Representations of Nature-Based Tourism, Development Conflict, and Sustainable Development Futures in Chilean Patagonia

Gabriel Inostroza Villanueva, Fabien Bourlon, Trace Gale-Detrich, and Heidi Blair

Abstract This chapter explored how *Modernization, Transformation*, and *Control* Sustainable Development (SD) imaginaries and trajectories interacted, nuanced, and mediated the approaches of neoliberal development initiatives and the conflicts which surrounded them. A collective case study approach was employed to better understand actors, their strategies, and perceptions about three large-scale development proposals, the Patagonia National Parks network, the HidroAysén hydroelectric project, and the Río Cuervo hydroelectric project. We identified six themes that characterized participants' preferences related to processes: a desire for greater proactiveness around *transparency*, a *binding participation* process of governance, *bottom-up* decision making, *re-empowerment* of local groups, *decentralization*, and *improved oversight* practices. Further, directed content analysis revealed numerous

G. Inostroza Villanueva (✉)
Universidad Austral de Chile, Patagonia Campus, Coyhaique, Chile
e-mail: gabriel.inostroza@uach.cl

F. Bourlon
Centro de Investigación en Ecosistemas de la Patagonia (CIEP), Sustainable Tourism Research Line, Human-Environmental Interactions Group, Coyhaique, Chile

Université Grenoble Alpes, Institute of Urban Planning and Alpine Geography - UMR 5194, Grenoble, Isère, France
e-mail: fabienbourlon@ciep.cl; fabien.bourlon@umrpacte.fr

T. Gale-Detrich
Centro de Investigación en Ecosistemas de la Patagonia (CIEP), Sustainable Tourism Research Line, Human-Environmental Interactions Group, Coyhaique, Chile

Cape Horn International Center (CHIC), Puerto Williams, Magallanes & Chilean Antarctica Region, Chile
e-mail: tracegale@ciep.cl

H. Blair
University of Montana, College of Forestry and Conservation, Missoula, MT, USA
e-mail: heidi.blair@mso.umt.edu

229

manifestations of the three SD futures trajectories and several examples of interactions and overlap between the SD imaginaries. Seemingly, some development actors navigated between SD imaginaries and tactics to maneuver between local and national actors, agendas, and decision-making processes. We have termed these tactics as *SD agility*, which are initially defined as: "the strategic ability to maneuver between SD imaginaries and trajectories to achieve strategic SD outcomes." SD agility tactics merit additional study as we believe they may represent an important capacity for SD futures trajectories.

Keywords Patagonia · Futures of sustainability · Development trajectories · Neoliberal development · Patagonia National Park · HidroAysén · Río Cuervo

10.1 Introduction

The Aysén Region of Chilean Patagonia is a biogeographic region with an abundance of relatively-well conserved natural resources. In recent decades, different visions of sustainability have led to tensions and conflict in the area with respect to proposals for territorial development. Since 1990, when Chile returned to democracy, numerous territorial conflicts have arisen around large privately driven development projects, generally driven by foreign investors and firms and/or large-scale exportation opportunities, like mining, aquaculture, timber harvesting, and energy production. For example, for over a decade beginning in 2005, intentions to implement large hydroelectric projects within the region triggered backlash from many of the region's citizens and groups. They preferred less invasive and extractive forms of development, like scientific or nature-based tourism (NBT) as they believed them to be more complementary with social and environmental conservation (Inostroza and Cànoves 2014). Other local actors have different views for the best path of regional development. For example, some groups have found recent nature conservation initiatives which have used private funding to influence the creation and reclassification of public protected areas (PAs) suspicious. In the past, the creation of PAs has meant the prohibition of many traditional local uses of PA lands and resources (e.g., animal grazing or small-scale cuts for firewood) by local communities (Blair et al. 2019; Inostroza 2007; Oltremari et al. 1994). And, even among NBT stakeholders, who often depend on PAs to develop their work, there has been a range of opinions about these projects, especially in relation to the processes that have been employed in their designation, and ongoing uncertainty about how new PAs and stricter PA designations will affect future NBT within the region (Gale and Ednie 2019).

Many believe that Aysén's territorial tensions and conflicts have their origin in the neoliberal economic development model that was implemented during the

1973–1990 dictatorship of Augusto Pinochet (Gentes and Policzer 2022; Harvey 2005; Latta and Aguayo 2012). This model is characterized by large national and transnational private sector actors who are motivated by the pursuit of economic profit and enabled by free markets operating under government guarantees (Harvey 2001, 2005; Latta and Aguayo 2012). Neoliberalism has remained the dominant economic model over the decades that have followed Chile's return to democracy. Numerous authors have written about the implications of this overarching model for Chile's environmental management and social well-being (Borrie et al. 2020; De Matheus e Silva et al. 2018; Gentes and Policzer 2022; Jones 2012; Latta and Aguayo 2012). These authors have described Chile as having a complex public/private oversight and governance system with limited regulatory capacity, that was purposefully designed through the implementation of the neoliberal model, to limit and weaken governmental authority over private business.

Latta and Aguayo (2012) noted some progress has been made in the years following the dictatorship to develop policy and oversight processes to facilitate debate with respect to development and protect Chilean environments and communities. For example, Chile's Environmental Impact Assessment System (SEIA) was implemented in 1997, requiring all new public and private projects to evaluate and declare their prospective environmental impacts, using either a Declaration of Environmental Impact (DEI) or a full Environmental Impact Assessment (EIA), depending on the scope of the project. Both are reviewed by the appropriate regulatory government agencies. In the case of an EIA, additional rigor requires a public comment period and a plan detailing mitigation, remediation, or compensation measures. Finally, the Environmental Evaluation Service (SEA) deliberates and eventually approves or rejects project proposals based on the outcomes of these processes (Chilean Environmental Evaluation Service 2022; Latta and Aguayo 2012; OECD 2016).

The SEIA's mission stresses sustainable development (SD), environmental and social well-being, citizen participation, transparency, mitigation, and compensation; yet it promises the efficiency facilitated by a neoliberal approach to economic development:

> To contribute to sustainable development, the preservation and conservation of natural resources, and the quality of life of the country's inhabitants, through the management of the Environmental Impact Assessment System, ensuring a transparent and efficient technical and environmental evaluation, in coordination with State agencies, while encouraging and facilitating citizen participation in the evaluation processes, with the purpose of mitigating, compensating and/or repairing significant environmental impacts (Chilean Environmental Evaluation Service 2022. para. 1).

As SD is emphasized as first and foremost within Chile's SEA mission, it seems appropriate and necessary to understand how this concept nuances and mediates conflicts around different development proposals that have advanced within the Aysén Region of Chilean Patagonia.

10.2   Theoretical Framework

Sustainable development and neoliberalism   Since the early years of the twenty-first century, academic discourse has advanced two points of view in parallel: the domination of neoliberal development agendas that worsen social inequities and the dominance of sustainable development (SD) principles that empower democracy, social justice, and the environment (Baldwin et al. 2019; Kashwan et al. 2019; Khan 2015; Kumi et al. 2014; Raco 2005; Slocum et al. 2019; Torres et al. 2015; Wolters 2022). For example, Raco (2005) pointed out the irony in these parallel conceptions, suggesting that SD has become an all-inclusive "meta-narrative" for development throughout the world; yet the spread of neoliberalism and its inherent market-based demands for enhanced capital accumulation, have augmented social inequities and resource exploitation, resulting in a development that is anything but "sustainable." Throughout the years, researchers have called for increased focus to address the many equity issues that have arisen from global interventions at local levels and the development of mechanisms to address the unfair power relations that frame many neoliberal development conflicts (Kashway et al. 2019; Khan 2015; Kumi et al. 2014; Slocum et al. 2019). Others have criticized neoliberal approaches and their conceptions of nature as being too narrow, focused solely on its commodification and economic value (Baldwin et al. 2019; Wolters 2022).

In recent years, new neoliberal forms of conservation-based development (CBD), like private protected areas (PAs) and NBT, seem to better align with SD objectives; yet, these types of development have also been criticized for predatory land grabbing, displacement, and practices that reduce social embeddedness of nature (Blair et al. 2019; Borrie et al. 2020; Büscher and Davidov 2016; De Matheus e Silva et al. 2018; Fletcher 2011; Holmes 2012, 2014). For example, Borrie et al. (2020) identified, "in order to improve long-term support and integration of PPAs and NBT, greater attention needs to be given to social well-being outcomes (including equity and justice concerns), building of social capital, and the preservation of local identities and histories" (p.1). Neoliberal ideology suggests that corporate social responsibility (CSR) and corporate community development (CCD) tactics can effectively address the social impacts of large-scale capitalist development projects within local communities, including those proposed by Borrie et al. (2020). Yet, debate continues to evaluate CSR/CCD as a replacement for the wider social role played by governments in SD, through the protection of the public good (Babidge 2013; Gentes and Policzer 2022; McLennan and Banks 2019).

Social Conflict   Reboratti (2019) defined conflict as involving, "situations in which concrete facts are viewed and interpreted in dissimilar and opposing ways by different groups and sectors of society, who adopt actions according to those visions and interpretations"(p. 2). The causes of conflicts originate from the different representations of social groups around nature, land use, and the use of natural resources (Sæþórsdóttir 2012). Williams (2002) argued that conflicts can be expected when multiple representations of a particular place occur. Jacobsen and Linnell (2016) identified that, "The extent to which stakeholders regard a management system as being just and fair is a key social dimension of conflict" (p.197). Torres et al. (2015) contextualized growing social conflict in Chile around neoliberal development, highlighting the important role that Chilean student movements have played in resistance movements and discourse since 2011. Discourses have a role in the construction of social realities, serving the interests of specific social groups. It is within this difference of interests that conflicts arise (Sæþórsdóttir 2012).

Sustainability Social Change Imaginaries Adloff and Neckel (2019) proposed three trajectories of social change related to sustainability that evolve from distinct imaginaries: Modernization, Transformation, and Control. According to Adloff and Neckel (2019), these three trajectories do not operate in a vacuum; rather, they are likely interrelated and may even operate incrementally, with one form contributing to the evolution of another. They have called for researchers to consider these trajectories within their research about SD and conflict, through consideration of the actors, practices, and structures involved in these imaginations, evidence that they are already manifesting and interacting, and evidence of how these imaginations might be able to provoke societal change on larger scale.

According to the authors, an imaginary based on sustainability through *Modernization* involves transforming economies and institutions around *"green"* and sustainable models of growth (e.g., CBD, circular economies). They explained:

> Programs dedicated to a sustainable modernization intend to improve the ecological balance of modern societies by means of technological and social innovations, so that the earth's capacities are no longer overstrained. These programs do not intend to fundamentally alter existing structures—such as liberal democracy and market capitalism—or crucial elements of the modern lifestyle—such as individualism, consumption, prosperity, and mobility—but only to adapt these to the changed conditions, characterized by ecological constraints (Adloff and Neckel 2019, p. 1018).

In stark contrast, a sustainability through *Transformation* imaginary proposes a fundamental societal paradigm shift, redefining well-being and eliminating the centrality of economic growth within development (e.g., *buen vivir*, or good living in English; post-capitalism), as this hinders SD potential. This social imaginary seeks an end to competitive, growth-based social orders and a fundamental change in the human–nature relationship. While there are many debates about specific concepts and manifestations, there are also numerous intersections. Adloff and Neckel (2019) attributed this to a common reference point: "the insight that the natural and social foundations of life on earth will not be protected by means of a further economization of sustainability" (p.1020).

The *Control* social imaginary of SD takes a third path, which is decidedly more extreme. It is rooted in assumptions that the future will be marked by increasing ecological emergencies (e.g., pandemics, floods, pollution, wars, drought) and collapse. Thus, this imaginary envisions an authoritarian sustainability with markedly different paths for wealthy elites and the rest of humanity. Enclaves would emerge to protect those with means, while the rest of society would be increasingly exposed to ecological disasters. Adloff and Neckel (2019) explained, "Under the conditions of ecological emergencies, sustainability as control refers to a world of resilience rather than one of genuine sustainability…It refers to forms of coping with crises and adapting to emergencies once they have occurred" (p.1021). In practice, this imaginary operates through *particularlist ethics*, and a belief that the population as a whole will not be capable of enhancing their resilience equally; thus, regulating and restraining practices (e.g., strongholds, exclusions, dispossession, banishment, monitoring, force) will be more appropriate than preventative practices.

This chapter responds to Adloff and Neckel's (2019) call for research using a collective case study approach to explore the ways in which SD imaginaries and trajectories have manifested for actors associated with the NBT sector during three development conflicts within the Aysén region of Chilean Patagonia. We were especially interested in examining how these SD imaginaries interact, nuance, and mediate the approaches of neoliberal development initiatives and the conflicts which surround them. Thus, our case study was guided by two overarching research questions:

- How did regional actors' perceptions, tensions, and conflicts compare with respect to the three development contexts?
- How did sustainability imaginaries manifest within the perspectives expressed by actors about development projects?

10.3   Methodology

10.3.1   Research Design and Justification

The exploratory nature of our research lent itself to qualitative methods, allowing us to engage with the many layers of actors' perceptions and imaginaries (Schutt and Chambliss 2013; Yin 2016). We employed a case study approach, focused on the complexity of development conflict in Aysén and the potential ways in which Adloff and Neckel's (2019) sustainability imaginaries nuance and mediate conflict trajectories (Creswell 2014; Simons 2009). Our approach considered Stake's (2003) ontology, which called for the recognition of reality as subjective, based on elements that are intrinsically linked. We also applied his guidance on the election of a collective case study approach as appropriate for "investigating a phenomenon, population, or general condition" (p.136). Thus, we purposefully selected three recent large-scale development proposals: the Patagonia National Parks Network CBD project, the HidroAysén hydroelectric project, and the Río Cuervo hydroelectric project, to represent a spectrum of development approaches and outcomes in our analysis of SD imaginaries (Fig. 10.1). Common factors between these cases involved their neoliberal driven frameworks, their large scale, and their potential to affect regional NBT operators. We sought to identify and better understand how SD

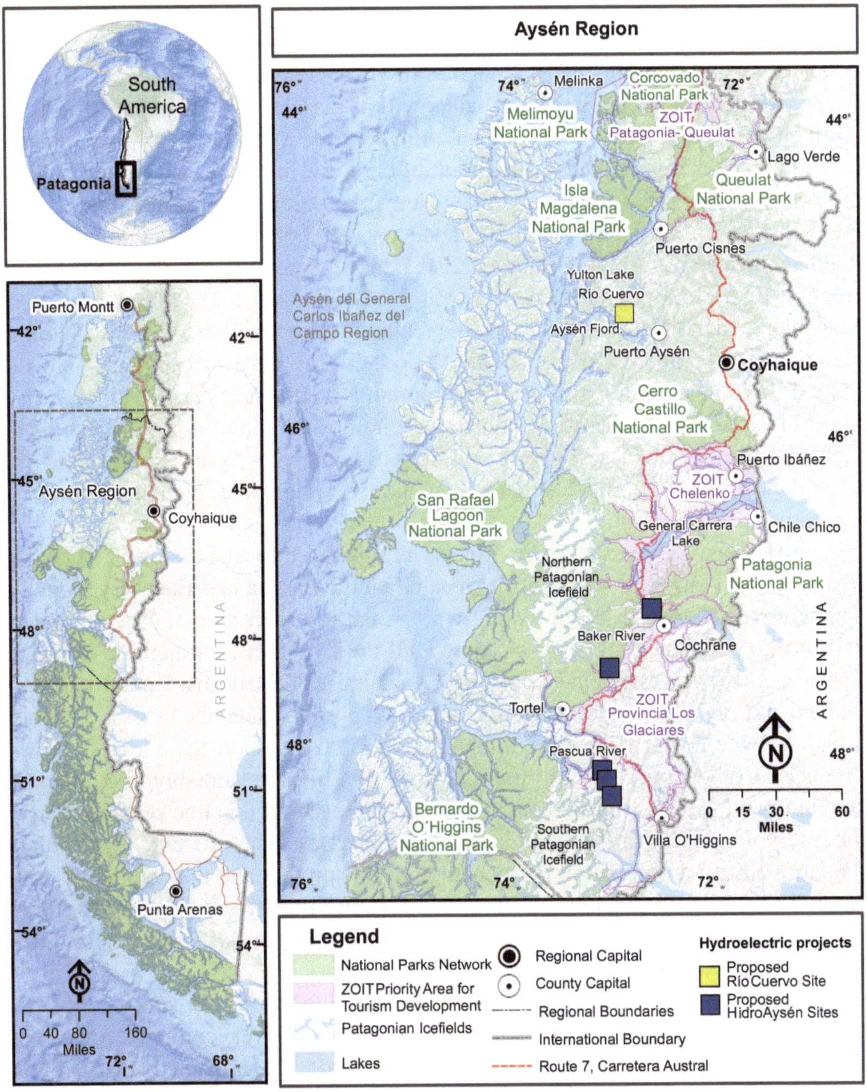

Fig. 10.1 Location of the three large-scale development proposals analyzed within the current study

imaginaries manifested within the perspectives and narratives expressed by actors related to these development contexts and the conflicts they provoked.

The Patagonia National Parks Network CBD project involved the work carried out in Chile by businessman and ecologist Douglas Tompkins, through the creation of private PAs. In 2015, a group of foundations linked to the Tompkins family presented their formal proposal to the Chilean Government to donate 4076.25 km² to the Treasury, for the creation of a "Network of National Parks in Chilean Patagonia"

(Chilean Ministry of the Interior and Public Security 2018). In exchange, the Chilean government had to incorporate at least double that surface area (more than 9000 fiscal km^2) within the National System of government Protected Wildlife Areas (SNASPE), protecting these lands within the category of national park. In April of 2017, the Chile's Council of Ministers for Sustainability, approved the creation, expansion, and recategorization of eight PA initiatives within the context of the new Patagonia National Parks Network, accepting the donation of the 4076.25 km^2 initially committed and annexing 9493.68 km^2 of new public lands (Chilean Ministry of the Environment 2017). In the Aysén Region, this included the creation of Melimoyu National Park, Patagonia National Park, the expansion of Magdalena Island National Park, and the reclassification of Cerro Castillo National Reserve as a National Park. In total, 13,569.93 km^2 were incorporated into the SNASPE (Chilean National Forestry Corporation 2018). Adding the already existing protected area, the network consists of more than 45,000 km^2, distributed in eight national parks across the Los Lagos Region, the Aysén Region and the Magallanes Region of Chilean Patagonia (Chilean National Forestry Corporation 2018).

The HidroAysén hydroelectric project, as submitted to SEIA, consisted of five hydroelectric dams and power plants located on the Pascua and Baker rivers in the southern reaches of the Aysén region, with a total reservoir area of 59.1 km^2 and a capacity to generate 2750 MW of energy. Of the five dams, the highest would be 114 m and the lowest would be 40 m. Its capital came from two main partners: Colbún and Endesa. In addition to the dams and power plants, the project proposed a high-voltage power line of more than 2000 km, to transport the energy that was produced to the central and northern regions of Chile. Presumably, the company sought to address the environmental and social impacts of this line separately from the reservoirs, as this line was not included in the assessment process. As a result of the large citizen mobilization in opposition to this project and the complaints filed by civil society organizations, on June 10, 2014, the HidroAysén hydroelectric project was rejected by the Chilean Council of Ministers for Sustainability (Chilean Ministry of the Environment 2014).

The Río Cuervo hydroelectric project, as submitted to SEIA, consisted of two hydroelectric dams and a power plant on the Cuervo River, commune of Aysén, 26 km northwest of Puerto Aysén (communal capital). The projected reservoir area was 131.66 km^2, with two adjacent dams, 55 and 57 m high, respectively. The project was proposed by Energía Austral, a subsidiary of the Xstrata Group. Its installed capacity would have been 600 MW (Pramar Ambiental Consultores 2007). After several legal proceedings, in 2018 the Third Environmental Court of the Chilean Republic reversed earlier SEIA approval for the project and the company finally abandoned its plans (Chilean Third Environmental Court 2018). In addition to the reservoirs, the project proposed to transport the energy to the central region of Chile via lines that would be buried in the fiords, along the coast. Presumably, as in the case of HidroAysén, the transmission line proposal would be submitted to the SEIA as a separate project.

10.3.2 Data Collection

The data collection process and format differed for each of the three contexts of interest in our work; nevertheless, as recommended by Stake (2003), our process employed many common practices and approaches that would help ensure a holistic treatment of the intrinsic links and subjective realities that emerged. Thirty-seven participants from the NBT, environment, energy, and public-service sectors contributed perspectives about the development initiative proposals and approaches through interviews and commentaries during a public panel (Table 10.1).

For each of the contexts, primary data collection began with an invitation to share their work related to NBT in Aysén and their overall perspectives about the region and its development. These accounts were followed by questions oriented to explore participants' perceptions, imaginaries, and lived experiences regarding the particular project and their perspectives about its potential impacts for regional NBT. All the participants were asked to share their preferred future scenarios for the project and NBT, and to reflect on the possibility that both could co-exist and thrive.

With respect to the Patagonia National Parks Network, a forum-debate was organized in August 2018 in the city of Coyhaique, entitled, *Tourism and Patagonia Parks Network: opportunities or obstacles to development ?* during which the creation of the network was officially announced. This forum was organized and convened by a civil society organization linked to NBT, the Association of Tourism Professionals of Chilean Patagonia (APROT), and a political authority of the region, a regional NBT leader and a leader of the conservation sector, all from the Aysén region, were invited as panelists. Representatives from the mining sector were also invited, as this sector was one of the big opponents to the creation of Patagonia National Park, but they declined to participate. As an introduction to the forum-debate, the Chilean National Forestry Corporation (CONAF) was invited to give a presentation on the Patagonia National Parks Network.

The event was attended by 85 people including the three panelists and participants from CONAF. Three questions were posed to the panelists at the forum to characterize the dimensions of conflicts and explore perceptions and values related

Table 10.1 Study participant profiles

| Development initiative | Number of interviews/panelists by sector | | | | |
	Nature-based tourism sector	Environment sector	Energy sector	Public sector	Total
Patagonia National Parks Network conservation-based development project	1	2	Not applicable	1	4
HidroAysén hydroelectric development project	9	4	2	10	25
Río Cuervo hydroelectric development project	2	3	1	2	8
Total	**12**	**9**	**3**	**13**	**37**

to the project and sustainable imaginaries. The first question involved understanding whether the panelists considered the Patagonia Parks Network CBD project to be an opportunity or a threat to the territory. The second asked panelists to reflect on the adequacy of the citizen participation process employed in the establishment of the Patagonia Parks Network CBD project. The third asked panelists to consider how a range of development projects, including CBD, NBT, ranching, mining, forestry, and salmon farming—among others—might harmoniously coexist in the Aysén region. Panelist answers were recorded and transcribed verbatim, to form the primary dataset for this development context.

In order to develop a more complete and well-rounded understanding of conflict and SD trajectories, we augmented the four primary data accounts from the forum with secondary data obtained from Chilean popular press articles, and press releases from core actors and interest groups, that presented perspectives on the *Patagonia National Parks Network* (Stake 2003; Boblin et al. 2013). We obtained an initial set of results by employing the Google Search Engine, with custom search settings for time frame (2017–2019) and type of result (news). From the initial list, we reviewed place of publication, titles, and summary paragraphs for all results, maintaining those that were published in Chilean outlets, and offered new perspectives and opinions of the initiative. If multiple articles shared the same basic information and outlook, only one was maintained. A total of 17 articles were added to the dataset in this manner. For each article, we reviewed the full text and extracted all of the relevant meaning units within. These meaning units included: "words, sentences, paragraphs, graphs, maps, illustrations, and other graphical alternatives to text, which are related through content and context" (Gale et al. 2019, p.12). Meaning units from the 17 articles were added to the existing dataset obtained from the forum transcripts.

For the HidroAysén and Río Cuervo projects, in-depth interviews were conducted with various NBT and conservation stakeholders, representatives of the energy projects, and public sector employees, between 2012 and 2018. In total, 33 accounts were collected. NBT sector participants included NBT micro-entrepreneurs and union leaders. The environmental sector involved representatives of environmental NGOs. The energy sector consisted of company representatives driving the two hydroelectric projects. The public sector included representatives of government and public institutions. Most of the participants were persons living and working within the Aysén Region, and the rest were national level actors, who were interviewed in Santiago. The interview sought to characterize tensions and dimensions of conflicts, and explore perceptions and values related to SD imaginaries and trajectories. For example, participants were asked to reflect on NBT in the Aysén region and how it might evolve in the future, both with and without the implementation of the particular hydroelectric project of interest. Then, they were asked to visualize and describe the compatibility of NBT with other economic sectors, like the hydroelectric sector. Participants also reflected on the role of the NBT sector in the environmental tensions and conflicts provoked by the hydroelectric project. Finally, participants were asked to share their perceptions of the process and outcomes reached between and the energy sector, the NBT sector, and the environmental sector. We also complemented the primary data with secondary data in these two contexts; using a similar search process related to conflicts occurring between 2012 and 2018, and the following search strings: *HidroAysén* and *Río Cuervo* (Stake 2003; Boblin et al. 2013). As before, we reviewed the full text and extracted all of the relevant meaning units from the 15 secondary sources that were included through this process, adding these to the existing dataset obtained from the forum transcripts.

10.3.3 Data Processing and Analysis

For data processing and analysis, we sought an approach that would permit us to integrate data across the three case contexts, opting for a combination of directed content analysis (Gale et al. 2019) and inductive coding methods. Although interviews and events occurred in different contexts and timeframes, the Spanish audio recordings and participant-observer notes were transcribed and joined within a master database, to form a unique dataset for this study. For the first research question, we sought to understand the underlying meaning, or latent context, within the dataset, with respect to tensions and conflicts regarding the development contexts. We began with an open coding process that resulted in 12 sub-themes, grouped into two primary categories. This stage was followed by triangulation and member-checking, to resolve points of confusion and ensure the consistency of the final themes and categories. For the second research question, we employed a consistent system of deductive coding to identify and analyze theory-driven codes across the dataset (Fereday and Muir-Cochrane 2006). We used the definitions established by Adloff and Neckel (2019) for the three SD imaginaries to guide our codebook, and the codification of textual phrases within the three trajectories: *Modernization, Transformation*, and *Control*. Each of the members of the research team conducted an initial coding process, and then a second stage of analysis focused on triangulation of the original coding among the three-member research team for this paper. During both of these phases of analysis, the researchers accepted participant comments at face value, although when the context or comment was unclear, elaborations were sometimes requested to clarify confusion that arose during coding and/or triangulation stages.

10.4   Results

The following sections provide an overview of the study results, beginning with the characterization and descriptions of the actors and their interests, and the two categories and 12 sub-themes that arose from latent meaning analysis of the data associated with tensions and conflicts for the three development conflicts. Next, we present the results of the directed content analysis of the three SD trajectories: *Modernization,*

Transformation, and *Control*. This section closes with a summary of the results with respect to the overlap and/or interaction of SD trajectories within the dataset.

10.4.1 Contexts, Actors, Interests, and Underlying Dynamics of Tensions and Conflicts

Figure 10.2 provides an overview of those from the Aysén Region who are in support and in opposition to each of the three development contexts, along with their main criticisms and acknowledgements. Major NBT associations and groups within the region expressed their support for each of the three initiatives, although the sector was divided for the HidroAysén hydroelectric development project, with some groups showing support while others opposed. With respect to the two hydroelectric development projects (HidroAysén, Río Cuervo), opposition was driven by broad coalitions consisting of environmental groups, outdoor enthusiasts, outdoor educators, some tourism professional organizations, and several foundations related to Doug and Kris Tompkins. With the Patagonia National Parks Network CBD project, opposition was driven by actors within the mining and ranching sectors, located in the buffer zones around the proposed area of Patagonia National Park.

While all three projects were criticized for technical reasons (e.g., anthropogenic impacts of excessive tourism, ecosystem damage from dams, landscape impacts, potential for disaster in event of earthquakes), they were also criticized for process and governance issues. Of the three projects, the Río Cuervo hydroelectric development project received less criticism in this aspect; in fact, many participants commented that the project approached stakeholders proactively to discuss concerns and develop mitigation plans. The benefits noted for all three projects related primarily to their economic upside potential for local communities, although the CBD project was also viewed as having the potential to *green* Chile's image and contribute to international conservation commitments.

Inductive coding of the latent meanings of participants' discussions of the conflicts and tensions around development processes for these three contexts helped to dig deeper into participants perceptions and values around the development process and governance. This analysis resulted in 12 themes that were grouped around two main categories: Trust (Fig. 10.3) and Distrust (Fig. 10.4). While technical and ideological differences were central to tensions and conflict in all three cases, these categorical perceptions of process and governance around development contexts in Aysén were expressed throughout the dataset, from actors in support and in opposition to the contexts.

The Trust category (Fig. 10.3) was comprised of six main themes, which defined participants' desires for regional development processes: a desire for greater proactivity around *transparency*, a *binding participation* process of governance, *bottom-up* decision making, *re-empowerment* of local groups, *decentralization*, and

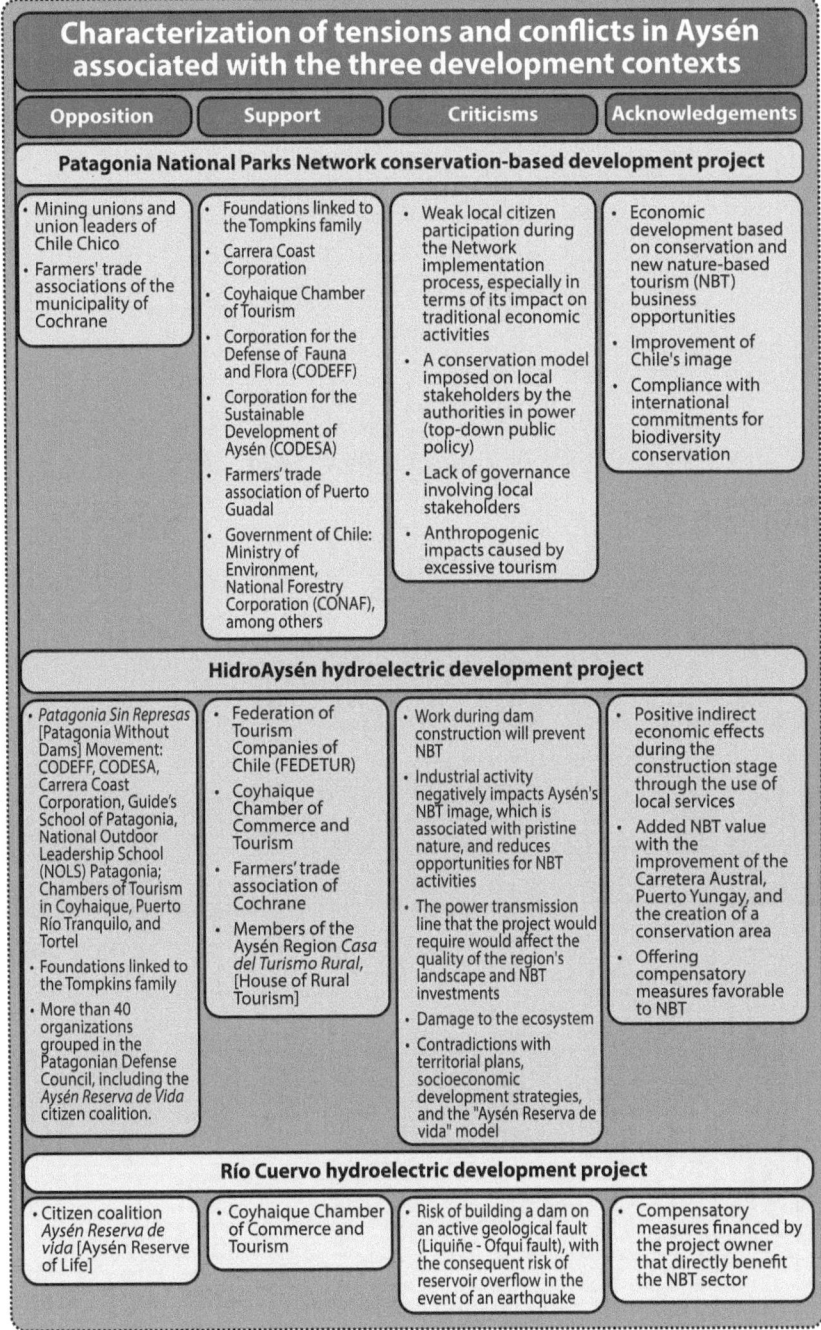

Fig. 10.2 Overview of actors and interests with respect to the three development contexts

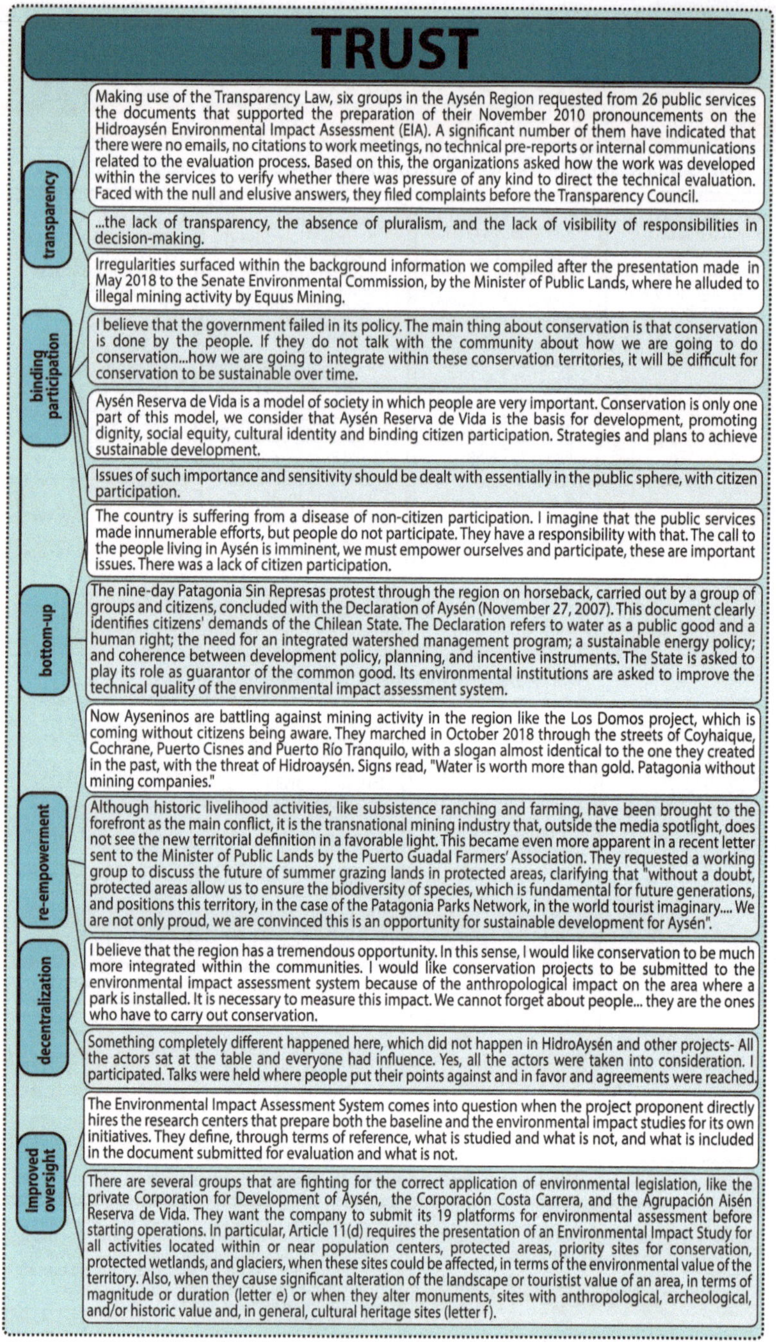

TRUST

transparency

Making use of the Transparency Law, six groups in the Aysén Region requested from 26 public services the documents that supported the preparation of their November 2010 pronouncements on the Hidroaysén Environmental Impact Assessment (EIA). A significant number of them have indicated that there were no emails, no citations to work meetings, no technical pre-reports or internal communications related to the evaluation process. Based on this, the organizations asked how the work was developed within the services to verify whether there was pressure of any kind to direct the technical evaluation. Faced with the null and elusive answers, they filed complaints before the Transparency Council.

...the lack of transparency, the absence of pluralism, and the lack of visibility of responsibilities in decision-making.

Irregularities surfaced within the background information we compiled after the presentation made in May 2018 to the Senate Environmental Commission, by the Minister of Public Lands, where he alluded to illegal mining activity by Equus Mining.

binding participation

I believe that the government failed in its policy. The main thing about conservation is that conservation is done by the people. If they do not talk with the community about how we are going to do conservation...how we are going to integrate within these conservation territories, it will be difficult for conservation to be sustainable over time.

Aysén Reserva de Vida is a model of society in which people are very important. Conservation is only one part of this model, we consider that Aysén Reserva de Vida is the basis for development, promoting dignity, social equity, cultural identity and binding citizen participation. Strategies and plans to achieve sustainable development.

Issues of such importance and sensitivity should be dealt with essentially in the public sphere, with citizen participation.

The country is suffering from a disease of non-citizen participation. I imagine that the public services made innumerable efforts, but people do not participate. They have a responsibility with that. The call to the people living in Aysén is imminent, we must empower ourselves and participate, these are important issues. There was a lack of citizen participation.

bottom-up

The nine-day Patagonia Sin Represas protest through the region on horseback, carried out by a group of groups and citizens, concluded with the Declaration of Aysén (November 27, 2007). This document clearly identifies citizens' demands of the Chilean State. The Declaration refers to water as a public good and a human right; the need for an integrated watershed management program; a sustainable energy policy; and coherence between development policy, planning, and incentive instruments. The State is asked to play its role as guarantor of the common good. Its environmental institutions are asked to improve the technical quality of the environmental impact assessment system.

Now Ayseninos are battling against mining activity in the region like the Los Domos project, which is coming without citizens being aware. They marched in October 2018 through the streets of Coyhaique, Cochrane, Puerto Cisnes and Puerto Río Tranquilo, with a slogan almost identical to the one they created in the past, with the threat of Hidroaysén. Signs read, "Water is worth more than gold. Patagonia without mining companies."

re-empowerment

Although historic livelihood activities, like subsistence ranching and farming, have been brought to the forefront as the main conflict, it is the transnational mining industry that, outside the media spotlight, does not see the new territorial definition in a favorable light. This became even more apparent in a recent letter sent to the Minister of Public Lands by the Puerto Guadal Farmers' Association. They requested a working group to discuss the future of summer grazing lands in protected areas, clarifying that "without a doubt, protected areas allow us to ensure the biodiversity of species, which is fundamental for future generations, and positions this territory, in the case of the Patagonia Parks Network, in the world tourist imaginary.... We are not only proud, we are convinced this is an opportunity for sustainable development for Aysén".

decentralization

I believe that the region has a tremendous opportunity. In this sense, I would like conservation to be much more integrated within the communities. I would like conservation projects to be submitted to the environmental impact assessment system because of the anthropological impact on the area where a park is installed. It is necessary to measure this impact. We cannot forget about people... they are the ones who have to carry out conservation.

Something completely different happened here, which did not happen in HidroAysén and other projects- All the actors sat at the table and everyone had influence. Yes, all the actors were taken into consideration. I participated. Talks were held where people put their points against and in favor and agreements were reached.

Improved oversight

The Environmental Impact Assessment System comes into question when the project proponent directly hires the research centers that prepare both the baseline and the environmental impact studies for its own initiatives. They define, through terms of reference, what is studied and what is not, and what is included in the document submitted for evaluation and what is not.

There are several groups that are fighting for the correct application of environmental legislation, like the private Corporation for Development of Aysén, the Corporación Costa Carrera, and the Agrupación Aisén Reserva de Vida. They want the company to submit its 19 platforms for environmental assessment before starting operations. In particular, Article 11(d) requires the presentation of an Environmental Impact Study for all activities located within or near population centers, protected areas, priority sites for conservation, protected wetlands, and glaciers, when these sites could be affected, in terms of the environmental value of the territory. Also, when they cause significant alteration of the landscape or touristic value of an area, in terms of magnitude or duration (letter e) or when they alter monuments, sites with anthropological, archeological, and/or historic value and, in general, cultural heritage sites (letter f).

Fig. 10.3 Themes and evidence related to the Trust category that emerged within the conflict-related data for the three development contexts

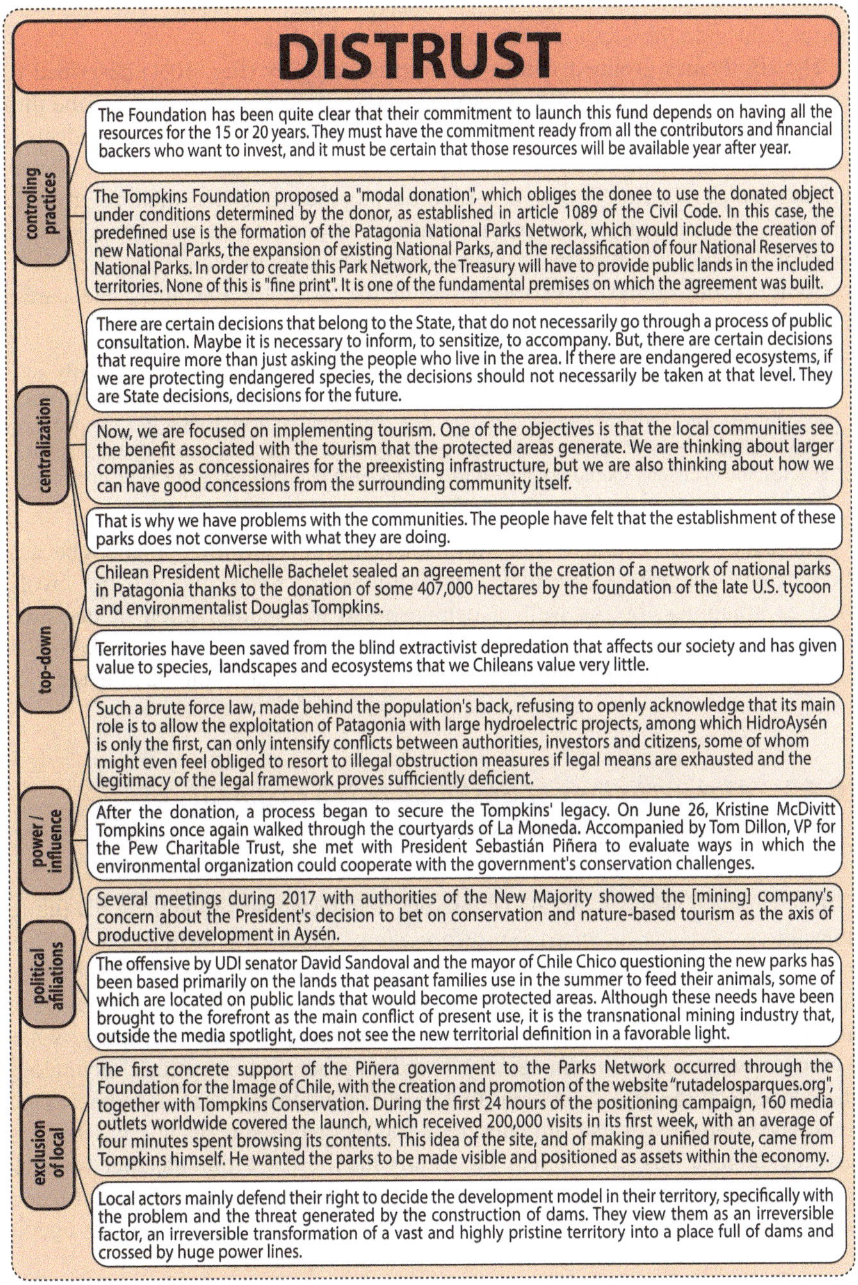

DISTRUST

controling practices

The Foundation has been quite clear that their commitment to launch this fund depends on having all the resources for the 15 or 20 years. They must have the commitment ready from all the contributors and financial backers who want to invest, and it must be certain that those resources will be available year after year.

The Tompkins Foundation proposed a "modal donation", which obliges the donee to use the donated object under conditions determined by the donor, as established in article 1089 of the Civil Code. In this case, the predefined use is the creation of the Patagonia National Parks Network, which would include the creation of new National Parks, the expansion of existing National Parks, and the reclassification of four National Reserves to National Parks. In order to create this Park Network, the Treasury will have to provide public lands in the included territories. None of this is "fine print". It is one of the fundamental premises on which the agreement was built.

centralization

There are certain decisions that belong to the State, that do not necessarily go through a process of public consultation. Maybe it is necessary to inform, to sensitize, to accompany. But, there are certain decisions that require more than just asking the people who live in the area. If there are endangered ecosystems, if we are protecting endangered species, the decisions should not necessarily be taken at that level. They are State decisions, decisions for the future.

Now, we are focused on implementing tourism. One of the objectives is that the local communities see the benefit associated with the tourism that the protected areas generate. We are thinking about larger companies as concessionaires for the preexisting infrastructure, but we are also thinking about how we can have good concessions from the surrounding community itself.

That is why we have problems with the communities. The people have felt that the establishment of these parks does not converse with what they are doing.

top-down

Chilean President Michelle Bachelet sealed an agreement for the creation of a network of national parks in Patagonia thanks to the donation of some 407,000 hectares by the foundation of the late U.S. tycoon and environmentalist Douglas Tompkins.

Territories have been saved from the blind extractivist depredation that affects our society and has given value to species, landscapes and ecosystems that we Chileans value very little.

Such a brute force law, made behind the population's back, refusing to openly acknowledge that its main role is to allow the exploitation of Patagonia with large hydroelectric projects, among which HidroAysén is only the first, can only intensify conflicts between authorities, investors and citizens, some of whom might even feel obliged to resort to illegal obstruction measures if legal means are exhausted and the legitimacy of the legal framework proves sufficiently deficient.

power / influence

After the donation, a process began to secure the Tompkins' legacy. On June 26, Kristine McDivitt Tompkins once again walked through the courtyards of La Moneda. Accompanied by Tom Dillon, VP for the Pew Charitable Trust, she met with President Sebastián Piñera to evaluate ways in which the environmental organization could cooperate with the government's conservation challenges.

Several meetings during 2017 with authorities of the New Majority showed the [mining] company's concern about the President's decision to bet on conservation and nature-based tourism as the axis of productive development in Aysén.

political afiliations

The offensive by UDI senator David Sandoval and the mayor of Chile Chico questioning the new parks has been based primarily on the lands that peasant families use in the summer to feed their animals, some of which are located on public lands that would become protected areas. Although these needs have been brought to the forefront as the main conflict of present use, it is the transnational mining industry that, outside the media spotlight, does not see the new territorial definition in a favorable light.

exclusion of local

The first concrete support of the Piñera government to the Parks Network occurred through the Foundation for the Image of Chile, with the creation and promotion of the website "rutadelosparques.org", together with Tompkins Conservation. During the first 24 hours of the positioning campaign, 160 media outlets worldwide covered the launch, which received 200,000 visits in its first week, with an average of four minutes spent browsing its contents. This idea of the site, and of making a unified route, came from Tompkins himself. He wanted the parks to be made visible and positioned as assets within the economy.

Local actors mainly defend their right to decide the development model in their territory, specifically with the problem and the threat generated by the construction of dams. They view them as an irreversible factor, an irreversible transformation of a vast and highly pristine territory into a place full of dams and crossed by huge power lines.

Fig. 10.4 Themes and evidence related to the Distrust category that emerged within the conflict-related data for the three development contexts

improved oversight practices. Sample excerpts from interviews and secondary sources illustrate the depth of each theme within the data.

The six themes grouped within the Distrust category (Fig. 10.4) described the issues participants described with respect to the processes employed for the three development contexts: *controlling practices, centralization, top-down* decision making, improper use of *power* and *influence, political affiliations* and alliances, and the *exclusion of local people* within development decision-making processes and plans. Again, sample excerpts from interviews and secondary sources illustrate some of the ways in which these themes emerged within the data.

Some of the themes overlapped within the data; for example, one participant stated:

> I think that there are certain decisions that belong to the State, that do not necessarily go through a process of consulting the public, maybe it is necessary to inform, to sensitize, to accompany. But there are certain decisions that require more than asking just the people who live in the area. If there are endangered ecosystems, if we are protecting endangered species, the decisions should not necessarily be taken at that level, they are State decisions that have to be taken, decisions for the future.

This excerpt was characterized within the theme of controlling practices because the narrative suggests a controlled decision-making process that did not involve local communities; but as well, it demonstrated the centralization of decisions, through the president and the Tompkins, who developed their negotiations in Santiago, Chile's capital, some 1505 km from the area that would be affected.

10.4.2 Manifestations of Sustainable Development Futures Trajectories

Directed content analysis revealed numerous manifestations of the three SD futures trajectories —*Modernization; Transformation; Control*— within the dataset. Manifestations of the SD futures trajectories were related to all three development contexts (Patagonia National Parks Network; HidroAysén; Río Cuervo). Figure 10.5 illustrates the core concepts of the *Modernization* SD future trajectory, which worked from an adaptation imaginary, based on practical innovations, technology, and concepts of the *green economy*, that would push toward more sustainable practices. The sample excerpts from interviews and secondary sources in Fig. 10.5 illustrate the range of ways in which this trajectory manifested within the interviews and secondary sources, including NBT, new financing mechanisms (i.e., Project Finance for Permanence [PFP]), the *greening* of Chile's image, a pro-business energy agenda based in non-conventional renewable energies, among others.

Figure 10.6 illustrates the core concepts of the Transformation SD future trajectory, which involved imaginaries that involved fundamental social transformation, along a new development model that involved non-consumptive values, *buen vivir* (living well, in English), utopias, and an overall alignment with the earth system. While the evidence was less prominent for this SD trajectory, several excerpts

Modernization Sustainable Development Trajectory

Imaginaries that innovate around environmental challenges; circular economy, green economy; faith in technological progress; view markets and capitalism as effective instruments of sustainability. Based on an evolution, step-by-step, of existing socio-economic infrastructure and logic.

It is not only an unprecedented preservation effort; it is also an invitation to imagine without further delay ways to rationally occupy our land, to create other economic activities, to use natural resources without depredating them. In other words, to generate sustainable development, the only profitable development in the long term.

We are advancing by leaps and bounds and the international community is already watching with attention the process that Chile is undergoing to become a green country, a world reference in conservation and environmental issues.

Protected Patagonia is already attracting great tourist interest: the New York Times ranked it sixth on the list of 52 places to visit in 2018. These parks will help cement an image of a green country that will benefit from the global growth of low-impact nature tourism.

The PFP [Project Finance for Permanence] is the result of the efforts of a group of conservationists, former bankers, and consultants who employed Wall Street practices to develop a new way of protecting large ecosystems. It's an approach that could also work in large-scale social projects. This program has already been successfully applied in Canada, Brazil, and Costa Rica.

We always considered that it [HidroAysén] was a bad project for the region and the country. There were other alternatives like non-conventional renewable energies, but, they said this was a utopia. It turns out that last year Chile registered the lowest price for solar energy ever known in the world. Historically, development in Patagonia has been based on extensive ranching and fishing, which have already reached their potential and are in decline. Tourism is one of the few sectors that is growing at double-digit rates, and nature tourism offers the greatest growth potential. So, the creation of national parks as a conservation tool is transformed into a lever for sustainable economic development, with a direct benefit to the surrounding communities. That is why we want to position this network of parks as the "The Patagonia Parks Route".

I do not agree with dams coming, but if they are going to come, they have to respect those of us who are here and what exists and help those of us who are here. Because whether we want development or not, it is part of sustainability. We are very lacking in things, development must come for the community. Their idea was to support tourism, they were going to improve the port, those things. Without the project, maybe they will still be done, but it will be slower, and we need them.

Faced with a pro-business line in the Energy Agenda, even a group like Matte may be willing to think about losing HidroAysén, on the condition that it ensures them a long-term sustainable energy matrix policy, which gives them economic certainty. Therefore, it is more important and productive for them, because this reform will act as an adjustment valve. My impression is that they expect this to serve as a central element for addressing accumulation patterns, lowering the energy cost in production. If that is resolved, then Matte and Angellini are willing to compromise on HidroAysén because in the long term their energy costs will be lower.

Chile is experiencing an opportunity to use electric energy to solve major environmental challenges like urban pollution. We are seeing a transition and that is why in January of this year we signed an agreement commiting not to develop new coal-fired plants.

We have to look at renewables in the broad sense. The opportunity for renewable energy is already a fact, the question is when. We have to reconcile economic, technical, and environmental viability. By solving these challenges, electricity will be 100% renewable. But what will be truly renewable is when renewable electricity replaces fossil fuels in the rest of the economy.

It is crucial to generate a public-private financial sustainability mechanism that allows us to face the costs of caring for these parks, ensuring that they do not become a liability for local, regional, and national economies; on the contrary: an opportunity for the country brand and the regions, as one of the last great protected natural places in the world. There are successful experiences to learn from like Costa Rica, the protected areas of the Brazilian Amazon and the parks in Bhutan, initiatives in which WWF has collaborated with governments, communities, and other NGOs, generating a model that could support the challenges that the great news of the Patagonia Parks brings us.

Fig. 10.5 Manifestations of Modernization Sustainable Development Trajectory within the conflict dataset for the three development contexts

emerged from interviews and secondary sources, as highlighted in Fig. 10.6. *Transformation* SD imaginaries seemed to contribute to some of the actors' actions with respect to the tensions and conflicts that emerged in the development contexts. Two movements arose, in particular, as representative of the *Transformation* SD trajectory: the *Aysén Reserva de Vida* (ARV; Aysén Reserve of Life in English) and *Patagonia Sin Represas* (Patagonia without Dams) movements.

Although perhaps not to the degree described by Adloff and Neckel (2019), the *Control* SD futures trajectory also manifested within the dataset, as shown in the sample excerpts from interviews and secondary sources, provided in Fig. 10.7. Manifestations pertained to expressions of narratives and imaginaries relating to the anticipation of future socio-ecological disturbances, forms of State controls,

Transformation Sustainable Development Trajectory

Transformation: imaginaries of fundamental social transformation; beyond capitalism: buen vivir=living well; real utopias; reduced consumption; reduced materialism; new community structures, aligned with the earth system.

The words development and progress are not well explained to people and communities. Development does not necessarily have to do with the intervention of large industries in the territory. Development has to do with something much more integral and holistic and unfortunately governments have failed to see this. Not everything is based on the generation of money and accelerating consumerism as it is exacerbated today on the planet. Here in Aysén, I believe we have a historic opportunity to do something different. We must change the paradigm. We have entered a neoliberal model of only producing and working in slavery and we are not living, we are surviving.

I have never understood...compete with whom? Why do we need to compete if the Aysén region can implement a different model of development? Let us compete with ourselves, to be more responsible, sustainable, to live more on a human scale. We have to move towards autonomy, to decide on our own development. Okay, so we are part of a national or international framework, we can use them as references, but we are building our own rules; not complying with competitiveness or indexes at national level.

Aysén Reserva de Vida has a bio-relevant territorial vision. It seeks binding citizen participation and advocacy. It seeks respect for all human rights, including the right to water.

The conflict surrounding the Hidroaysén Project indicates a need for change in the relationship between civil society and the State. It's synonymous with a deficit in the governance model, and a lack of control in the economic sphere on the part of governments and citizens. The Patagonia Sin Represas social movement has configured a repertoire of actions which have brought us national and international visibility, and enabled us to question the legitimacy of what is happening with the democratic and economic system in the country.

The movement constitutes a territorial struggle that has employed new practices of political action and the use of judicial resources to transcend the local space. We've used the global space to claim the right to maintain the environment of this pristine region and the possibility of creating parks, of the land, and for the land. The social mobilization against this project has been growing. Its consequences concern not only the inhabitants of Patagonia. The problem is national and of humanity, insofar as Aysén is recognized as a "reserve of life", a "sanctuary of nature", "heritage of humanity", a "source of life", and other varied representations of Patagonia. Today, these expressions are not exclusively coming from the people of Aysén, nor from those who defend the southern region of the country. This local discourse is recognized globally, and it is through collective action that it is unified. Diverse social activities are providing it with content.

The marches brought together more than 120,000 demonstrators, giving birth to a social movement that shouted "enough". A true prelude to a cycle of mobilizations that put environmental issues, and also educational issues, at the center of political debate.

Fig. 10.6 Manifestations of Transformation Sustainable Development Trajectory within the conflict dataset for the three development contexts

Control Sustainable Development Trajectory

Imaginaries build on an anticipation of socioecological disturbances and particularist ethics. Geoengineering practices, state and military control; segregation, surveillance; Earth system control structures; new infrastructures of technological, military and state control.

This network was an old idea that Douglas had already expressed to President Sebastián Piñera after he took office in 2010. Both knew each other; the North American had even advised him on the creation of Tantauco Park in Chiloé. However, his government decided to prioritize the creation of Yendegaia Park in 2013, due to the high costs that the network project meant for the country. But after Kristine's offer to President Bachelet, which adjusted the terms of the donation, La Moneda sealed a protocol agreement with the Tompkins Foundation in March 2017, that finally came to fruition this year with the firming of the largest public-private national park donation in history.

Unfortunately territorial zoning is not binding. We have Territorial Zoning Plans and Coastal Area Zoning Plans but they are just preferences, so they are useless. An ordinary citizen may not be able to relocate, but if you have money you can. The only things that are supposedly more restrictive are natural protected areas, monuments etc. …. But that's as far as it goes… In Quitralco Nature Sanctuary there are nine salmon concessions… They were installed before it was declared, but they should still be removed. And at the time, CONAF also justified HidroAysén.

The electricity concessions bill, promoted by the government, backed by a UDI-DC agreement, and currently in its third reading in Congress, is broadly speaking, a highly convenient attempt for Endesa and Colbún to make the management of electricity projects more expeditious by legalizing national heritage abuse, both natural (by reducing protection for parks and nature reserves) and legal (by reducing the rights of landowners and other affected parties). Approval of these terms would greatly facilitate the materialization of HidroAysén.

The Tompkins Foundation proposed a "modal donation", which obliges the donee to use the donated object under conditions determined by the donor, as established in article 1089 of the Civil Code. In this case, the predefined use is the formation of the Patagonia National Parks Network, which would include the creation of new National Parks, the expansion of existing National Parks, and the reclassification of four National Reserves to National Parks. In order to create this Park Network, the Treasury will have to provide public lands in the included territories. None of this is "fine print". It is one of the fundamental premises on which the agreement was built.

The State needs to generate incentives and a regulatory framework that guarantees the conservation of these places in the long term. There is nothing to be gained by a private individual announcing to protect a certain area, if that desire to protect rests only on his personal will or later on that of his heirs. Protection is either forever or it is not protection.

Today, much more than before, the political class understands the importance of nature conservation. I do not want to mention anyone in particular, so as not to leave anyone out, but undoubtedly, former President Ricardo Lagos has strongly incorporated global environmental changes in his discourse. But from our personal experience, today we can say with satisfaction that we have created national parks with three Chilean presidents: President Lagos, President Piñera and President Bachelet, who knew how to understand the relevance of these historical legacies.

The Cerro Castillo community is very sensitive to their environment and believes that tourism and the use of their natural spaces is a way to improve their economic activity. I believe that the government failed in its policy…If they do not talk with the community about how they are going to do conservation, it will be difficult for conservation to be sustainable over time. If conservation is not done by the people, this conservation is seen as imposed. This network of parks is seen as a central decision among elite groups that decided that a network of parks would be beneficial for us.

The Environmental Impact Assessment System comes into question when the project proponent directly hires the research centers that prepare both the baseline and the environmental impact studies for its own initiatives. They define, through terms of reference, what is studied and what is not, and what is included in the document submitted for evaluation and what is not.

In April 2012 the Chamber of Deputies approved a 417-page report by its Human Rights Commission, that was immediately dismissed by the company. It documents in detail, "serious conflicts of interest in authorities participating in the [project approval] process" and the "modification, or alteration, of technical reports and sectoral pronouncements in favor of HidroAysén". Irregularities occurred in so many public services. It is difficult to conceive of an explanation that does not involve a systematic and deliberate policy of undue intervention by the company.

I did not say that people should not be considered, but there are certain decisions, which are State decisions, and they have to be made regardless of whether they can be socialized or discussed.

Fig. 10.7 Manifestations of Control Sustainable Development Trajectory within the conflict dataset for the three development contexts

particularistic ethics, and earth system control structures like the proposed dams and the PA network.

While the examples provided in the previous results represented meaning units that were coded to a single SD futures trajectory, several of the meaning units that were extracted from the interviews and secondary sources were coded to multiple SD trajectories, suggesting overlaps and/or interactions in the ways these trajectories manifest in territories or development contexts (Fig. 10.8).

Overlap and/or Interaction between Sustainable Development Trajectories

Modernization - Transformation - Control

Adequate legislation that prioritizes environmental protection, an alert and mobilized citizenry, and a State that provides oversight. These are conditions that are fundamental to avoid harmful projects. But nothing will change if we do not modify the production and consumption patterns that have plunged us into the current socio-environmental crisis. In this sense, the creation of national parks has proven to be a successful instrument for the conservation of biodiversity and to contain, in part, this crisis.

The views of the environmental sector is influenced by the opinions of more radical environmentalist groups, like Tompkins, who think that nature is pristine and that it cannot be touched, and nothing can be done. That is a selfish view from the point of view of development, because in our opinion, it is not possible to have a territory and not take advantage of the natural resources. We believe that it is possible to take advantage of the resources, through sustainable and rational use of the natural resources, and provide what is needed for the future. To leave a piece of land untouched is a utopia. We believe that it is a very selfish view and that it is not possible to do it here. That is why some more radical environmental foreign capitalists buy large extensions of land and block them and seal them. Metaphorically, they put a padlock on them and who wins with that? Only a group of people who think that way. The communities that need a more harmonious development do not understand it and do not accept it.

The objective is not to develop tourism, but to focus on the environment and its protection.. Tourism may be involved, but it is not really the priority. We are in the tourism business out of necessity, not out of desire.

The mobilizations against Hidroaysén have provided the main environmental movements in the country with a taste of triumph. The fact is that the blow they have dealt is even greater than thought. Two important business groups, among them the Matte Group, reversed their positions. This was directly influenced by the pressure of the mobilizations and a successful campaign that managed to even influence rejection of the project at the international level. This coincided with a historic moment in which the environmental organizations were joined by thousands of people in the streets. The social strength achieved in the opposition campaign broke the very transitional prior scheme, in which area NGOs were the protagonists. This time it was the communities and citizens who came to the forefront.

Today , the NGOs that previously led the environmental world fulfilled the role of having positioned the issue, but I have the impression that these groups, as well as the political parties, have suffered in terms of public sentiment of their representativeness. And to the extent that the institutional system loses legitimacy, citizens and companies are now in direct dialogue with the government. What they both are skipping is the political system.

I have the impression that the Matte Group gave up the HidroAysén project some time ago. They are focusing their efforts on the Energy Agenda currently being promoted by the government, where market regulation and the introduction of competition will mark a new phase for the sector. This long-term reform is more central for these groups than the HidroAysén project itself.

Could the development of HidroAysén affect the development of other renewable energy sources? It would literally block the development of non-conventional renewable energy (NCRE) in the Central Interconnected System (SIC) by monopolizing the market for years. HidroAysén is sinking in a past where 'huge was beautiful'. The truth is that 'the mega', the 'large-scale economy', is functional only to the electric oligopoly imposed in Chile. It is clearly dysfunctional for the country. Citizens reject this abusive format and demand a radically different energy model. First, Chile must develop energy efficiency, savings, and energy conservation. Then, it must deploy distributed generation systems, 'smart grids', by macro-zone, combining the different NCRE offered by the regions throughout the country. This is modern, socially, and ecologically sustainable.

Fig. 10.8 Overlap and/or Interaction between Sustainable Development Trajectory within the conflict dataset for the three development contexts

10.5 Discussion

The results of our study identified several commonalities between actors and groups with respect to the tensions and conflicts around development contexts proposed for the Aysén region of Chile. While technical aspects of the projects varied, all three implied large-scale territorial land-use change that would have an intense local impact. Many of the outcomes that the projects emphasized would provide benefits on a larger scale (e.g., national energy supply, national and global conservation goals, national and global carbon offsetting, national and global NBT). Thus, while there were fundamental differences between actors and groups with respect to the ways they valued nature and its use, several commonalities arose with respect to tensions and conflicts around process and governance. Similar tendencies have arisen in other studies of NBT actors and development conflict within the Aysén territory (Bourlon 2020; Inostroza 2016).

The two main categories (Trust and Distrust) that evolved in our coding of latent themes portrayed the underlying factors that emerged around these process and governance related tensions and conflicts. They portray the stark contrasts between what local community actors sought from development contexts and what they rejected. In all three development contexts, actors sought greater proactivity from developers with respect to *transparency* about their intentions and actions, a *binding participation* process of planning and governance, *bottom-up* decision making about how things would occur in their territory, *re-empowerment* of local groups, *decentralization*, and *improved oversight* practices, especially with respect to Chile's EIA process. In contrast, they rejected actors who employed *controlling practices*, especially when these controls were imposed through *top-down* decision making at a *central* level, with repeated *exclusion of local people* within development decision-making processes and plans. Distrust also manifested through perceptions (and proof) of improper uses of *power* and *influence, political affiliations* and alliances.

Chile's EIA process stresses SD as its first and foremost mission. Thus, as all large-scale public and private development projects must participate in this process, one must assume that the creators of Chile's large energy projects, like HidroAysén and Energía Austral, were operating within the logics and parameters of a sustainability trajectory. In fact, several Control SD tactics were observed in their approaches; especially in the case of HidroAysén, who in recognition of future water scarcity, sought geoengineering on a massive scale, to facilitate the requirements of mining, at the expense of local communities and their needs. Nevertheless, our analysis suggested stakeholder recognition and rejection of these Control SD trajectories, initially at a local level and later on, with the successful consolidation of opposition movements and coalitions, through widespread national and international manifestations. One of the main issues with these Control SD trajectory projects was their large scale, and their design for national (not local) needs. Moreover, as the energy produced by the dams would largely support extractive industry and big business (i.e., mining in northern Chile), the irreversible changes in Aysén were considered an unacceptable sacrifice.

In contrast, the intensity of opposition in the Río Cuervo hydroelectric development context was much lower. Conflicts gathered fewer actors, coalitions were not obtained, and there was less media coverage. The Río Cuervo project threatened a smaller geographic sector of Aysén (i.e., the Puerto Aysén commune), with indirect impacts on NBT and the artisanal fishing sector, as the dam and power plant would have much less impact on viewsheds, as compared with the HidroAysén proposal, and the transmission lines would run under the fjords. To a certain extent, the project was even recognized as potentially having positive impacts related to cruise ship tourism, as its construction and transmission infrastructure would necessitate improvements to the ports in Puerto Chacabuco, that would also facilitate this mass tourism model. Several factors likely contributed to the differences we noted in opposition intensity. For example, our data reinforced positive aspects about the approach employed by the company for positioning and advancing the initiative at the local level. In the Río Cuervo context, Energía Austral representatives purposefully assembled actors and proactively worked on plans to mitigate the project's impacts. For example, one of our study participants described:

> Something completely different happened here, which did not happen in HidroAysén and other projects, because all the actors sat at the table and everyone had influence; all the actors and yes, they were taken into consideration, I do not know the extent, but I participated, talks were held where people put their points against and in favor and agreements were reached.

Environmental groups were the main opponents to the Río Cuervo hydroelectric development context, steadfastly maintaining their views that the project would create irreparable damage to local ecosystems and risks for local community safety; thus, did not represent SD. The perceptions that surfaced from participants representing these two groups illustrated clear examples of the Control and Modernization SD imaginaries. For example, in the following excerpt, representatives for Energía Austral, the company behind the Río Cuervo project, described the Control SD approaches they perceived from environmental groups, emphasizing their own approaches through Modernization SD imaginaries, that were, in their opinion, more representative of the best interests of local territories:

> The views of the environmentalist sectors are influenced by the opinions of more radical environmentalist groups, such as Tompkins, who thought that nature was pristine and that it cannot be touched, and nothing can be done. That is a selfish view from the point of view of development, because in our opinion, it is not possible to have a territory and not take advantage of the natural resources. We believe that it is possible to take advantage of the resources, making a sustainable and rational use of the natural resources and providing for the future what is needed. To leave a piece of land untouched is a utopia, we believe that it is a very selfish view, and it is not possible to do it here, that is why some more radical environmental foreign capitalists buy large extensions of land and block them and seal them. Metaphorically, they put a padlock on them and who wins with that? Only a group of people who think that way, but the communities that need a more harmonious development do not understand it and do not accept it.

This excerpt was one of several examples of interactions and overlap that emerged in our study between the SD imaginaries (Fig. 10.8). Although the Energía

Austral representative's representation of Tompkins' vision portrayed a Control SD mindset, it also accurately captured the Transformation SD trajectory that the couple and their foundations employed at the off set of their projects in Chile. Based on principles of Deep Ecology and Doug's vision of a "next economy" that would intimately intertwine economic sustainability and the restoration of natural systems (Tompkins 2012), the couple's initial approach to private conservation was centered on a Transformational SD futures trajectory (Bourlon 2017). Along the trajectory of their work in Patagonia they experienced significant tensions with local and national stakeholders. Thus, as the following excerpt from our dataset reinforces, the Tompkins realized that the cost of a Transformation SD trajectory, might be the failure of achieving the large-scale change they sought:

> ...the objective is not to develop the tourism offer, but to focus on the environment and protection... tourism may be involved, but it is not really the priority... we are in the tourism business out of necessity, not out of desire.

In 2017, Kris Tompkins shared her perspective of what is needed to achieve SD in the future (González Isla 2017). Her statement reaffirms the overlaps and interactions between SD imaginaries that we observed in other aspects of the Tompkins long-term approach. The first part of her comments describes the need for Control SD approaches: "Adequate legislation that prioritizes environmental protection, an alert and mobilized citizenry, and a State that oversees, are fundamental conditions to avoid harmful projects." Her next statement moves toward Transformative SD imaginaries, "But nothing will change if we do not modify our production and consumption patterns, which have plunged us into the current socio-environmental crisis." And the final part of her excerpt returns to a Modernization SD trajectory, focused on incremental improvements, "In this sense, the creation of national parks has proven to be a successful instrument for the conservation of biodiversity and to contain, in part, this crisis."

While the Tompkins maintained their Transformation SD ideology throughout the trajectory of their work in Chilean Patagonia, our data suggests that they adapted their strategies to work within the political and social settings they encountered. Seemingly, they navigated between Modernization and Control SD imaginaries and tactics to maneuver between local and national actors, agendas, and decision-making processes. These tactics, which we have termed, *SD agility*, merit additional study, as they may represent an important capacity for SD futures trajectories. Within the context of our study, we have established an initial definition of SD agility as: "the strategic ability to maneuver between SD imaginaries and trajectories in order to achieve strategic SD outcomes."

We hypothesize that SD agility may work in similar manners as other capacities (e.g., *boundary concepts, translation*), that have been successfully used to bridge differences between interdisciplinary and transdisciplinary actors (De Witt and Hedlund 2017; Gale and Ednie 2019; Gale et al. 2021; Hedlund-de Witt 2013). These practices orient communication strategies to resonate with stakeholders' values. They *meet people where they are* by interacting with them in ways that will resonate with their values (Kohl and McCool 2016, p. 271). Perhaps this concept also works in political and legislative concepts, requiring an agility to recognize and maneuver between SD trajectories, depending on the imaginaries of receptors and the ways in which these imaginaries are enacted within policy, legislation, and practice.

A second example of the SD agility concept involves the *Aysén Reserva de Vida* (ARV) movement. This movement surfaced within each of the three development contexts of our study in several different contexts and roles. For example, ARV frequently arose in association with an environmental defense stance that operated from an SD Control imaginary focused on limiting the forms of growth and change that occurred in the Aysén Region. Nevertheless, as representatives of the movement described, they obtained their desired outcomes through an SD agility that included agents operating from a range of SD imaginaries:

> The first thing we learned many years ago is that we must make alliances... from time to time you end up allying with the devil, but in this case it was the lesser of two evils...Do not play David against Goliath, rather, play Gulliver, and use everyone to tie them up. The key is long campaigns. You have to try to buy time, because they end up collapsing for improbable reasons.

In his 2019 book, Peter Hartmann, the ARV movement's founder, described the movement based on a Transformation SD-based imaginary and purpose, saying, "the citizens' mega-proposal 'Aisén Reserva de Vida' emerged in 1990 as a search for a sustainable regional development alternative and in response to the traditional centralist, destructive, and predatory model" (p. 2).

One of our interview participants explained that over the years, the ARV movement has been positioned as,

> ...a citizen proposal built by a wide diversity of Ayseninos and Ayseninas, organizations, and actors from other parts of the country and the world, who work to protect the heritage of this region in Patagonia, and promote the desire to live in a different model of society, based on economic, social and cultural rights, so as to improve the quality of life for, and by, the inhabitants of this territory, and thus, contribute to a more sustainable world.

As such, ARV offers a fourth proposed development context for Aysén, which arose within our case study, through push-back and support for elements of the other three proposed development contexts.

During the early years of the ARV movement, efforts focused on aligning citizens who were operating from a range of SD imaginaries and trajectories around promoting the Aysén Region as being a reservoir of life that must be protected and conserved. For example, many early participants were professionals who had arrived in the region during the late 1980s and early 1990s, to work with technocratic institutions that were ramping up in Coyhaique, the region's capital, with the transition back to democracy (Blanco Wells 2009). Many had studied abroad during the 1970s and 1980s, with disciplinary training geared toward natural resources, parks and reserves, wildlife, outdoor education, and other aspects of conservation and ecology. As the historic photos in Fig. 10.9 illustrate, early ARV efforts aligned with the work of these professionals. Photo A was taken during the second Latin American Congress of National Parks and other PAs, which occurred in 2008. In this session, professionals associated with ARV shared the movement's objectives,

Fig. 10.9 Historic photos documenting key events in the early years of the Aysén Reserva de Vida movement

principals, and strategies, with the goal of extending the movement throughout Chilean and Argentine Patagonia. Photo B was taken during the same year, at the launch of the movement's declaration statement, which is featured in the large, framed pictures shown in front of the participant members standing for the picture.

The ARV declaration shared environmental, cultural, economic, social, and political principles for citizen activism and engagement, proposed and shared by people who self-ascribed to the ARV movement. The declaration was developed by members during a project financed by the Arena Foundation, which sought to strengthen ties between social, cultural, environmental, and NBT organizations and entrepreneurs from Chilean and Argentine Patagonia around the pursuit of sustainability. Framed versions of the declaration statement were displayed within the businesses, offices, and homes of participants. Considering the momentum of the movement, its resonance with professionals in Aysén technocratic institutions, the Arena Foundation project outcomes, and growing interest in public participation, it is not surprising that, a few years later, the ARV concept and slogan were incorporated into the 2010 regional development strategy put forth by the Regional Government of Aysén (Ilpes/Cepal-Gore Aysén 2010), even though the plan largely maintained Chile's dominant neoliberal paradigm and a Modernization SD futures trajectory.

Since then, the movement has been increasingly associated with overlapping and interacting SD trajectories. For example, in recent years, several Chilean geographers have associated the ARV movement with larger Modernization SD imaginaries that have manifested in Aysén (Miranda Cabaña 2016; Núñez et al. 2018, 2020), identifying State policy and discourse which has encouraged CBD, and especially NBT (Aliste et al. 2018; Miranda Cabaña 2016; Núñez and Aliste 2017; Núñez et al. 2014, 2018, 2019, 2020). Their research proposes that recent land redistribution in Aysén represents a new form of *green* colonization within the territory, and that this reterritorialization process has been framed and enabled by the State, through the positioning of a new social narrative of the Aysén region, as a *Reserva of Vida* (Miranda Cabaña 2016; Núñez et al. 2018, 2020).

Nevertheless, our study reinforced the ARV movement's self-expressed commitment to a Transformation SD trajectory, even though members sometimes employed Control SD tactics in conflicts with other development paradigms, and/or supported the Modernization SD trajectory expressed within the Patagonia National Parks Network CBD project. For example, although one of the panelists (who has been a long-standing leader within the ARV movement), advocated for the Patagonia National Parks Network CBD project and the NBT opportunities it could bring for local communities in Aysén, they also described:

> Not everything is based on the generation of money and accelerating consumerism as it is exacerbated today on the planet. Here in Aysén I believe we have a historic opportunity to do something different. We must change the paradigm. We have entered a neoliberal model of only producing and working in slavery and we are not living, we are surviving.

This second example of SD agility seemed to work differently from what we observed within the first example involving the Tompkins' SD trajectory. Perhaps, since the ARV coalition is citizen-based rather than private, and dependent on a broader base of participants who likely operate along a range of SD imaginaries, their work has required an inherent SD agility in order to resonate with their participants' values. Thus, ARV may be willing to *meet people where they are*, providing spaces and opportunities in which they may advance toward the ARV Transformation SD imaginary. Alternatively, perhaps the attractiveness of Aysén Reserva de Vida, as a slogan, has been appropriated by other groups, and used to position a new CBD narrative and reterritorialization process based on Modernization SD imaginaries that differ in substance from the Transformative SD trajectory that the original movement has proposed. Clearly, additional research is warranted to better understand the interactions and overlaps that occur when SD imaginaries and futures trajectories are implemented at a territorial level, including the ways in which they evolve and adapt over time.

10.6   Conclusions

The collective case study we have presented within this chapter characterized the tensions and conflicts that arose in the Aysén Region of Chilean Patagonia, in response to three large-scale development proposals. We identified dimensions of Trust and Distrust that affected local actors' perceptions of these projects and the conflicts that surrounded them. Further, we have identified that Adloff and Neckel's (2019) SD imaginaries and futures trajectories manifested within the debates that surrounded all three of the proposed large-scale development projects. Frequently, we found examples of Modernization and Control SD imaginaries in the narratives, strategies, and tactics employed by the development projects and by many of their opponents. These tendencies seem logical, especially as Chile is a unitary republic with a purposefully developed, overarching neoliberal economic model, that is reinforced by an overly complex public/private oversight and governance system with limited regulation capacity, that effectively limits and weakens governmental authority over private business (Borrie et al. 2020; De Matheus e Silva et al. 2018; Gentes and Policzer 2022).

For example, given their size and complexity, both the HidroAysén and Río Cuervo hydroelectric projects were submitted for full EIA under Chile's SEA system. Both projects advanced for assessment as project-level EIAs, as neither project bundled their reservoirs and power plant proposals with the proposals for their transmission lines. This type of approach has been criticized as having inherent limitations for four main reasons: (1) it is reactive, and prevents an effective anticipation and planning around sites that are environmentally sensitive; (2) project-level EIAs are normally financed by the proponent of the project and have often been shown as biased against the environment; (3) often occur after key decisions in an overall strategy have been taken, thus no serious alternative assessment occurs; and (4) project-level EIAs do not effectively consider project subcomponents, ancillary developments, or the cumulative impacts of multiple projects (United Nations Environment Programme 2018).

The HidroAysén hydroelectric project proposal was initially approved through the SEA system (May 2011), under the government of Sebastian Piñera, but was subsequently rejected (June 2014) under the presidency of Bachelet. Both during and after the EIA process, the project was criticized for legal and procedural irregularities, but was ultimately rejected for technical reasons, including the absence of a community resettlement plan. The Río Cuervo hydroelectric project was also initially approved by the SEA (May 2012), in a decision that was subsequently ratified by the Committee of Ministers for Sustainability. After numerous appeals, it was also ultimately rejected by the Third Environmental Court of the Republic in Valdivia for technical reasons related to its proposed location above the active Liquiñe-Ofqui Faultline. Interestingly, despite the massive scale of the Patagonia National Parks Network, which spans three Chilean administrative regions, and the well-documented impacts of NBT in PAs and communities around the globe, there has been no mention of evaluating the environmental impacts (positive and/or negative), that could arise as a result of the project, or the subsequent development of the Route of the Patagonia Parks.

In our process to understand the actors involved and positions for and against the three development contexts, we identified latent themes related to the components of trust and distrust that merit future research and consideration within SD development trajectories and processes. We believe that improving Chile's current Environmental Impact Assessment System (SEIA) represents an important process improvement that could positively affect several of the themes that we identified related to the Trust category. For example, currently the SEIA focuses on evaluating extractive projects and does not evaluate others, such as the Patagonia National Park Network, despite their potential social, economic, and environmental impacts. In light of the growth of CDB in Chile, and especially Patagonia, it seems necessary to consider how to adjust the SEIA so that it can effectively evaluate development projects based on conservation.

Further, Chile's SEIA process warrants change to improve its capacity to accurately consider the SD related impacts of traditional large-scale development projects. Comparing Chile's environmental impact assessment process to systems in Brazil, Spain, and Canada, a recent evaluation identified four main weaknesses in the current Chilean system: high national-level centralization, a lack of scoping requirements, the non-binding nature of the EIA process, and the lack of consideration of project alternatives (Rodríguez-Luna et al. 2021). To address these weaknesses and improve the capacity of Chile's SEIA regulation, this study recommended modifications to the current system, including decentralization, mandating design alternatives within projects, the incorporation of scoping, a register of baseline information reviewers, and an improved public information process during and post-evaluation. Our research, and especially the underlying factors of trust and distrust that surfaced through the latent themes, supports the recommendations made by Rodríguez-Luna et al. (2021), as important steps in improving Chile's SD evaluation and oversight process.

We hope that this research helps to advance the agenda outlined by Adloff and Neckel (2019), providing further evidence that supports their hypothesis about the interrelated nature of these SD trajectories. Our results identified several interactions and interrelations between Modernization, Transformation, and Control SD imaginaries and trajectories, demonstrating how they often overlap and interact as development contexts unfold. The two examples we described within our discussion seem to strongly support another of Adloff and Neckel's (2019) hypotheses, that some SD forms can impulse or contribute to the evolution of others. We would extend that hypothesis, through consideration of SD agility. Our results suggest that SD agility may represent an important capacity within large-scale SD transformations, and as such, may be an important component of understanding SD territorial transitions. Further research is warranted around all these hypotheses.

We conclude by adding one final possible explanation for the pluralism and interactions observed within the work of ARV. Perhaps the SD agility we observed within their strategies has arisen through the collaborative initiatives and alliances that have formed during their many campaigns and projects, in order to be heard and listened to within the current limits of Chile's environmental and social oversight capacity. For example, in the introduction to Peter Hartmann's (2019) book about the movement, he explained:

> In our Chile, individual views are not considered, unless one is famous or influential, which forces us to act through organizations. We have learned that organizing is very important and that in this way incredible goals can be achieved…In general we think that it is necessary to work jointly or in alliance with people and organizations related to the objectives we want to achieve. (p. 2–3)

Acknowledgements This work was supported by Chile's National Research and Development Agency (ANID) under ANID's Regional Program R17A10002; the CIEP R20F0002 project; ANID PAI79170138; ANID FONDECYT Regular 1230020, and the CHIC-ANID PIA/BASAL PFB210018. We are grateful to Aldo Farias Herrera, Universidad Austral de Chile, for his support with the elaboration of the cartography in Fig. 10.1, and Patricio Segura Ortiz, journalist and director of the private Corporation for the Development of Aysén, for his permission to share the photographs in Fig. 10.9.

References

F. Adloff, S. Neckel, Futures of sustainability as modernization, transformation, and control: A conceptual framework. Sustain. Sci. **14**, 1015–1025 (2019) https://doi.org/10.1007/s11625-019-00671-2

E. Aliste, M. Folchi, A. Núñez, Discourses of nature in new perceptions of the natural landscape in southern Chile. Front. Psychol. **9**, Article 1177 (2018) https://doi.org/10.3389/fpsyg.2018.01177

S. Babidge, "Socios": The contested morality of "partnerships" in indigenous community-mining company relations, northern Chile. J. Latin American Caribbean Anthropol **18**(2), 274–293 (2013) https://doi.org/10.1111/jlca.12020

C. Baldwin, G. Marshall, H. Ross, J. Cavaye, J. Stephenson, L. Carter, C. Freeman, A. Curtis, G. Syme, Hybrid neoliberalism: Implications for sustainable development. Soc. Nat. Resour. **32**(5), 566–587 (2019) https://doi.org/10.1080/08941920.2018.1556758

H. Blair, B. Keith, T. Gale, Protected areas, tourism, and rural transition in Aysén, Chile. Sustainability **11**(24), Article 7087 (2019) https://doi.org/10.3390/su11247087

G. Blanco Wells, The social life of regions. Salmon farming and the regionalization of development in Chilean Patagonia. Wageningen University, 2009

S.L. Boblin, S. Ireland, H. Kirkpatrick, K. Robertson, Using Stake's qualitative case study approach to explore implementation of evidence-based practice. Qual. Health Res. **23**(9), 1267–1275 (2013) https://doi.org/10.1177/1049732313502128

B. Borrie, T. Gale, K. Bosak, Privately protected areas in increasingly turbulent social contexts: Strategic roles, extent, and governance. J. Sustain. Tour. **30**(11), 2631–2648 (2020) https://doi.org/10.1080/09669582.2020.1845709

F. Bourlon, La bio-geografía de Douglas Tompkins, una mirada comprensiva de la conservación privada en la Patagonia chilena [The bio-geography of Douglas Tompkins, a comprehensive view of private conservation in Chilean Patagonia]. Revista de Aysenología **4**, 86–98 (2017)

F. Bourlon, Voyager en Patagonie: Les usages touristiques de la nature [Travelling in Patagonia: The tourist uses of nature]. L'Harmattan, (2020)

B. Büscher, V. Davidov, Environmentally induced displacements in the ecotourism-extraction nexus. Area **48**(2), 161–167 (2016) https://doi.org/10.1111/area.12153

Chilean Third Environmental Court, Pronunciamiento del Tercer Tribunal Ambiental, República de Chile sobre el caso de Corporación Privada para el Desarrollo de Aysén, Corporación Pro Defensa de la Flora y Fauna y Hugo Díaz Márquez v. la Resolución Exenta N° 0914 del 5 de agosto 2016, del Servicio de Evaluación Ambiental adoptado por el Comité de Ministros [Pronouncement of the Republic of Chile's Third Environmental Court on the case of Corporación Privada para el Desarrollo de Aysén, Corporación Pro Defensa de la Flora y Fauna and Hugo Díaz Márquez v. Exempt Resolution N° 0914 del August 5, 2016, of the Environmental Assessment Service adopted by the Committee of Ministers]. Tercer Tribunal Ambiental de Chile (2018)

Chilean Environmental Evaluation Service, ¿Cuál es el proceso de evaluación de impacto ambiental? [What is the environmental impact assessment process?]. Servicio de Evaluación Ambiental de Chile (2022). Retrieved 15 Jan 2023, from https://sea.gob.cl

Chilean Ministry of the Environment, Consejo de Ministros para la Sustentabilidad ratifica creación de la Red de Parques de la Patagonia [Council of Ministers for Sustainability ratifies the creation of the Patagonian Parks Network]. Ministerio del Medio Ambiente (April 10, 2017). https://mma.gob.cl/consejo-de-ministros-para-la-sustentabilidad-ratifica-creacion-de-la-red-de-parques-de-la-patagonia/

Chilean Ministry of the Environment, Comité de Ministros acoge reclamaciones presentadas por la ciudadanía y decide rechazar proyecto HidroAysén [Committee of Ministers accepts claims presented by citizens and decides to reject HidroAysén project]. Ministerio del Medio Ambiente (June 10, 2014). https://mma.gob.cl/comite-de-ministros-acoge-reclamaciones-presentadas-por-la-ciudadania-y-decide-rechazar-proyecto-hidroaysen/

Chilean Ministry of the Interior and Public Security, Leyes, reglamentos, decretos y resoluciones de Orden General. Normas Generales. CVE 1508732. 5 p (2018)

Chilean National Forestry Corporation, *Red de Parques Nacionales de la Patagonia [National Parks Network of Patagonia]* (Unpublished, 2018)

J. Creswell, *Research Design: Qualitative, Quantitative, and Mixed Methods Approaches* (SAGE Publications Inc, Thousand Oaks, 2014)

L.F. De Matheus e Silva, H.M. Zunino, V. Huiliñir Curío, El negocio de la conservación ambiental: Cómo la naturaleza se ha convertido en una nueva estrategia de acumulación capitalista en la zona andino-lacustre de Los Ríos, sur de Chile [The business of environmental conservation: How nature has become a new strategy of capitalist accumulation in the Andean-lacustrine zone of Los Ríos, southern Chile.]. Scripta Nova **22** (2018). https://doi.org/10.1344/sn2018.22.19021

A. De Witt, N. Hedlund, Toward an integral ecology of worldviews, in *The Variety of Integral Ecologies: Nature, Culture, and Knowledge in the Planetary Era*, ed. by S. Mickey, S. Kelly, (SUNY Press, Albany, 2017), pp. 305–344

J. Fereday, E. Muir-Cochrane, Demonstrating rigor using thematic analysis: A hybrid approach of inductive and deductive coding and theme development. Int J Qual Methods **5**(1), 80–92 (2006) https://doi.org/10.1177/160940690600500107

R. Fletcher, Sustaining tourism, sustaining capitalism? The tourism industry's role in global capitalist expansion. Tour. Geogr. **13**(3), 443–461 (2011) https://doi.org/10.1080/1461668 8.2011.570372

T. Gale, A. Ednie, K. Beeftink, Acceptability and appeal: How visitors' perceptions of sounds can contribute to shared learning and transdisciplinary protected area governance. J. Outdoor Recreat. Tour. **35**, 1–13 (2021) https://doi.org/10.1016/j.jort.2021.100414

T. Gale, A. Ednie, Can intrinsic, instrumental, and relational value assignments inform more integrative methods of protected area conflict resolution? Exploratory findings from Aysén, Chile. J. Tour. Cult. Chang. **18**(6), 690–710 (2019) https://doi.org/10.1080/14766825.2019.1633336

T. Gale, A. Ednie, K. Beeftink, Worldviews, levels of consciousness, and the evolution of planning paradigms in protected areas. J. Sustain. Tour. **27**(11), 1609–1633 (2019) https://doi.org/10.108 0/09669582.2019.1639720

I. Gentes, P. Policzer, Weakness by design: Neoliberal governance over mining and water in Chile. Territory, Politics, Governance (2022) Advanced online publication. https://doi.org/10.108 0/21622671.2022.2134196

C. González Isla, Kristine McDivitt Tompkins: "La conservación no es ni debe ser privativa solo de los ricos" [Kristine McDivitt Tompkins: Private conservation shouldn't be just for the rich"]. La Tercera (April 21, 2017). https://www.latercera.com/noticia/kristine-mcdivitt-tompkins-la-conservacion-no-privativa-solo-los-ricos/#:~:text=las%20200%20hect%C3%A1reas.-,La%20conservaci%C3%B3n%20no%20es%20ni%20debe%20ser%20privativa%20solo%20de,por%20vivir%20en%20este%20planeta%22

P. Hartmann Samhaber, Aisén reserva de vida. Testimonio de un arquitecto activista ambiental formado en la Facultad de Arquitectura y Urbanismo de la Universidad de Chile [Aisén reserve of life. Testimony of an environmental activist architect trained at the Faculty of Architecture and Urbanism of the University of Chile]. Universidad de Chile Facultad de Arquitectura y Urbanismo (2019)

D.L. Harvey, *A Brief History of Neoliberalism* (Oxford University Press, Oxford, 2005)

D. Harvey, *Spaces of Capital: Towards a Critical Geography* (Routledge, New York, 2001)

A. Hedlund-de Witt, *Worldviews and the Transformation to Sustainable Societies: An Exploration of the Cultural and Psychological Dimensions of our Global Environmental Challenges* (Vrije Universiteit, Amsterdam, 2013). https://doi.org/10.13140/RG.2.1.4492.8406

G. Holmes, Biodiversity for billionaires: Capitalism, conservation and the role of philanthropy in saving/selling nature. Dev. Chang. **43**(1), 185–203 (2012) https://doi.org/10.1111/j.1467-7660.2011.01749.x

G. Holmes, What is a land grab? Exploring green grabs, conservation, and private protected areas in southern Chile. J. Peasant Stud. **41**(4), 547–567 (2014) https://doi.org/10.1080/0306615 0.2014.919266

Ilpes/Cepal-Gore Aysén, *Aysén: Matices de una identidad que asoma [Aysén: Nuances of a looming identity]* (Ocho Libros Editores, 2010)

G. Inostroza, Vinculación turística entre áreas silvestres protegidas y comunidades campesinas aledañas. Caso Parque Nacional Chiloé, sector Cucao, Chile [Tourism linkage between protected wild areas and neighboring rural communities. Case: Chiloé National Park, Cucao sector, Chile]. Universidad Austral de Chile (2007)

G. Inostroza, G. Cànoves, Turismo sostenible y proyectos hidroeléctricos: Contradicciones en la Patagonia Chilena [Sustainable tourism and hydroelectric projects: Contradictions in Chilean Patagonia]. Cuadernos de Turismo **34**, 115–138 (2014)

G. Inostroza, Turismo sostenible y conflicto por el uso de los recursos. Estudio de caso: Patagonia chilena, Región de Aysén [Sustainable tourism and resource use conflict. Case study: Chilean Patagonia, Aysén Región]. Universitat Autònoma de Barcelona (2016). http://hdl.handle.net/10803/386418

K.S. Jacobsen, J.D.C. Linnell, Perceptions of environmental justice and the conflict surrounding large carnivore management in Norway — Implications for conflict management. Biol. Conserv. **203**, 197–206 (2016) https://doi.org/10.1016/j.biocon.2016.08.041

C. Jones, Ecophilanthropy, neoliberal conservation, and the transformation of Chilean Patagonia's Chacabuco valley. Oceania **82**(3), 250–263 (2012)

P. Kashwan, L.M. MacLean, G.A. García-López, Rethinking power and institutions in the shadows of neoliberalism: (An introduction to a special issue of world development). World Dev. **120**, 133–146 (2019) https://doi.org/10.1016/j.worlddev.2018.05.026

M.A. Khan, Putting 'good society' ahead of growth and/or 'development': Overcoming neoliberalism's growth trap and its costly consequences. Sustain. Dev. **23**(2), 65–73 (2015) https://doi.org/10.1002/sd.1572

E. Kumi, A.A. Arhin, T. Yeboah, Can post-2015 sustainable development goals survive neoliberalism? A critical examination of the sustainable development-neoliberalism nexus in developing countries. Environ. Dev. Sustain. **16**(3), 539–554 (2014) https://doi.org/10.1007/s10668-013-9492-7

J. Kohl, S. McCool, *The Future Has Other Plans: Planning Holistically to Conserve Natural and Cultural Heritage* (Fulcrum Publishing, Golden, 2016)

A. Latta, B.E.C. Aguayo, Testing the limits neoliberal ecologies from Pinochet to Bachelet. Latin American Perspect **39**(185), 163–180 (2012) https://doi.org/10.1177/0094582X12439050

S. McLennan, G. Banks, Reversing the lens: Why corporate social responsibility is not community development. Corp. Soc. Responsib. Environ. Manag. **26**(1), 117–126 (2019) https://doi.org/10.1002/csr.1664

F. Miranda Cabaña, Políticas del estado y la incorporación de espacios en la geografía del capitalismo: El caso de Patagonia Aysén [State policy and the incorporation of spaces in the geography of capitalism: The case of Aysén Patagonia]. BEGEO **4**, 50–70 (2016)

A. Núñez, E. Aliste, Discursos ambientales y procesos de fronterización en Patagonia-Aysén (Chile): De los paisajes de la mala hierba a los del bosque sagrado. [Environmental discourses and frontier processes in Patagonia-Aysén (Chile): From weed-filled landscapes to sacred forests]. Fronteiras: J. Soc. Technol. Environ. Sci **6**(1), 198–218 (2017). https://doi.org/10.2166/4/2238-8869.2017v6i1.p198-218

A. Núñez, E. Aliste, Á. Bello, El discurso del desarrollo en Patagonia-Aysén: La conservación y la protección de la naturaleza como dispositivos de una renovada colonización. Chile, siglos XX-XXI [The development discourse in Patagonia-Aysén: Conservation and nature protection as devices of a renewed colonization. Chile, XX-XXI centuries]. Scripta nova **18**(46) (2014)

A. Núñez, E. Aliste, A. Bello, J. Astaburuaga, Eco-extractivism and nature discourses in Patagonia-Aysén: New geographical imaginaries and renewed processes of territorial control. Revista Austral de Ciencias Sociales **35**, 133–153 (2018)

A. Núñez, M. Benwell, E. Aliste, Interrogating green discourses in Patagonia-Aysén (Chile): Green grabbing and eco-extractivism as a new strategy of capitalism? Geogr. Rev. **112**(5), 688–706 (2020) https://doi.org/10.1080/00167428.2020.1798764

A. Núñez, F. Miranda, E. Aliste, S. Urrutia, Conservacionismo y desarrollo sustentable en la geografía del capitalismo: Negocio ambiental y nuevas formas de colonialidad en Patagonia-Aysén [Conservationism and sustainable development in the geography of capitalism: Environmental business and new forms of coloniality in Patagonia-Aysén], in *(Las) otras geografías en Chile: perspectivas sociales y enfoques críticos [(The) Other Geographies in Chile: Social Perspectives and Critical Approaches]*, ed. by A. Núñez, E. Aliste, R. Molina, (Lom, 2019), pp. 23–46

OECD, Regulatory policy in Chile: Government capacity to ensure high-quality regulation, OECD (2016). Retrieved 18 Jan 2023, from https://doi.org/10.1787/9789264254596-en

J. Oltremari, V. Leyton, J. Ignacio, T. Tomic, Memorias del taller FAO/CEGADES: Modelos de gestión integral del Sistema Nacional de Áreas Silvestres Protegidas en Chile [Proceedings of the FAO/CEGADES workshop: Models for integrated management of the National System of Protected Wildlife Areas in Chile]. FAO (1994)

Pramar Ambiental Consultores, Estudio de impacto ambiental central hidroeléctrica río Cuervo [Environmental impact study río Cuervo hydroelectric power plant]. Unpublished (2007)

M. Raco, Sustainable development, rolled-out neoliberalism and sustainable communities. Antipode **37**(2), 324–347 (2005) https://doi.org/10.1111/j.0066-4812.2005.00495.x

C. Reboratti, Tensiones geográficas: Controversias y conflictos ambientales en Argentina [Geographical tensions: Environmental controversies and conflicts in Argentina]. Investigaciones Geográficas **100**, 1–9 (2019). https://doi.org/10.14350/rig.60015

D. Rodríguez-Luna, N. Vela, F.J. Alcalá, F. Encina-Montoya, The environmental impact assessment in Chile: Overview, improvements, and comparisons. Environ. Impact Assess. Rev. **86**, Article 106502 (2021) https://doi.org/10.1016/j.eiar.2020.106502

A. Sæþórsdóttir, Tourism and power plant development: An attempt to solve land use conflicts. Tour. Plan. Dev **9**(4), 339–353 (2012) https://doi.org/10.1080/21568316.2012.726255

R.K. Schutt, D.F. Chambliss, Qualitative data analysis, in *Making Sense of the Social World: Methods of Investigation*, ed. by D.F. Chambliss, R.K. Schutt, (SAGE, Thousand Oaks, 2013), p. 248–283

H. Simons, *Case Study Research in Practice*, 1st edn. (SAGE Publications Inc, Los Angeles, 2009)

R.E. Stake, Case studies, in *Strategies of Qualitative Inquiry*, 2nd edn., (SAGE Publications Inc, Thousand Oaks, 2003)

S.L. Slocum, D.Y. Dimitrov, K. Webb, The impact of neoliberalism on higher education tourism programs: Meeting the 2030 sustainable development goals with the next generation. Tour. Manag. Perspect. **30**, 33–42 (2019) https://doi.org/10.1016/j.tmp.2019.01.004

D. Tompkins, The next economy. Transitions from globalization to eco-localism, in *Explorando las nuevas fronteras del turismo*, ed. by F. Bourlon, P. Mao, M. Osorio, T. Gale, (Ediciones Ñire Negro, 2012), pp. 107–124

R. Torres, G. Azócar, J. Rojas, A. Montecinos, P. Paredes, Vulnerability and resistance to neoliberal environmental changes: An assessment of agriculture and forestry in the Biobio region of Chile (1974-2014). Geoforum **60**, 107–122 (2015) https://doi.org/10.1016/j.geoforum.2014.12.013

United Nations Environment Programme, Assessing environmental impacts - A global review of legislation. UN Environmental Programme (2018). Retrieved 25 Jan 2023, from https://europa.eu/capacity4dev/unep/documents/assessing-environmental-impacts-global-review-legislation

D.R. Williams, Social construction of Artic wilderness: Place meanings, value pluralism, and globalization, in *Wilderness in the Circumpolar North: Searching for Compatibility in Ecological, Traditional, and Ecotourism Values*, ed. by A.E. Watson, L. Alessa, J. Sproull, (United States Dept. of Agriculture, Forest Service, Rocky Mountain Research Station, 2002), pp. 120–132

T. Wolters, Why is ecological sustainability so difficult to achieve? An in-context discussion of conceptual barriers. Sustain. Dev. **30**(6), 2025–2039 (2022) https://doi.org/10.1002/sd.2326

R.K. Yin. *Qualitative Research from Start to Finish*. (2nd Edition). New York. The Guilford Press. (2016)

Chapter 11
Identification of Causal Chains for Sustainable Tourism Development Within Two Chilean Patagonia National Parks: Cerro Castillo and Torres del Paine

Andrés Adiego, Germaynee Vela-Ruiz Figueroa, Fiorella Repetto Giavelli, and Trace Gale-Detrich

Abstract The Austral Macrozone of Chilean Patagonia (Aysén and Magallanes Regions) is home to 80% of the total area of the 106 protected areas (PAs) of the National System of Wildlife Protected Areas (SNASPE), and many of its PAs are experiencing intense socio-environmental challenges related to the balance between effective conservation and growing tourism development, in the face of uncertainty and change. This chapter takes an in-depth look at the causal chains for sustainable tourism development within the Cerro Castillo and Torres del Paine National Parks, using Ante Mandić's (Environ Syst Decis 40(4):560–576, 2020) conception of the *Drivers, Pressures, State, Impact, and Response (DPSIR) model* for advancing the sustainability of PAs that are managing nature-based tourism growth. Outcomes of

---

The original version of the chapter has been revised. A correction to this chapter can be found at https://doi.org/10.1007/978-3-031-38048-8_19

A. Adiego
Centro de Investigación en Ecosistemas de la Patagonia (CIEP), Sustainable Tourism Research Line, Human-Environmental Interactions Group, Coyhaique, Chile

Universidad de Zaragoza, Department of Geography and Territorial Planning, Zaragoza, Spain
e-mail: andres.adiego@ciep.cl

G. Vela-Ruiz Figueroa
Vela-Ruiz Consultorías Ambientales SpA, Punta Arenas, Chile

F. Repetto Giavelli
Torres del Paine Legacy Fund, Punta Arenas, Chile
e-mail: fiorella@supporttdp.org

T. Gale-Detrich (✉)
Centro de Investigación en Ecosistemas de la Patagonia (CIEP), Sustainable Tourism Research Line, Human-Environmental Interactions Group, Coyhaique, Chile

Cape Horn International Center (CHIC), Puerto Williams, Chile
e-mail: tracegale@ciep.cl

© The Author(s) 2023, Corrected Publication 2023
T. Gale-Detrich et al. (eds.), *Tourism and Conservation-based Development in the Periphery*, Natural and Social Sciences of Patagonia,
https://doi.org/10.1007/978-3-031-38048-8_11

the study represent an important first step for developing a better understanding of the causal chains related to the economic, social, and environmental dynamics of tourism in PAs within Chilean Patagonia and validate the value of moving forward with Mandić's (Environ Syst Decis 40(4):560–576, 2020) to advance understanding of tourism's effects on their conservation and management and thus, improve their potential for sustainability.

Keywords Patagonia · Tourism · Protected areas · Drivers, Pressures, State, Impact, and Response (DPSIR) model · Causal chains · Sustainable development

11.1 Introduction

The sustainable development and provision of nature-based tourism experiences provoke various socio-environmental challenges for protected areas (PAs) that require effective planning, management, and coordination with a range of actors, in a context that is increasingly complex and unpredictable (Gale et al., 2019; Kohl, 2018; Mandić, 2020; McCool, 2009; Spenceley et al., 2018). This situation becomes especially critical in the Austral Macrozone of Chilean Patagonia (Aysén and Magallanes Regions), which houses more than 80% of the total area protected within the 106 PAs of the National System of Wildlife Protected Areas (SNASPE for its acronym in Spanish, Fig. 11.1).

The two regions of this macrozone contain vast territories with low degrees of human intervention. Patagonian PAs protect fragile ecosystems, including highly biodiverse ocean and fjords, mountain environments, glaciers, wetlands, grasslands, and steppe, all of which are highly sensitive to climate change, and provide vital habitat for a range of species, including the emblematic puma (*Puma concolor*) and the huemul (*Hippocamelus bisulcus*) (Chilean National Forestry Corporation 2007, 2009b). Their uniqueness and fragility also make them very attractive settings for nature-based tourism (Barrena et al., 2019), which is an important axis of regional and local development that requires adequate planning and management to ensure sustainability.

11.1.1 Evaluation of the Linkages Between Tourism Use and PA Conservation

One of the tools that has been used to understand causal chains is the *Drivers, Pressures, State, Impact, and Response (DPSIR) model*, a framework which facilitates understanding of the complex cause and effect relationships generated between human activities, the environment and society (Gari et al., 2015; Patrício et al., 2016). Within the model, *drivers* are defined as the social, economic, and/or

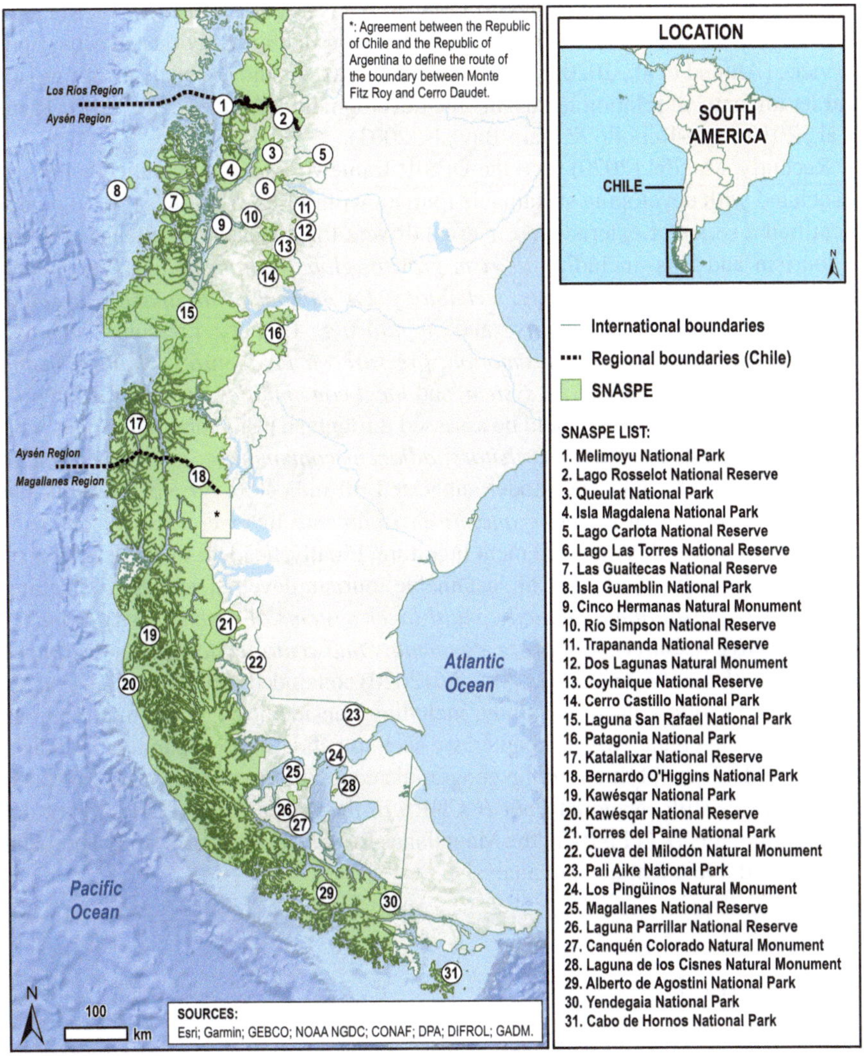

Fig. 11.1 Units of the National System of Wildlife Protected Areas (SNASPE) in the Aysén and Magallanes Regions

ecological factors that cause changes in the system, determining human production and consumption activities, which exert *pressures* on the environment. These pressures generate *impacts* that affect the health of ecosystems and human beings, shaping their *state*. To correct or mitigate the negative impacts on the state, society generates *responses*, which can be directed toward any part of this causal chain (EEA, 2003). The DPSIR model has been widely used in socioecology, facilitating the analysis of factors influencing coastal and marine ecosystems (Delgado et al., 2021; Gari et al., 2015; Lewison et al., 2016), processes affecting marine and terrestrial PA conservation (Eklund & Cabeza, 2017; López & Pardo, 2018; Mandić, 2020; Ojeda-Martínez et al., 2009), the effects of productive activities on ecosystem services (Ahmed et al., 2020; Delgado & Tironi, 2019), and the analysis of tourism and its impacts, in relation to sustainable development (Mustika et al., 2017; Ruan et al., 2019; Vera-Rebollo & Ivars-Baidals, 2003).

Recently, Mandić (2020) used the DPSIR framework to structure the challenges associated with developing sustainable tourism within PAs. Mandić's (2020) model identified a series of external and internal drivers for increasing challenges related to tourism and PAs, including *tourism growth, globalization, political, economic and socio-cultural environments, technology, PA management capacity, strategic and organizational adaptability,* and *monitoring.* Primary pressures included *increased visitation, access, promotion, pressure on PA features, changes in the position and strength of the PA system,* and *local communities.* Mandić (2020) proposed that the current state could be assessed through an evaluation of changes with respect to the *key PA features, visitors, adjacent communities,* and the *economic development of the PA.* The model suggested 30 indicators, grouped around the *ecological, socio-cultural,* and *economic environments,* that were drawn from leading international and PA management literature. Finally, he identified five responses to help address the challenges of sustainable tourism development in PAs, calling for *improved institutional capacity, multi-layer systems of planning and management, effective resource monitoring, community and visitor education,* and *stronger alliances with local communities.* Mandić (2020) concluded calling for a geographically focused follow up to his study, including consideration of the interrelations between causal chains. In this chapter, we seek to validate and expand on Mandić's (2020) model by applying it to the geographic context of two Patagonian SNASPE PAs: Cerro Castillo National Park (CCNP), in the Aysén Region, and Torres del Paine National Park (TPNP), in the Magallanes Region, using a case study approach and directed document content analysis.

11.2 Methods

The exploratory nature of this research drove the decision for case study methodology, which is particularly suitable for qualitative inquiry, where the researcher has little control over the object of study (Ebneyamini & Sadeghi Moghadam, 2018; Stake, 1994). Case study research can facilitate a deep and holistic consideration of phenomena occurring in a given context, even when variables blur the boundaries between the phenomenon and its context (Boblin et al., 2013; Harrison et al., 2017; Yin, 2009). Two Patagonian SNASPE PAs within the Austral Macrozone were chosen for the research (Fig. 11.2): Cerro Castillo National Park (CCNP), in the Aysén Region, and Torres del Paine National Park (TPNP) in the Magallanes Region. Case study sites were chosen based on their importance for both tourism and conservation, and for the contrasts they presented through differences in their stages of

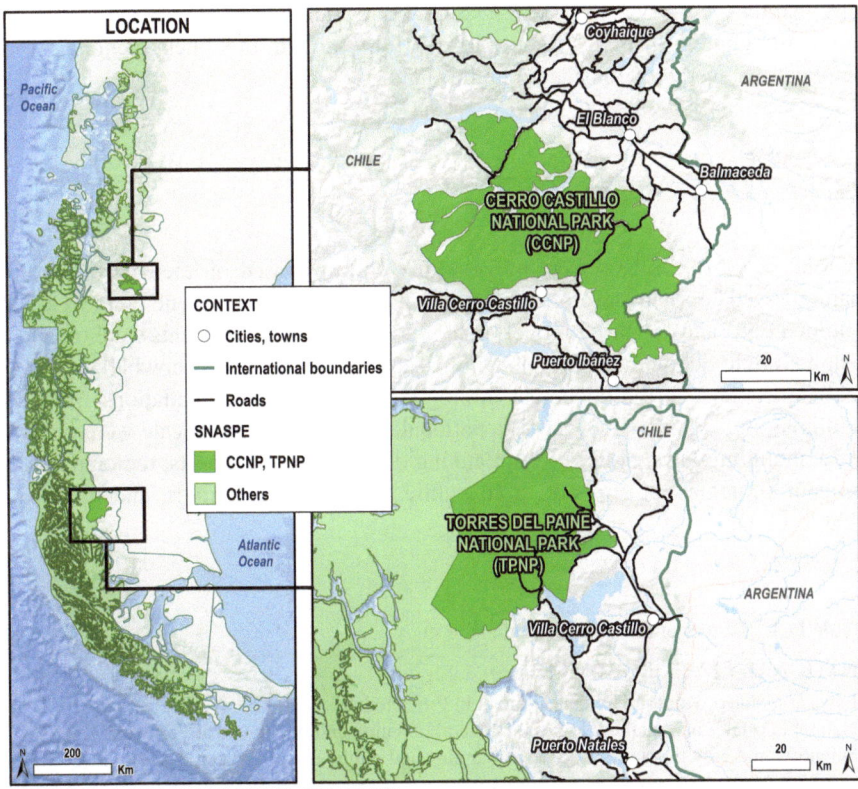

Fig. 11.2 Localization of Cerro Castillo National Park (CCNP) and Torres del Paine National Park (TPNP)

tourism development, or destination maturity. Both parks protect ecosystems of great ecological value and are surrounded by localities where tourism is one of the main economic activities (Rivas & Rojas, 2020). Both PAs are experiencing continual growth in tourism demand for their varied tourism offers that include 1-day and multi-day trails, climbing, and camping, among other activities, all of which generate economic income for local entrepreneurs. The TPNP is an example of a consolidated, mature nature-based tourism destination, with a strong influence over the regional and national economy. In contrast, the CCNP represents an emerging tourism destination with much lower visitation than the TPNP. Currently, its economic impacts are more local, but it has been prioritized nationally as one of Chile's emblematic PAs, and as such, has received significant national and international promotion as a must-see tourism destination.

For both parks, the case studies considered salient aspects of the evolution of tourism planning and use from 2000 to 2021. Given that at the time of writing this chapter the impacts that the SARS Cov-2 pandemic will generate in the socio-economic context of both PAs are not yet fully known, this phenomenon and its implications were not considered in the analysis.

11.2.1 Data Collection

A total of 78 documents were purposefully chosen, based on their pertinence to nature-based tourism development, planning, and/or management of the CCNP, and/or TPNP, and the chosen timeframe of 2000 to 2021. Documents were obtained using Google Scholar and Google Online search engines, and a snowball sampling technique, through the document bibliographies, and recommendations from PA administrators. Documents across a national, regional and local scale were considered, including laws, policies, PA planning documents and reports, regional development strategies, and reports from public and private programs, and initiatives (Table 11.1).

Table 11.1 Sources of information used in the study

NATIONAL LEVEL DOCUMENTS	
Laws associated with tourism development in Protected Areas (4)	
National policies and methodologies on general planning and public use of Protected Areas (7)	
Scientific and gray literature on general planning and public use of Protected Areas (10)	
REGIONAL DOCUMENTS—PATAGONIA	
Scientific and gray literature related to tourism and Protected Areas (5)	
REGIONAL AND/OR LOCAL LEVEL DOCUMENTS—AYSÉN	**REGIONAL AND/OR LOCAL LEVEL DOCUMENTS—MAGALLANES**
Regional policies and strategies related to tourism and/or Protected Areas (9)	Regional policies and strategies related to tourism and/or Protected Areas (8)
Scientific and gray literature related to tourism and Protected Areas (5)	Scientific and gray literature related to tourism and Protected Areas (1)
GENERAL AND TOURISM PLANNING PROTECTED AREA DOCUMENTS	
Cerro Castillo National Park (CCNP)	**Torres del Paine National Park (TPNP)**
Protected Area management tools (4)	Protected Area management tools (1)
Specific reports on general planning and public use of the CCNP (1)	Specific reports on general planning and public use of the TPNP (4)
Scientific and gray literature related to the tourism development of the CCNP and/or its immediate surroundings (2)	Scientific and gray literature related to the tourism development of the TPNP and/or its immediate surroundings (17)
INTERVIEWS/COMMUNICATIONS	
CCNP	**TPNP**
CONAF personnel, CCNP (2)	CONAF personnel, TPNP (3)

Note: The number of documents/interviews reviewed are in parentheses

11.2.2 Data Analysis

The directed content analysis employed a deductive approach to validate and extend the existing DPSIR model proposed by Mandić (2020). The authors read through the 78 documents, or units of analysis (Graneheim & Lundman, 2004), extracting meaning units (e.g., words, phrases, paragraphs, or graphics with a related context or content) associated with nature-based tourism and PA management within the PAs of interest (Gale et al., 2019). Meaning units were categorized according to the five constructs of the DPSIR framework, using a process of individual coding, followed by triangulation to achieve consensus. When doubts existed about the meaning units, the research team complemented their review with semi-structured interviews with PA administrators to help clarify the context and meaning. These sessions were carried out in-person, online, or through email. Next, the intermediate DPSIR frameworks for each PA were compared to the conceptual model proposed by Mandić (2020). Mandić's constructs were validated through the data that arose in the directed content analysis and when new concepts emerged, the investigators returned to the indicator documents proposed by Mandić (2020), to ground their findings with indicators and concepts within these reference documents (IUCN and WCPA, 2017; OECD, 2004; United Nations, 2007; WTO, 2004). Additionally, national sustainability indicators for PAs and tourism destinations were considered (De la Maza et al., 2014; Rivas, 2014), and in some cases, semi-structured interviews with PA administrators were retaken to clarify perspectives.

11.3 Results and Discussion

This section presents the results for the two case studies. Each case study begins with a brief description of the PA, followed by an overview of the main concept groupings that emerged from the data, organized around the DPSIR model. Finally, each of the main concept groupings is discussed and contextualized within Mandić's (2020) model.

11.3.1 Cerro Castillo National Park Case Study

The CCNP is located in the Aysén Region of Chile, (Fig. 11.2), about 1400 km south of Santiago. It has an area of 1435 km^2 and protects several vegetation formations including temperate Andean evergreen and deciduous forest/scrub, Andean anti-boreal desert and temperate Patagonian steppe; and emblematic fauna including puma (*Puma concolor*), the Magellanic woodpecker (*Campephilus magellanicus*), the guanaco (*Lama guanicoe*), and the huemul (*Hippocamelus bisulcus*), which is in danger of extinction (Chilean National Forestry Corporation, 2021). Surrounding population centers include Coyhaique (57 km, 49,667 inhabitants, regional capital), Balmaceda (35 km, 405 inhabitants), El Blanco (20 km, 250 inhabitants), Puerto Ingeniero Ibáñez (30 km, 764 inhabitants), and Villa Cerro Castillo (5 km, 376 inhabitants; Chilean National Statistics Institute, 2019). The CCNP is accessible by the Carretera Austral, the main north to south transportation route for the region, which crosses the PA for a 25 km stretch. CCNP received 13,350 registered visitors in 2019, placing it above the average of the other SNASPE units in the region, and is likely to have received thousands more that did not register, as the park is accessible around most of its perimeter (Chilean Subsecretary of Tourism, 2021). Figure 11.3 provides the conceptual DPSIR model that emerged from the data for the challenges associated with developing sustainable tourism within the CCNP.

11.3.1.1 Drivers

The primary external driver encountered in the CCNP data involved Mandić's (2020) *international and regional political and economic environment* category, specifically, a disconnect between high levels of public and private nature tourism promotion investment (Blair et al., 2019; Aysén Regional Government, 2014; Chilean National Tourism Service & Guazzinni Consultores, 2017; Chilean National Tourism Service, 2017), compared with extremely low conservation investment (Petit et al., 2018; Repetto-Giavelli et al., 2018; Toledo, 2017). Disparities are further complicated by the lack of a sufficient legal framework for PA management (Petit et al., 2018; Praus et al., 2011). Chile is currently advancing the Biodiversity and Protected Areas Service (SBAP, for its acronym in Spanish) bill, which will create an institution under the Ministry of the Environment to unify the administration of Chile's public, private, terrestrial, and marine PAs. The bill was unanimously approved in the Senate in 2019 (Chilean Senate, 2019), but the enactment of the law has not yet been finalized (Manzur, 2021).

Other external drivers included Mandić's (2020) *globalization,* and *international and regional tourism growth* categories, evidenced by growing visitation, and the high priority that the Aysén regional government has placed on nature-based tourism as an economic driver (Aysén Regional Government, 2009a, b, 2014; Chilean National Tourism Service, 2017). Finally, Mandić's (2020) *climate change* external driver category emerged in the data describing stresses arising from changing environmental conditions.

Internal drivers emerged in the CCNP data related to all three categories identified by Mandić (2020): *protected area management effectiveness and efficiency, effective flexibility and adaptability of strategic document,* and *organization structure and monitoring.* The park's management plan was recently updated in 2021, with a methodology that includes adaptive management capacity (Chilean National Forestry Corporation 2017b, 2021). The preceding management plan was developed in 2009 and based on rational comprehensive planning principles (Chilean

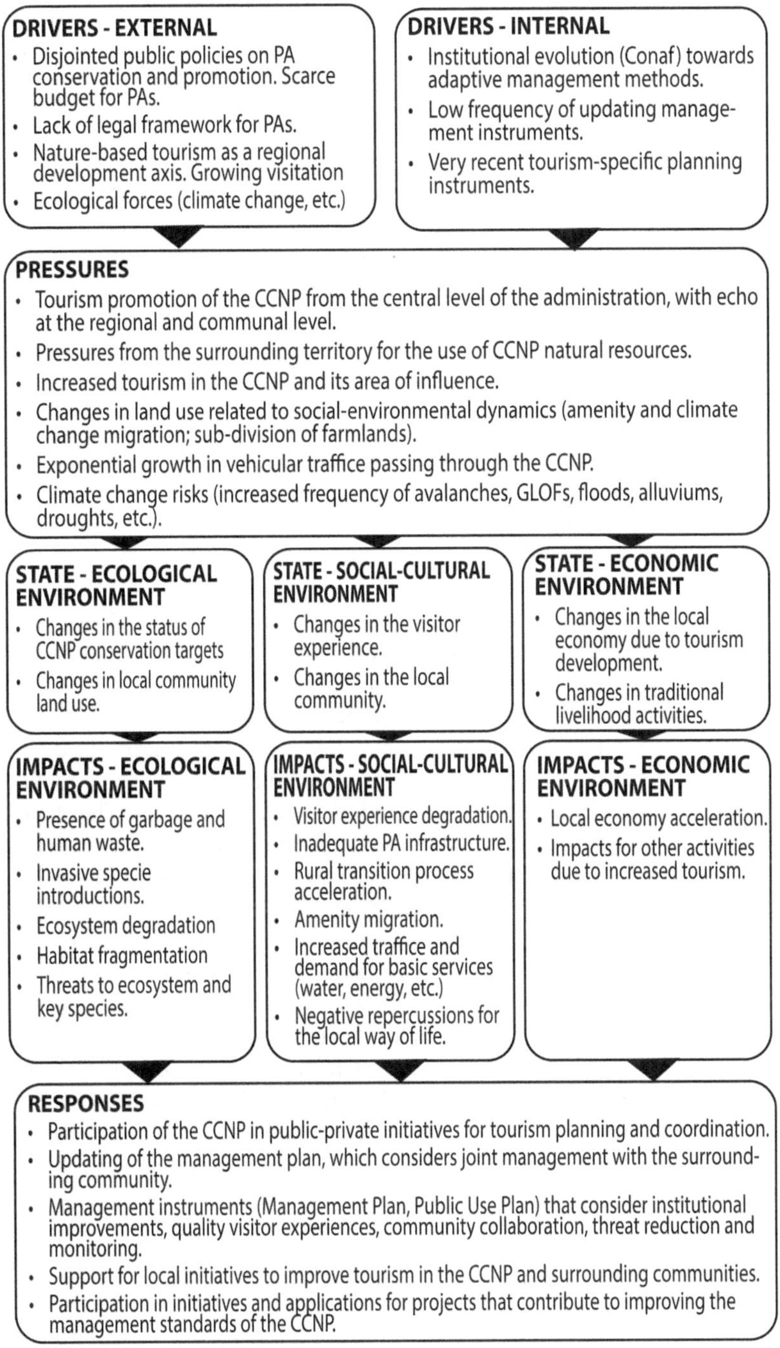

DRIVERS - EXTERNAL
- Disjointed public policies on PA conservation and promotion. Scarce budget for PAs.
- Lack of legal framework for PAs.
- Nature-based tourism as a regional development axis. Growing visitation
- Ecological forces (climate change, etc.)

DRIVERS - INTERNAL
- Institutional evolution (Conaf) towards adaptive management methods.
- Low frequency of updating management instruments.
- Very recent tourism-specific planning instruments.

PRESSURES
- Tourism promotion of the CCNP from the central level of the administration, with echo at the regional and communal level.
- Pressures from the surrounding territory for the use of CCNP natural resources.
- Increased tourism in the CCNP and its area of influence.
- Changes in land use related to social-environmental dynamics (amenity and climate change migration; sub-division of farmlands).
- Exponential growth in vehicular traffice passing through the CCNP.
- Climate change risks (increased frequency of avalanches, GLOFs, floods, alluviums, droughts, etc.).

STATE - ECOLOGICAL ENVIRONMENT
- Changes in the status of CCNP conservation targets
- Changes in local community land use.

STATE - SOCIAL-CULTURAL ENVIRONMENT
- Changes in the visitor experience.
- Changes in the local community.

STATE - ECONOMIC ENVIRONMENT
- Changes in the local economy due to tourism development.
- Changes in traditional livelihood activities.

IMPACTS - ECOLOGICAL ENVIRONMENT
- Presence of garbage and human waste.
- Invasive specie introductions.
- Ecosystem degradation
- Habitat fragmentation
- Threats to ecosystem and key species.

IMPACTS - SOCIAL-CULTURAL ENVIRONMENT
- Visitor experience degradation.
- Inadequate PA infrastructure.
- Rural transition process acceleration.
- Amenity migration.
- Increased traffice and demand for basic services (water, energy, etc.)
- Negative repercussions for the local way of life.

IMPACTS - ECONOMIC ENVIRONMENT
- Local economy acceleration.
- Impacts for other activities due to increased tourism.

RESPONSES
- Participation of the CCNP in public-private initiatives for tourism planning and coordination.
- Updating of the management plan, which considers joint management with the surrounding community.
- Management instruments (Management Plan, Public Use Plan) that consider institutional improvements, quality visitor experiences, community collaboration, threat reduction and monitoring.
- Support for local initiatives to improve tourism in the CCNP and surrounding communities.
- Participation in initiatives and applications for projects that contribute to improving the management standards of the CCNP.

Fig. 11.3 Conceptual diagram of the Drivers, Pressures, State, Impact, and Response (DPSIR) model for the Cerro Castillo National Park (CCNP)

National Forestry Corporation, 2009b). Also, in 2017, the CCNP developed its first public use plan (PUP), dedicating significant attention to planning tourism use and establishing scenarios and protocols for ongoing monitoring and adaptive management (Chilean National Forestry Corporation, 2017c).

11.3.1.2 Pressures

The data illuminated several pressures related to sustainable tourism development within the CCNP, in alignment with the dynamics identified by Mandić (2020). A primary pressure involved Mandić's (2020) *deterioration of the local community's position*. Central government initiatives prioritizing conservation and nature-based tourism development within and around the CCNP have impacted rural transition dynamics, resulting in conflicts and low levels of trust for park administration within the communities surrounding the park. In the 1970s, when the Cerro Castillo National Reserve was established, conflicts arose over land ownership, use, purchase, and sale (Blair et al., 2019). Many of these conflicts were resolved in an authoritarian manner and have left long-standing scars which manifest as resistance to CCNP conservation ordinances and rules. For example, the CCNP experiences continual pressures from the surrounding territory related to livestock use within its boundaries, firewood extraction, and the interaction of domestic animals with wildlife (Chilean National Forestry Corporation, 2017c).

Other Mandić's (2020) pressures categories, including *facilitated accessibility and visibility* and *strengthening of the protection system*, also arose in the CCNP data. In 2018, the status of the PA was elevated from National Reserve to National Park, which implies a higher level of protection and use restrictions. The change has been accompanied by increased national, regional, and local level tourism promotion and emphasis, including global press coverage and focus. Promotion and the development of tourism services and infrastructure around the PA have captured the interest of investors, manifesting in rapid acquisition of rural properties in order to subdivide them and offer them as plots for amenity and climate-change migration. These changes and increased attention have added new fuel to previous conflict and trust issues, creating further divisions in local communities, with respect to the most appropriate uses of the CCNP (CESPA, 2019; Chilean National Forestry Corporation, 2009a; Municipality of Coyhaique, 2014; Paralelo 47, 2019; Chilean Subsecretary of Tourism, 2021). Moreover, the CCNP has experienced a pronounced *increase in number of arrivals* (Mandić, 2020), from 1033 visitors in 2009 to 13,350 in 2019 (Chilean Subsecretary of Tourism, 2021) and *increasing pressure on national park features* (Mandić, 2020), through exponential growth in vehicular traffic passing through the park, via the main regional highway, the Carretera Austral (Chilean National Forestry Corporation, 2017c; Gale et al., 2018). While visitation dropped by a little more than 50% during COVID-19 (6080 registered annual visits), it has returned with strength in 2022. Finally, as Mandić (2020) observed, there have been a number of *negative effects on protected resources due to extreme weather conditions*. In the CCNP, the following are related to the effects produced by climate change including an increase in the frequency of avalanches, sudden glacial lake outbursts (GLOFs), floods, alluviums, and droughts, over recent decades, which have impacted tourism-related activities and are predicted to affect vegetation formations in the medium- and long-term (Chilean National Forestry Corporation, 2017c).

11.3.1.3 Status

The data obtained for the CCNP documented a series of changes resulting from the drivers and pressures affecting sustainable tourism development within the CCNP (Chilean National Forestry Corporation 2017c, 2021). With respect to the *ecological environment* (Mandić, 2020) changes related to the CCNP's conservation targets and land use in the surrounding areas. For the *socio-cultural environment* (Mandić, 2020) changes affected both the visitor experience and the well-being of the local population of the communities around the park. *Economic environment* (Mandić, 2020) changes involved an increase in the local economy due to tourism development and changes in traditional productive activities.

11.3.1.4 Impacts

CCNP data revealed *ecological environment* impacts related to park conservation targets and the proper functioning of park ecosystems (Chilean National Forestry Corporation 2017c, 2021). Inappropriate visitor behavior has led to degradation in the *state of the resource and of species* (Mandić, 2020), augmenting the presence of garbage and human waste, water contamination, introduction of invasive species, and harassment of wildlife. Although the CCNP views tourism as an engine for local development (CEQUA, 2017b), shortcomings in the availability of human and financial resources have led to a partial implementation of management and PUPs (CEQUA, 2017b), overloads for the CCNP staff, and a lack of appropriate infrastructure to meet demand. Some of the impacts of these conditions include insufficient drinking water, trails in poor condition, soil erosion and compaction, increased hazards and risks due to lacking maintenance, furtive campfires, and the existence of micro-dumps (CEQUA, 2017b; Chilean National Forestry Corporation, 2017c). While the Carretera Austral has greatly facilitated access to the park and connectivity for the region, it has fragmented important habitat within the CCNP, and speeding has caused deaths, injuries, and behavioral changes for wildlife key species (CEQUA, 2017b; Chilean National Forestry Corporation 2017c, 2021).

Socio-cultural environment impacts included *tourist and visitor satisfaction with the overall experience* (Mandić, 2020), resulting from crowding, a lack of adequate infrastructure, and a lack of informational and educational materials (Chilean National Forestry Corporation, 2017a). The *local population bordering the park* (Mandić, 2020) has also experienced impacts. In recent years, as nature-based tourism has assumed a larger role in territorial development, CCNP management has developed new local concessions and links with the community in order to improve tourism and recreational management (Chilean National Forestry Corporation, 2017c; Inostroza & Rovira, 2020). The *number of single-day and multi-day visitors* has increased, which has impacted the *ratio of tourists/visitors to residents*, and the *percent of men and women employed in the tourism sector* (Mandić, 2020). And, growing awareness and information about the park have accelerated rural transitions and amenity migration (Blair et al., 2019), increasing the local population and demand for basic services (water, electricity, cleaning, access roads, etc.), augmenting the presence of domestic animals and vehicular traffic, and contributing to the disruption of biological corridors (Blair et al., 2019; Paralelo 47, 2019; Sepúlveda & Lara, 2021). Although some of the local population have aligned with tourism initiatives, several documents (e.g., Paralelo 47, 2019; Sepúlveda & Lara, 2021; Chilean National Tourism Service & Guazzinni Consultores, 2017) evidenced *dissatisfaction with tourism in the destination* (Mandić, 2020), citing negative repercussions for the local way of life and a fundamental difference in land values. For example, the traditional ranching culture, which is still very present, values the natural environment as a space for productive exploitation, rather than an asset to be preserved for the proper functioning of ecosystems, or for tourism and recreation (Chilean National Forestry Corporation, 2017a; Gale & Ednie, 2020; Paralelo 47, 2019). From an *economic environment* perspective, tourism development has led to *economic development of the area around the park* (Mandić, 2020). A growth of companies, businesses, and sales volumes linked to tourism is observed (Chilean National Tourism Service, 2017). Gender impacts have also been observed; for example, studies have suggested that local tourism around the CCNP employs approximately twice as many women as men, while agriculture and cattle ranching, involved approximately twice as many men as women (Paralelo 47, 2019).

11.3.1.5 Responses

The data in the CCNP indicated a number of initiatives taking place in, and around the CCNP, that align with the five Mandić (2020) responses. For example, in terms of *multi-layered PA planning and management,* CONAF has recently updated the CCNP management plan with a vision that explicitly mentions the development of basic infrastructure for the sustainable enjoyment of the natural environment and joint management with the surrounding community (Chilean National Forestry Corporation, 2021). That plan is layered with the CCNP's first PUP and complements recent community tourism plans that involved joint work with the municipality, the community and CONAF, and where the CCNP is presented as a key piece (Paralelo 47, 2019). The CCNP's management plan and PUP emphasize strategies to improve *institutional capacity building, indicator development and key resource monitoring, community and visitor education,* and *positive impacts for local communities* (Chilean National Forestry Corporation 2017c, 2021). As a tangible step in achieving these strategies, the CCNP is forming alliances with NGOs and universities to strengthen planning, management, and monitoring capacity, which have facilitated additional resources for management plan updates and grant development, social capital building with local communities, and technical support for the application to the IUCN Green List of Protected and Conserved Areas (Programa Austral Patagonia, 2022; M. Sepúlveda, 23 de Mayo 2022, "personal communication"). Being part of this list is associated with effective and transparent management, which guarantees the conservation of nature and considers social, economic, and cultural benefits for the territories where they are inserted (IUCN & WCPA, 2017).

11.3.2 Torres del Paine National Park Case Study

The TPNP, with an area of 1814 km^2, is located at 50°57′ south latitude; 73°8′ west longitude, 2700 km south of Santiago, Chile, in the Magallanes Region (Fig. 11.2), and contains a private 44 km^2 PA, the Las Torres Reserve, within its boundaries. Since its creation in 1959, the TPNP has been considered *an ensemble of scenic beauty of exceptional tourist value* (BCN, 1962). The park includes part of the Southern Patagonia Ice Field and protects diverse ecosystems, including Magellanic deciduous forest, pre-Andean scrub, Patagonian steppe, and Andean desert (Pisano, 1974). The TPNP preserves its paleontological, archaeological, historical, cultural heritage, and a great diversity of birds and mammals, like the huemul and puma (Vela-Ruiz & Reppetto-Giavelli, 2017). The main population centers surrounding the TPNP are Villa Cerro Castillo (130 inhabitants) and Puerto Natales (19,023 inhabitants) (Chilean National Statistics Institute, 2019). The TPNP is accessed from the cities of Puerto Natales (100 km) and Punta Arenas (400 km). Figure 11.4 provides the conceptual DPSIR model that emerged from the data for the challenges associated with developing sustainable tourism within the TPNP.

11.3.2.1 Drivers

The data related to the TPNP suggested that one of the main external drivers is CONAF's limited ability to enforce regulations and manage the parks, which is constrained by severe underfunding of the SNASPE (less than US$2 per 10,000 km^2, per year) (Toledo, 2017), and the lack of an integrated regulatory and institutional body for PAs (Petit et al., 2018; Praus et al., 2011; Vela-Ruiz et al., 2018). In addition, the TPNP is affected by the *international and regional political and economic environment* (Mandić, 2020), and the disconnections that exist between local, regional, and national policy regarding the management, development, and promotion of tourism in the TPNP. A third external driver that emerged in the data was related to the *socio-cultural environment* (Mandić, 2020). Historically, CONAF has limited community involvement in the PA's management (CONAF, 2004); however, social manifestations have demanded greater participation in decisions involving PAs, and CONAF has increasingly moved toward including local communities in governance (Chilean National Forestry Corporation, 2020a). Other observed external drivers in TPNP are related to *globalization* and *international and regional*

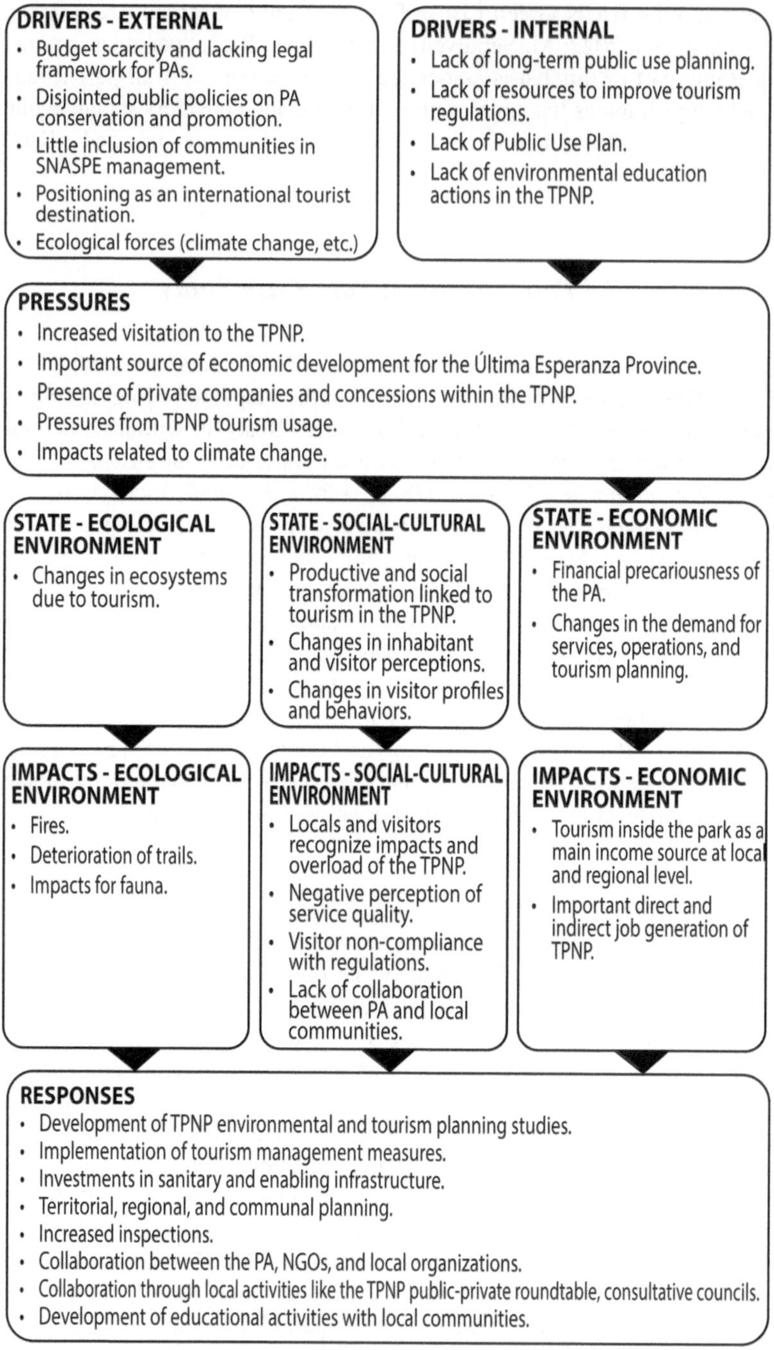

DRIVERS - EXTERNAL
- Budget scarcity and lacking legal framework for PAs.
- Disjointed public policies on PA conservation and promotion.
- Little inclusion of communities in SNASPE management.
- Positioning as an international tourist destination.
- Ecological forces (climate change, etc.)

DRIVERS - INTERNAL
- Lack of long-term public use planning.
- Lack of resources to improve tourism regulations.
- Lack of Public Use Plan.
- Lack of environmental education actions in the TPNP.

PRESSURES
- Increased visitation to the TPNP.
- Important source of economic development for the Última Esperanza Province.
- Presence of private companies and concessions within the TPNP.
- Pressures from TPNP tourism usage.
- Impacts related to climate change.

STATE - ECOLOGICAL ENVIRONMENT
- Changes in ecosystems due to tourism.

STATE - SOCIAL-CULTURAL ENVIRONMENT
- Productive and social transformation linked to tourism in the TPNP.
- Changes in inhabitant and visitor perceptions.
- Changes in visitor profiles and behaviors.

STATE - ECONOMIC ENVIRONMENT
- Financial precariousness of the PA.
- Changes in the demand for services, operations, and tourism planning.

IMPACTS - ECOLOGICAL ENVIRONMENT
- Fires.
- Deterioration of trails.
- Impacts for fauna.

IMPACTS - SOCIAL-CULTURAL ENVIRONMENT
- Locals and visitors recognize impacts and overload of the TPNP.
- Negative perception of service quality.
- Visitor non-compliance with regulations.
- Lack of collaboration between PA and local communities.

IMPACTS - ECONOMIC ENVIRONMENT
- Tourism inside the park as a main income source at local and regional level.
- Important direct and indirect job generation of TPNP.

RESPONSES
- Development of TPNP environmental and tourism planning studies.
- Implementation of tourism management measures.
- Investments in sanitary and enabling infrastructure.
- Territorial, regional, and communal planning.
- Increased inspections.
- Collaboration between the PA, NGOs, and local organizations.
- Collaboration through local activities like the TPNP public-private roundtable, consultative councils.
- Development of educational activities with local communities.

Fig. 11.4 Conceptual diagram of the Drivers, Pressures, State, Impact, and Response (DPSIR) model for the Torres del Paine National Park (TPNP)

tourism growth (Mandić, 2020). For example, the TPNP received UNESCO designation as "Man and the Biosphere Reserve" in 1978 (Chilean National Forestry Corporation, 2020b), Trip Advisor recognized TPNP as the "Eighth Wonder of the World" in 2013, and it was elected as one of the most important tourist images of Chile, and areas inside and surrounding the park were designated as a Zone of Tourism Interest (ZOIT, for its acronym in Spanish) in 2019 (Chilean Subsecretary of Tourism, 2018). These situations increased tourism promotion and visibility at national and international levels. Another important external driver affecting the sustainability of tourism in the TPNP involved Mandić's (2020) category of *climate change*. Variations in temperature and precipitation, loss of biodiversity and occurrence of extreme phenomena are threats to tourism activity in Chile (UNDP et al., 2019), as they put at risk the natural attractions of PAs (MINECON & MMA, 2019).

Similar to the situation in the CCNP, the primary internal drivers that arose in this case study were related to all three categories identified by Mandić (2020): *protected area management effectiveness and efficiency, effective flexibility and adaptability of strategic document and organization structure*, and *monitoring*, especially with respect to the absence of long-term public use planning. While there have been efforts to advance tourism planning and several actions has been applied to improve the tourism management (AMBAR, 2004; Vela-Ruiz et al., 2018), the TPNP does not currently have enough resources or planning instruments (e.g., PUP), to effectively regulate this activity. Further, the TPNP still needs to improve regulations, tourism impacts monitoring, and environmental education actions (CEQUA, 2017a; J. Linnebrink, 9 August 2021, "personal communication"; Vela-Ruiz et al., 2018).

11.3.2.2   Pressures

The TPNP case study surfaced a number of pressures on tourism sustainability within the park. With respect to socioeconomic pressures, the TPNP has experienced a significant *increase in the number of arrivals* (Mandić, 2020), with an average growth of 12%, between 2009 and 2019. In 2019, 304,947 visitors were registered (Chilean Subsecretary of Tourism, 2021). In 2020, visitation dropped to 142,881, as a result of the COVID-19 pandemic; but, by the end of 2021, it has returned to pre-pandemic levels. To manage increased demand, the TPNP has become the SNASPE unit at national level with the highest number of tourism concessions, and there are also tourism services in the Las Torres Reserve. The TPNP is a fundamental income source within the Última Esperanza Province (Vela-Ruiz & del Delgado, 2010); and, as such, local actors apply significant pressure to influence tourism management decisions within the TPNP (Vela-Ruiz et al., 2018). This situation, as Mandić's model (2020) identified in a broader context, often influences *the park's ability to achieve conservation objectives*. In terms of ecological pressures, the evidence supported climate change impacts (Mandić, 2020), including the retreat of mountain glaciers and the loss of glacial mass in the Southern Icefield, affecting important TPNP tourist attractions, like the Grey Glacier (Weidemann et al., 2018).

11.3.2.3 State

Similar to the CCNP case study and the Mandić (2020) conceptual model, TPNP data documented changes related to the *ecological, socio-cultural, and economic environments*, as a result of tourism development and activity. Several scientific studies documented *ecological environment* changes that have taken place in the TPNP, and the effects these changes have had for the park conservation; for example, the ecosystem impacts generated by forest fires (Vidal et al., 2012), trails deterioration (Torres et al., 2018), and changes in structure and behavior of birds and carnivores due to food supplementation (Cabello-Cabalín, 2017). For example, in the *socio-cultural environment*, TPNP events and cycles have contributed to a constant productive and social transformation of tourism in the region, with evidence of service diversification and growth in recent years (Vela-Ruiz & del Delgado, 2010), tourist behavior changes, and changing perceptions of TPNP tourism from both local residents and visitors. From an *economic environment* perspective, tourism linked to the TPNP has been one of the main development factors for the Última Esperanza Province, especially for Puerto Natales, transforming productive activities and revitalizing social organization (Ferrer, 2003; Vela-Ruiz & del Delgado, 2010). According to Valverde (2020), in 2017, TPNP revenues from tourism services located within the park, entrance fees, and concession income, generated \$US 56,284,000 in revenues directly related to tourism, and \$US 182,931,000 in additional indirect revenues. This is equivalent to 3% and 10% of Magallanes' GDP, respectively.

11.3.2.4 Impacts

With respect to the ecological environment, the *state of resources and species* (Mandić, 2020) within the TPNP has been strongly affected in recent decades, due to three mega fires caused by poor tourist practices (years 1985, 2005 and 2011), that have significantly changed the attributes of the forests (Vidal et al., 2012), and the presence of invasive exotic species of flora in camping and trail sectors (Vidal et al., 2015). *Land degradation* (Mandić, 2020) has also been linked to tourism and the poor condition of mountain trails (Farrell & Marion, 2001; Torres et al., 2018), which is evidenced by changes in vegetation and soil, erosion, exposure of roots and loose rocks, habitat fragmentation due to the presence of multi-trails, and the widening of trails (Repetto & Cabello, 2015; Torres et al., 2018). Regarding fauna, Cabello-Cabalín (2017) and Repetto and Cabello (2015) point out tourism impacts related to the structure and behavior of birds and carnivores, due to food supplementation generated by poor *waste management and runoff from sewage treatment plants* (Mandić, 2020). There is also evidence of impacts from the harassment of fauna and from the use of anticoagulant poisons in lodging areas, affecting native rodents and birds of prey Cabello-Cabalín (2017). In terms of the socio-cultural environment, evidence supported Mandić's (2020) proposition to consider *resident satisfaction with tourism in the destination* and the *impacts of tourism on the destination's destiny,* as well as *tourist and visitor satisfaction with the overall experience.* According to Fernández Génova et al. (2020), local stakeholders are critical of the way tourism has developed in TPNP and the impacts it has generated. Despite specific efforts, the TPNP has not advanced a long-term strategy to link stakeholders and establish governance that is recognized and validated by local authorities and community stakeholders. They criticize the expansion of concessions as damaging to the park, with negative consequences for the conservation and sustainability of the destination. They also question the concession model, which has not prioritized local communities (Barrena et al., 2019; Sepúlveda & Lara, 2021). In addition, several studies have negatively evaluated the quality of services provided within the TPNP (CEQUA, 2016; Chilean National Forestry Corporation and EES Ingeniería, 2019; Chilean National Forestry Corporation & Search Consultores, 2014). Trail and infrastructure crowding issues, linked to the high tour operator demand for the *W'* and *Macizo Paine* circuits (CEQUA, 2017a), which has led to overcrowding of full day trails (Torres et al., 2019). Moreover, visitor experiences have been affected by visitor behavior problems, including the use of informal campsites, use of fire in unauthorized places, and bad practices to avoid paying entrance fees and services in the park (Barrena et al., 2019). Evidence of the economic environment impacts centered on the regional dependence on TPNP related tourism as a major factor of the GNP, which relates to Mandić's (2020) identification of impacts produced through the *economic development of the park area.* In relation to *direct tourism employment* (Mandić, 2020), 6700 direct jobs were estimated in relation to the TPNP, for 2017, in the Magallanes Region and other parts of Chile (Valverde, 2020). Additionally, Valverde (2020) estimated that 10,850 indirect jobs were generated during this same year.

11.3.2.5   Responses

Some of Mandić's (2020) five responses to improve the sustainability of tourism, in and around PAs, manifested within the data for the TPNP. Nevertheless, external and internal drivers limited the PA's advances with respect to a local governance that supports the tourism management and conservation of the park, the generation of spaces for dialogue, and the collaboration between public and private stakeholders in the territory. The evidence was present to support Mandić's (2020) conceptual response about the importance of *multi-layered PA planning and management tactics.* Although the TPNP currently lacks a PUP, important actions have focused on resolving urgent gaps, prioritizing the construction of enabling infrastructure for workers, and adequate sanitation systems. Between 2014 and 2017, a tourism management system (TMS) was proposed for the TPNP through a participatory and collaborative planning process with a wide range of public and private stakeholders (Fernández Génova et al., 2020). This included a micro-zoning of public use and an action plan to improve management and visitor flow management, in addition to proposing impact mitigation measures for trails and lodging sites through the implementation of technological innovations (Vela-Ruiz & Reppetto-Giavelli, 2017; Vela-Ruiz et al., 2018). These park planning advances have been complemented by advances in territorial planning, which *improve institutional capacity* (Mandić, 2020) for CONAF, by adding new legal tools, financial resources, management strategies and technological innovations, trying to improve tourism management (entrance payment, visitors registration system, reservations for accommodation), and to mitigate environmental problems generated by tourism activity (trail designs, fauna monitoring, sewage treatment plants). Between 2017 and 2018, Torres del Paine was declared a ZOIT destination, with an associated tourism management action plan that responds to several gaps identified in the park (Chilean Subsecretary of Tourism, 2018). In parallel, the municipalities of Torres del Paine and Natales have made progress in updating provincial tourism plans and Torres del Paine has updated tourism ordinances and established an environmental ordinance (Municipality of Natales and Municipality of Torres del Paine, 2021; Municipality of Torres del Paine, 2015, 2020a, b). In addition, the State has increased its oversight of tour operators (Martínez, 2016), and sanctions for inappropriate behavior by visitors have been more frequent and publicized (e.g., the expulsion of six tourists in 2022 for irresponsible behavior; two for improper use of cookstoves; another involved a guide charged with reckless endangerment for employing bad practices when photographing a puma in 2021). With respect to *indicator and key resource management,* CONAF has established important alliances with academia and with national and international NGOs, to improve the condition of trails and infrastructure, and work in partnership with local businesses and visitors, who collaborate with funding and concrete conservation actions (https://www.tdplegacyfund.org/). With respect to *creating positive impacts for local communities* (Mandić, 2020), important participatory efforts have been undertaken during recent years, including updating of the Torres del Paine Biosphere Reserve management plan (Chilean National Forestry Corporation, 2020b), the definition of conservation targets, the declaration of the Torres del Paine ZOIT (Chilean Subsecretary of Tourism, 2018), the TMS (Fernández Génova et al., 2020), the functioning of the park's Consultative Council, and the intermittent establishment of public-private roundtables. Finally, CONAF and other institutions are focusing efforts to improve and expand *community and visitor education* (Mandić, 2020).

11.4 Conclusions

The two case studies developed in this chapter supported Mandić's (2020) conceptual DPSIR model for advancing the sustainability of PAs that are managing nature-based tourism growth. The holistic, systems thinking DPSIR approach taken by Mandić (2020) facilitated a triple bottom line approach to conceptualizing the challenges associated with sustainable tourism development in the CCNP and TPNP. Most of the categories identified by Mandić (2020) arose in the data related to the CCNP and TPNP, and missing categories were also helpful, as they helped illuminate geographic specific dynamics that might emerge in the future. For example, while Mandić (2020) emphasized *technological innovations* as a driver, the case studies identified little to no evidence to support this driver with respect to tourism development and management in and around the CCNP, but technological innovations have surged in TPNP as a response. The five responses Mandić (2020) posed as challenges for the sustainable development of PAs resonated well with studies and needs that emerged in the CCNP and TPNP cases, even though each of the PAs were experiencing very different contexts and levels of tourism destination development. In both cases, these responses validated current efforts to implement the new SBAP legal framework and strategy for PA management in Chile, which prioritize initiatives consistent with *improved institutional capacity, multi-layer systems of planning and management, effective resource monitoring, community and visitor education,* and *stronger alliances with local communities*. This paper expands on Mandić's (2020) conceptual work by adding a geographic focus; however, it falls short of considering the interrelations between the proposed parameters. As well, we have employed a qualitative approach, rather than taking Mandić's (2020) suggestion about "measuring each parameter and testing their mutual correlations" (p.13). While these limitations are important to consider, we believe this chapter sets the course for future work along these lines and represents an important first step for developing a better understanding of the causal chains related to the economic, social, and environmental dynamics of tourism in PAs within Chilean Patagonia, as it has validated the value of moving forward with Mandić's (2020) work, and thus, may help understanding of tourism's effects on their conservation and management and, therefore, improve the potential for sustainability.

Acknowledgements We are grateful for the contributions of all the local actors in both case studies, which facilitated the understanding of the territory and made it possible to obtain concrete data. In particular, the contributions of Hernán Velásquez, Carlos Hochstetter, and Álvaro Salín, from CONAF Región de Aysén; and José Linnebrink, Mauricio Ruiz, Michael Arcos, Rodrigo Rodríguez, Cristian Ruiz, and Verónica Osorio, from CONAF Región de Magallanes, who provided important information for the preparation of the chapter. Andrés Adiego and Trace Gale-Detrich received support from Chile's National Agency for Research and Development (ANID) under the ANID Regional Program R17A10002; CIEP project R20F0002. Trace Gale-Detrich also received support from the CHIC-ANID PIA/BASAL PFB210018. We thank the Innova Corfo Bienes Públicos para la Competitividad Nacional project, Code. 14BPC4-28654, for sharing data.

References

S.N. Ahmed, L.H. Anh, P. Schneider, A DPSIR assessment on ecosystem services challenges in the Mekong delta, Vietnam: Coping with the impacts of sand mining. Sustainability (Switzerland) **12**(22), 1–29 (2020). https://doi.org/10.3390/su12229323

AMBAR, *Desarrollo e implementación de un sistema de gestión del uso público en los parques Torres del Paine y Bernardo O'Higgins [Development and Implementation of a Public Use Management System in Torres del Paine and Bernardo O'Higgins Parks]* (EuroChile, CORFO, 2004)

Aysén Regional Government, *Estrategia regional de desarrollo de Aysén [Regional Development Strategy for Aysén]* (GORE-Aysén, 2009a). Retrieved 1 Dec 2022, from https://www.goreaysen.cl/controls/neochannels/neo_ch112/appinstances/media42/EDR_AYSEN.pdf

Aysén Regional Government, *Política regional de turismo de Aysén [Aysén Regional Tourism Policy]* (GORE-Aysén, 2009b). Retrieved 1 Dec 2022, from https://www.goreaysen.cl/controls/neochannels/neo_ch95/appinstances/media204/01_AysenPoliticaTurismo.pdf

Aysén Regional Government, *Plan especial de desarrollo de zonas extremas: Región de Aysén [Special Development Plan for Extreme Zones: Aysén Region]* (GORE-Aysén, 2014). Retrieved 1 Dec 2022, from http://www.goreaysen.cl/controls/neochannels/neo_ch237/appinstances/media1423/PEDZE_FINAL_25AGO14.pdf

J. Barrena, M. Lamers, S. Bush, G. Blanco, Governing nature-based tourism mobility in National Park Torres del Paine, Chilean Southern Patagonia. Mobilities **14**(6), 745–761 (2019). https://doi.org/10.1080/17450101.2019.1614335

BCN, *Decreto 1050, amplía la extensión del parque nacional de turismo "Lago Grey" y declara que en lo sucesivo se denominará parque nacional de turismo "Torres del Paine" [Decree 1050 extends the extension of the "Lago Grey" National Tourism Park and Declares that It Will Henceforth Be Called "Torres del Paine" National Tourism Park]* (Biblioteca del Congreso Nacional, 1962). http://bcn.cl/2ha5j

H. Blair, K. Bosak, T. Gale, Protected areas, tourism, and rural transition in Aysén, Chile. Sustainability (Switzerland) **11**(24), 1–22 (2019). https://doi.org/10.3390/su11247087

S.L. Boblin, S. Ireland, H. Kirkpatrick, K. Robertson, Using Stake's qualitative case study approach to explore implementation of evidence-based practice. Qual. Health Res. **23**(9), 1267–1275 (2013). https://doi.org/10.1177/1049732313502128

J. Cabello-Cabalín, La fauna del Parque Nacional Torres del Paine y la importancia de su conservación [The fauna of Torres del Paine National Park and the importance of its conservation], in *Guía de conocimiento y buenas prácticas para el turismo en el Parque Nacional Torres del Paine [Guide to Knowledge and Best Practices for Tourism in Torres del Paine National Park]*, ed. by G. Vela-Ruiz, F. Repetto-Giavelli, (Ediciones CEQUA, 2017), pp. 78–101

CEQUA, *Informe final de análisis de las encuestas sobre calidad de la experiencia turística a los visitantes y disposición a regular el caudal en los circuitos montañosos del Parque Nacional Torres del Paine – Estancia Cerro Paine [Final Analysis Report of the Surveys on the Quality of the Tourist Experience for Visitors and Willingness to Regulate the Flow in the Mountain Circuits of Torres del Paine National Park – Estancia Cerro Paine]* (Ediciones CEQUA, 2016)

CEQUA, *Diagnóstico de empresas operadoras de actividades turísticas de aventura en los circuitos de montaña del Parque Nacional Torres del Paine [Diagnosis of Operating Companies of Adventure Tourism Activities in the Mountain Circuits of Torres del Paine National Park]* (Innova Corfo 14BPC4-28654, 2017a)

CEQUA, *Informe final: Encuesta sobre planificación y uso turístico en las áreas protegidas públicas y privadas [Final Report: Survey on Tourism Planning and Use in Public and Private Protected Areas]* (Innova Corfo 14BPC4-28654, 2017b)

CESPA, *Plan de desarrollo comunal de Río Ibáñez, 2018–2026 [Community Development Plan of Río Ibáñez, 2018–2026]* (Municipalidad de Río Ibáñez, 2019). Retrieved 25 Nov 2022, from https://rioibanez.cl/wp-content/uploads/2022/10/Pladeco-rio-ibanez-2018-2026.pdf

Chilean National Forestry Corporation, *Política para el desarrollo del ecoturismo en áreas silvestres protegidas del estado y su entorno [Policy for the Development of Ecotourism in Protected Wild Areas of the State and Its Surroundings]* (Corporación Nacional Forestal, 2004). Retrieved 5 Nov 2022, from https://www.conaf.cl/wp-content/uploads/2018/05/Pol%C3%ADtica-Ecoturismo-SNASPE.pdf

Chilean National Forestry Corporation, *Plan de manejo Parque Nacional Torres del Paine [Torres del Paine National Park Management Plan]* (Corporación Nacional Forestal, 2007). Retrieved 20 Dec 2022, from https://parquetorresdelpaine.cl/biblioteca/plan-de-manejo/

Chilean National Forestry Corporation, *Estadística visitantes unidad SNASPE, año 2008 [SNASPE Unit Visitor Statistics, Year 2008]* (Corporación Nacional Forestal, 2009a). Retrieved 20 Dec 2022, from https://www.conaf.cl/wp-content/files_mf/1385733281totalvisitantes2008.pdf

Chilean National Forestry Corporation, *Plan de manejo de la Reserva Nacional Cerro Castillo [Cerro Castillo National Reserve Management Plan]* (Corporación Nacional Forestal, 2009b). Retrieved 20 Dec 2022, from https://www.conaf.cl/wp-content/files_mf/1382467911RNCerroCastillo.pdf

Chilean National Forestry Corporation, *Anexo, informe de las actividades de participación pública en el desarrollo del plan de uso público de la Reserva Nacional Cerro Castillo [Annex, Report on Public Participation Activities in the Development of the Public Use Plan for the Cerro Castillo National Reserve]* (Corporación Nacional Forestal, 2017a)

Chilean National Forestry Corporation, *Manual para la planificación del manejo de las áreas protegidas del SNASPE [Manual for Management Planning of Protected Areas of SNASPE]* (Corporación Nacional Forestal, 2017b). Retrieved 12 Jan 2023, from https://www.conaf.cl/wp-content/files_mf/1515526054CONAF_2017_MANUALPARALAPLANIFICACI%C3%9 3NDELASAREASPROTEGIDASDELSNASPE_BajaResoluci%C3%B3n.pdf

Chilean National Forestry Corporation, *Plan de uso público de la Reserva Nacional Cerro Castillo [Cerro Castillo National Reserve Public Use Plan]* (Corporación Nacional Forestal, 2017c)

Chilean National Forestry Corporation, *Marco de acción para la participación de la Comunidad en la gestión del SNASPE [Framework of Action for Community Participation in the Management of the SNASPE]* (Corporación Nacional Forestal, 2020a). Retrieved 12 Nov 2022, from https://www.conaf.cl/wp-content/files_mf/1605703679MARCODEACCIONPARALAPARTICIPAC IONCONAF.2020.pdf

Chilean National Forestry Corporation, *Reserva de Biosfera Torres del Paine: Desafíos de un nuevo territorio [Torres del Paine Biosphere Reserve: Challenges of a New Territory]* (Maval SPA, 2020b)

Chilean National Forestry Corporation, *Plan de manejo Parque Nacional Cerro Castillo [Cerro Castillo National Park Management Plan]* (Corporación Nacional Forestal, 2021)

Chilean National Forestry Corporation, EES Ingeniería, *Encuesta de percepción de áreas silvestres protegidas. Informe final de resultados Parque Nacional Torres del Paine [Protected Wildlife Areas Perception Survey. Final Results Report Torres del Paine National Park]* (Corporación Nacional Forestal, 2019)

Chilean National Forestry Corporation, Search Consultores, *Evaluación de la percepción de los visitantes respecto de la calidad del servicio ofrecido en las áreas silvestres protegidas del estado 2014: Parque Nacional Torres del Paine [Evaluation of Visitors' Perception of the Quality of Service Offered in State Protected Areas 2014: Torres del Paine National Park]* (Corporación Nacional Forestal, 2014)

Chilean National Statistics Institute, *Ciudades, pueblos, aldeas y caseríos 2019 [Cities, Towns, Villages and Hamlets 2019]* (INE, 2019). Retrieved 10 Dec 2022, from https://geoarchivos.ine.cl/File/pub/Cd_Pb_Al_Cs_2019.pdf

Chilean National Tourism Service, *Anuario de turismo Región de Aysén 2017 [Tourism Yearbook Aysén Region 2017]* (Sernatur, 2017)

Chilean National Tourism Service, Guazzinni Consultores, *Plan de acción para la gestión participativa de Zonas de Interés Turístico (ZOIT): Zona de interés turístico Chelenko [Action Plan for the Participatory Management of Zones of Tourist Interest (ZOIT): Chelenko Area of Tourist Interest]* (Sernatur, 2017). Retrieved 11 Nov 2022, from http://www.subturismo.gob.cl/wp-content/uploads/2015/10/Plan-Accion-ZOIT-Chelenko.pdf

Chilean Senate, *Servicio de biodiversidad y áreas protegidas (SBAP). 25 de Julio 2019. N°175/ SEC/19* (Senado de Chile, 2019). Retrieved 14 Nov 2022, from https://mma.gob.cl/wp-content/uploads/2019/09/PL-SBAP-Aprobado-en-Senado-1.pdf

Chilean Subsecretary of Tourism, *Plan de acción Zona de Interés Turístico (ZOIT) "Destino Torres del Paine" [Action Plan for the Zone of Tourist Interest (ZOIT) "Torres del Paine Destination"]* (Sernatur, 2018). Retrieved 15 Nov 2022, from http://www.subturismo.gob.cl/wp-content/uploads/2015/10/Plan-de-Acci%C3%B3n-ZOIT-Destino-Torres-del-Paine.pdf

Chilean Subsecretary of Tourism, *Cuadros de visitas a parques, monumentos, reservas y áreas protegidas nacionales [Tables of Visits to Parks, Monuments, Reserves and National Protected Areas]* (Sernatur, 2021)

C.L. De la Maza, C. Cerda, G. Cruz, G. Mancilla, J.P. Fuentes, C. Estades, F. Medrano, E. Aliste, Á. Piroska, *Manual para aplicar indicadores de sustentabilidad en áreas protegidas: Ámbito biofísico [Manual for Applying Sustainability Indicators in Protected Areas: Biophysical Setting]* (Universidad de Chile, Corfo, Sernatur, 2014)

L.E. Delgado, A. Tironi, Sistemas socio-ecológicos y servicios ecosistémicos: Modelos conceptuales para el humedal del Río Cruces (Valdivia, Chile) [Social-ecological systems and ecosystem services: Conceptual models for the Río Cruces wetland (Valdivia, Chile)], in *Naturaleza en sociedad: Una mirada a la dimensión humana de la conservación de la biodiversidad [Nature in Society: A Look at the Human Dimension of Biodiversity Conservation]*, ed. by C. Cerda, E. Silva-Rodríguez, C. Briceño, (Ocho Libros, 2019), pp. 177–205

L.E. Delgado, C.C. Zúñiga, R.A. Asún, R. Castro-Díaz, C.E. Natenzon, L.D. Paredes, D. Pérez-Orellana, D. Quiñones, H.H. Sepúlveda, P.M. Rojas, G.R. Olivares, V.H. Marín, Toward social-ecological coastal zone governance of Chiloé Island (Chile) based on the DPSIR framework. *Sci. Total Environ.* **758**, 143999 (2021). https://doi.org/10.1016/j.scitotenv.2020.143999

S. Ebneyamini, M.R. Sadeghi Moghadam, Toward developing a framework for conducting case study research. *Int. J. Qual. Methods* **17**(1), 1–11 (2018). https://doi.org/10.1177/1609406918817954

EEA, *Environmental Indicators: Typology and Overview* (EEA, 2003). Retrieved 15 Nov 2022, from http://costabalearsostenible.es/PDFs/AMYKey%20References_Indicators/EEA%202003.pdf

J. Eklund, M. Cabeza, Quality of governance and effectiveness of protected areas: Crucial concepts for conservation planning. *Ann. N. Y. Acad. Sci.* **1399**(1), 27–41 (2017). https://doi.org/10.1111/nyas.13284

T. Farrell, J.L. Marion, Trail impacts and trail impact management related to visitation at Torres del Paine National Park, Chile. *Leisure/Loisir* **26**(1–2), 31–59 (2001)

M. Fernández Génova, G. Vela-Ruiz Figueroa, F. Repetto-Giavelli, J. Torres Mendoza, N. Recabarren Traub, I. González Ruiz, R. López Márquez, Local stakeholders' perception as a contribution to the identification of negative impacts on protected areas: A case study of Torres del Paine National Park, in *Socio-ecological Studies in Natural Protected Areas*, ed. by A. Ortega-Rubio, (Springer International Publishing, Cham, 2020), pp. 215–242

D. Ferrer, *Conservación de la naturaleza y territorio en Chile. El Parque Nacional de Torres del Paine y su área de influencia socioeconómica [Nature Conservation and Territory in Chile. Torres del Paine National Park and Its Area of Socioeconomic Influence]* (Universidad Autónoma de Madrid, 2003)

T. Gale, A. Ednie, Can intrinsic, instrumental, and relational value assignments inform more integrative methods of protected area conflict resolution? Exploratory findings from Aysén, Chile. *J. Tour. Cult. Chang.* **18**(6), 690–710 (2020). https://doi.org/10.1080/14766825.2019.1633336

T. Gale, A. Adiego, A. Ednie, A 360° approach to the conceptualization of protected area visitor use planning within the Aysén Region of Chilean Patagonia. *J. Park. Recreat. Adm.* **36**(3), 22–46 (2018). https://doi.org/10.18666/jpra-2018-v36-i3-8371

T. Gale, A. Ednie, K. Beeftink, Worldviews, levels of consciousness, and the evolution of planning paradigms in protected areas. *J. Sustain. Tour.* **27**(11), 1609–1633 (2019). https://doi.org/10.1080/09669582.2019.1639720

S.R. Gari, A. Newton, J.D. Icely, A review of the application and evolution of the DPSIR framework with an emphasis on coastal social-ecological systems. *Ocean Coast. Manag.* **103**, 63–77 (2015). https://doi.org/10.1016/j.ocecoaman.2014.11.013

U.H. Graneheim, B. Lundman, Qualitative content analysis in nursing research: Concepts, procedures and measures to achieve trustworthiness. Nurse Educ. Today **24**(2), 105–112 (2004). https://doi.org/10.1016/j.nedt.2003.10.001

H. Harrison, M. Birks, R. Franklin, J. Mills, Case study research: Foundations and methodological orientations. Forum Qual. Sozialforsch. **18**(1) (2017). https://doi.org/10.17169/fqs-18.1.2655

G. Inostroza, A. Rovira, Sistema de gobernanza para la gestión turística en áreas silvestres protegidas del estado, Aysén, Chile [Governance system for tourism management in state protected wilderness areas, Aysén, Chile], in *Turismo sustentable y áreas silvestres protegidas en Patagonia, Chile [Sustainable Tourism and Protected Wilderness Areas in Patagonia, Chile]*, ed. by P. Szmulewicz, (Ediciones Kultrún, 2020), pp. 113–123

IUCN, WCPA, *IUCN Green List of Protected and Conserved Areas: Standard, Version 1.1. IUCN* (IUCN, WCPA, 2017). Retrieved 20 Jan 2023, from https://portals.iucn.org/library/sites/library/files/resrecrepattach/IUCN%20Green%20List%20Standard%20Version%201.1%20%2025%20September%202018%20update_0.pdf

J. Kohl, Big-picture perspective: A plan means little if it doesn't get implemented. That's why planners need to take a holistic view of the process. Planning **84**(10), 22–25 (2018)

R.L. Lewison, M.A. Rudd, W. Al-Hayek, C. Baldwin, M. Beger, S.N. Lieske, C. Jones, S. Satumanatpan, C. Junchompoo, E. Hines, How the DPSIR framework can be used for structuring problems and facilitating empirical research in coastal systems. Environ. Sci. Policy **56**, 110–119 (2016). https://doi.org/10.1016/j.envsci.2015.11.001

I. López, M. Pardo, Socioeconomic indicators for the evaluation and monitoring of climate change in national parks: An analysis of the Sierra de Guadarrama National Park (Spain). Environments – MDPI **5**(2), 1–16 (2018). https://doi.org/10.3390/environments5020025

A. Mandić, Structuring challenges of sustainable tourism development in protected natural areas with driving force–pressure–state–impact–response (DPSIR) framework. Environ. Syst. Decis. **40**(4), 560–576 (2020). https://doi.org/10.1007/s10669-020-09759-y

M.I. Manzur, *Informativo 14, proyecto de ley que crea el servicio de biodiversidad y áreas protegidas [Information Brief 14, Draft Law Creating the Biodiversity and Protected Areas Service]* (Chile Sustentable, 2021). Retrieved 5 Dec 2022, from https://www.chilesustentable.net/wp-content/uploads/2021/01/Informativo-SBAP-14-de-ChS.pdf

R. Martínez, *Un 78% de inspecciones a locales en Torres del Paine motivó sumarios* (El Mercurio, 2016). http://www.economiaynegocios.cl/noticias/noticias.asp?id=301147

S.F. McCool, Constructing partnerships for protected area tourism planning in an era of change and messiness. J. Sustain. Tour. **17**(2), 133–148 (2009). https://doi.org/10.1080/09669580802495733

MINECON, MMA, *Plan de adaptación al cambio climático del sector turismo en Chile [Climate Change Adaptation Plan for the Tourism Sector in Chile]* (Ministerio del Medio Ambiente, Subsecretaría de Turismo, Sernatur, 2019). Retrieved 25 Nov 2022, from https://mma.gob.cl/wp-content/uploads/2020/01/Plan-de-Adaptacion-al-Cambio-Climatico-del-sector-Turismo-en-Chile.pdf

Municipality of Coyhaique, *Plan de desarrollo comunal: PLADECO 2014–2018 [Community Development Plan: PLADECO 2014–2018]* (Municipalidad de Coyhaique, 2014). Retrieved 7 Dec 2022, from https://www.coyhaique.cl/portalmunicipalidad/files/pladeco20142018/PLADECO2014_2018.pdf

Municipality of Natales & Municipality of Torres del Paine, *Plan de desarrollo turístico intercomunal destino Natales-Torres del Paine 2021–2024 [Intercommunal Tourism Development Plan Natales-Torres del Paine Destination 2021–2024]* (Municipalidad de Natales, Municipalidad de Torres del Paine, 2021). Retrieved 11 Dec 2022, from https://www.munitorresdelpaine.cl/Portal%20TA/Otros%20Antecedentes/PLADETUR%202021-2024(RESUMEN%20EJECUTIVO).pdf

Municipality of Torres del Paine, *Ordenanza municipal de turismo comuna Torres del Paine [Municipal Tourism Ordinance of the Torres del Paine Commune]* (Municipalidad de Torres del Paine, 2015)

Municipality of Torres del Paine, *Modificación a la ordenanza municipal de turismo de la comuna Torres del Paine [Modification of the Municipal Tourism Ordinance of the Torres del Paine Commune]* (Municipalidad de Torres del Paine, 2020a)

Municipality of Torres del Paine, *Ordenanza municipal medio ambiental de la comuna de Torres del Paine [Municipal Environmental Ordinance of the Commune of Torres del Paine]* (Municipalidad de Torres del Paine, 2020b)

P.L.K. Mustika, R. Welters, G.E. Ryan, C. D'Lima, P. Sorongon-Yap, S. Jutapruet, C. Peter, A rapid assessment of wildlife tourism risk posed to cetaceans in Asia. J. Sustain. Tour. **25**(8), 1138–1158 (2017). https://doi.org/10.1080/09669582.2016.1257012

OECD, *OECD Key Environmental Indicators* (OECD, 2004). Retrieved 1 Dec 2022, from https://www.oecd.org/env/indicators-modelling-outlooks/31558547.pdf

C. Ojeda-Martínez, F. Giménez Casalduero, J.T. Bayle-Sempere, C. Barbera Cebrián, C. Valle, J. Luis Sanchez-Lizaso, et al., A conceptual framework for the integral management of marine protected areas. Ocean Coast. Manag. **52**(2), 89–101 (2009). https://doi.org/10.1016/j.ocecoaman.2008.10.004

Paralelo 47, *Plan de desarrollo turístico para Villa Cerro Castillo [Tourism Development Plan for Villa Cerro Castillo]* (Municipalidad de Río Ibáñez, 2019). Retrieved 28 Nov 2022, from https://www.perturismoaysen.com/wp-content/uploads/2019/01/INFORME-PLADETUR-FINAL-2-002.pdf

J. Patrício, M. Elliott, K. Mazik, K.N. Papadopoulou, C.J. Smith, DPSIR-Two decades of trying to develop a unifying framework for marine environmental management? Front. Mar. Sci. **3**(9) (2016). https://doi.org/10.3389/fmars.2016.00177

I.J. Petit, A.N. Campoy, M.J. Hevia, C.F. Gaymer, F.A. Squeo, Protected areas in Chile: Are we managing them? Rev. Chil. Hist. Nat. **91**(1) (2018). https://doi.org/10.1186/s40693-018-0071-z

E. Pisano, Estudio ecológico de la región continental sur del área Andino-Patagónica [Ecological study of the southern continental region of the Andean-Patagonian area]. An. Inst. Patagon. **1–2**, 59–104 (1974)

S. Praus, M. Palma, R. Domínguez, *La situación jurídica de las actuales áreas protegidas de Chile [The Legal Status of Chile's Current Protected Areas]* (PNUD-GEF, 2011)

Programa Austral Patagonia, ¿Qué es la Lista Verde de UICN y por qué esta semana estaremos hablando de ella? [What is the IUCN Green List and why will we be talking about it this week?] *Programa Austral Patagonia* (2022, May 16). https://programaaustralpatagonia.cl/que-es-la-lista-verde-de-uicn/

F. Repetto, J. Cabello, Potencial de restauración ecológica en zonas de uso público en el Parque Nacional Torres del Paine [Ecological restoration potential in public use areas of Torres del Paine National Park]. An. Inst. Patagon. **43**(3), 115–121 (2015)

F. Repetto-Giavelli, G. Vela-Ruiz, M. Díaz-Beros, M. Fernández Génova, R. Lopez, B. Gonzalez, Diagnóstico sobre la planificación e implementación del uso público en el Sistema Nacional de Áreas Silvestre Protegidas del Estado de Chile [Diagnosis of the planning and implementation of public use in the National System of Protected Wildlife Areas of the State of Chile]. Biodiversidata **7**, 116–124 (2018)

H. Rivas, *Destinos turísticos sustentables: Propuestas para un sistema de distinción [Sustainable Tourist Destinations: Proposals for a System of Distinction]* (Universidad Andrés Bello, CORFO, 2014)

H. Rivas, C. Rojas, Gobernanza para el desarrollo sustentable del turismo en la Patagonia chilena. Desde la macro visión de los planes maestros a la focalización en las zonas de interés turístico [Governance for the sustainable development of tourism in Chilean Patagonia. From the macro vision of master plans to the focus on areas of tourist interest]. Estud. Perspect. Tur. **29**(4), 1232–1254 (2020)

W. Ruan, Y. Li, S. Zhang, C.H. Liu, Evaluation and drive mechanism of tourism ecological security based on the DPSIR-DEA model. Tour. Manag. **75**, 609–625 (2019). https://doi.org/10.1016/j.tourman.2019.06.021

C. Sepúlveda, M. Lara, *Comunidades y áreas protegidas de la Patagonia chilena [Communities and Protected Areas in Chilean Patagonia]* (Editorial Andros, 2021)

A. Spenceley, G. Hvenegaard, R. Bushell, Y. Leung, S. McCool, P. Eagles, The impacts of protected area tourism, in *Tourism and Visitor Management in Protected Areas: Guidelines for Sustainability*, ed. by C. Groves, (IUCN, 2018), pp. 9–26

R.E. Stake, Case studies, in *Handbook of Qualitative Research*, ed. by N.K. Denzin, Y.S. Lincoln, (SAGE, Thousand Oaks, 1994), pp. 236–247

C. Toledo, *Análisis económico de los ingresos y egresos del Sistema Nacional de Áreas Silvestres Protegidas del Estado (SNASPE) [Economic Analysis of the Income and Expenditures of the National System of State Wildlife Protected Areas (SNASPE)]* (Publicaciones Fundación Terram, 2017)

J. Torres, F. Repetto-Giavelli, F. Quezada, M. Sánchez, B. González, F. Vela-Ruiz, Evaluación de senderos de montaña y medidas de manejo para mitigar impactos en el Parque Nacional Torres del Paine, Chile [Evaluation of mountain trails and management measures to mitigate impacts in Torres del Paine National Park, Chile]. Áreas Nat. Prot. Scr. 4(2), 7–28 (2018). https://doi.org/10.18242/anpscripta.2018.04.04.02.0001

J. Torres, G. Vela-Ruiz, C. Olave, F. Repetto, M. Sánchez, Análisis de la dinámica de visitantes en los senderos de montaña del Parque Nacional Torres del Paine [Analysis of visitor dynamics on the mountain trails in Torres del Paine National Park]. Biodiversidata **8**, 87–94 (2019)

UNDP, Chilean Ministry of the Environment, Chilean National Tourism Service, *Elaboración de diagnóstico de vulnerabilidad ante el cambio climático para el sector turismo [Elaboration of a Climate Change Vulnerability Assessment for the Tourism Sector]* (PNUD, Ministerio del Medio Ambiente, Sernatur, 2019). Retrieved 10 Dec 2022, from https://procurement-notices.undp.org/view_file.cfm?doc_id=186016

United Nations, *Indicators of Sustainable Development: Guidelines and Methodologies* (United Nation Publications, 2007)

J. Valverde, *Bases para una estrategia económica sustentable en la Reserva de la Biosfera Cabo de Hornos [Basis for a Sustainable Economic Strategy in the Cape Horn Biosphere Reserve]* (IEB, UMAG, UNT, UAB, UCI, 2020)

G. Vela-Ruiz, M.M. del Delgado, Contribución del enfoque de desarrollo territorial rural a la comprensión de los procesos generados en torno a áreas protegidas en la Patagonia chilena [Contribution of the rural territorial development approach to the understanding of the processes generated around protected areas in Chilean Patagonia]. Rev. Chil. de Estud. Reg. **1**, 83–96 (2010)

G. Vela-Ruiz, F. Reppetto-Giavelli, *Guía de conocimiento y buenas prácticas para el turismo en el Parque Nacional Torres del Paine [Guide to Knowledge and Best Practices for Tourism in Torres del Paine National Park]* (Ediciones CEQUA, 2017)

G. Vela-Ruiz, F. Repetto-Giavelli, M. Fernandez, J. Torres, N. Recabarren, J. Cabello, et al., *Informe final: Sistema de manejo turístico para el Parque Nacional Torres del Paine [Final Report: Tourism Management System for Torres del Paine National Park]* (Innova Corfo 14BPC4-28654, 2018)

J. Vera-Rebollo, J. Ivars-Baidals, Measuring sustainability in a mass tourist destination: Pressures, perceptions and policy responses in Torrevieja, Spain. J. Sustain. Tour. **11**(2–3), 181–203 (2003). https://doi.org/10.1080/09669580308667202

O.J. Vidal, C. San Martín, S. Mardones, V. Bauk, C.F. Vidal, The orchids of Torres del Paine Biosphere Reserve: The need for species monitoring and ecotourism planning for biodiversity conservation. Gayana Bot. **69**(1), 136–146 (2012)

O.J. Vidal, M. Aguayo, R. Niculcar, N. Bahamonde, S. Radic, C.S. Martín, A. Kusch, J. Latorre, J. Félez, Invasive plants in Torres del Paine National Park (Magallanes, Chile): Current knowledge, post-fire distribution and implications for ecological restoration. An. Inst. Patagon. **43**(1), 75–96 (2015)

288 A. Adiego et al.

S.S. Weidemann, T. Sauter, P. Malz, R. Jaña, J. Arigony-Neto, G. Casassa, C. Schneider, Glacier mass changes of lake-terminating Grey and Tyndall glaciers at the southern Patagonia icefield derived from geodetic observations and energy and mass balance modeling. Front. Earth Sci., 6 (2018). https://doi.org/10.3389/feart.2018.00081

WTO, *Indicators of Sustainable Development for Tourism Destinations: A Guidebook* (World Tourism Organization, 2004)

R. Yin, *Case Study Research: Design and Methods*, 4th edn. (SAGE Ltd, Thousand Oaks, 2009)

Chapter 12
Visual Dimensions of Conservation Landscapes: An Exploration of Patagonian Fjordic Landscapes from the Perspective of Prospective Chilean Tourists

Andrea Báez Montenegro, Trace Gale-Detrich, and Laura Rodríguez

Abstract This chapter explores Chilean tourists' valuation of tourism experiences in and around the Chilean village of Puyuhuapi. In 2020, local stakeholders were concerned about the impacts that salmon aquaculture infrastructure within the fjords surrounding Puyuhuapi might have on tourism. Research was undertaken to examine how prospective national tourists might perceive visible salmon aquaculture infrastructure within Puyuhuapi's landscapes. Two hypothetical experience scenarios were designed, with short texts describing possible experiences and visual cues portraying typical destination landscapes. The scenarios were differentiated by the presence of movable floating sea cages. The scenarios were presented to potential tourists through an online survey ($n = 804$ responses). Results supported current nature-based tourism experience positioning, suggesting it was well received with and without the presence of salmon aquaculture infrastructure. The visible presence of the movable floating sea cages did not provoke significant differences in the valuation of the landscape for prospective tourists; nevertheless, several significant interactions occurred between

A. Báez Montenegro
Universidad Austral de Chile, Institute of Statistics, Valdivia, Chile

Centro de Investigación en Ecosistemas de la Patagonia (CIEP), Human-Environmental Interactions Group, Coyhaique, Chile
e-mail: abaez@uach.cl

T. Gale-Detrich (✉)
Centro de Investigación en Ecosistemas de la Patagonia (CIEP), Human-Environmental Interactions Group, Coyhaique, Chile

Cape Horn International Center (CHIC), Puerto Williams, Chile
e-mail: tracegale@ciep.cl

L. Rodríguez
Universidad Austral de Chile, Institute of Architecture and Urban Planning, Faculty of Architecture and Arts, Edificio Ernst Kasper, Valdivia, Chile
e-mail: lrodriguez@uach.cl

tourism experience attributes and socio-demographic characteristics, including population density, level of education, and sex. Discussion focuses on defining a series of hypotheses to inform future research and the importance of expanding understanding of Chilean perspectives and imaginaries of Patagonia and its abundant natural settings and values.

Keywords Patagonia · Conservation landscapes · Nature-based tourism · Tourism experience · Viewsheds

12.1   Introduction

Landscape is a multidimensional concept; the European Landscape Convention of 2000 defined it as "an area, as perceived by people, whose character is the result of the action and interaction of natural and/or human factors" (United Kingdom Secretary of State for Foreign and Commonwealth Affairs, 2012, p. 4); therefore, landscape can be identified as the set of interrelations derived from the interaction between geomorphology, climate, soil, vegetation cover, water, and anthropic modifications (Dunn, 1974). Over the past 20 years, the concept of landscape has acquired great relevance in Chile, based on an understanding that landscapes are natural resources that are easily depreciated and difficult to renew (Muñoz et al., 1993, 2000, 2012; Muñoz, 2004a, b). Landscapes represent complex systems composed of various interacting attributes and are therefore difficult to repair. Thus, authors have emphasized the importance of landscape preservation and conservation measures within land-use planning (Muñoz et al., 1993, 2012; Muñoz, 2004a, b). Landscape evaluation and planning contribute to the preservation and conservation of protected areas (PAs), as well as small localities that share the components characteristic of PAs, by applying a systemic approach to evaluating the appropriate management measures for the set of natural and human elements present in a place.

Landscape typologies include the *natural landscape*, which can be considered as one of the main elements for territories promoting nature-based tourism activities, as they play a central role in creating the destination image, or *positioning*, used to promote and sell tourism within the national and global markets (Aguilar et al., 2015). According to Saqib (2019), "destination image can be defined as a tourist's general impression of a destination, that is, the 'sum of beliefs, ideals, and impressions' that a visitor has toward a certain place" (p.133). Creating a tourism destination or experience image involves a strategic positioning process that associates a destination with certain characteristics (e.g., naturalness, sustainability, uniqueness) and certain uses (e.g., relaxation, adventure, learning). Landscape represents a fundamental component of destination positioning strategies because a high landscape value can contribute to an image that motivates visitation and destination choice and sets the stage for particular types of tourism experiences (Aguilar et al., 2015; Saqib, 2019). For example, a destination that focuses on nature-based tourism is likely to position itself within the marketplace through advertising and imagery that features natural landscapes that will be attractive to potential visitors (Saqib, 2019). It will likely develop and emphasize products (e.g., tourism experiences) that allow these visitors to experience its natural landscapes in manners that they will consider appealing, unique, and personally relevant.

Environmental awareness is another core concept associated with landscape valuation. In Chile, the use of public PAs for recreation has become more widespread, as a result of increased free time and the rise in the standard of living for some segments of the population (Ednie et al., 2020; Gale et al., 2018). Muñoz (2004a) observed that, for a number of reasons, average Chilean citizens are increasingly internalizing a kind of *environmental awareness* that results in a new appreciation of natural spaces and their ecosystems, associating this appreciation with growing citizen resistance to the loss of spaces of high tourist, landscape, and recreational value. Gale and Ednie's (2019) study supported earlier work in this area, finding aesthetic and recreational value dimensions to be among the highest rated by national visitors to PAs in the Aysén region of Chile.

This chapter explored Chilean tourists' valuation of tourism experiences in and around the village of Puyuhuapi, in Chilean Patagonia, based on visual and descriptive product prompts. Puyuhuapi is considered a nature-based tourism destination, with a high natural landscape value that is important to conserve. In 2020, when this research was conducted, many tourism providers within this locality were concerned about the installation of salmon aquaculture infrastructure (movable floating sea cages) within the surrounding fjords and the impacts this might have for tourism (Rozas, 2021). Research was undertaken to help stakeholders understand how prospective national tourists might perceive visible salmon aquaculture infrastructure within Puyuhuapi's landscapes. Specifically, research explored (a) whether current nature-based tourism experience positioning was well received and relevant, with and without the presence of salmon aquaculture infrastructure (movable floating sea cages) in the Puyuhuapi bay, (b) whether the visible presence of the movable floating sea cages provoked significant differences in the valuation of the landscape for prospective tourists, and (c) whether there were socio-demographic characteristics that acted as determinants.

The study employed a *product concept testing* methodology, combining visual and descriptive cues, which comes from marketing research and is based on the development of an experiential tourist product (Dahan & Srinivasan, 2000). Two hypothetical experience scenarios were designed: both employing typical destination landscapes and differentiated by the presence (or not) of movable floating sea cages. The scenarios were presented to potential tourists (Chilean professionals) through an online survey. The results of the study will inform local tourism planning and contribute to local tourism policy and program development.

12.2   Theoretical Framework

12.2.1   Landscape from the Observer's Perspective

Scientific knowledge, and in particular geography, has approached the study of landscape as a way of integrating natural and social science (Arts et al., 2017). For example, Benson and Roe (2000) posed that the identification of landscapes, and needs related to landscapes, has arisen in response to the cultural emphasis that has been given to territorial studies in recent years. For scholars, landscape must be defined and protected to maintain the sustainability of the planet for future generations (Benson & Roe, 2000).

Landscape ecology has traditionally characterized landscapes as systems using the concept of *land units* (Campos-Campos et al., 2018; Zonneveld, 1989). A land unit is an ecologically homogeneous land tract that can be surveyed and characterized according to a series of attributes, which typically include soil, vegetation, and landform (Zonneveld, 1989). Landscape units are especially helpful in territorial planning and management as their delineation helps identify spatial patterns for the physical and social dynamics of territories (Campos-Campos et al., 2018).

Landscape is considered to be a polysemic concept that—historically—was viewed differently within the arts and academia (Klonk, 1993). During the Romantic Period, the idea of landscape arose as an object of artistic practice and was proposed from a perceptual standpoint. In contrast, academics of the period (e.g., Vidal de la Blache in France; Carl Sauer in the United States) used landscape to refer to the ways in which the surface of the earth integrates physical, geographic, and cultural elements (Jones, 2003).

Contemporary landscape studies have shifted towards the arts' conceptualization, focused on understanding the perceptions of a landscape rather than the evaluation of the natural elements and conditioning factors that compose it. This facilitates a differentiation between the landscape that is observed—which is subject to perception—and the environment that facilitates the perception (Fariña & Solana, 2007). With these elements in mind, it is possible to determine the *viewshed*, defined as "the part of the territory that is visible from a point in it" (Fariña & Solana, 2007, p. 263). As Ramírez and López (2015) indicated, "landscape is not a natural reality independent of who observes it, but rather it is the meaning that human beings give to materialized nature. It is the surface of the Earth, seen and interpreted" (p. 72). Therefore, landscape is subject to society's frameworks of understanding and in constant transformation.

Thus, landscape no longer comprises only the natural components of the earth's surface but also incorporates constructed, symbolic, and even abstract components derived from the viewer's lived experience, worldview, and cultural background. As Rodríguez (2021) indicated:

> Landscape presents the world as it is, but at the same time it is a construction, a composition of this world, a way of seeing and feeling. It represents the cultural projection of a society in a given space, admitting, however, that this is something dynamic (p. 176).

And Ramírez and López (2015) emphasized the importance of the symbolic aspect of landscapes, noting that individuals frequently symbolize spaces and objects by assigning meanings to them. Following the argument reflected by Nogué (2007), "landscape can be understood as a social product, resulting from the cultural projection of a society in a given space" (p. 12). These perspectives have helped evolve the study of landscape, incorporating a close relationship with human beings rather than an approach that is alien to them.

Ramírez and López (2015) posed that original landscapes have been modified by society, transitioning them from natural landscapes to cultural landscapes that assume societal or individual significance and symbolism. Consequently, landscapes express many different types of thoughts, ideas, and emotions; they are conceived through different ways of seeing and interpreting. They are cultural constructions that "reflect particular ideologies that convey particular ways of appropriating space" (Nogué, 2007, p.12). The visual, or perceived landscape, includes the aesthetics, the observer's sensory perception of the landscape, and the effect of a given landscape on the observer. Although all five senses are involved, the visual sense is considered to be the most relevant. As such, landscape is a physical reality that is experienced according to the perceptive capacity and cultural background of the observer (Center for Environmental Agrarian Studies—CEA, 2022; Zube et al., 1982).

12.2.2 Natural Landscapes

The research presented in this chapter focused on the natural landscape. The *naturalness* of a landscape can be defined by measuring the number of natural landscape units (without human intervention) within a territory, as compared to the number of landscape units with anthropic (human caused) modification. Natural landscapes have very little human construction, and if it exists, it is dispersed and does not monopolize the landscape. Natural landscapes can be subdivided into two main categories: *undisturbed* and *intervened*. Undisturbed natural landscapes include continuous land units with no, or very little, human intervention (e.g., pristine desert landscapes, scrublands, steppe, forests, ocean, rivers, lakes, glaciers). Intervened natural landscapes have a natural background that is fragmented by landscape units that have been, or are currently, subjected to anthropogenic activities (e.g., residential settlements, agricultural valleys, aquacultural installations, or transportation and communication elements such as roads, highways, bridges, power lines or antennas).

12.2.3 Chilean Legislation and Environmental Awareness

In Chile, the landscape is considered to be a component of the environment and is an object of protection within the Chilean Environmental Impact Assessment System (SEIA), when it is located in an area with unique natural or cultural landscape value (Chilean National Commission on the Environment 1993, 1994; Chilean Environmental Evaluation Service, 2019). As such, possible impacts for landscape uniqueness are included within Chile's environmental impact studies, as instructed within the General Environmental Law (Chilean Law No. 19,300). Nevertheless, landscape regulation in Chile is vague. For example, Chile has ratified some international treaties, including the Ramsar Convention and the Convention on the Protection of Heritage, that imply landscape protection but do not establish a definition or set clear parameters for protection (Sandoval, 2021). Chile's current constitution makes no reference to landscape, and analysis of related constitutional guarantees and obligations (e.g., the right to live in an environment free of pollution, the duty of the State to protect the preservation of nature) have concluded that landscape protection is currently limited and restricted, observing that the constitution regulates the environment as a guarantee granted to people and, therefore, its notion is directly anthropocentric (Bermúdez, 2014; Femenías, 2017; Sandoval, 2021).

Chile does, however, have more clearly articulated strategies for promoting environmental awareness regarding natural landscape protection. For example, the National Biodiversity Strategy (2017–2030) is based on achieving the following vision:

> Chilean society understands, values, respects and integrates biodiversity and the country's ecosystem services as a source of its own well-being, stopping their loss and degradation, restoring them, protecting them, using them in a sustainable way and distributing the benefits of biodiversity in a fair and equitable manner, maintaining the possibilities of satisfying the needs of future generations. (Chilean Ministry of the Environment, 2016, p.49)

The strategy's objectives include developing awareness, participation, and information about biodiversity and ecosystem services, as a basis for human welfare and Chile's sustainable development (Chilean Ministry of the Environment, 2016).

12.2.4 Evaluating Perceptions of Landscape Naturalness

Bernáldez (1985) and Bourassa (1990) established that landscapes contain and emit a series of signs through which they communicate their identity. Understanding human reactions to landscapes is complex, especially when it comes to understanding the components of their appreciation and/or preferences. Landscape preferences are understood as the valuation of the perceived scenic quality of the visual environment and landscape. They are the result of a complex system of innate and acquired factors, based on biological, social, and personal reactions to the figurative or symbolic character of particular scenic elements (De Fuente de Val et al., 2004).

In the past, measurement of the visual landscape (Calvin et al., 1972; Daniel & Vining, 1983; Dunn, 1974; Fines, 1968; Zube et al., 1982) did not employ a standardized methodology. Studies, generally undertaken by experts, concentrated on descriptions and subsequent classification within categories. Contemporary research has evolved a range of methods for visual landscape measurement, including the direct valuation of the visual landscape, through measures of quality, fragility, and use capacity (Muñoz et al., 2000; Muñoz, 2004a). Other methods have focused on the landscape value of a locality or territorial unit, considering landscape as an environmental component that humans perceive through the visual sense, and models that reflect all the *visual* landscape qualities of a territory (Zambrano & González, 2002). In Chile, landscape preferences have been evaluated through PA visitor surveys and mixed methods approaches that included direct assessment of representative subjectivity and subsequent indirect analysis of characteristics and components (De Fuente de Val & De Lucio, 2003; Muñoz et al., 2012).

12.3   Materials and Methods

12.3.1   Study Area

This study focused on the territory in and around Puerto Puyuhuapi, a small village located in the Cisnes Commune, in the Aysén region of Chilean Patagonia. Puyuhuapi is one of the gateway communities for Queulat National Park, Magdalena Island National Park, and the Patagonian fjords (Fig. 12.1). Located at the northern end of the Puyuhuapi Fjord, the village is set against a backdrop of calm waters,

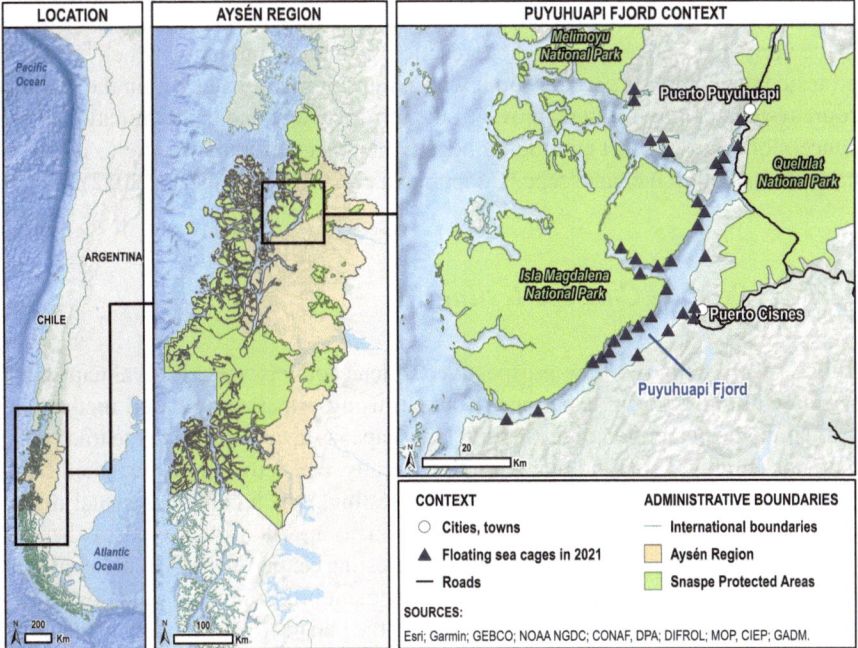

Fig. 12.1 Study area and territorial context, within the Aysén Region and Chile

surrounded by mountains with abundant vegetation. Puyuhuapi was settled in 1935 by four German families that relocated to the area before the arrival of war in Europe. Taking advantage of Chilean land grants, the families founded a settlement based on farming, livestock, fishing, and other pursuits, including the manufacture of fine fabrics and rugs (Ludwig Winkler, 2013). The German settlers employed carpenters and other trades from the Island of Chiloe, who brought their culture and traditions, resulting in a unique heritage and cultural mix, reflected in the local architecture, gastronomy, and craftsmanship.

According to the Cisnes Commune Development Plan (Municipality of Cisnes, 2018), Puerto Puyuhuapi has a population of around 1,000 persons. The area's most important economic activity is tourism, followed by artisan fishing, and commerce. The plan notes a growth in private investment in the area, noting a high correlation between these investments and tourism. Other, less integral economic activities include small-scale ranching and the extraction of forest resources.

According to the popular press, in 1999 the Chilean Marine Subsecretary granted the Salmones de Chile company aquaculture concessions within the Puyuhuapi Bay area, located in the viewshed directly in front of the town (Rozas, 2021). The company operated the concession with salmon infrastructure present between 2001 and 2008, and again between 2012 and 2013, but had not operated in the location since. In 2020, the company faced losing the concession if it did not reactivate use, and as such, planned to reinitiate activities amidst the complications of the COVID-19 pandemic. Without communicating to the community in advance, they installed a series of movable floating sea cages in the Puyuhuapi Bay at the end of August 2020 with the intention of installing 64,000 rainbow trout fingerlings. These interventions in the viewshed provoked immediate reaction with the community, which has since protested the company's tactics, both within the community and within the Chilean court system, calling for the removal of the infrastructure and the termination of the concession (Rozas, 2021), citing its negative impacts on tourism, due to a loss of naturalness within the landscape (Cooperativa.cl, 2021; Greenpeace, 2022).

12.3.2   Product Concept Testing

This research explored how prospective Chilean tourists valued Puyuhuapi landscapes within a tourism experience context, using a marketing-based method for evaluating the perception of the natural landscape, as an alternative to the traditional methods employed for land planning. Specifically, the study employed a methodology known as *proof of concept*, or concept testing, which combines visual depictions and textual descriptions (Dahan & Srinivasan, 2000) and exposes them to potential "customers" using an A/B (or split testing) experiment. A/B concept testing typically accompanies a product's development process by periodically exploring the feasibility of a product or service from the perspective of the target audience. This approach incorporates feedback gained along the way to maximize a product's utility and chances of success (Dahan & Srinivasan, 2000; Wang, 2007). In this study, A/B product concept testing was used to test the visual impact of the floatable sea cages in the Puyuhuapi fjord. Participants included prospective Chilean tourists, with a focus on professionals. Two hypothetical "products" were created (Product A, Product B), using scenes and experiential descriptors typically found in current tourism materials and websites for Puerto Puyuhuapi (e.g., Ruta de los Parques, Chile es Tuyo, Go Chile, Trip Advisor). Next, A/B testing employed online surveying to expose the products to two homogeneous groups of potential tourists, who were asked to evaluate different aspects (attributes) of the hypothetical products, including visual aspects.

12.3.3 Survey Procedures

A closed (online) A/B test survey was conducted between September and November 2021. The target population included potential national visitors to the Puyuhuapi area. Participant recruitment was conducted using a database of emails of Chilean professionals containing over 100,000 records. An algorithm built in the Qualtrics platform was used to assign prospective participants to survey group 1 or 2, and then sent an invitation email with a differentiated link (described in detail below). Finally, a sample of 402 responses was obtained for each of the two product surveys (Product A, Product B), for a total sample of 804 surveys.

12.3.4 Questionnaire Design

Each survey instrument was divided into three parts. The first part consisted of the tourism experience analysis, where participants were presented with narrative and visual cues, and asked to respond to six questions. The textual description included the following positioning statement and a bullet-point list of *don't miss* experiences:

Visit Puyuhuapi, the portal community for Queulat National Park, Magdalena Island National Park and the Patagonian fjords. Located at the northern end of the Puyuhuapi fjord, surrounded by mountains of incredible beauty with abundant vegetation, in this quiet town you can still breathe the traditions of the first settlers of German and Chiloé origin.

- Relax in the hot springs near Puyuhuapi
- Explore Queulat National Park
- Tour the village plaza to learn about the history of Puyuhuapi through illustrative panels.
- Guided kayaking in the fjord (boat and bike excursions are also available)
- Follow an interpretative trail from the tourist information kiosk, passing by buildings that represent the German architectural heritage of the village.
- Photograph the heritage of Puyuhuapi, represented through the houses located on the waterfront of the fjord.
- Hike the Los Canelos trail on the edge of the town.

The photo montages represented typical landscapes of the area that were frequently shown in tourism promotional literature. The background of the photos within the montage were the same for Product A (No aquaculture interventions) and Product B (Visible aquaculture interventions). The only difference between the two montages was that in Product A, the natural landscapes did not include aquaculture interventions, while in Product B, the photos included the movable floating sea cages (see Fig. 12.2). Study participants only saw one of the photo montages, according to which product group they were randomly assigned.

Following the descriptive and visual product prompts, participants were asked to respond to the following six questions (they were able to return to the description and photo montage throughout the survey if desired). The Likert-scale questions were identical for participants in Group 1, who were exposed to Product A (No Aquaculture interventions), and Group 2, who were exposed to Product B (Visible aquaculture interventions):

- Initial reaction: What is your initial reaction to this proposed tourism experience?
- Uniqueness: How unique is this proposed tourism experience compared to other products currently available?
- Attractiveness: How attractive is this proposed tourism experience compared to other products currently available?
- Appeal: How much do you like or dislike this proposed tourism experience?
- Sustainability: How sustainable is this proposed experience?
- Personal Relevance: How relevant is this proposed tourism experience for you personally?

The second part of the questionnaire sought clarification and context about participants' responses to the tourist experience presented, by asking them to write down the proposed visual and experiential aspects they liked most and least. The final section asked participants to share socio-demographic characteristics, including sex, age, income, education, place of residence, belonging to an Indigenous people, marital status, and professional situation in the last 3 months.

12.3.5   Statistical Analysis

Descriptive statistical analysis was performed to determine means and standard deviations for quantitative variables, frequencies, and percentages for categorical variables. Comparative analysis employed hypothesis tests for mean differences, using the Likert scale in quantitative form. Although we would have preferred to conduct multiple A/B experiments within the original list of over 100,000 records, the available project resources allowed us to conduct a single A/B experiment from the larger database. Thus, *bootstrapping* was performed. Bootstrapping is a statistical procedure used to approximate characteristics of the distribution in a sample through simulation. It generates a large number of samples by

Product A - NO aquaculture interventions

Product B - VISIBLE aquaculture interventions

Fig. 12.2 Supporting photo montages, representing landscapes that are frequently pictured within tourism promotional messages, without (Product A—top) and with (Product B—bottom) aquaculture interventions

resampling the original sample (with replacement). Each of these new samples are used to calculate estimates of a parameter, which are then combined to form a sampling distribution (Kung et al., 2020). Based on these distributions, parametric and non-parametric hypothesis tests were used to determine the difference in means or medians between the two groups (Product A, Product B) for each question associated with the experience. In addition, an analysis of variance and/ or Friedman's test was performed to test for interaction between socio-demographic characteristic variables and the experience ratings. For the interactions that were significant, Tukey's multiple comparisons tests were performed. To determine the effect of socio-demographic characteristic variables on the alternatives (Product A, Product B), a bivariate analysis of variance was performed between the assigned *tourism experience alternatives* factor (Product A—No aquaculture interventions, and Product B—Visible aquaculture interventions) and the following socio-demographic characteristic variables: sex, education, income, age, and population density. Population density was obtained according to the population and surface area of the region of origin of the respondents, which was categorized as low (<40 persons/km^2), medium (40–99 persons/km^2), medium high (100–400 persons/km^2), or high (>400 persons/km^2). Data analysis was performed with RStudio statistical software.

12.4 Results

12.4.1 Study Sample

As shown in Table 12.1, the sample characteristics for Group 1 and Group 2 were homogeneous, as sought within A/B testing, with a majority of participants living in Chile's central zone (74.0 and 67.4%, respectively). This zone includes the Metropolitan, Valparaiso, and O'Higgins regions, together representing 56% of the general population of Chile (Chilean National Institute of Statistics (INE), 2019). The next highest concentration of participants resided in Chile's southern zone (20.0% and 19.2%, respectively), which includes the regions of Maule, Ñuble, Bío Bío, Araucanía, Los Ríos, and Los Lagos, representing 29.9% of Chile's total population (INE, 2019). Much lower percentages of study participants resided in the northern (Arica-Parinacota, Tarapacá, Antofagasta, Atacama, and Coquimbo Regions) and southernmost (Aysén and Magallanes Regions) zones of Chile, which together account for 14.1% of the Chilean population (12.6 and 1.5%, respectively). For both groups, the participants skewed slightly more female than the national average (53.5 versus the national average of 51%), with the majority of participants (87.4%) between 25 and 54 years old. Approximately 50% of the combined sample reported a family income above 1.5 million pesos, which seems accurate considering the study focused on Chilean

Table 12.1 Sample distribution

Variables	Categories	Group 1 Product A	Group 2 Product B
Zone of Chile	North	3.9	9.1
	Central	74.0	67.4
	South	20.0	19.2
	Austral	2.1	4.3
Population density in place of residence	Low density (<40 persons/km^2)	20.6	28.4
	Medium-low density (40–99 persons/km^2)	15.2	11.4
	Medium-high density (100–400 persons/km^2)	10.2	10.7
	High density (>400 persons/km^2)	54.0	49.5
Sex	Male	46.5	47.5
	Female	53.5	52.9
Age	18–24	0.3	0.3
	25–34	19.2	20.5
	35–44	38.3	39.8
	45–54	29.9	23.2
	55–64	8.7	10.1
	65 or more	3.6	6.1
Civil status	Married	44.9	43.7
	Living with a partner	23.5	24.8
	Widowed	1.2	1.6
	Divorced/separated	10.5	7.9
	Other	2.5	3.8
Children at home	No	47.3	50.6
	Yes	52.7	49.4
Highest level of education attained	High School (or lower)	6.3	5.4
	Technical associate degree	18.8	16.4
	University bachelor's degree or higher	75.0	78.2
Monthly Income	<$CLP300,000	5.1	4.0
	$CLP301,000–600,000	7.9	6.9
	$CLP601,000–900,000	13.7	8.9
	$CLP901,000–1,500,000	24.8	26.1
	>$CLP1,500,000	48.6	54.1

professionals and that 68.5% of the participants were either married or living with a partner. According to the INE (2022), in 2021, the average monthly salary for employed workers in Chile was $681.039 CLP, and 16% received an income equal or greater than 1.0 million pesos per month.

12.4.2 Responses to Current Nature-Based Tourism Experience Positioning with and without the Presence of Aquaculture Infrastructure

In general, study participants in both groups 1 and 2 found the tourism experiences they were presented in Puyuhuapi to be highly attractive, appealing, and relevant. Approximately 66% of the respondents indicated an extremely positive initial reaction to the experience they viewed, and another 29% indicated a somewhat positive first reaction to the proposed experience (Fig. 12.3a). Similarly, more than 60% indicated they found the offering extremely appealing, and another 33.8% agreed to finding it moderately or somewhat appealing (Fig. 12.3b). And the majority of study respondents indicated that the product they were exposed to in Puyuhuapi was very attractive (56.7%), followed by 28% who indicated that it was somewhat attractive (Fig. 12.3c). More than 90% expressed that the Puyuhuapi tourism offerings were personally relevant at some level, with more than 71.6% expressing moderate or extreme levels of personal relevance (Fig. 12.4a). Participants expressed less enthusiasm about the uniqueness of their proposed experience; for example, less than half of participants (43%) found the Puyuhuapi experience they viewed to be very or extremely unique (Fig. 12.4b). And, while slightly more than a quarter of participants (26.1%) evaluated their experience as being extremely sustainable; the majority of participants perceived at least some level of risk regarding the sustainability of the tourism experience they were provided (66.1%), with 1.1% evaluating their experience as having some degree of unsustainability (Fig. 12.4c).

12.4.3 Comparative Analysis Between Tourism Experience Alternatives with and without the Presence of Visible Aquaculture Infrastructure

Figure 12.5 shows the results of the hypothesis tests (T-tests) to see if there were differences between Group 1 and Group 2 mean ratings for the tourism product experience attributes (Initial reaction, Uniqueness, Attractiveness, Appeal, Sustainability, Personal relevance). Surprisingly, the mean values and confidence intervals for the two groups were very similar, suggesting high levels of agreement around the product experience attributes, regardless of which product was evaluated (see also Table 12.2).

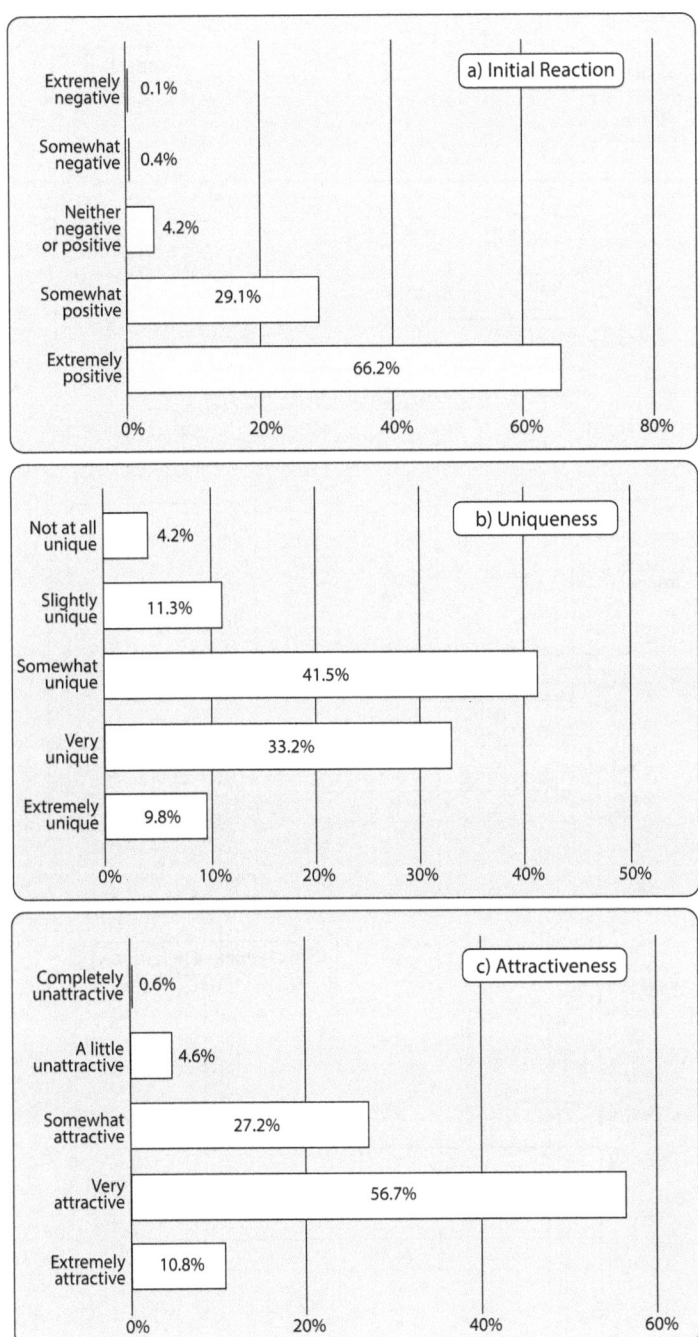

Fig. 12.3 Frequency distributions showing participant perceptions of the tourism experience described within the survey [(**a**) Initial Reaction; (**b**) Uniqueness; (**c**) Attractiveness]

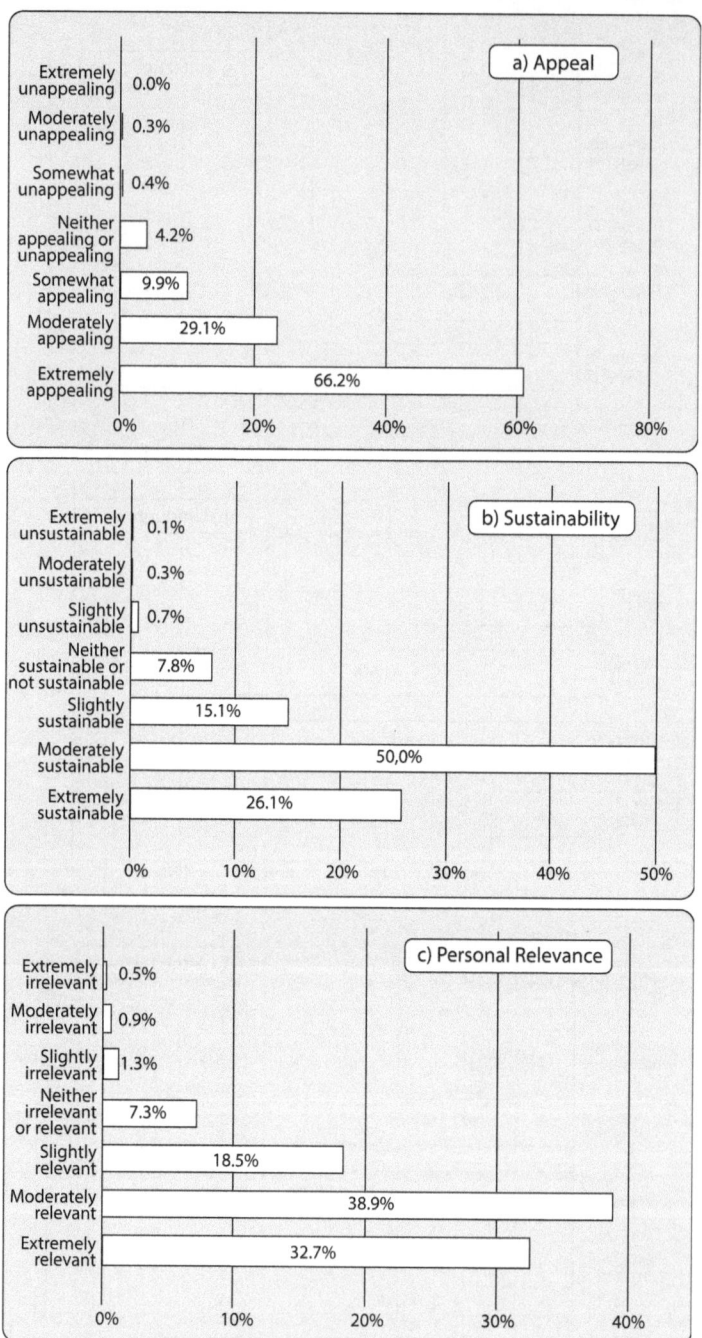

Fig. 12.4 Frequency distributions showing participant perceptions of the tourism experience described within the survey [(**a**) Appeal; (**b**) Sustainability; (**c**) Personal Relevance]

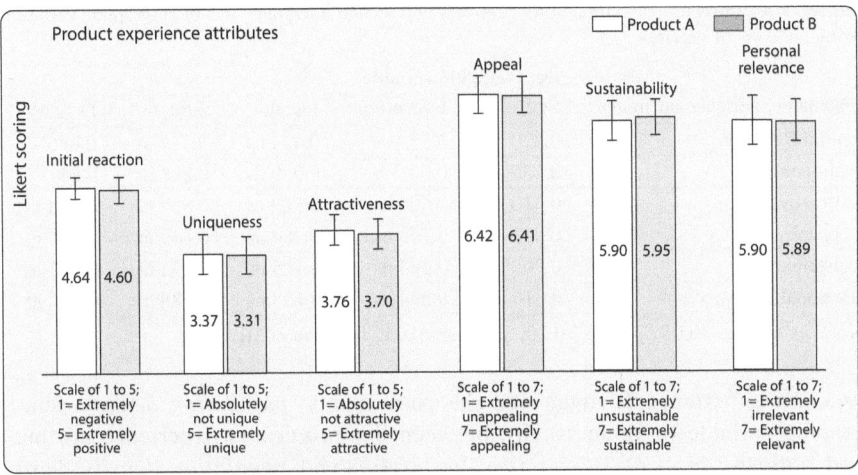

Fig. 12.5 Mean scores and confidence intervals of the groups for Product A (Group 1) and Product B (Group 2) experience attribute ratings

Table 12.2 Means, bootstrap confidence intervals, and T-tests for Product A (Group 1) and Product B (Group 2) experience attribute ratings

Product experience attributes[c]	Product concept		Bootstrap[a] 95% confidence interval				T-test for equality of means[b]
			Inferior		Superior		
Product:	A	B	A	B	A	B	P-value
Initial reaction	4.64	4.60	4.58	4.54	4.69	4.66	0.36 ns
Uniqueness	3.37	3.31	3.26	3.22	3.46	3.40	0.40 ns
Attractiveness	3.76	3.70	3.69	3.62	3.83	3.77	0.25 ns
Appeal	6.42	6.41	6.32	6.32	6.51	6.50	0.96 ns
Sustainability	5.90	5.95	5.80	5.86	5.99	6.04	0.40 ns
Personal relevance	5.90	5.89	5.79	5.78	6.00	6.00	0.85 ns

[a]Unless otherwise stated, results are based on 1000 bootstrap samples
[b]ns p-value > 0.05
[c]The Initial Reaction, Uniqueness, and Attractiveness variables used a 5-point Likert scale. The Appeal, Sustainability, and Personal relevance variables used a 7-point Likert Scale

12.4.4   The Effect of Socio-demographic Characteristic Variables on Experience Attribute Ratings with and without the Presence of Visible Aquaculture Infrastructure

Table 12.3 presents the p-value and significance of mean product experience attributes ratings across socio-demographic characteristic variable groupings. Several statistically significant differences were found between the *initial*

Table 12.3 Summary of interactions between experience attributes and demographic variables using analysis of variance

Product experience attributes	Demographic variables				
	Sex	Education	Income	Age	Density
Initial Reaction	0.820 ns	0.539 ns	0.444 ns	0.747 ns	0.066[†]
Uniqueness	0.435 ns	0.023[*]	0.909 ns	0.268 ns	0.021[*]
Attractiveness	0.610 ns	0.609 ns	0.649 ns	0.935 ns	0.118 ns
Appeal	0.092[†]	0.778 ns	0.984 ns	0.849 ns	0.886 ns
Sustainability	0.705 ns	0.294 ns	0.125 ns	0.973 ns	0.682 ns
Personal Relevance	0.310 ns	0.294 ns	0.638 ns	0.589 ns	0.516 ns

Note: ns p-value >0.05; [†]p-value<0.10; [*]p-value <0.05; [**]p-value <0.01

reaction experience attribute and the participants' *population density* demographic variable (p-value <0.10); between the *uniqueness* experience attribute and both the *level of studies* (p-value <0.05) and *population density* demographic variables (p-value <0.05); and between the *appeal* experience attribute and the *sex* demographic variable (p-value <0.10). There were no significant differences observed between any of the socio-demographic characteristic variables and the questions about tourism experience attractiveness, sustainability, and relevance.

To determine where the interaction occurred as a function of population density, Tukey's a posteriori multiple comparisons analysis (DHS) was performed. Graphs of significant interactions between experience attributes and demographic variables are shown in Figs. 12.6 and 12.7. For example, people residing in a high-density region of Chile (>400 persons/km^2) did not present significant differences in the mean score for their *initial reactions* (Product A = 4.61; Product B = 4.62); however, for persons residing in areas with low population density (<40 persons/km^2), there was a significant difference in the mean score (Product A = 4.65; Product B = 4.49; $p \leq 0.10$). For *uniqueness*, the mean scores were significantly different across all four population density levels ($p \leq 0.05$), with similar interactions to the *initial reaction* results; persons living in areas of lower density perceived Product A as being more unique (Product A = 3.45; Product B = 3.27), and persons in living in high density areas perceived them as less unique (Product A = 3.35; Product B = 3.29). There were also significant differences in the mean values for *uniqueness* and *levels of education*, in that participants with lower level of education perceived Product B to be significantly more unique than Product A, while persons with undergraduate degrees or higher evaluated Product A to be significantly more unique than Product B. Finally, significant differences were observed between the mean ratings of women and men for the *appeal* experience attribute, with women rating both of the products with significantly higher *appeal* (Product A = 6.58; Product B = 6.46) than men (Product A = 6.23; Product B = 6.33). Nevertheless, it is interesting to note that women rated Product A more appealing than Product B, whereas men found Product B to be more appealing.

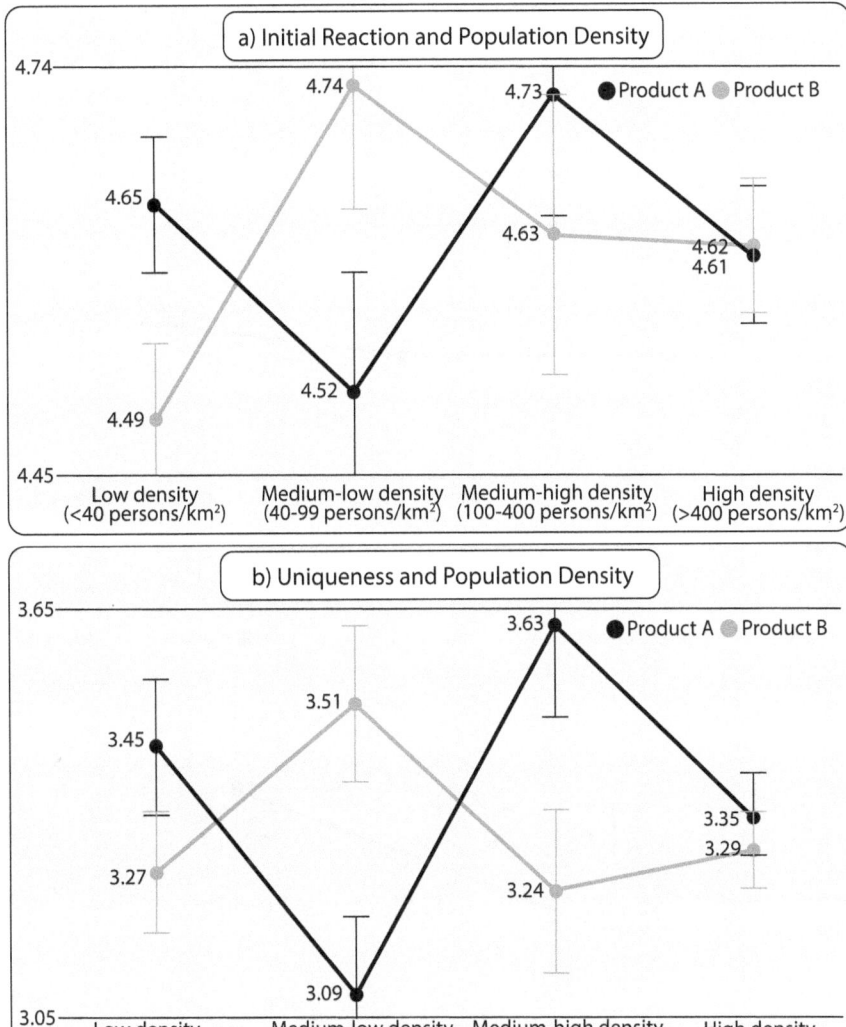

Fig. 12.6 Graphs of significant interactions between experience attributes and demographic variables [(**a**) Initial Reaction and Population Density; (**b**) Uniqueness and Population Density]

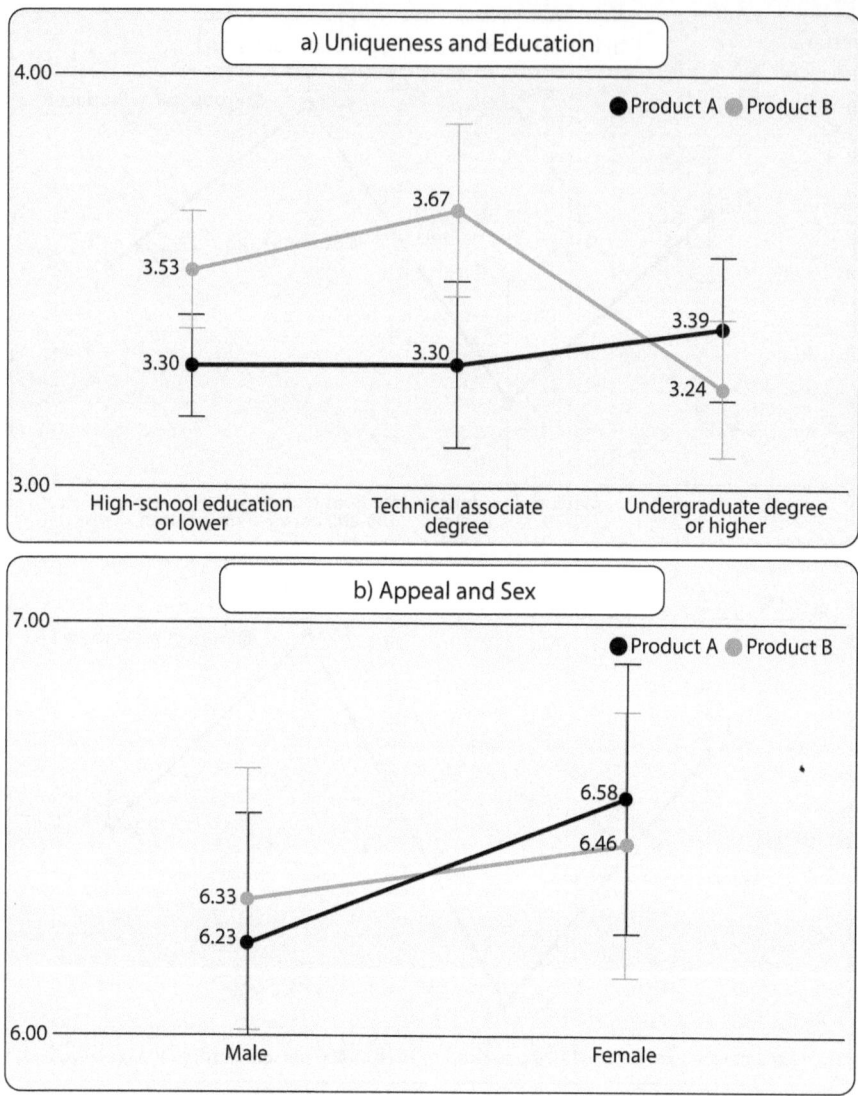

Fig. 12.7 Graphs of significant interactions between experience attributes and demographic variables [(**a**) Uniqueness and Education; (**b**) Appeal and Sex]

12.5   Discussion

This research sought to understand the perceptions and preferences of Chilean professionals with respect to tourism experiences being offered in and around Puyuhuapi, Chile, and the potential impacts of the aquaculture infrastructure within the natural and cultural landscapes. To understand the potential of the current study, it is helpful to contextualize the sample characteristics with recent Chilean internal tourism tendencies. For example, longitudinal studies realized by the Chilean Subsecretary of Tourism (2022) indicated that in 2019, Chileans made more than 22-million overnight tourism trips and another 17-million full-day excursions within national territory. More than 40% of the trips taken by Chilean national tourists originated from the Metropolitan region of Chile, and specifically, the Maipú Commune (Chilean Subsecretary of Tourism, 2022). The principal Chilean regional destinations for national tourism included Valparaíso, the Metropolitan region, Biobio, the Araucanía, and O'Higgins, which received a combined total of 13.6 million national tourists in 2019, which was slightly lower than the prior year (Chilean National Tourism Service, 2020). For these regions, around a quarter of visitors were from the same region, while the other three-quarters lived in other regions. In contrast, the Aysén region of Chile received 180,381 national tourists in 2019, down 36.5% from the prior year (Chilean National Tourism Service, 2020). While less information has been documented about national tourism tendencies and behaviors in Aysén, a recent search of Trip advisor emphasized all terrain tours along the Carretera Austral and nature tours as being the most popular experiences in Aysén. And 5 of the top 25 regional excursions, according to tripadvisor.cl, focused on the area of Puyuhuapi, and especially, visits to Queulat National Park.

The recent tourism context within Chile and the Aysén Region supports the importance of understanding the perceptions and preferences of the national tourism sector for choosing tourism experiences and landscapes. And, the demographic profile of the current study, with its high incidence of participants from Chile's more developed, urban zones, and of participants with higher levels of education and income, provides informative input for the local tourism sector, in and around Puyuhuapi, as they seek to understand the potential of their current tourism experience positioning and possible impacts resulting from the presence of aquaculture infrastructure in their landscapes.

12.5.1   Understanding the Potential for Current Tourism Experience Positioning with Prospective Chilean Tourists

Participant responses to current nature-based tourism experience positioning were extremely positive, regardless of which product they viewed. They expressed positive initial reactions to the descriptions and photography of the destination and its natural and cultural landscapes, finding them to be attractive, appealing, and personally relevant, with or without the presence of movable floating sea cages. On one hand, these results should encourage the tourism sector about the viability of its current tourism positioning and experiences and reassure them that all is not lost if the salmon infrastructure remains in the landscape. However, tendencies with respect to some of the other experience variables (e.g., Uniqueness, Sustainability) suggest some underlying concerns that should be addressed through further research and planning.

For example, less than 10% of the combined sample perceived the tourism experience they were presented to be *extremely unique*. Rather, more than half of the participants (57%) found the proposed experience and its associated landscapes, to be *somewhat, slightly or not at all unique*. This should be concerning to destination managers, who are seeking to differentiate their positioning within marketing and promotional materials, especially considering the remote and hard-to-reach nature of the Puyuhuapi destination. Further, Chilean legislation protected landscapes that have unique natural or cultural landscape values. Study results do not support building this argument for the landscapes of Puyuhuapi, based on the perceptions of prospective Chilean tourists. Future research, focused on defining unique aspects of the Puyuhuapi destination and landscapes, from the vantagepoint of prospective and actual tourists, could help improve both destination positioning and experience offerings, and perhaps, contribute to defining strategies for protecting territory landscapes.

Future research should also be oriented toward understanding how prospective Chilean tourists define the concept of sustainability for tourism landscapes. The majority of study participants (73.9%) responded that the Puyuhuapi landscapes included in the tourism experiences (both with and without salmon aquaculture infrastructure) were not *extremely sustainable*. One possible explanation for these scores could be that participants perceived the landscapes within the experiences as vulnerable. Maybe they would prefer to visit the destination if protections could ensure their permanence over time. These outcomes would certainly correlate with the apprehension that residents have expressed regarding the threats coming from the development and the impact they could have on the conservation of the natural landscapes, and the ecosystems of the area. Nevertheless, this hypothesis must be validated through future research before a tourism-based justification for landscape protection can be developed.

12.5.2 Comparative Analysis Between Tourism Experience Alternatives with and without the Presence of Visible Aquaculture Infrastructure

The lack of significant differences between the tourism experience alternatives (Product A, Product B) were surprising. Nevertheless, it is important to remember that participants were only exposed to one or the other alternatives, without the possibility to compare between the two. Returning to the landscape studies research provides several possible explanations that serve as hypotheses for future work (Nogué, 2007; Ramírez & López, 2015; Rodríguez, 2021).

For example, Nogúe (2007), and later Rodríguez (2021), proposed that landscapes represent the cultural projection of societies. Moreover, as multiple landscape researchers have observed, landscape experiences reflect observers' perceptive capacities and cultural backgrounds (Center for Environmental Agrarian Studies—CEA, 2022; Zube et al., 1982). The majority of Chileans live and work in a very urban cultural setting dominated by urban and industrial landscapes (Gale et al., 2021; Janubová & Greš, 2016). This may help explain why participants who were exposed to Product B were so accepting of the industrial components within otherwise largely natural landscapes. For them, the natural components of the landscape may have sufficiently dominated the overall visual aspect of the product experience; thus, the viewsheds presented in Product B (Visible aquaculture interventions), aligned with the nature-based tourism description provided.

Alternatively, perhaps their responses to Product B reflected a complacency about the presence of salmon aquaculture infrastructure. Results from Ednie and Gale (2021) identified similar complacency tendencies with respect to urban dwellers and natural sounds. Urban dwellers were exposed to anthropogenic sounds within the natural settings with which they felt most connected and those who reported hearing them more often, found them as more acceptable within a natural setting. Ednie and Gale (2021) warned that this type of urban complacency about anthropogenic components in nature created risks with respect to the reliability of social norm data for the conservation of natural areas. Results of the current study seem to align with these findings, reinforcing the authors' call for continued research around urban frames of reference and perceived naturalness (e.g., Groulx et al., 2017).

Ramírez and López's (2015) research, which proposed that modified landscapes may hold cultural significance and symbolism for observers, also seems relevant for this discussion of our results. Perhaps the largely urban sample of our study assigned a positive significance to the industrial development portrayed through the salmon aquaculture infrastructure. Chile is the second largest salmon producer in the world, with a reported 26% of the production worldwide, and an average reported annual growth of more than 10% over the past 3 years (Cuéllar, 2022; Riedemann Fuentes et al., 2021). According to Cuéllar (2022), 99% of Chile's salmon is produced within the three Patagonian regions of Los Lagos (42.2%), Aysén (40.5%), and Magallanes (17%). Chilean salmon is sold throughout the country and exported around the world. Given the importance of salmon production within the Chilean economy and the professional, urban nature of our sample population, it seems plausible that a natural landscape that has been modified with the presence of floatable sea cages, may be perceived as *interesting, appealing, attractive, personally relevant*, and even *sustainable*.

Although Muñoz (2004a) observed increasing levels of environmental awareness among Chilean citizens, it is also plausible that the majority of participants in this study were not aware or knowledgeable, or perhaps, were apathetic about the value and importance of the natural ecosystems in and around Puyuhuapi. They may also lack awareness or knowledge about the difficulties (e.g., the impacts of the ISA virus between 2007 and 2009; red tide events after massive dead salmon dumping in 2016; the sinking of the Seikongen well boat, with its toxic cargo of oil and decomposed salmon in 2017; numerous large-scale salmon escapes, and an increased presence of industrial trash along the coast) associated with salmon aquaculture production in Patagonia (Riedemann Fuentes et al., 2021). Thus, the focus on building environmental awareness within Chile's National Biodiversity Strategy (2017–2030) seems both appropriate, and vitally important for communities like Puyuhuapi, who hope to build a tourism offering based on natural landscape valuation.

12.5.3 Determination of the Effect of Socio-demographic Characteristic Variables on the Product Alternatives with and without the Presence of Visible Aquaculture Infrastructure

Understanding the interactions that occurred between product experience attributes and socio-demographic characteristic variables adds important context that may inform future research needs and strategies for building environmental awareness as well as destination experience positioning. For example, earlier we discussed landscape studies research that has drawn out the importance of cultural backgrounds, projections, and observers' perceptive capacities (Center for Environmental Agrarian Studies—CEA, 2022; Nogué, 2007; Rodríguez, 2021; Zube et al., 1982). While the current study was exploratory, with a limited scope, our analysis showed significant interactions between several of the tourism experience attributes (i.e., *initial reaction, uniqueness, appeal*), and demographic variables (*population density, levels of education, sex*) that can orient future cultural and landscape perception research. Focused future research to understand the frames of reference for people living in distinct population density situations in Chile may help illuminate the impacts of living within heavily industrialized settings, and the types of environmental awareness efforts that are required for reaching an urban society that is often concentrated far away from the remote areas where natural environments are conserved.

The interactions between *uniqueness* and *level of education* were also interesting, suggesting the need for differentiation strategies, both in environmental awareness building efforts, and for tourism positioning and experience design. It seems possible that the differences in preferences associated with education may be related to career possibilities and options. For example, the salmon aquaculture sector in Chile has provided important sources of work through a range of jobs for persons with technical degrees, or with high-school degrees or less, within the hatchery, farming, and processing phases, specifically in roles that are associated more directly with infrastructure sites and installations (Riedemann Fuentes et al., 2021). Perhaps, these groups of participants rated *uniqueness* higher for the natural landscapes that were modified with the presence of floatable sea cages, because they represented innovations that could lead to jobs and increased economic opportunities. More research is warranted to understand these differences and their implications for environmental awareness and tourism.

12.6 Conclusions

This chapter explored some of the tensions that can arise for conservation-based development in Chilean Patagonia, responding to 2020 concerns from Puyuhuapi tourism providers about potentially negative impacts for viewsheds affected by the installation of salmon aquaculture infrastructure within the surrounding fjords. Our research sought to help these stakeholders understand the potential impact of this salmon aquaculture infrastructure on potential tourists, by exploring whether current nature-based tourism experience positioning was well received and relevant, with and without the presence of salmon aquaculture infrastructure (movable floating sea cages) in the Puyuhuapi bay. Our results suggested that the current destination and experience positioning used in Puyuhuapi were relevant to participants. Moreover, the visible presence of the movable floating sea cages did not provoke significant differences in the valuation of the landscape for these prospective tourists. Nevertheless, there were socio-demographic characteristics (i.e., population density, level of education, sex) that acted as determinants for respondents' characterization of the initial reaction, uniqueness, and appeal of the two hypothetical products.

Results provided interesting initial insights around the complex challenge of balancing conservation and development in a way that permits sustainability and quality of life for the long-term. It raised a number of interesting hypotheses and research questions that warrant future research to inform conservation-based development strategies and outcomes. Specifically, the lack of significant differences between participants perceptions of the landscapes, with and without the presence of salmon aquaculture infrastructure (movable floating sea cages) in the Puyuhuapi bay, suggests the importance of expanding understanding of Chilean perspectives and imaginaries of Patagonia, and its abundant natural settings and values.

Acknowledgements This work was supported by Chile's National Research and Development Agency (ANID) under ANID's Regional Program R17A10002; the CIEP R20F0002 project; and the CHIC-ANID PIA/BASAL PFB210018. We would like to thank Andrés Adiego, the Centro de Investigación en Ecosistemas de la Patagonia (CIEP) and the Department of Geography and Territorial Planning/Management of the Universidad de Zaragoza, for his assistance with the development of the map for our study area (Fig. 12.1).

References

A. Aguilar, A. Palafox, J. Anaya, El turismo y la transformación del paisaje natural [Tourism and the transformation of natural landscape]. Noésis. Rev. de Cienc. Soc. **24**(47-1), 19–30 (2015). https://doi.org/10.20983/noesis.2015.12.2

B. Arts, M. Buizer, L. Horlings, V. Ingram, C. Oosten, P. Opdam, Landscape approaches: A state-of-the-art review. Annu. Rev. Environ. Resour. **42**(1), 439–463 (2017)

J.F. Benson, M.H. Roe, *Landscape and Sustainability* (Tylor & Francis, 2000). https://doi.org/10.4324/9780203995785

J. Bermúdez, *Fundamentos de Derecho Ambiental [Fundamentals of Environmental Law]* (Ediciones Universitarias de Valparaíso, 2014)

F.G. Bernáldez, *Invitación a la ecología humana. La adaptación afectiva al entorno [Invitation to Human Ecology. The Affective Adaptation to the Environment]* (Tecnos, Madrid, 1985)

S. Bourassa, A paradigm for landscape aesthetics. Environ. Behav. **22**(6), 787–812 (1990). https://doi.org/10.1177/0013916590226004

J. Calvin, J. Dearinger, M. Curtin, An attempt at assessing preferences for natural landscapes. Environ. Behav. **4**, 447–470 (1972)

O. Campos-Campos, G. Cruz-Cárdenas, R.J.C. Aquino, R. Moncayo-Estrada, M.A.V. Machuca, L.A.Á. Meléndez, Historical delineation of landscape units using physical geographic characteristics and land use/cover change. Open Geosci. **10**(1), 45–57 (2018). https://doi.org/10.1515/geo-2018-0004

Center for Environmental Agrarian Studies – CEA, *Estudios del paisaje visual [Visual Landscape Studies]* (2022). Retrieved 21 Nov 2022, from https://www.ceachile.cl/paisaje/#tipos

Chilean Environmental Evaluation Service, *Guía de evaluación de impacto ambiental del valor paisajístico en el SEIA [Guide to Environmental Impact Assessment of Landscape Value in SEIA]* (Santiago, 2019)

Chilean Ministry of the Environment, *Estrategia nacional de biodiversidad: 2017–2030 [Chilean National Biodiversity Strategy 2017–2030]* (Santiago, 2016)

Chilean National Commission on the Environment, *Directrices para la evaluación del impacto ambiental de los proyectos de inversión [Guidelines for the Evaluation of the Environmental Impact of Investment Projects]*. Presidential instructions. (Chilean National Commission on the Environment, Chilean Government, Santiago de Chile, 1993)

Chilean National Commission on the Environment, *Manual de evaluación de impacto ambiental: conceptos y antecedentes básicos [Manual on Environmental Impact Assessment: Basic Concepts and Background]* (CONAMA, Santiago, 1994)

Chilean National Institute of Statistics, *Ciudades, pueblos, aldeas y caseríos 2019 [Cities, Towns, Villages, and Hamlets 2019]* (2019). https://geoarchivos.ine.cl/File/pub/Cd_Pb_Al_Cs_2019.pdf

Chilean National Institute of Statistics, *El ingreso laboral promedio mensual en Chile fue de $681.039 en 2021 [Average Monthly Labor Income in Chile Was $681,039 in 2021]* (INE, 2022, July 21). https://www.ine.gob.cl/prensa/detalle-prensa/2022/07/21/el-ingreso-laboral-promedio-mensual-en-chile-fue-de-$681.039-en-2021

Chilean National Tourism Service, *Cuenta pública participativa 2019 [Participatory Public Account 2019]* (2020). Retrieved 14 Dec 2022, from https://www.sernatur.cl/wp-content/uploads/2020/05/CuentaPublicaSernatur2019_15mayo.pdf

Chilean Subsecretary of Tourism, *Big Data para el Turismo Interno [Big Data for Internal Tourism]* (2022). Retrieved 14 Dec 2022, from https://www.sernatur.cl/dataturismo/big-data-turismo-interno/

Cooperativa.cl, *Coyhaique: Rechazaron recurso de protección contra salmonera en Puyuhuapi [Coyhaique: Appeal for Protection Against Salmon Farm in Puyuhuapi Rejected]* (Cooperativa, 2021, October 23). https://cooperativa.cl/noticias/pais/region-de-aysen/coyhaique-rechazaron-recurso-de-proteccion-contra-salmonera-en-puyuhuapi/2021-10-23/092235.html

J.A. Cuéllar, *Salmon Production in the World: Current Outlook* (Veterinaria Digital, 2022, January 12). https://www.veterinariadigital.com/en/articulos/salmon-production-in-the-world-current-outlook/

E. Dahan, V. Srinivasan, The predictive power of internet-based product concept testing using visual depiction and animation. J. Prod. Innov. Manag. **17**(2), 99–109 (2000). https://doi.org/10.1111/1540-5885.1720099

T.C. Daniel, J. Vining, Methodological issues in the assessment of landscape quality, in *Behavior and the Natural Environment*, ed. by I. Altman, J.F. Wohlwill, (Springer US, Boston, 1983), pp. 39–84. https://doi.org/10.1007/978-1-4613-3539-9_3

G. De la Fuente de Val, J.V. De Lucio, *La importancia de considerar las expectativas y pref-erencias paisajísticas de visitantes, gestores y expertos ambientales en la gestión de espa-cios naturales del mediterráneo [The Importance of Considering the Landscape Expectations and Preferences of Visitors, Managers and Environmental Experts in the Management of Mediterranean Natural Areas]* (Centro de Investigaciones Ambientales de la Comunidad de Madrid, Madrid, 2003)

G.J. De la Fuente de Val, J.A. Atauri Mezquida, J.V. De Lucio Fernández, El aprecio por el paisaje y su utilidad en la conservación de los paisajes de Chile Central [Appreciation for the land-scape and its usefulness in the conservation of Central Chile's landscapes]. Ecosistemas **13**(2), 82–89 (2004)

M.C. Dunn, *Landscape Evaluation Techniques: An Appraisal and Review of the Literature* (Centre for Urban and Regional Studies, University of Birmingham, Birmingham, 1974)

A. Ednie, T. Gale, Soundscapes and protected area conservation: Are noises in nature mak-ing people complacent? Nat. Conserv. **44**, 177–195 (2021). https://doi.org/10.3897/natureconservation.44.69578

A. Ednie, T. Gale, K. Beeftink, A. Adiego, Connecting protected area visitor experiences, wellness motivations, and soundscape perceptions in Chilean Patagonia. J. Leis. Res. **53**(3), 377–403 (2020). https://doi.org/10.1080/00222216.2020.1814177

J. Fariña, J. Solana, El análisis del paisaje [Landscape analysis], in *La ciudad y el medio natural*, ed. by J. Fariña, (Akal, Madrid, 2007), pp. 258–280

J. Femenías, La culpabilidad en la responsabilidad por daño ambiental y su relación con el sistema de evaluación de impacto ambiental [Guilt in liability for environmental damages and its con-nection with the Chilean environmental impact assessment system]. Rev. Derecho de la Pontif. Univ. Catól. de Val. **48**(1), 233–259 (2017)

K.D. Fines, Landscape evaluation: A research project in East Sussex. Reg. Stud. **2**(1), 41–55 (1968). https://doi.org/10.1080/09595236800185041

T. Gale, A. Ednie, Can intrinsic, instrumental, and relational value assignments inform more inte-grative methods of protected area conflict resolution? Exploratory findings from Aysén, Chile. J. Tour. Cult. Chang. **18**(6), 690–710 (2019). https://doi.org/10.1080/14766825.2019.1633336

T. Gale, A. Adiego, A. Ednie, A 360° approach to the conceptualization of protected area visitor use planning within the Aysén Region of Chilean Patagonia. J. Park. Recreat. Adm. **36**, 22–46 (2018). https://doi.org/10.18666/JPRA-2018-V36-I3-8371

T. Gale, A. Ednie, K. Beeftink, Thinking outside the park: Connecting visitors' sound affect in a nature-based tourism setting with perceptions of their urban home and work soundscapes. Sustainability **13**(12), 6572 (2021). https://doi.org/10.3390/su13126572

Greenpeace, *Greenpeace denuncia concesión ilegal de empresa salmonera en la bahía de Puyuhuapi [Greenpeace Condemns Salmon Company's Illegal Concession in Puyuhuapi Bay]* (Greenpeace, 2022, November 4). https://www.greenpeace.org/chile/noticia/uncategorized/greenpeace-denuncia-operaciones-ilegales-de-la-empresa-salmon-concesiones-xi-region-s-a/

M. Groulx, C.J. Lemieux, J.L. Lewis, S. Brown, Understanding consumer behaviour and adapta-tion planning responses to climate-driven environmental change in Canada's parks and pro-tected areas: A climate futurescapes approach. J. Environ. Plan. Manag. **60**(6), 1016–1035 (2017). https://doi.org/10.1080/09640568.2016.1192024

B. Janubová, M. Greššs, Urbanization of poverty and the sustainable development of urban areas in Chile. Theor. Empir. Res. Urban Manag. **11**(4), 17–29 (2016)

M. Jones, The concept of cultural landscape: Discourse and narratives, in *Landscape Interfaces: Cultural Heritage in Changing Landscapes*, ed. by H. Palang, G. Fry, (Springer Netherlands, Dordrecht, 2003), pp. 21–51. https://doi.org/10.1007/978-94-017-0189-1_3

C. Klonk, *Science and the Perception of Nature: British Landscape Art in the Late Eighteenth and Early Nineteenth Centuries*. (Doctoral thesis) (1993). https://doi.org/10.17863/CAM.11575

S.C. Kung, T.W. Chien, Y.T. Yeh, J.C.J. Lin, W. Chou, Using the bootstrapping method to verify whether hospital physicians have different h-indexes regarding individual research achievement: A bibliometric analysis. Medicine **99**(33), e21552 (2020). https://doi.org/10.1097/ MD.0000000000021552

L. Ludwig Winkler, *Puyuhuapi: curanto y kuchen, historia oral de un pueblo de Aysén [Puyuhuapi: Curanto and Kuchen, Oral History of an Aysén Village]* (Ediciones Kultrún, 2013)

Municipality of Cisnes, *Comuna de Cisnes Plan de Desarrollo Comunal, 2018–2028 [Cisnes Commune Plan for Development of the Cisnes Commune, 2018–2028]* (Puerto Cisnes, 2018). https://municipalidadcisnes.cl/wp-content/uploads/2022/09/PLADECO-MUNICIPALIDAD-CISNES-FINAL-II.pdf

A. Muñoz, La evaluación del paisaje: Una herramienta de gestión ambiental [Landscape evaluation: A tool for environmental management]. Chil. J. Nat. Hist. **77**, 139–156 (2004a)

A. Muñoz, Los humedales del río Cruces y la convención de Ramsar: Un intento de protección fallido [The wetlands of the Cruces river and the Ramsar convention. A failed attempt at protection]. Gest. Ambient. **10**, 11–26 (2004b)

A. Muñoz, A. Badilla, H. Rivas, Evaluación del paisaje en un humedal del sur de Chile: el caso del río Valdivia (X Región) [Evaluation of landscape in a wetland of the south of Chile: The case of Valdivia river]. Rev. Chil. Hist. Nat. **66**, 403–417 (1993)

A. Muñoz, J. Moncada, A. Larraín, Variación de la percepción del recurso paisaje en el sur de Chile [Variation in the perception of the landscape resource in southern Chile]. Rev. Chil. Hist. Nat. **73**, 681–690 (2000)

A. Muñoz, J. Moncada, L. Gómez, Evaluación del paisaje visual en humedales del Río Cruces, sitio Ramsar de Chile [Evaluation of the visual landscape in wetlands of the Río Cruces, a Ramsar site in Chile]. Rev. Chil. Hist. Nat. **85**(1), 73–88 (2012). https://doi.org/10.4067/ S0716-078X2012000100006

J. Nogué, *La construcción social del paisaje. Paisaje y teoría [The Social Construction of Landscape. Landscape and Theory]* (Biblioteca Nueva, Madrid, 2007)

B. Ramírez, L. López, *Espacio, paisaje, región, territorio y lugar: la diversidad en el pensamiento contemporáneo [Space, Landscape, Region, Territory and Place: Diversity in Contemporary Thought]* (UNAM, Instituto de Geografía: UAM, Xochimilco, México, 2015)

A. Riedemann Fuentes, T. Bansal, F. Pardo Núñez, *Informe Industria Salmonera en Chile y Derechos Humanos. Evaluación de impacto sectorial [The Salmon Industry and Human Rights in Chile: Sector-Wide Impact Assessment]* (Instituto Nacional de Derechos Humanos, Instituto Danés de Derechos Humanos, 2021)

L. Rodríguez, Paisajes del trauma: una cinta de Moebius [Landscapes of trauma: A Moebius tape], in *Historia, Trauma, Memoria*, ed. by C. Balbontín, L. Rodríguez, (Libros del amanecer, Santiago, 2021), pp. 173–196

J. Rozas, *La comunidad está en pie de guerra contra la instalación de balsas jaulas de la empresa Salmones de Chile frente a la pequeña localidad de la Región de Aysén [The Community Is Up in Arms Against the Installation of Cage Rafts of the Company Salmones de Chile in Front of the Small Town in the Aysén Region]* (Diario Acuícola, 2021, September 17). https:// www.diarioacuicola.cl/noticia/actualidad/2021/09/vecinos-de-puyuhuapi-interpusieron-recurso-de-proteccion-por-instalacion-de-salmonera#:~:text=Un%20recurso%20de%20 protecci%C3%B3n%20ante,de%20la%20comuna%20de%20Cisnes

M. Sandoval, Medio ambiente y paisaje: una aproximación desde el Sistema de Evaluación de Impacto Ambiental [Environment and landscape: An approach from the environmental impact assessment system]. Eco-Reflexiones **1**(1), 1–6 (2021). http://dacc.udec.cl/wp-content/ uploads/2021/04/Eco-Reflexi%C3%B3n-N%C2%B01.pdf

N. Saqib, A positioning strategy for a tourist destination, based on analysis of customers' perceptions and satisfactions: A case of Kashmir, India. J. Tour. Anal. **26**(2), 131–151 (2019). https:// doi.org/10.1108/JTA-05-2019-0019

United Kingdom Secretary of State for Foreign and Commonwealth Affairs, *European Landscape Convention. Treaty Series No. 36* (United Kingdom, 2012)

Y. Wang, *Product Development and Management* (Tsinghua University Press, Beijing, 2007)

M. Zambrano, V. González, *La valoración en el ordenamiento territorial [Valuation in Land Use Planning]* (Cuenca, 2002)

I.S. Zonneveld, The land unit—A fundamental concept in landscape ecology, and its applications. Landsc. Ecol. **3**(2), 67–86 (1989). https://doi.org/10.1007/BF00131171

E.H. Zube, J.L. Sell, J.G. Taylor, Landscape perception: Research, application and theory. Landsc. Plan. **9**(1), 1–33 (1982). https://doi.org/10.1016/0304-3924(82)90009-0

Part III
Building Resilience and Sustainability

Part III offers strategies for building resilience and sustainability, contextualized through Patagonia's varied and volatile circumstances of climate change, environmental and human health, and geographic periphery.

Chapter 13
Employing Local Tourism Councils to Improve Protected Area Tourism Development and Governance in the Aysén Region of Chile

Adriano Rovira, Gabriel Inostroza Villanueva, Guillermo Sebastián Pacheco Habert, and Pablo Szmulewicz

Abstract This chapter presents a recent regional project developed to improve tourism governance in and around the protected areas (PAs) administered by the National System of State Natural Protected Areas (SNASPE) within the Aysén Region of Chile. The project focused on the design of a participatory multi-scale governance and management system that would enable local communities to work with SNASPE PAs and improve tourism services for visitors, both within PAs and within the surrounding communities. As part of the project, a proposal for the creation of Local Tourism Councils (LTC) was developed. In this chapter, we discuss the validation of the proposed local governance model through a pilot implementation of the LTC concept, within the Cerro Castillo National Park (CCNP) and two of its primary gateway communities: Cerro Castillo Village and Puerto Ingeniero Ibáñez. Early outcomes for the CCNP and its gateway communities seem to support the potential for the LTC model. We present a series of enabling factors observed during the project that may inform the creation of other LTCs in other areas. Achieving this outcome would help stimulate local economies and improve the potential for tourism development to be compatible with the conservation of natural and cultural heritage.

Keywords Patagonia · Nature-based tourism · Collaborative governance · Protected areas · Gateway communities

A. Rovira (✉) · G. S. Pacheco Habert · P. Szmulewicz
Universidad Austral de Chile, Faculty of Economics and Administrative Sciences – Institute of Tourism, Valdivia, Chile
e-mail: arovira@uach.cl; Guillermo.pacheco@uach.cl; pszmulew@uach.cl

G. Inostroza Villanueva
Universidad Austral de Chile, Campus Patagonia, Coyhaique, Chile
e-mail: gabriel.inostroza@uach.cl

13.1   Introduction

The Aysén Region is located in the southernmost tip of Chile, between 43°38′ and 49°16′ south latitude, in the heart of Chilean Patagonia. It is one of the largest and least populated regions of Chile, with a territory of 108,494 km² and a population density of just one person per square kilometer (DNA Expertus, 2016). The natural landscapes of the region are valued for their scenic beauty, their uniqueness, and for their perceived pristineness (DNA Expertus, 2016). Currently, approximately 50% of the region's territory is under protection within the National System of State Natural Protected Areas (SNASPE). Regional SNASPE protected areas (PAs) currently include seven National Parks, eight National Reserves and two Natural Monuments, which together cover approximately 51,620 km².

Natural PAs are particularly attractive for nature enjoyment and recreation, helping to satisfy a number of human needs. For example, recreational and touristic use of PAs has been shown to foster and strengthen connections with nature, and in many cases, levels of support for conservation (Figueira, 2011; Gale & Ednie, 2019). Nevertheless, recreation and tourism use produce impacts for PAs and can negatively affect their objectives related to ecosystem and biodiversity conservation (Hummel et al., 2019), especially when use is improperly managed or controlled. Prior to COVID-19, global tourism trends showed a substantial increase in demand for nature tourism, which manifested in Aysén through an increase in the number of tourists seeking nature, adventure, and extreme sports. Visits to SNASPE PAs grew during this timeframe, averaging a 7% increase per year between 2010 and 2019 (Szmulewicz & Aedo, 2020). This increase in visitation has been even more drastic in the Aysén Region, which saw a 28.7% increase per year between 2012 and 2018 (Pacheco & Boldt, 2020).

Aysén's SNASPE PAs are managed by the Chilean National Forestry Corporation (CONAF), a private legal entity, which reports to the Chilean Ministry of Agriculture. CONAF's responsibilities include regulating and defining appropriate use in these areas, in balance with the principles of conservation and nature protection. Historically, CONAFs management approach has not included participatory governance strategies; however, increasing social and economic interest in SNASPE PAs and their management has fostered increasing recognition and support for these types of approaches. For example, from 2016 to 2018, CONAF implemented a 3-year initiative within Aysén to improve visitor experiences and build management capacity. Gale et al. (2018) observed that,

> Rather than viewing the protected areas as islands connected to the center by better infrastructure and roads, the project placed emphasis on better integrating them within their own local areas, and the implementation of adaptive models for maintaining conservation priorities through the growth and increasing recreation experience demands. (p. 26)

Tourism stakeholders have also noted the need for better coordination and a shared governance structure for SNASPE PAs. For example, analysis conducted by the region's strategic nature-based tourism development program, PER Turismo Aysén, determined that the lack of a coordinated system of governance for the different actors linked to nature-based tourism in the region represented a weakness for sustainable tourism growth (DNA Expertus, 2016). Furthermore, Aedo (2020) noted that, based on their proximity, all of the towns and cities of the Aysén Region can be associated with at least one SNASPE PA. For these reasons, a regional gateway communities' approach to nature-based recreation and tourism management has been gaining interest and support within the Aysén region. The gateway strategy would concentrate on tourism services within these communities (lodging, food services, guides, etc.), rather than within the PAs themselves (Aedo, 2020).

This chapter presents the lessons learned from a recent 2-year regional project that was developed to improve the governance of tourism development in and around Aysén SNASPE PAs. The project was financed by CORFO, the Chilean economic development agency, through funding designed to enhance regional economic competitiveness. The objective of this project was to design a participatory multi-scale governance system that would enable local communities to work with PAs and improve tourism services for visitors, both within PAs and their surrounding communities. This would be achieved by facilitating direct dialogue between SNASPE PA administration, other public sector institutions, the private sector, academic institutions, and the social and commercial organizations of the Aysén Region's PA gateway communities. Improving these linkages, it was believed, would help stimulate local economies and improve the potential for tourism development to be compatible with the preservation of natural and cultural heritage.

The project consisted of four main stages: (1) a review of participatory governance best practices, (2) an analysis of the existing governance structure in the region of Aysén, (3) the development of a participatory governance proposal for PAs and their gateway communities in Aysén, and (4) a pilot implementation of the proposed system within one area of the region. The following sections will share highlights for the first three phases as well as an in-depth review of phase four. Our aim is to provide insights about how local community participatory mechanisms can be integrated into multilevel governance systems in order to improve conservation area planning and support locally sustainable tourism.

13.2 Project Phases

13.2.1 Phase 1: Participatory Governance Best Practices

The first stage of the project focused on a review of participatory governance theory and best practice. The concept of governance originated from strategic alliances between the public and private sectors (i.e., Mayntz, 2005), a conceptualization that is very present in Chile as evidenced by the tourism governance of Chilean destinations (Pacheco et al., 2015). In recent years, the concept of governance has branched to include new approaches designed to address changing societal needs and demands, including participatory, collaborative, adaptive, polycentric, public, and multilevel governance (Casady et al., 2020; Cejudo et al., 2018; Christensen et al.,

2020; Cunill-Grau & Leyton, 2016; Pacheco & Henríquez, 2018; Vella et al., 2015). Differences between these styles of governance often center around principles of corporate governance, including the processes, rules, and practices for directing and controlling a system or institution. Traditionally, *top-down* hierarchical governance processes were employed by governments and private entities (Lovrić et al., 2018; Chang & Watanabe, 2019; Maestre-Andrés et al., 2018), in contrast to the *bottom-up* governance approaches employed by communities (Pacheco & Henríquez, 2018). Bottom-up approaches emphasize coordination between actors at the local, regional, and national scales through multilevel and polycentric governance platforms that are focused and designed to address the environment and sustainable management of nature (Vella et al., 2015; Matson et al., 2016; Urquiza et al., 2019).

Although a close relationship is maintained between governance and management, collaborative governance systems separate the two in order to allow actors to move beyond the simple technical aspects of PA management and address more complex social issues like local community acceptance and participation in decision-making. Borrini-Feyerabend and Hill (2019) highlight the differences between the two concepts, specifying that management pertains to the actions completed to achieve proposed objectives, while governance encompasses the decision-making process itself and who participates in forming the objectives and their associated strategies. In short, governance defines who is responsible for making decisions on financing, monitoring, and evaluating management, among other duties.

The governance of public PAs begins with the process and decisions related to their creation and delineation, or the way in which they are to be managed. These processes can occur through a number of different frameworks or approaches, ranging from exclusive government or privately driven processes to those driven by communities. Traditionally, PA management has largely been the responsibility of national governments or private actors and entrepreneurs, who govern these PAs without participation from the local and/or Indigenous communities that live within or nearby these areas (Brenner, 2019; Lovrić et al., 2018; Major et al., 2018; Mardones, 2018; Maretti et al., 2019; Niedziałkowski et al., 2018). As with the rest of the world (i.e., national parks in Africa or Asia), participatory mechanisms in Chilean and Latin American PAs remain scarce (Koy et al., 2019; Chang & Watanabe, 2019).

Nevertheless, through the synergies and shared challenges facing conservation and nature-based tourism, new participatory mechanisms have emerged that make it possible for communities to co-manage these territories to various degrees (Tseng et al., 2019; Bello et al., 2016; Islam et al., 2017; Pacheco, 2014; Pacheco & Szmulewicz, 2013). Best-practice governance frameworks are increasingly being designed to share decision-making authority between the public entities responsible for the PA's administration and other actors present in the territory where the PA is located (Pacheco & Boldt, 2020; Rovira et al., 2020; Worboys & Trzyna, 2019). These new forms of PA governance make up an emerging and inclusive conservation paradigm that is unique to the twenty-first century, promoting participation, transparency, pluralism, and other democratic features and practices that strengthen nature conservation measures (Brenner & De la Vega, 2014; Sanz & Torres, 2006; Stoll-Kleemann et al., 2010). Community participation is a key component in these governance models, which require that local communities have the authority to influence decision-making in a binding way. For example, new public governance approaches, like the *Whole of Government Approach* and the *Joint Government Approach*, have infused the idea of integrating more actors in the deliberation and decision-making processes of public institutions. In general, these approaches focus mainly on addressing discoordination within the public sector by incorporating the private sector when necessary (Christensen et al., 2020).

This greater coordination among actors through governance systems is not yet prevalent in Chile, where governance approaches still focus on forms of *New Public Management* that place greater emphasis on strategic alliances between public and private institutions. Public participation is also an important component of Chile's new public management and governance and has gradually infused a critical look at Chilean institutionalism (Casady et al., 2020; Howlett et al., 2017). Through these emerging processes for political organization, public and private actors, as well as local Chilean communities, are adapting to new structures in which binding mechanisms of participation are central to the governance process. While these approaches have not been implemented in the SNASPE PAs of Aysén, current trends suggest that their incorporation would be both timely and possible. Thus, it seems possible that binding participation mechanisms could also be incorporated within the governance and administration of Chilean PAs and related public policy.

The review of participatory governance theory and best practice undertaken in this initial stage of the project provided guiding principles for the remaining three stages. Within the context of this project, governance was conceptualized as a collective process of deliberation in which a diverse set of actors generate agreements and make decisions. This same group of actors subsequently control the implementation of these agreements and actions through accountability and compliance mechanisms (Casady et al., 2020; Cejudo, 2011; Cejudo et al., 2018; Cunill-Grau & Leyton, 2016; Christensen et al., 2020; Howlett et al., 2017; Osborne, 2010; Rhodes, 2015). The project sought to inform governance systems in PA gateway communities that face sustainability and management challenges by improving local organization and tapping into the knowledge and practices of these gateway communities (Ostrom, 2000; Pacheco, 2018). Furthermore, the project recognized participatory governance as being directly linked to multilevel and polycentric governance through the integration of civil society in the planning processes of conservation areas and associated public policies.

13.2.2 Phase 2: Analysis of the Governance Process in the Region of Aysén

Informed by the guiding principles outlined above, the second phase of the project evaluated current governance practices within Aysén SNASPE PAs. This was accomplished through a series of meetings with regional tourism and SNASPE PA authorities, in which a conceptual map was developed to help understand the current structure and relations between nature-based tourism and PA stakeholders, and document the dynamics and challenges associated with nature-based tourism governance in and around SNASPE PAs within the Aysén Region (Fig. 13.1).

Participants identified that, at the time of the project, the core of PA governance rested with the SNASPE, which was administered by CONAF. SNASPE governance comprised of a *Consejo Consultivo* (Advisory Council) for each PA. Most regional PA Advisory Councils were long-standing organizations, composed mainly of local ranchers, educational stakeholders, and tourism concessionaires, working within the PAs. They did not necessarily link the PAs with a broad range of PA stakeholders or with actors who were concerned with tourism development issues within the surrounding localities. Furthermore, while the basic function of the PA Advisory Council was to participate in PA planning, management, and administration, their input was considered "advice" rather than a branch of PA governance with genuine authority.

In parallel to SNASPE PA governance, a formal system of tourism governance existed, represented by the left-hand column in Fig. 13.1. Tourism governance was composed of the governmental agencies in charge of the tourism sector. The regional representative for the Chilean Ministry of Agriculture, which currently oversees CONAF, participated in this governance system, though coordination with CONAF and/or PA administrators was intermittent. The formal tourism governance system included local-level representatives of the region's three officially designated Zones of Tourist Interest (ZOIT) and representatives of the Municipal, commune-level tourism workgroup.

Finally, participants identified a number of local tourism stakeholders and service providers within the communities directly related to the SNASPE PAs. At the time of the stakeholder mapping process, these stakeholders and providers had limited interaction with and/or voice within PA administration and policy creation. Regional tourism authorities, speaking on behalf of local tourism stakeholders and service providers, expressed frustration with the current governance process, expressing a need for new mechanisms to address issues affecting tourism experience development and management associated with the PAs. The tourism sector did not feel represented within PA Advisory Councils. While they acknowledged that CONAF had realized some participatory processes within local communities—especially during management and visitor use planning processes—many perceived these processes to be more consultative or informative, rather than binding (Pacheco & Boldt, 2020).

The best-practice principles summarized in the first phase of the project carry important implications for improving the PA nature-based tourism governance structures outlined by these conceptual mapping exercises. Numerous researchers have described the problems that arise within exclusionary PA governance systems (or that only involve consultative or informative forms of stakeholder participation) including deteriorating trust between different stakeholder groups and an erosion in the legitimacy of PA planning and administration (Davies et al., 2018; Gale & Ednie, 2019; Kohl & McCool, 2016; Soliku & Schraml, 2018). Studies specific to Aysén examining stakeholder perceptions and conflict in and around PAs cite a

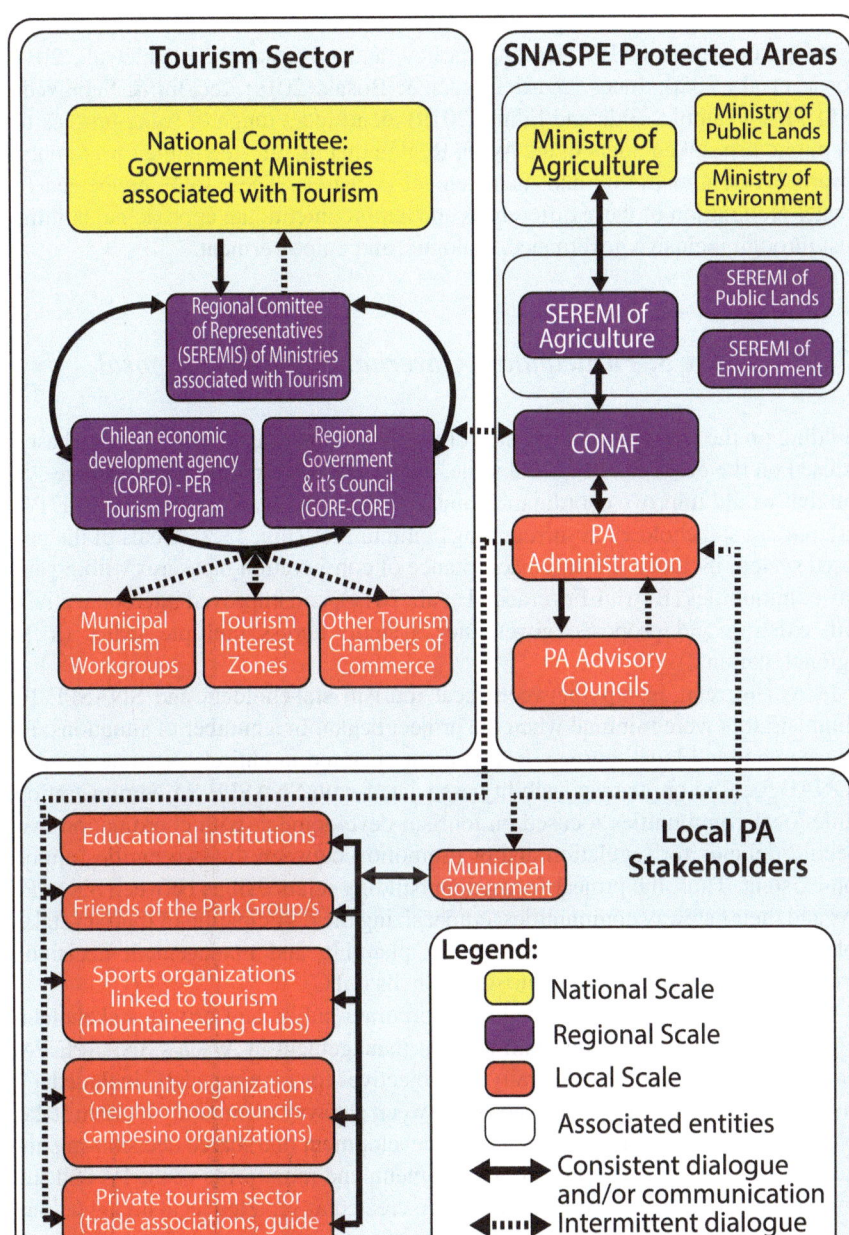

Fig. 13.1 Current structure of actors and their relations with respect to nature-based tourism and SNASPE PAs in Aysén

range of challenges, including low levels of transparency, coordination, and meaningful participation in PA planning, policy, and governance (Blair et al., 2019; Borrie et al., 2020; Jones, 2012; Louder & Bosak, 2019; Tecklin & Sepúlveda, 2014). For example, Gale and Ednie (2019) identified a range of value profiles for PA public use stakeholders in the Aysén Region that manifested in their preferences toward PA administration and management decisions. Their study advocated for greater recognition of these differences and a more intentional approach to building trust through inclusive governance, dialogue, and empowerment.

13.2.3   Phase 3: Participatory Governance Model Proposal

Building on the concepts and findings of the first two phases of the project, Phase 3 focused on the evolution of a local scale, nature-based tourism/PA governance system that would improve coordination and interaction between Aysén SNASPE PAs and tourism stakeholders in surrounding communities (Fig. 13.2). Goals of the proposed system included increased acceptance of conservation measures within gateway communities (Borrini-Feyerabend et al., 2014), and improved interconnectivity with existing and proposed governance systems across multiple scales (local, regional, national).

In Aysén, relationships between local tourism stakeholders and SNASPE PA administrators were minimal when the project began. In a number of situations, PA management and local tourism development operated in entirely separate spheres. CONAF focused on its responsibility associated with SNASPE PA administration, while local communities focused on tourism development, with coordination over specific policies for regulation and/or promotion between the two entities almost non-existent. Thus, the project focused on building relationships between SNASPE PAs and their gateway communities, emphasizing the participation of tourism stakeholders in these communities within PA planning and management decisions, through the creation of Local Tourism Councils (LTC).

The LTC concept was designed to incorporate public and private stakeholders that have a direct relationship with tourism management in Aysén's SNASPE PAs and gateway communities. Generally, the objectives for forming a LTC included: (1) establishing a collaborative relationship between gateway community tourism stakeholders and PA administration, and (2) the development of a shared decision-making platform for tourism development, management, and monitoring of the PA and surrounding area. Nevertheless, LTCs would be created as separate legal organizations, each with their own declared purposes, specific objectives, operational plans, and bylaws that may vary depending on the specific needs and situations of each PA/ gateway community group. LTC members could include PA Advisory Councils and/ or *Friends Groups*, private tourism sector organizations (trade associations, guide associations), sports organizations linked to tourism (mountaineering clubs), local schools and educational institutions, territorial organizations (neighborhood

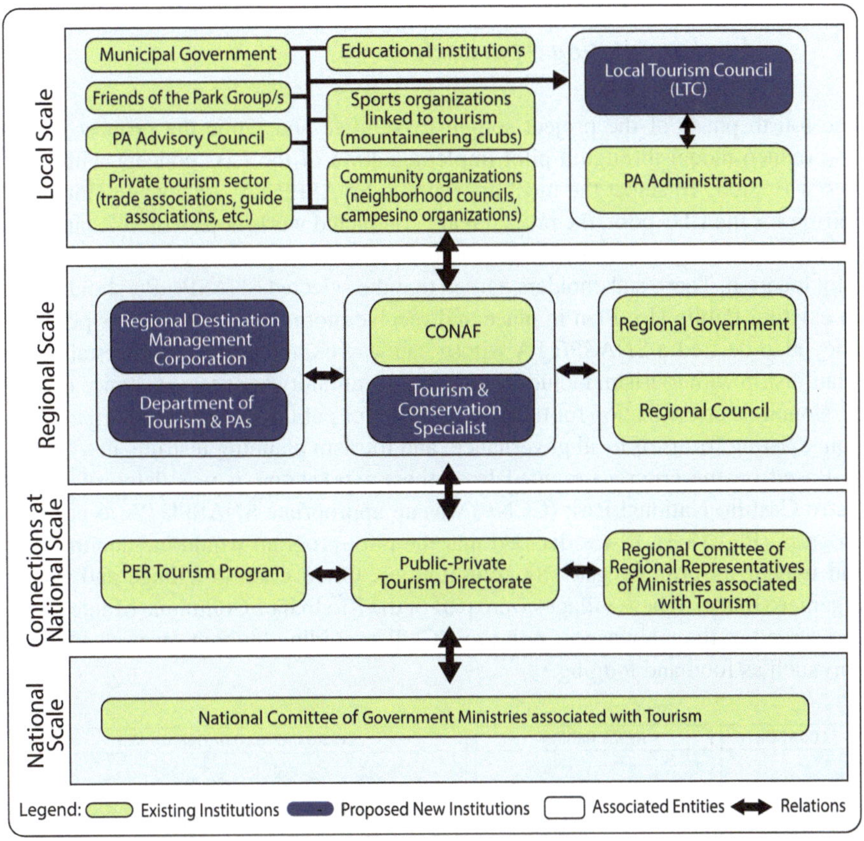

Fig. 13.2 Proposed structure at regional scale

councils, campesino organizations), Municipal commune-level governments, public service representatives, and potentially, SNASPE PA administrators.

In addition to the design of LTCs, the project also developed a set of recommendations to further strengthen public-private coordination and linkages for tourism in and around Aysén PAs at a regional scale. These included a proposal to create a new public-private regional corporation that would serve as a destination management organization (DMO) for the Aysén Region. The proposed DMO would involve the regional government and the regional council, public agencies in charge of tourism, CONAF, representatives of tourism associations, and NGOs. This regional corporation would include a regional tourism and conservation specialist from CONAF and a specialized tourism and PAs department. It would be responsible for setting general nature-based tourism guidelines and specific guidelines for tourism within regional SNASPE PAs, as well as act as a liaison between the nature-based tourism sector and regional/national authorities.

13.2.4   Phase 4: Pilot Participatory Governance Model Implementation Approach

The fourth phase of the project sought to validate and refine the proposed local governance model through a pilot implementation of the LTC concept within the Aysén Region. To select the most appropriate SNASPE PA—gateway community pairing for the LTC pilot, the project team conducted workshops with relevant public and private regional tourism stakeholders to define the selection criteria for the pilot location. These stakeholders agreed that the selected SNASPE PA should have an existing Public Use Plan in place and receive more than 2500 visitors per year. They also desired a SNASPE PA whose gateway communities had an established relationship with tourism, including gateway community services, existence of concessionaires, prioritization for tourism, priority for public investment, tourism tradition, existing forms of local governance, and tourism planning instruments.

Based on the criteria provided from these workshops, it was determined that Cerro Castillo National Park (CCNP) was an appropriate SNASPE PA to pilot the program (Fig. 13.3). It was decided that the pilot program would include the Park and two of its primary gateway communities, Cerro Castillo Village and Puerto Ingeniero Ibáñez. These villages were part of the Río Ibáñez Commune Municipality and served as the main access point to CCNP, providing tourism services and support such as food and lodging.

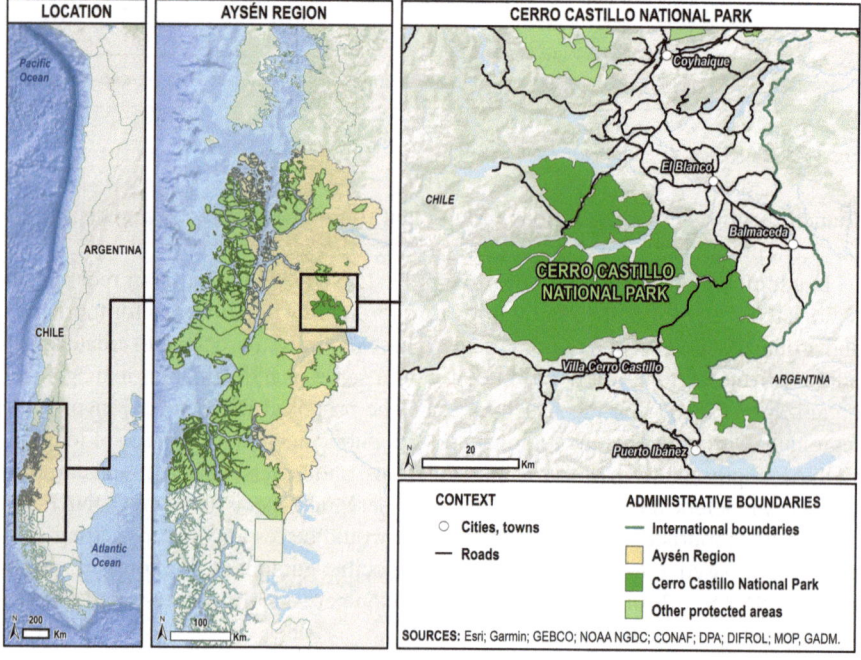

Fig. 13.3 Cerro Castillo National Park within the Aysén regional context of the National System of State Natural Protected Areas

Pilot implementation began with a series of trust-building meetings with two of the key stakeholders: CONAF and the commune-level government (Municipality of Río Ibáñez). These meetings focused on developing a better understanding of each entity's priorities and challenges, on finding common ground around tourism and protected area management issues, and on identifying key tourism stakeholders within the territory. Following these meetings, local tourism stakeholders were convened to discuss the local tourism context, tourism-park relations, and the feasibility of implementing a LTC, which was formally established on July 10, 2019, as the "Cerro Castillo National Park Local Tourism Development Corporation." Its primary objective was to promote the sustainable tourism development of CCNP and its surrounding communities improving coordination between CONAF, local organizations, and the commune government, among others. The CCNP LTC included members of the pre-existing "Local Council" of CCNP and representatives of 16 member organizations: *Cerro Castillo Andean Social and Campesino Sports Club, Aysén Climbing Club, Coyhaique Chamber of Tourism, Villa Cerro Castillo Neighborhood Council No. 3, The Roses of Villa Cerro Castillo Senior Citizens Club, Entre Cerros de Castillo Committee of Young Farmers, El Bosque Trade Association, Friends of the Cerro Castillo National Reserve Group, Farmer's Committee of Entre Cerros of Castillo, El Bosque Association of Friends of the Cerro Castillo National Reserve*, the *Farmer's Committee of Upper North Río Ibáñez Rivera Norte*, the *Bicentennial Cerro Castillo Rural High School*, the *Patagonia Aysén Regional Guides Association*, the *Andean Patagonian Mountaineering Club*, and the *Association of Mountain Ski Guides of Patagonia*.

From September 2018 to June 2019, an additional seven meetings and two workshops were held with the LTC members to finalize the LTC structure, including the types of members it included, its operating regulations, and the scope of LTC functions and commitments, which focused on tourism destination planning and management and the development of quality tourist offerings with a strong local identity. These workshops also provided the forum for creation of the LTC's operational management plan (Inostroza & Rovira, 2020).

LTC members were provided with an array of tools and information, including the guidelines of the CCNP Public Use Plan, recent academic research and literature, and other local tourism planning instruments (e.g., Tourism Development Plan for Villa Cerro Castillo; Management Plan for the Chelenko Zone of Tourist Interest [ZOIT]). A tourism supply and demand inventory was developed during this phase of the project to further inform planning efforts (Rovira et al., 2020). Informed by these inputs, stakeholder analysis identified facilitating and limiting factors associated with the local tourism system, tourism-park relations, and priorities for future LTC work. They determined that tourism's potential was growing exponentially in and around the CCNP as a result of improved regional and local access, increased national promotion of CCNP, and a recent decision to increase the park's protection category from national reserve to national park (Blair et al., 2019; Gale et al., 2018). Despite these promising developments, there were also a number of factors that were limiting local tourism, including low service quality, lack of technical preparation, and a scarcity of innovative tourism products within the gateway community of Villa Cerro Castillo (Inostroza & Rovira, 2020).

An operational management plan was assembled to address these points, and then, the COVID-19 pandemic occurred. What was initially a huge setback for tourism and the LTC project, turned out to be a major argument for the need of the LTC concept. The COVID-19 pandemic required the tourism sector to pivot with agility in order to survive. During 2020 and 2021, the CCNP LTC implemented several projects to assist the tourism community in response to COVID-19, facilitating a much faster recovery than most other gateway communities of the Aysén Region. For example, the CCNP LTC prioritized the improvement of public health and safety conditions for CCNP circuits and procured funding to implement their plans from the Chilean Development Corporation (CORFO), an institution that supports projects of economic relevance for the country. This project enabled LTC members to improve safety protocols and infrastructure within CCNP tourism routes. Tactics included the design of a website (www.rutacerrocastillo.cl) with information about the CCNP and its surrounding area, including COVID-19 protocols and public health information to help prospective visitors safely plan their trips. The project also facilitated the design of an information kiosk, tourist signage, and the construction of four sanitary station prototypes inside the CCNP.

Another CCNP LTC project, financed by Chile's Technical Cooperation Service (SERCOTEC) within the Ministry of Economy, Development and Tourism, enabled the purchase of sanitation equipment and materials for local tourism microentrepreneurs and the LTC offices. It provided training on sanitary protocols for tourism entrepreneurs, facilitated development of linkage actions with the CCNP related to tourism sector recovery, and funded the development of a CCNP LTC logo and press kit. As COVID-19 recovery has continued during 2022, the CCNP LTC has maintained its efforts with new actions that bode well for future impact, including the hiring of a council manager and the establishment of regular meetings.

Thus far, outcomes of the CCNP pilot project appear to support the proposed LTC governance format. The pilot program has shown it has the capacity to improve local public-private coordination between PA administrators, Municipal Governments, and local tourism stakeholders in PA gateway communities in the Aysén Region of Chile. Tourism stakeholders and local authorities within the CCNP gateway communities supported the creation of the LTC and agreed that the LTC system would contribute to local development for the inhabitants surrounding CCNP. Furthermore, both CCNP administrators and members of the regional CONAF PA management team actively participated in the LTC formation process, expressing their confidence about the potential for increased community involvement to yield better relations and coordination between the CCNP and its gateway communities. Importantly, creation of the CCNP LTC verified the legal feasibility and compatibility of this type of organization within the existing institutional framework of Aysén. And finally, the sustained interest and participation of tourism stakeholders, local authorities, and CONAF, suggests potential for this approach to persist over time.

The following section outlines some of the key lessons learned during the LTC pilot program that enable successful collaborative local governance of tourism in and around SNASPE PAs.

13.3   Lessons Learned

13.3.1   Lesson 1: Shared Commitment to Collaborative Governance

Shared understanding of the importance of collaborative governance for local tourism within and around the CCNP was an enabling factor for the success of the CCNP LTC pilot initiative. CONAF's CCNP Public Use Plan (2017) prioritized communication with local stakeholders during its design and implementation, calling for the development of a collaborative governance model that would include local entities and achieve a greater integration with other territorial management initiatives and groups (Chilean National Forestry Corporation, 2017; Gale et al., 2018). CONAF's formal acknowledgment of its commitment to work with local tourism stakeholders and authorities to create and manage public use inside the park and to support sustainable tourism development of the surrounding area set the tone for advances within the pilot LTC territory. As a result, there was a willingness on the part of LTC participants to align aspects of their mission and objectives with the CCNP Public Use Plan and other tourism planning instruments for the territory surrounding the park, including the Tourism Development Plan of Villa Cerro Castillo and similar Municipal tourism development plans (Inostroza & Rovira, 2020). This alignment of goals has continued through the CCNP LTC training, technical assistance, and promotion initiatives developed during 2020 and 2021, and it has been reinforced through the coordinated management that has occurred with different stakeholders—particularly CONAF and the Municipality of Río Ibáñez—who have both repeatedly expressed the importance of the local community—CCNP relationship through their words and actions.

13.3.2   Lesson 2: Obtaining Legal Status for the Local Tourism Council

Another enabling factor for the success of the CCNP LTC was the ability of the council to obtain a legal status within Chile. This was critical for the proposal because it recognized the council as a legitimate authority on tourism within the territory, giving the LTC the capacity to coordinate with other tourism and territorial actors and manage public and private resources through formal competitive projects (grants). Obtaining this legal status was largely because of the council's creation under the auspices of the Municipality of Río Ibáñez, whose active involvement and sponsorship from the beginning of the process provided the CCNP LTC with legal backing and political support for its plans and actions. Although this legal authority did not extend to actions within SNASPE PA (e.g., CCNP), the Municipality's support facilitated coordination with CONAF and improved the potential of the LTC to influence and/or form binding agreements for PA planning and management decisions.

13.3.3   Lesson 3: Structure of the Organization

The organizational structure employed for the CCNP LTC was a third enabling factor for the success observed to date; specifically, the decision to employ a technical resource person who serves as the corporation's manager. This person is in charge of managing resources, formulating projects, and coordinating with different stakeholders. Having this resource has enabled the LTC to maintain proactivity within the territory and enhance communications and articulation between stakeholders, the LTC, and the rest of the existing tourism governance structure. The CCNP LTC structure aligns well with the existing tourism governance system, strengthening local capacity and coordination. In turn, this enhances articulation between the local commune-level government (Municipality), the regional government (regional branch of the National Tourism Service, SERNATUR), and the national government (Chilean Subsecretary of Tourism) that has the potential to lead to concrete local outcomes with regional/national support. For example, recognition of the potential for strengthening tourism at the local level has resulted in support throughout the tourism governance system around the implementation of a new "Tourism Hosts" program that links the concept of host community with the concept of gateway community, through an alliance with the Universidad Austral de Chile (UACh).

13.3.4   Lesson 4: Collaboration with Other Civil Society Actors

It is important to highlight other collaborations that have evolved between the CCNP LTC and civil society actors as another enabling factor that has contributed to early successes. Both the Patagonia Aysén Foundation and the Balloon Latam Foundation have offered support for the Council's initiatives to increase the social capital of this territory and contribute to the continuous improvement of the services offered and the quality of the visitors' experience. All of this will likely generate better conditions for local economic development.

As mentioned above, there is an agreement between the Municipality and the Chilean Subsecretary of Tourism that has enabled the development of a tourism training and advisory program called "Tourism Hosts." This program, which has been developed in part by members of the Patagonia Campus of the Universidad

Austral de Chile, has provided small businesses and start-ups with training and advice on tourism matters and raised local community awareness about the tourism and development opportunities that exist due to their proximity to the CCNP. The Municipality and the UACh have also implemented joint collaboration agreements to support conservation and tourism initiatives, connected with the university administered Austral Patagonia Program (https://programaaustralpatagonia.cl/), which is funded through grants from the PEW Foundation. This program leads the development of the gateway community strategy in Patagonia, whose purpose is to increase the links between PAs and surrounding local towns. The program helps communities identify themselves as *gateways* in order to gain access to the economic benefits associated with local tourism services, and the ecological benefits associated with reducing pressure on the PA by transferring tourism services to local communities (e.g., camping areas, hotels, retail outlets), rather than locating them within the PA itself. Currently, the Austral Patagonia Program is supporting two important initiatives as part of this agreement: development of the gateway community's strategy, and the declaration of a Protected National Asset for the Río Ibáñez Commune.

Lastly, it is important to mention that the regional components of the project's proposal have not moved forward. These components proposed mechanisms to improve coordination among regional institutions for tourism development in the Aysén Region; specifically, the formation of a Regional Destination Management Corporation with a Tourism and PA Department, that would coordinate with CONAF, SERNATUR, the Regional Government, and its Council. While all of these entities have expressed support for this proposal, and agreement about the need for a regional coordination body to manage everything related to tourism in the Region, the creation of such an entity would involve overcoming significant legal, political, and funding challenges. Nevertheless, recent changes in the legislation governing regional governments augur more favorable conditions for the creation of bodies such as the one proposed in the coming years. This seems logical in a region that is committed to nature tourism as one of the most important activities for the regional and local economy, and which has more than 50,000 km^2 under state protection.

13.3.5 Lesson 5: The Importance of Grassroots Support

Finally, the most relevant enabling factor for the CCNP LTC has undoubtedly been the support of grassroots territorial organizations, whose participation lends credibility and ownership to the LTC process. Communication and cooperation between the organizations and their trust in the Municipal Commune government have been instrumental to early successes. Nevertheless, it is important to note that during the period of the pilot, the CCNP LTC board was chaired by the commune's mayor. Inostroza and Rovira (2020) noted that the formal leadership of the mayor in this initial stage was positive, given the important role played by the commune

Municipality throughout the process. However, they warned that continued leadership over a prolonged period could lead to negative impacts for the sustainability of the Council, as it could lead to risk of a co-optation of local interests by the political authority in office. The authors recommend continued participation and commitment from the commune-level government but suggest that, in the future, the CCNP LTC chair position be maintained by a neutral party. This would demonstrate maturity of the corporation and the empowerment of local leaders.

13.4 Closing Reflections

The project presented within this chapter was implemented to advance toward a fully functioning, participatory, multi-scale, governance, and management system for tourism in and around the SNASPE PAs of the Aysén Region of Chile. It focused on the design and pilot testing of a new governance structure that encouraged direct dialogue between public sector institutions, the private sector, academia, and the social and commercial organizations of the Aysén Region's SNASPE PA gateway communities. Such a structure would enable local communities to work with PAs and improve tourism services for visitors, both within SNASE PAs and their surrounding communities. This was founded on the belief that achieving this outcome would help stimulate local economies and improve the potential for tourism development to be compatible with the conservation of natural and cultural heritage.

Early outcomes for the CCNP and its gateway communities seem to support the potential for the LTC model of local shared governance. The pilot program achieved a number of positive results during the past few years through the efforts and enabling factors discussed in earlier sections. However, perhaps the most important impact of the project extends beyond the pilot implementation and exposes the potential for creating LTCs in other areas of the Aysén Region and Chilean Patagonia. The analysis conducted during the early phases of the project indicated a growing sentiment that the current advisory councils of SNASPE PAs are not an effective method for managing and governing tourism within PAs and their gateway communities. The advisory councils do not have a legal authority and are comprised of a diversity of actors with different interests (not all of them tourism-related) and broad territorial coverage. In contrast, the LTC approach offers a formal, locally focused institution centered around tourism development, management, and sustainability that has demonstrated positive, tangible results. The legal formation is relatively simple, as petitions and decisions about LTC legality can be decided at the local Municipality (commune level), and this avoids the greater bureaucracy and complications that accompany regional, or national organizations. With the support of the local municipal mayor and PA administrator (CONAF), there is a good chance that the LTC will be successful in its formation and petition for legal recognition. Furthermore, our project analysis indicated high levels of interest in participatory local governance among local tourism stakeholders, who expressed interest in being part of the decision-making process regarding tourism development in their territory and commitment to collaborate on developing effective shared governance mechanisms.

Acknowledgments We would like to thank Andrés Adiego, of the Centro de Investigación en Ecosistemas de la Patagonia (CIEP), and the Department of Geography and Territorial Planning, Universidad de Zaragoza, for his assistance with the development of the map for our study area (Fig. 13.3).

References

E. Aedo, Evaluación del sistema regional de turismo vinculado a áreas silvestres protegidas en Aysén [Evaluation of the regional tourism system linked to wildlife protected areas in Aysén], in *Turismo sustentable y áreas silvestres protegidas en Patagonia, Chile [Sustainable Tourism and Protected Wilderness Areas in Patagonia, Chile]*, ed. by P. Szmulewicz, (Ediciones Kultrun, 2020), pp. 74–79

F. Bello, B. Lovelock, N. Carr, Enhancing community participation in tourism planning associated with protected areas in developing countries: Lessons from Malawi. Tour. Hosp. Res. **18**(3), 309–320 (2016). https://doi.org/10.1177/1467358416647763

H. Blair, K. Bosak, T. Gale, Protected areas, tourism, and rural transition in Aysén, Chile. Sustainability (Switzerland) **11**(24), 1–22 (2019). https://doi.org/10.3390/su11247087

B. Borrie, T. Gale, K. Bosak, Privately protected areas in increasingly turbulent social contexts: Strategic roles, extent, and governance. J. Sustain. Tour. **13**(11), 2631–2648 (2020). https://doi.org/10.1080/09669582.2020.1845709

G. Borrini-Feyerabend, R. Hill, Governance for nature conservation, in *Governance and Management of Protected Areas*, ed. by G.L. Worboys, M. Lockwood, A. Kothari, S. Feary, I. Pulsford, (Editorial Universidad El Bosque, ANU Press, 2019), pp. 175–214

G. Borrini-Feyerabend, P. Bueno, T. Hay-Eddie, B. Lang, A. Rastogi, T. Sandwith, Primer on governance for protected areas, in *Governance Thematic Strand of the 2014 IUCN World Parks Congress*, (IUCN, Gland, 2014)

L. Brenner, Multi-stakeholder platforms and protected area management: Evidence from El Vizcaíno Biosphere Reserve, Mexico. Conserv. Soc. **17**(2), 147–160 (2019). https://doi.org/10.4103/cs.cs_18_63

L. Brenner, A.C. De la Vega, Participatory governance of natural protected areas: The case of El Vizcaíno biosphere reserve. Reg. Soc. **26**(59), 183–213 (2014)

C.B. Casady, K. Eriksson, R.E. Levitt, W.R. Scott, (Re)defining public-private partnerships (PPPs) in the new public governance (NPG) paradigm: An institutional maturity perspective. Public Manag. Rev. **22**(2), 161–183 (2020). https://doi.org/10.1080/14719037.2019.1577909

G. Cejudo, La nueva gestión pública. Una introducción al concepto y a la práctica [The new public management. An introduction to the concept and practice], in *Nueva gestión pública [New Public Management]*, ed. by G. Cejudo, (Siglo XXI Editores, 2011), pp. 17–47

G. Cejudo, P.J. May, H. Saetren, P. Hupe, S. Winter, Investigación sobre implementación y gobernanza: Direcciones y retos para el futuro [Implementation and governance research: Directions and challenges for the future]. Gest. y Polít. Pública **27**(1), 269–283 (2018)

L. Chang, T. Watanabe, The mutual relationship between protected areas and their local residents: The case of Qinling Zhongnanshan UNESCO global geopark, China. Environments **6**(5), 49 (2019). https://doi.org/10.3390/environments6050049

Chilean National Forestry Corporation, *Plan de uso público de la Reserva Nacional Cerro Castillo [Cerro Castillo National Reserve public use plan]* (Corporación Nacional Forestal, 2017)

T. Christensen, K. Yamamoto, S. Aoyagi, Trust in local government: Service satisfaction, culture, and demography. Adm. Soc. **52**(8), 1268–1296 (2020). https://doi.org/10.1177/0095399719897392

N. Cunill-Grau, C. Leyton, La provisión privada de servicios de protección social. Confrontando el debate anglosajón con los imperativos de América Latina [The private provision of social protection services. Confronting the Anglo-Saxon debate with Latin American imperatives]. Reforma y Democr. (66), 35–66 (2016)

I.P. Davies, R.D. Haugo, J.C. Robertson, P.S. Levin, The unequal vulnerability of communities of color to wildfire. PLoS One **13**(11), e0205825 (2018). https://doi.org/10.1371/journal.pone.0205825

DNA Expertus, *PER turismo Aysén: Marco estratégico y hoja de ruta [PER Turismo Aysén: Strategic Framework and Roadmap]* (PER Turismo, 2016). Retrieved 1 Dec 2022, from https://www.perturismoaysen.com/wp-content/uploads/2017/09/doc_resumen_ejecutivo.pdf

V. Figueira, Turismo y visitas a las áreas protegidas. Breve referencia al Portugal continental [Tourism and visits to protected areas. Brief reference to mainland Portugal]. Estud. Perspect. Tur. **20**, 1214–1232 (2011)

T. Gale, A. Ednie, Can intrinsic, instrumental, and relational value assignments inform more integrative methods of protected area conflict resolution? Exploratory findings from Aysén, Chile. J. Tour. Cult. Chang. **18**(6), 690–710 (2019). https://doi.org/10.1080/14766825.2019.1633336

T. Gale, A. Ednie, A. Adiego, A 360° approach to the conceptualization of protected area visitor use planning within the Aysén Region of Chilean Patagonia. J. Park. Recreat. Adm. **36**, 22–46 (2018). https://doi.org/10.18666/JPRA-2018-V36-I3-8371

M. Howlett, A. Kekez, O. Poocharoen, Understanding co-production as a policy tool: Integrating new public governance and comparative policy theory. J. Comp. Policy Anal.: Res. Pract. **19**, 1–15 (2017). https://doi.org/10.1080/13876988.2017.1287445

C. Hummel, D. Poursanidis, D. Orenstein, M. Elliott, M.C. Adamescu, C. Cazacu, G. Ziv, N. Chrysoulakis, J. Van der Meer, H. Hummel, Protected area management: Fusion and confusion with the ecosystem services approach. Sci. Total Environ. **651**, 2432–2443 (2019). https://doi.org/10.1016/j.scitotenv.2018.10.033

G. Inostroza, A. Rovira, Sistema de gobernanza para la gestión del turismo en áreas silvestres protegidas por el Estado, Aysén, Chile [Governance system for tourism management in state protected wilderness areas, Aysén, Chile], in *Turismo sustentable y áreas silvestres protegidas en Patagonia, Chile [Sustainable Tourism and Protected Wilderness Areas in Patagonia, Chile]*, ed. by P. Szmulewicz, (Ediciones Kultrún, 2020), pp. 113–123

M. Islam, L. Ruhanen, B. Ritchie, Adaptive co-management: A novel approach to tourism destination governance? J. Hosp. Tour. Manag. **37**, 97–106 (2017). https://doi.org/10.1016/j.jhtm.2017.10.009

C. Jones, Ecophilanthropy, neoliberal conservation, and the transformation of Chilean Patagonia's Chacabuco valley. Oceania **82**, 250–263 (2012)

J.M. Kohl, S.F. McCool, *The Future Has Other Plans: Planning Holistically to Conserve Natural and Cultural Heritage* (Fulcrum Publishing, Golden, 2016)

J.K. Koy, A.M.M. Ngonga, D.A. Wardell, Moving beyond the illusion of participation in the governance of Yangambi Biosphere Reserve (Tshopo Province, Democratic Republic of Congo). Nat. Conserv. **33**, 33–54 (2019). https://doi.org/10.3897/natureconservation.33.30781

E. Louder, K. Bosak, What the Gringos brought: Local perspectives on a private protected area in Chilean Patagonia. Conserv. Soc. **17**(2), 161–172 (2019). https://doi.org/10.4103/cs.cs_17_169

N. Lovrić, M. Lovrić, W. Konold, A Grounded Theory approach for deconstructing the role of participation in spatial planning: Insights from nature park Medvednica, Croatia. Forest Policy Econ. **87**, 20–34 (2018). https://doi.org/10.1016/j.forpol.2017.11.003

S. Maestre-Andrés, L. Calvet-Mir, E. Apostolopoulou, Unravelling stakeholder participation under conditions of neoliberal biodiversity governance in Catalonia, Spain. Environ. Plan. C: Politics Space **36**(7), 1299–1318 (2018). https://doi.org/10.1177/2399654417753624

K. Major, D. Smith, A.B. Migliano, Co-managers or co-residents? Indigenous peoples' participation in the management of protected areas: A case study of the Agta in the Philippines. Hum. Ecol. **46**, 485–495 (2018). https://doi.org/10.1007/s10745-018-0007-x

G. Mardones, El aislamiento social de la conservación de la naturaleza en el bosque templado del sur de Chile. Caso de estudio: Parque Nacional Alerce Andino y Reserva Nacional Llanquihue [The social isolation of nature conservation in the temperate forest of southern Chile. Case study: Alerce Andino National Park and Llanquihue National Reserve]. CUHSO Cult.-Hombre-Soc. **28**(2), 141–169 (2018). https://doi.org/10.7770/0719-2789.2018.cuhso.05.a03

C.C. Maretti, A.R. Leão, A.P. Prates, E. Simões, R.B.A. Silva, K.T. Ribeiro, et al., Marine and coastal protected and conserved areas strategy in Brazil: Context, lessons, challenges, finance, participation, new management models, and first results. Aquat. Conserv. Mar. Freshwat. Ecosyst. **29**(S2), 44–70 (2019). https://doi.org/10.1002/aqc.3169

P. Matson, W.C. Clark, K. Andersson, *Pursuing Sustainability: A Guide to the Science and Practice* (Princeton University Press, Princeton, 2016)

R. Mayntz, Nuevos desafíos de la teoría de la gobernanza [New challenges of governance theory], in *La gobernanza hoy: 10 textos de referencia [Governance Today: 10 Reference Texts]*, ed. by A. Cerrillo, (INAP, 2005), pp. 83–98

K. Niedziałkowski, E. Komar, A. Pietrzyk-Kaszyńska, A. Olszańska, M. Grodzińska-Jurczak, Discourses on public participation in protected areas governance: Application of Q methodology in Poland. Ecol. Econ. **145**, 401–409 (2018). https://doi.org/10.1016/j.ecolecon.2017.11.018

S.P. Osborne, The (new) public governance: A suitable case for treatment, in *The New Public Governance? Emerging Perspectives on the Theory and Practice of Public Governance*, ed. by S.P. Osborne, (Routledge, London, 2010), pp. 1–16

E. Ostrom, *La gobernanza de los bienes comunes. La evolución de las instituciones de acción colectiva* (C. Iturbe, & A. Sandoval, Trans.) *[Governing the commons. The evolution of institutions for collective action]* (Universidad Nacional Autónoma de México, 2000). Original work published 1990

G. Pacheco, *Indicadores de paisaje a escala humana: Elementos para la elaboración de un modelo teórico metodológico [Human Scale Landscape Indicators: Elements for the Development of a Theoretical and Methodological Model]* (CEDER, Universidad de Los Lagos, 2014)

G. Pacheco, Turismo comunitario y procesos de gobernanza en Chile: Un análisis comparativo con experiencias brasileñas [Community-based tourism and governance processes in Chile: A comparative analysis with Brazilian experiences]. Gest. Tur. **30**, 54–85 (2018). https://doi.org/10.4206/gest.tur.2018.n30-03

G. Pacheco, J. Boldt, Gestión turística en las Áreas Silvestres Protegidas (ASP) en la Región de Aysén, Patagonia, Chile [Tourism management in wildlife protected areas (ASP) in the Aysén Region, Patagonia, Chile], in *Turismo sustentable y áreas silvestres protegidas en Patagonia, Chile [Sustainable Tourism and Protected Wilderness Areas in Patagonia, Chile]*, ed. by P. Szmulewicz, (Ediciones Kultrun, 2020), pp. 63–74

G. Pacheco, C. Henríquez, El turismo de base comunitaria y los procesos de gobernanza en la comuna de Panguipulli, sur de Chile [Community-based tourism and governance processes in the commune of Panguipulli, southern Chile]. Gest. Tur. **25**, 42–62 (2018). https://doi.org/10.4206/gest.tur.2016.n25-03

G. Pacheco, P. Szmulewicz, Sinergias y conflictos entre el desarrollo turístico y otros sectores económicos. El caso del turismo de intereses especiales en la región de Los Ríos [Synergies and conflicts between tourism development and other economic sectors. The case of special interest tourism in the Los Ríos region]. Gest. Tur. (20), 39–59 (2013). https://doi.org/10.4206/gest.tur.2013.n20-03

G. Pacheco, J. Vera, J.C. Castaing, La gestión de destinos en la Región de Los Lagos-Patagonia Chilena ¿Una disputa entre asociatividad y competitividad? [Destination management in the Los Lagos Region, Chilean Patagonia. A dispute between associativity and competitiveness?]. Rev. Interam. Ambient. Tur. **11**(2), 148–162 (2015). https://doi.org/10.4067/317

R.A.W. Rhodes, Recovering the craft of public administration. Public Adm. Rev. **76**(4), 638–647 (2015). https://doi.org/10.1111/puar.12504

A. Rovira, C. Salas, G. Pacheco, El rol de la presencia de un área silvestre protegida en el desarrollo local. Un estudio de caso en la Región de Aysén, Patagonia chilena [The role of the presence of a protected wilderness area in local development. A case study in the Aysén Region, Chilean Patagonia]. Terra. Rev. Desarrollo Local (7), 48–71 (2020). https://doi.org/10.7203/terra.7.17417

C. Sanz, A.J. Torres, Gobernabilidad en las áreas protegidas y participación ciudadana [Governance in protected areas and citizen participation]. Papers **82**, 141–161 (2006). https:// doi.org/10.5565/rev/papers.2053

O. Soliku, U. Schraml, Making sense of protected area conflicts and management approaches: A review of causes, contexts and conflict management strategies. Biol. Conserv. **222**, 136–145 (2018). https://doi.org/10.1016/j.biocon.2018.04.011

S. Stoll-Kleemann, A. De la Vega-Leinert, L. Schultz, The role of community participation in the effectiveness of UNESCO biosphere reserve management: Evidence and reflections from two parallel global surveys. Environ. Conserv. **37**(3), 227–238 (2010). https://doi.org/10.1017/ S037689291000038X

P. Szmulewicz, E. Aedo, Desafíos de la gestión del turismo en las áreas silvestres protegidas de la Región de Aysén, Patagonia, Chile [Challenges of tourism management in the protected wildlife areas of the Aysén Region, Patagonia, Chile], in *Turismo sustentable y áreas silvestres protegidas en Patagonia, Chile [Sustainable Tourism and Protected Wild Areas in Patagonia, Chile]*, ed. by P. Szmulewicz, (Ediciones Kultrun, 2020), pp. 7–13

D. Tecklin, C. Sepúlveda, The diverse properties of private land conservation in Chile: Growth and barriers to private protected areas in a market-friendly context. Conserv. Soc. **12**(2), 203–217 (2014). https://doi.org/10.4103/0972-4923.138422

M.L. Tseng, C. Lin, C.W.R. Lin, K.J. Wu, T. Sriphon, Ecotourism development in Thailand: Community participation leads to the value of attractions using linguistic preferences. J. Clean. Prod. **231**, 1319–1329 (2019). https://doi.org/10.1016/j.jclepro.2019.05.305

A. Urquiza, C. Amigo, M. Billi, J. Cortés, J. Labraña, Gobernanza policéntrica y problemas ambientales en el siglo 21: Desafíos de coordinación social para la distribución de recursos hídricos en Chile [Polycentric governance and environmental problems in the 21st century: Social coordination challenges for the distribution of water resources in Chile]. Pers. Y Soc. **33**(1), 133–160 (2019). https://doi.org/10.53689/pys.v33i1.258

K. Vella, N. Sipe, A. Dale, B. Taylor, Not learning from the past: Adaptive governance challenges for Australian natural resource management. Geogr. Res. **53**(4), 379–392 (2015). https://doi. org/10.1111/1745-5871.12115

G.L. Worboys, T. Trzyna, Protected area management and governance, in *Governance and Management of Protected Areas*, ed. by G.L. Worboys, M. Lockwood, A. Kothari, S. Feary, I. Pulsford, (Editorial Universidad El Bosque, ANU Press, 2019), pp. 215–262

Chapter 14
Key Resilience Factors in Four Patagonia Nature-Based Tourism Destinations in the Aysén Region of Chile

Cecilia Gutiérrez Vega, Adriano Rovira, and Pablo Szmulewicz

Abstract This chapter supports the idea that tourism destinations should be prepared for adversities and that this preparation is the responsibility of key tourism stakeholders in the territory. Key resilience factors are identified for tourism destinations associated with protected areas in Chilean Patagonia. We analyze the strategic planning aspects of four case studies: Aysén Patagonia Queulat, Coyhaique-Puerto Aysén-Cerro Castillo, Lago General Carrera, and *Provincia de los Glaciare*s (Province of the Glaciers). A measurement model was applied to assess three fundamental pillars of resilience (capabilities, ownership, and connections), incorporating natural risk assessment as a mechanism to relate resilience capacity to the territorial context. Results indicate positive evaluations for several resilience factors in three of the four tourism destinations. Nevertheless, all four destinations presented high levels of natural risks, with the *Provincia de los Glaciare*s destination as the most vulnerable. The discussion focuses on the implications for each of the study's destinations, suggesting that destinations with higher levels of natural risks should focus on strengthening their resilience factors. Thus, these destinations should develop strategies to build connections (relations with other networks), capacities (management of tools and knowledge, social and technical skills), and ownership (participation of local actors and managers in the governance of the territory).

Keywords Patagonia · Resilience · Nature-based tourism · Natural risk assessment · Vulnerability

C. Gutiérrez Vega (✉) · A. Rovira · P. Szmulewicz
Universidad Austral de Chile, Faculty of Economics and Administrative Sciences, Valdivia, Los Rios Region, Chile
e-mail: ceciliagutierrez@uach.cl; adriano.rovira@gmail.com; pszmulew@uach.cl

© The Author(s) 2023
T. Gale-Detrich et al. (eds.), *Tourism and Conservation-based Development in the Periphery*, Natural and Social Sciences of Patagonia,
https://doi.org/10.1007/978-3-031-38048-8_14

14.1 Introduction

Economic crises, natural disasters, and climate change pose complex and interrelated challenges for the planning and management of tourist destinations. Since 2019, Chile has been engulfed by crises, including social and political unrest and the COVID-19 pandemic. Resilience has emerged as a metric for gauging recovery from such crises and has been applied to a diverse range of fields, including ecology, economics, psychology, sociology, and development studies. This chapter applies the resilience perspective to four Patagonian tourism destinations in the Aysén Region of Chile to evaluate factors that influence their resilience.

Patagonian tourism destinations consist of a range of distinct characteristics, including pristine and isolated landscapes, changing natural resources, and disparate levels of human and territorial development that interact with crises in complex and uncertain ways. Tourism activity in Patagonia is founded on the use of natural resources of great international value, causing the region to be particularly sensitive and vulnerable to the ongoing social and environmental turbulence in the world. Tourism offerings within Patagonia are typically related to the natural features and characteristics of the territory's protected areas (PAs), many of which are administered within the National System of State Natural Protected Areas (SNASPE). By 2017, Patagonia's Aysén Region had the largest system of SNASPE PAs in Chile (52,000 km²), distributed in 18 territorial units, of which 5 were National Parks, 11 were National Reserves, and 2 were Natural Monuments (Aedo et al. 2020). This study compares resilience factors between tourist destinations that hold an abundance of SNASPE PAs in order to enrich our current knowledge about tourism resilience. This was done by considering how resilience factors interact with the diverse natural risks that affect tourism in these PAs and their surrounding territories. A better understanding of the strengths and vulnerabilities of Patagonian tourism destinations will help guide future planning and management strategies in the region.

This study applies a resilience measurement model to consider the effect of various natural risks on system resilience. Secondary information from the following four tourism destinations in Chilean Patagonia is analyzed, as defined by the Chilean National Tourism Service (2014): Aysén Patagonia Queulat; Coyhaique-Puerto Aysén-Cerro Castillo, Lago General Carrera, and the *Provincia de los Glaciares* (Province of the Glaciers). Secondary data consisted of documents generated by regional actors and/or organizations within the public, private, and civil sectors. Documents were analyzed for information related to resilience factors in the following areas: Economic/Touristic, Environmental, Policy/Instructional, and Sociocultural.

14.2 Theoretical Framework

The concept of resilience has been researched within a wide range of disciplines, including psychology (e.g., Scoville 1942) and biology (e.g., Rutter 1987). Yet, there is no agreed upon definition for resilience among the various researchers and disciplines, nor is it known exactly in which discipline the use of the concept began (Kalawski and Haz 2003). Nevertheless, resilience has been, and continues to be, widely applied to situations and contexts to explain the capacity to resist and react in the face of adverse events. The United Nations International Strategy for Disaster Reduction (United Nations 2009) defined resilience as:

> the ability of a system, community or society exposed to hazards to resist, absorb, accommodate to and recover from the effects of a hazard in a timely and efficient manner, including through the preservation and restoration of its essential basic structures and functions. (p. 24)

This global view of resilience, which has formed the basis for further international agreements and strategies, emphasizes the importance of strengthening capacities to respond to change and provides general guidelines for disaster risk management and reduction.

According to research conducted by the World Resources Institute (WRI) in 2009, resilience can be conceptualized as the capacity of a system to receive disturbances or alterations and recover. For humans, this means improving our capacity to learn, plan, and organize (Cuevas-Reyes 2010). Walker et al. (2006) and Folke et al. (2002) described ecological resilience as the level of disturbance that an ecosystem can absorb without crossing the threshold of a different ecosystem structure or state. Alterations which can affect an ecosystem are varied and can range from natural ones that are not controllable by humans (i.e., earthquakes, tsunamis, hurricanes) to anthropogenic alterations (e.g., logging, indiscriminate fishing, overexploitation of pastures) that are controllable (Ecoespaña and the WRI 2009).

Although resilience can be approached from a personal or individual perspective (e.g., Henderson 2007), experts working with resilience in vulnerable tourism destinations have identified that a collective resilience focus, looking at the territory as a destination, may have greater utility (Ecoespaña and WRI 2009; Tanana et al. 2019). A collective approach may facilitate efforts to strengthen weak areas and improve preparedness for natural, social, health, economic, and political disturbances, among others. Furthermore, Tanana et al. (2019) recommend a shared focus that combines an understanding of vulnerability with prevention management, training, and awareness building with tourism destination stakeholders, visitors, and the local community.

In a tourism context, resilience has generally been defined in terms of the options and response capacity of vulnerable subsectors of the tourism industry to cope with shocks and changes generated at the local, regional, and global levels (Biggs et al. 2012; Hall et al. 2003; Henderson 2007; Kontogeorgopoulos 1999; Ritchie and Crouch 2003). Tourism resilience research contributes to a better understanding of how the tourism industry and its businesses might effectively respond and positively adapt in the face of global disruptions, disturbances, or changes (Farrell and Twining-Ward 2004; Tyrrell and Johnston 2008).

Although there is a significant body of literature that has studied the societal resilience (e.g., Biggs et al. 2012; De Sausmarez 2007; Farrell and Twining-Ward 2004; Smith and Henderson 2008; Strickland-Munro et al. 2010), its applications to tourism systems are more recent and exploratory in nature, with less precision for tourism-specific conceptual models and tools (Chang 2009; Farrell and Twining-Ward 2004; Plummer and Armitage 2007; Stadel 2008). The most recent works focus on measuring resilience in destinations affected by natural disasters in order to facilitate recovery (Gutiérrez 2013; Miller et al. 2017; Min et al. 2020; Nakanishi et al. 2014). This is supplemented by resilience research in other fields, including planning (Becken and Khazai 2017; Gutiérrez 2013; Holladay and Powell 2016), business management (Biggs et al. 2012; Calgaro et al. 2014; Guo et al. 2018), and destination development (Farrell and Twining-Ward 2004; McKercher and Young 1999), among others.

14.2.1 Resilience Factors

Measuring resilience requires the definition of resilience factors. Carpenter et al. (2012) synthesized six enabling conditions for resilience from the literature: *diversity*, which includes the breadth of reactions to changes or shocks, cultural diversity, and the heterogeneity of socio-ecological systems in the landscape; *modularity*, which includes different peoples' problem-solving approaches and organizational diversity; *openness*, which includes the strength of connections between socio-ecological systems; *reserves*, which include the capacities to re-mobilize system features that have been lost through disturbances, funding, recolonization, or social memory; *feedback*, which involves how ecosystems are enriched and/or networks of economic transactions; and *nesting*, which involves cross-scale governance systems made up of municipalities, provinces, and regions (Walker and Salt 2012; Levin 1999; Biggs et al. 2012). More recently, Herrera and Rodríguez (2016) proposed a series of factors for evaluating territorial resilience. Their matrix offered qualitative and quantitative factors within four dimensions (*economic, social, institutional*, and *infrastructure*). For each factor, they defined indicators and the impact they had on resilience.

Varghese et al. (2006) contended that levels of community resilience are affected by the degree of *local ownership* within the community, which they defined as including local autonomy and power, local flexibility in decision-making, and the distribution of local-level benefits. They pointed out that the extent of local ownership within a community affects the level of commitment and involvement of both local and external groups, and the forms of local ownership influenced their decision-making processes. Generally, the greater the commitment and involvement of workers, managers, and community members, the greater the possibility of setting goals that support local job creation, community programs, and long-term business viability.

In 2009, Ecoespaña and the WRI published a report which, among other things, supported the Varghese et al. (2006) concept of *local ownership*, as a community resilience factor. The Ecoespaña and the WRI (2009) report identified three key community resilience factors. First, they posed that *capacities* are based on the management of tools and knowledge that enable sustainable resource development. Capacities are linked to social, technical, and entrepreneurial skills used to manage resources and create enterprises and may include support skills that help build capacity and influence. The second resilience factor proposed in the 2009 report involved *ownership*, which was linked to the strong involvement of local development actors and managers and an enabling environment, including favorable public policies, a non-discriminatory fiscal and regulatory environment, and the commitment of government agencies. The third factor was *connections*, which were related to being articulated with others, not only horizontally but also vertically, in all areas in which a company/organization can be linked (i.e., both public and private spheres), to improve access to learning, support, and commercial networks and associations.

Additionally, several tools have been developed for measuring community resilience. For example, the international humanitarian organization, GOAL (2015), developed a tool to measure the components of community resilience to disasters. This tool grouped resilience components into five thematic areas: *governance, risk assessment, knowledge and education, risk management and vulnerability reduction*, and *disaster preparedness and response*. And the United Nations Office for Disaster Risk Reduction (2017a, b, 2021) has provided a series of self-assessment tools and local urban indicators to address the concept of resilient cities. Their resilience self-assessment tools address ten essential aspects: organizing for resilience, identifying risks, strengthening financial capacity for resilience, promoting resilient urban design and development, protecting natural buffer zones, strengthening institutional capacity, understanding and strengthening social capacity, increasing the resilience of vital infrastructure, ensuring an effective response to disasters, and accelerating the recovery process/building back better.

While all of these frameworks and tools are helpful and increase understanding of resilience in a range of contexts, their primary focus is on resilience elements that are internal to the territories, with little to no consideration for external elements that may also affect it. Thus, their conceptual use in tourism may be limited, as the tourism sector incorporates physical, environmental, human, governmental, internal, and external elements, among others. Therefore, Gutiérrez (2013) built on prior work, especially the community resilience factors proposed by Ecoespaña and the WRI (2009), integrating these factors with relevant tourism literature (e.g., Becken and Khazai 2017; Biggs et al. 2012; Calgaro et al. 2014; Farrell and Twining-Ward 2004; Guo et al. 2018; Holladay and Powell 2016; McKercher and Young 1999; Pearce et al. 2016; Varghese et al. 2006). Further, Gutiérrez (2013) identified four spheres of action particularly relevant to tourism destinations (economic—touristic, environmental, political-institutional, and sociocultural), and tourism-specific indicators for each of the three resilience community resilience factor domains (capabilities, ownership, and connections).

Within the Gutiérrez (2013) model, the economic-touristic sphere of action focused on tourism within the local economy, tourism dynamics, public and private investment, and tourism equipment and infrastructure. The environmental sphere of action focused on understanding and managing the environmental repercussions and impacts of tourism and of disasters and crises that could affect the environment and nature-based tourism destinations. The political-institutional sphere of action included aspects of tourism governance, institutional linkage, and crisis/disaster management. Finally, the sociocultural sphere of involved actions related to local communities and tourism, including employment, education, training, well-being, and the repercussions of any negative impacts.

14.2.2 Risks and Tourism

Recovery capacity varies depending on the degree of disturbance to which a territory is subjected. For example, disaster resilience evaluates the potential for disasters within a territory and the capacity for measures to protect and enhance infrastructure performance, thus reducing potential losses when extreme events occur (Tierney and Bruneau 2007). According to research developed by Blake and Sinclair (2003), the global tourism industry involves the movement of travelers between territories around the world; as such, the diversity of potential risks and disasters is unlimited. And, the contribution of tourism is so important, both for industrialized and developing countries, that any crisis that affects aspects of social life should be considered a concern by the tourism sector. Nevertheless, they found tourism resilience planning measures to be lacking at the time of their research, pointing out that the industry's general response to declining tourism demand was to pressure governments to implement policies to counteract the crisis, rather than taking preemptive or responsive action to improve resilience on its own.

The 2020 version of the Global Risks Report, published by the World Economic Forum (WEF), indicated that global economic risk, which dominated public discourse during the 2007–2015 period, had been overshadowed by environmental risks in recent years (2016–2020), with an emphasis on events related to climate change and natural disasters (e.g., tornadoes, floods, forest fires, droughts, and heat waves). Additionally, they noted that cyberattacks have caused severe damage around the world in recent years. In 2021, the WEF published follow-up research that specifically addressed global tourism resiliency, adding scenarios that reflected a greater long-term impact, including infectious diseases, failure in climate action, weapons of mass destruction, and the loss of biodiversity (Al-Khateeb 2021).

Many risk and crisis researchers have suggested that the tourism sector is particularly prone to disturbance because its permeability, dynamism, and dependence on other sectors of the economy make it especially vulnerable to crisis (e.g., Goeldner and Ritchie 2009; Lerbinger 1997; McKercher and Young 1999; Murphy and Price 2005; Muñiz and Brea 2010; Pennington-Gray et al. 2014; Richardson 1994). Murphy and Price (2005) noted that tourism development typically occurs in places with higher potential for natural and/or social risk, calling for greater consideration of this trend and the establishment of appropriate and timely information and security alerts. Muñiz and Brea (2010) observed that tourism was becoming more and more technologically driven and, as a result, tourism communication, information, and reservation systems were becoming more sophisticated, with greater automation and reliance on technology; yet, these required greater knowledge and responsibility on the part of travelers, making them more fragile and vulnerable. Goeldner and Ritchie (2009) posed that the nature of tourism, as a cross-cutting sector that influenced—and was influenced by—many other sectors, made it susceptible to a varied set of external factors that could generate a crisis. Henderson (2007) described the vulnerabilities associated with the fragmented nature of the tourism sector, writing: "a structure that offers products related to the experience and that are the result of the joint work of several suppliers who must face various problems of fragmentation and control" (p. 8).

Conversely, other researchers have noted that tourism's connection to other sectors of the economy does not necessarily weaken the industry (e.g., Glaesser 2003; Henderson 2007; Pennington-Gray et al. 2014; Pike 2004). For example, Pike (2004) observed that the tourism sector has proven itself to be resilient, through frequent tests that have demonstrated its capacity for quick recovery in the face of crisis, in many cases, much faster than other sectors. And after the 2010 earthquakes and tsunamis that affected destinations throughout Chile, Pennington-Gray et al. (2014) identified that the development of ongoing co-management approaches that would focus communities and destinations on bringing the diverse groups of actors involved in the tourism sector together through increased linkages, planning, and private-public coordination could contribute to increased resilience and effective disaster response/recovery. Glaesser (2003) and Henderson (2007) proposed a matrix that would help destinations identify the types of events that could affect tourism and analyze their possible impact, based on the level of surprise and the degree of control related to each event. According to these authors, greater levels of anticipation and control over events or shocks for tourism operators, tourists, and authorities would improve destination resilience (Glaesser 2003; Henderson 2007).

In recent years, territorial risk detention and reduction has advanced considerably as a result of the widespread adoption of the Sendai Framework for Disaster Risk Reduction 2015–2030 around the world. This framework is aligned with other global-level 2030 frameworks such as the Paris Agreement on Climate Change and the Sustainable Development Goals (United Nations 2015). The Sendai Framework offers several helpful concepts for evaluating and managing disaster risks and their potential impacts on the tourism sector. For example, it contains a roadmap of concrete actions to support its position about the primary role central governments should play in disaster risk reduction and the shared responsibility held by other public and private stakeholders—including local governments—for territorial assessment and resilience building (United Nations Office for Disaster Risk Reduction 2017b). One of the concrete tools that has been developed to assist with these efforts is the Quick Risk Estimation, which facilitates the identification and understanding of current and future risks, and exposure threats for human and physical assets (United Nations Office for Disaster Risk Reduction 2017a, b, 2021). This tool is notable as it employs citizen participation to identify risks and hazards associated with specific locations in smaller territories; thus, it may offer an interesting perspective for tourism destination risk evaluation efforts. And, while the QRD does not support large-scale risk assessment, the hazard indicators included in the QRE tool are aligned with the Sendai Framework for Disaster Risk Reduction 2015–2030 and the Sustainable Development Goals (United Nations Office for Disaster Risk Reduction 2017a, 2021).

14.3   Materials and Methods

14.3.1   Study Area

This chapter measures and compares key resilience factors in nature-based tourism destinations in the Aysén Region of Chilean Patagonia. The Aysén Region is one of the most remote areas of the Chilean territory, with a great wealth of natural resources and climatic variety, including Patagonian pampas, evergreen and deciduous forests, lakes, rivers, fjords, and glaciers. It is the region with the largest amount of fresh water in Chile and has the third largest extension of continental ice in the world. At the time of this research, approximately 50% of the territory was encompassed within SNASPE and managed under a range of Chilean mandates, including seven National Parks, eight National Reserves, and two Natural Monuments (Aedo et al. 2020). There were also several other forms of protected lands within the region, including private protected areas (PPAs), nature sanctuaries, national monuments, national conservation areas, marine protected areas, and municipal parks (Aedo et al. 2020).

In 2019, prior to a period of national social unrest and the subsequent COVID-19 pandemic, tourism in Aysén was at an all-time high, with a reported 217,711 tourist arrivals during the peak months of December 2018 through February 2019 (Chilean National Tourism Service 2019). For this same time period, the 11 SNASPE PAs that manage their entrances reported 88,158 visitors (Chilean National Tourism Service 2019). All four of the regional sub destinations chosen for the study focused their tourism offerings on nature and adventure, with PAs serving as important local attractions and settings. Each of the destinations had relevant strategic tourism planning instruments that guided tourism development, marketing, management, and governance (Pearce et al. 2016). Additionally, SNASPE PAs had relevant planning documents, including General Management Plans and Public Use Plans (Gale et al. 2018), and the Chilean National Emergency Office of the Ministry of the Interior (ONEMI) had designed various plans to deal with risk and disaster situations. Further, there was even a Regional Plan for Disaster Risk Reduction (Chilean National Emergency Office of the Ministry of the Interior 2018).

The Aysén Region of Chilean Patagonia can be divided into four sub destinations, which comprised the study area for the research presented in this chapter: Aysén Patagonia Queulat, Coyhaique-Puerto Aysén-Cerro Castillo, Lago General Carrera, and the *Provincia de los Glaciares* (Province of the Glaciers). The Aysén Patagonia Queulat sub destination is one of three officially declared Zones of

Interest for Tourism (ZOITs), along with Lago General Carrera and the Provincia de los Glaciares. ZOITs are officially designated territories within Chile, recognized for having special conditions that make them especially attractive for tourism. Along with other planning and management requirements, having a ZOIT designation implies that tourism development within the territory will be managed in a participatory manner (Chilean National Tourism Service 2017). Two of the sub destinations, Aysén Patagonia Queulat and the Provincia de los Glaciares, are located in more remote areas of the Aysén Region, while the other two, Coyhaique-Puerto Aysén-Cerro Castillo and Lago General Carrera, are closer to urban centers.

The Aysén Patagonia Queulat ZOIT is located within the northern part of the Aysén Region and includes several PAs (i.e., Queulat National Park, Melimoyu National Park, Isla Magdalena National Park, Lago Las Torres National Reserve, Lago Carlota National Reserve, and Lago Rosselot National Reserve), distributed within three communes: Guaitecas, Cisnes, and Lago Verde. The Coyhaique-Puerto Aysén-Cerro Castillo sub destination represents the most developed transect of the Aysén Region, including the regional capital of Coyhaique and the second largest city in the region, Puerto Aysén. PAs in this sub destination include large sections of the Cerro Castillo National Park, Río Simpson National Reserve, Coyhaique National Reserve, Dos Lagunas National Monument, and a number of private reserves and urban parks. The Lago General Carrera sub destination is also a ZOIT, under the name of Chelenko. The Chelenko ZOIT is located in the central zone of the Aysén Region, in the territory surrounding Lago General Carrera, which includes large sections of Cerro Castillo National Park, San Rafael Lagoon National Park, and the Jeinimeni Sector of Patagonia National Park, distributed within the Río Ibáñez and Chile Chico communes (Chilean National Tourism Service and Guazzinni Consultores 2017). The final sub destination in the study, located in the southernmost reaches of the Aysén Region, is the Provincia de los Glaciares, also designated as a ZOIT. This sub destination includes much of San Rafael Lagoon National Park, Patagonia National Park, and Bernardo O'Higgins National Park, distributed within three communes, Cochrane, Tortel, and O'Higgins (Chilean Subsecretary of Tourism 2017).

The Aysén Region and its sub destinations are exposed to a number of natural hazards including seismic (tectonism, mass movements associated with the presence of the Andes Mountains), tidal waves or tsunamis, volcanic eruptions, hydrometeorological events (i.e., river flooding, river erosion, snowfall, alluvium), and forest fires (Chilean National Emergency Office of the Ministry of the Interior 2018). The region has also experienced natural disasters of great magnitude, including the forest fires that occurred in the Guaitecas islands in 1998, the eruptions of the Hudson volcano in 1971, 1991, and 2011, forest fires in the sector of La Junta in 1996 affecting 50 km^2, forest fires in the sector of La Tapera in 2007 affecting 70 km^2, forest fires in the Pallavicini sector in 2009 affecting 80 km^2 in Chile and 120 km^2 in Argentina, a frontal system bringing blizzards to much of the region in 2010, and forest fires in Coyhaique, Balmaceda, and Puerto Chacabuco in 2014.

14.3.2   Evaluation Procedures for Resilience Factors and Risk

The research employed an exploratory, mixed-methods, descriptive approach to evaluate resilience factors and risk for the four Aysén tourism sub destinations mentioned: Aysén Patagonia Queulat, Coyhaique-Puerto Aysén-Cerro Castillo, Lago General Carrera, and the Provincia de los Glaciares.

For both resilience factors and risk, the research team identified documents containing information about plans, programs, natural and anthropogenic hazards and phenomena, and initiatives that could affect tourism resilience in each of the four sub destinations. Criteria used for the search included that the documents be available in digital format, accessible to the public, and published between the year 2000 and the time of the study. The final collection of 28 articles was authored by public institutions, business organizations, NGOs, universities, and national and regional research centers.

For the evaluation of tourism resilience factors, document analysis involved a systematic search of the texts and figures to identify content and imagery related to the Gutiérrez (2013) matrix, which included the Ecoespaña and the WRI (2009) resilience factors: *capacities*, *ownership*, and *connections* and four domains or spheres of action: *Economic/Touristic*, *Environmental*, *Political/Institutional*, and *Sociocultural* (Fig. 14.1).

When questions or doubts arose for the team, they contacted tourism, land planning, and risk/disaster experts within the Aysén Region and/or sub destination to clarify through guided discussions. This iterative approach of secondary and primary data collection continued until the research team was satisfied that they were informed adequately and could conduct a qualitative evaluation, based on rankings, for each of the sub destinations. This evaluation involved coding each of the factors and criteria with respect to three areas: (1) whether or not it was *observed* in the data, (2) their understanding of its *impact* on destination resilience (positive or negative), and (3) its *relevance* for destination recovery. Each member of the team coded individually, and then triangulation was conducted to arrive at a consensus around the final scores.

Natural and anthropogenic hazards were also identified and ranked according to *likelihood* and *severity*, using the definitions and process outlined by the QRE (United Nations Office for Disaster Risk Reduction 2017a, 2021). For example, the likelihood was assessed in terms of the potential for an event to occur as a result of a hazard, based on existing hazards, trends, and historical events, in comparison to other potential events within the same territory. Decisions were influenced by the team's understanding of the territory's exposure and vulnerability, as well as the current prevention measures and actions in place (scale of 1–10). Severity was rated according to the perceived impact and consequence that each of the hazards could have on the territory and sub destination (scale of 0–100). Lower scores were associated with decreased likelihood and severity, while higher scores were associated with increased likelihood and severity and therefore called for a more substantial response (United Nations Office for Disaster Risk Reduction 2017a, 2021).

Economic - Touristic Sphere of Action

CAPACITIES
- Tourist activities
- Public transport
- Tourist support services
- Use of Social Networks
- Contribution of tourism to income
- Contribution of tourism to employment
- Contribution to other economic activities
- Seasonality of tourism employment
- Tourism Signage

OWNERSHIP
- Tourist accommodation companies
- Gastronomic services companies
- Profitability of tourism companies
- Sports activities on offer
- Tourist excursions on offer

CONNECTIONS
- Communication channels and services
- Access to the destination and tourism sites
- Communication technologies

Environmental Sphere of Action

CAPACITIES
- Vehicle saturation capacity
- Destination cleanliness / ornamentation
- Contribution of tourism to noise quality
- Equipment and security of tourist sites
- Natural Disaster Prevention Programs

OWNERSHIP
- Potential tourist resources
- Offer of natural tourist attractions
- Trials and popular participation in the face of natural disasters

CONNECTIONS
- Infrastructure recovery programs
- Presence of environmental institutions
- Environmental actions
- Environmental awareness
- Tsunami warning mechanisms

Political - Institutional Sphere of Action

CAPACITIES
- Tourism planning tools
- Municipal tourism management
- Tourism management by business sector
- Investment actions in tourism promotion
- Local planning tools
- Tourism plans and programs
- Municipal tourism ordinances
- Programs to promote tourism entrepreneurship
- Public financial incentives in tourism
- Disaster Recovery Programs

OWNERSHIP
- Effectiveness of public financial incentives
- Team of professionals and technicians
- Level of tourism development
- Participation in tourism planning design and implementation

CONNECTIONS
- Technicians/professionals
- Tourism organizations
- Public sector institutions working in tourism
- Municipal management accountability
- Public/private sector coordination in tourism
- Municipal alliances for tourism development
- Liaison with local authorities

Socio-Cultural Sphere of Action

CAPACITIES
- Public health services for tourist use
- Quality customer service
- Community disposition towards tourism
- Contribution to curbing youth migration
- Contribution of tourism to overall image

OWNERSHIP
- Cultural tourist attractions
- Tourism training programs
- Education and training of workers
- Training and capacity building of entrepreneurs
- Impacts of tourism on the local population
- Recovery and promotion of culture/ heritage
- Prevention activity

Fig. 14.1 Resilience indicators by spheres of action for each of the three factors considered in the study

14.3.3 Analysis

To analyze the tourism destination resilience data, each member of the research team assigned a value from one to five for each indicator and sub destination, along a scale of one to five, based on the amount of evidence and clarity with which each indicator was addressed. Indicators that were not mentioned within any of the documents received a score of one. Those which received only casual mention within one or more of the documents received a score of two. Indicators that were explicitly addressed within one or more of the documents with a concrete mention within larger strategies or tactics received a three. Indicators for which there were specialized technical plans and strategies received a four. Finally, indicators with specialized technical plans and strategies that also had established monitoring actions received a five. A score of three or more signified that the indicator was *observed* at the destination. Once the individual rankings were complete, the team realized a process of triangulation to achieve a consensus for each indicator and sub destination. Then, the destination tourism resilience equation (Eq. 14.1), developed by Gutiérrez (2013), was employed to compare the findings for each of the four sub destinations:

$$F = \left\{ \left[\left(\text{PI}rO \times \omega i^r + \text{PI}nrO \times \omega i^{nr} \right) \times g \right] + \left[\left(\text{NI}rNO \times \omega i^r + \text{NI}nrNO \times \omega i^{nr} \right) \times h \right] \right\}$$
$$- \left\{ \left[\left(\text{NI}rO \times \omega i^r + \text{NI}nrO \times \omega i^{nr} \right) \times g \right] + \left[\left(\text{PI}rNO \times \omega i^r + \text{PI}nrNO \times \omega i^{nr} \right) \times h \right] \right\} \tag{14.1}$$

- "*F*" being the resilience factor analyzed (capabilities, ownership, connections)
- PI = Positive impact for resilience
- NI = Negative impact for resilience
- *r* = Highly relevant for recovery
- nr = Non-highly relevant for recovery
- *O* = Observed
- NO = Not observed
- ωi^r and ωi^{nr} = Weighting factor used to consider highly relevant and not highly relevant aspects for recovery in the measurement of the factor, $\sum \omega i^r + \sum \omega i^{nr} = 1$
- *g* and *h* = Weighting factor used to consider *observed* and *not observed* indicators, $g + h = 1$

During the resiliency indicator evaluation phase, relevancy differences emerged for the tourism destinations for some of the measures related to recovery considerations. Thus, a weighting was used (ωi^r and ωi^{nr}), which assigned a weight of 0.70 to the highly relevant factors that arose during a subjective determination by a panel of tourism stakeholders and experts, and a weight of 0.30 was assigned to non-highly relevant factors from this panel (Gutiérrez 2013). The objective of this weighting was to highlight aspects that tourism stakeholders and experts should view as decisive when facing adversities or crises (Gutiérrez 2013).

A second weighting was applied to the factors (*g* and *h*), based on the prior sensitivity analysis developed by García (2005), which established a weighting of 0.51

for resiliency factors that were observed by the panel (scores of 3–5) and 0.49 for factors that were not observed by the panel (scores of 1 or 2). The objective of this weighting measure was to assign a special value to the factors that would have an impact on the final resilience calculation for each of the destinations under study.

Once scores were achieved for each of the three factors, sub destination resiliency scores were calculated by adding the scores obtained for each of the three factors (capabilities, ownership, connections). Sub destination category scores for risk (geophysical, hydrological, meteorological, climatological, anthropogenic) were obtained by multiplying the average likelihood and severity values for the hazards that were observed in the data related to the respective category. Sub destination scores were calculated by multiplying the average likelihood and severity values for all of the hazards observed.

14.4 Results

14.4.1 Resilience Factors and Indicators

The majority of *Capabilities* factor indicators were evaluated as having a positive effect on destination resilience within all four sub destinations, with the exception of the *seasonality of tourism employment*, the *vehicle saturation capacity, destination cleanliness/ornamentation*, and the *contribution of tourism to noise quality* (Fig. 14.2). Nine highly relevant tourism resilience *capability* indicators were present for the sub destinations, including the *seasonality of tourism employment, equipment and security of tourist sites, natural disaster prevention programs, tourism planning tools, municipal tourism management, tourism management by business sector, tourism plans and programs, disaster recovery programs*, and *public health services for tourist use*. The main differences between sub destinations with respect to the capabilities factor manifested within the evaluation of whether or not indicators were observable within the data. For example, while capabilities relating to *the use of social networks* and *community disposition for tourism* were observed within the data for the Coyhaique-Puerto Aysén-Cerro Castillo, Lago General Carrera, and Provincia de los Glaciares sub destination, these indicators did not manifest within the data related to the Aysén Patagonia Queulat sub destination. And only the Lago General Carrera sub destination showed clear capacities with respect to the *vehicle saturation capacity*, the *equipment*, and the *security of tourist sites* indicators. Another notable difference was found in the *tourism plans and programs* and *contribution to curbing youth migration* indicators. For both of these indicators, all of the ZOIT destinations had clearly documented capacities; yet, they were not observed within the data for the Coyhaique-Puerto Aysén-Cerro Castillo sub destination.

All of the *Ownership* factor indicators were evaluated as having a positive effect on tourism resilience, with the exception of the *impacts of tourism on the local*

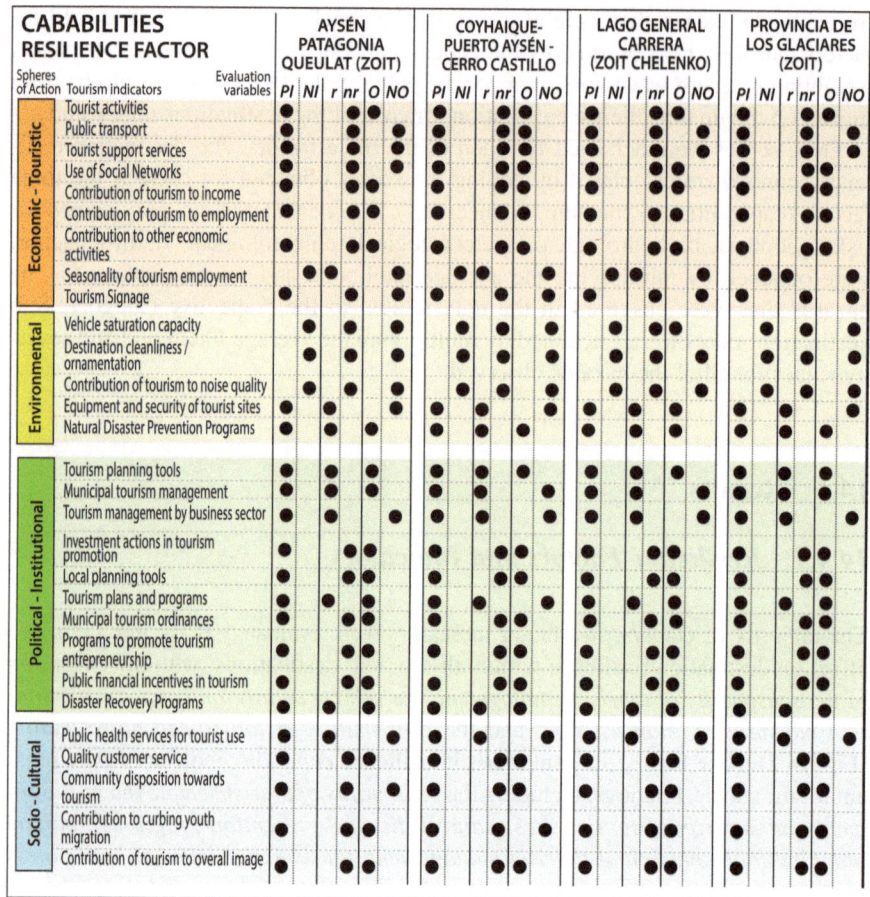

Fig. 14.2 Capabilities tourism resilience indicators scorecard for the four tourism sub destinations in Aysén

population (Fig. 14.3). Eight highly relevant indicators emerged within the literature associated with the sub destinations, including *trials and popular participation in the face of natural disasters, education and training of workers, training and capacity building of entrepreneurs, impacts of tourism on the local population, recovery and promotion of culture/heritage, prevention activity, effectiveness of public financial incentives*, and *levels of tourism development*. Of the three tourism resilience factors, the most differences between the sub destinations arose with respect to the observation of strategies and actions related to the indicators within the *Ownership* factor. In fact, differences resulted between the sub destinations for indicators across all four spheres of *Ownership* action. For example, *tourist accommodation companies* and *gastronomic services companies* were not observed in Provincia de los Glaciares. And evidence related to the *tourist excursions on offer*

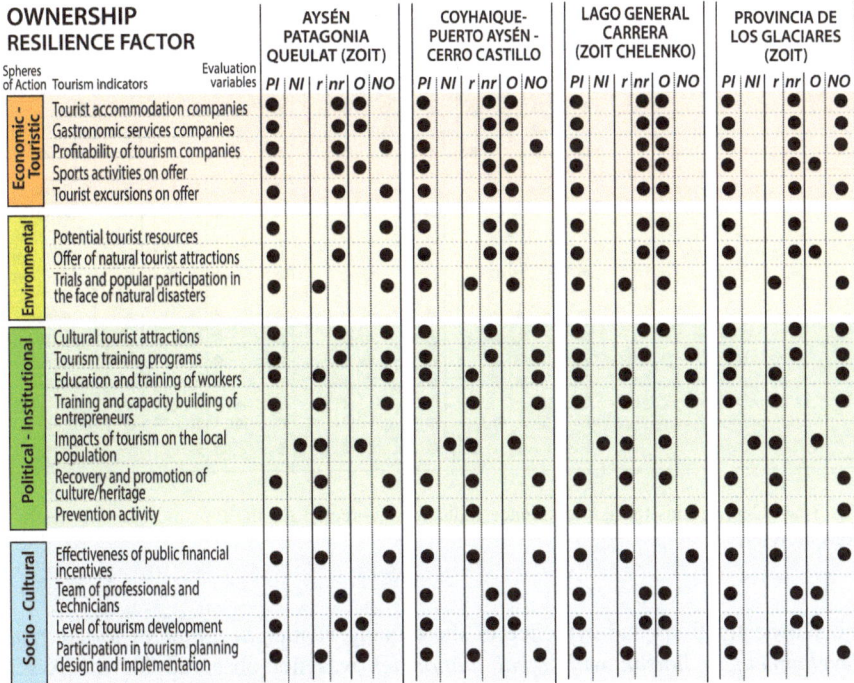

Fig. 14.3 Ownership tourism resilience indicators scorecard for the four tourism sub destinations in Aysén

indicator was observed in Coyhaique-Puerto Aysén-Cerro Castillo and Lago General Carrera, but not observed in Aysén Patagonia Queulat or Provincia de los Glaciares.

All of the *Connections* factor indicators were evaluated as having a positive effect on tourism resilience within the four sub destinations (Fig. 14.4). Six of the indicators were evaluated as being highly relevant for tourism recovery, including *communication channels and services, access to the destination and tourism sites, communication technologies, infrastructure recovery programs, public/private sector coordination in tourism,* and *municipal alliances for tourism development.* Differences between the sub destinations occurred across all three of the *Connections* spheres of actions, with respect to evidence of the indicators within the data. For example, evidence of *communication channels and services* arose within the Coyhaique-Puerto Aysén-Cerro Castillo and Lago General Carrera sub destinations but was absent in Aysén Patagonia Queulat and Provincia de los Glaciares. And, while the investigators observed clear evidence of *infrastructure recovery programs* and a *presence of environmental institutions,* in all of the sub destinations they did not observe evidence of *environmental actions* in Provincia de los Glaciares and only observed evidence of *environmental awareness* in Coyhaique-Puerto Aysén-Cerro Castillo. Similarly, for the *Political-Institutional* sphere of action, while all of

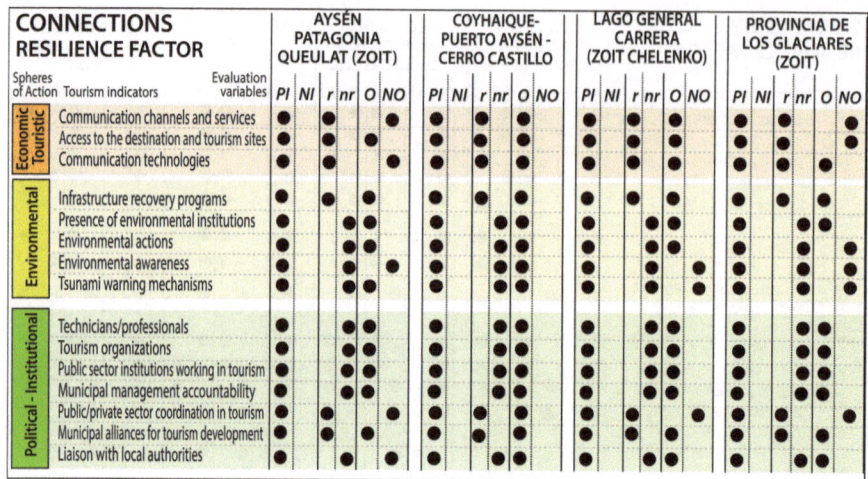

Fig. 14.4 Connections tourism resilience indicators scorecard for the four tourism sub destinations in Aysén

the sub destinations had evidence of there being *municipal alliances for tourism development*, a *liaison with local authorities* was not observed for the Aysén Patagonia Queulat sub destination, and evidence of *public sector institutions working in tourism* was only observed for the Coyhaique-Puerto Aysén-Cerro Castillo sub destination.

From the indicator evaluations, factor resilience scores were determined for the four sub destinations (Fig. 14.5). All four sub destinations scored positively for the *Capabilities* and *Connections* factors and negatively for the *Ownership* factor. The *Capabilities* factor of resilience was generally the strongest factor for sub destinations within the Aysén territory; however, for both the Aysén Patagonia Queulat and the Provincia de los Glaciares sub destinations, *Capability* and *Connections* factor scores were very similar; in fact, for the Provincia de los Glaciares, the *Connections* factor score was slightly stronger than the *Capabilities* factor score.

The Lago General Carrera sub destination scored most favorably across the factors, with scores of 0.420 and 0.369 for the *Capabilities* and *Connections* factors, respectively, and the most negative score for the Ownership factor (−0.326). The Aysén Patagonia Queulat sub destination scored least favorably across the factors, with scores under 0.1 for both *Capabilities* and *Connections* factors and a score of −0.449 for the *Ownership* factor. The other two sub destinations scored in between; the Coyhaique-Puerto Aysén-Cerro Castillo sub destination obtained a better rating than the Provincia de los Glaciares for both *Capabilities* and *Ownership*; yet, for *Connections*, the Provincia de los Glaciares ranked higher.

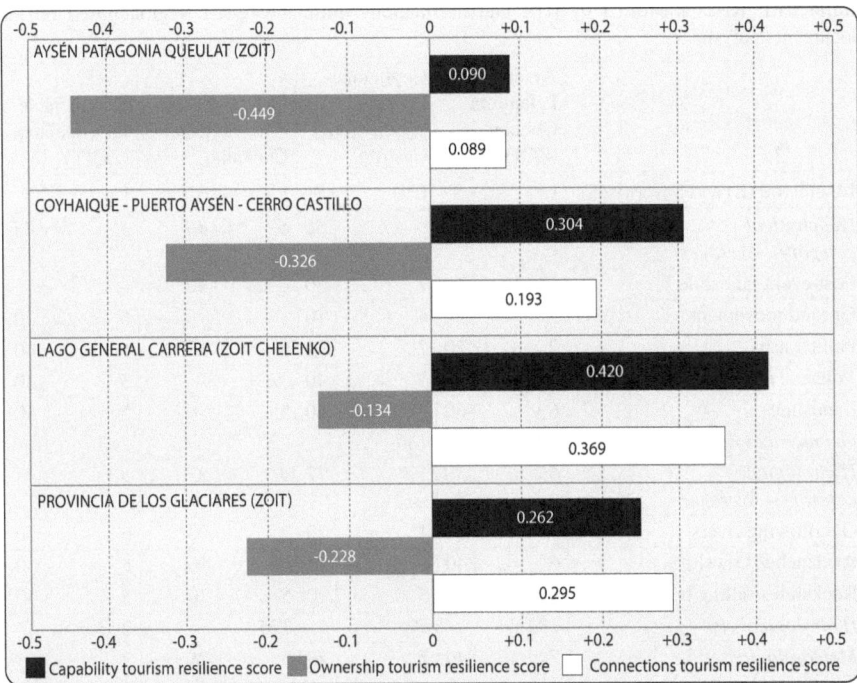

Fig. 14.5 Tourism resilience factor score for the four sub destinations in Aysén

14.4.2   Risk Analysis Results

The risk analysis for the four sub destinations (Table 14.1) showed several common risks across the four sub destinations, including *falling ash*, in the *Geophysical* category, and all of the phenomena within the *Hydrological* category of hazards (*overflowing rivers*, *avalanche/snowslides*, and *rockslides/falling boulders*), as well as several *Meteorological* risks, including *cold waves*, *snow/ice*, and *freezing* temperatures. *Anthropogenic* risks, including *traffic accidents* and *boating accidents*, were also common across the sub destinations.

The meteorological risk was evaluated as the highest category of risk, with an average score of 228 for the four sub destinations (Fig. 14.6). This average was heavily influenced by the Provincia de los Glaciares and the Aysén Patagonia Queulat sub destination scores (213 and 202, respectively). *Geophysical* risks were the next highest scoring category with an average risk of 206 between the four sub destinations. Within this category, the Lago General Carrera and Aysén Patagonia Queulat sub destinations received the highest scores (213 and 202, respectively).

Table 14.1 Risks identified by type and destination within the Aysén Region based on the document analysis

	Aysén Patagonia Queulat (ZOIT)		Coyhaique-Puerto Aysén-Cerro Castillo		Lago General Carrera (ZOIT Chelenko)		Provincia de los Glaciares (ZOIT)	
Likelihood (L) and Severity (S):	L	S	L	S	L	S	L	S
Geophysical category—Averages	*5*	*43*	*6*	*30*	*6*	*40*	*5*	*43*
Post-event landslide	6	50	7	30	6	40	–	–
Ground movement	–	–	4	10	–	–	6	30
Falling ash	2	20	7	30	6	50	6	60
Volcanic mudflows	–	–	5	40	–	–	3	30
Landslide	6	60	7	40	5	30	5	50
Geophysical risk	*202*		*180*		*227*		*213*	
Hydrological category—Averages	*6*	*30*	*5*	*17*	*4*	*27*	*5*	*47*
Overflowing rivers	6	10	4	20	2	20	6	40
Avalanche/snowslides	6	40	6	10	6	40	5	50
Rockslides/falling boulders	6	40	6	20	5	20	5	50
Hydrological risk	*180*		*89*		*116*		*248*	
Meteorological category—Averages	*7*	*40*	*7*	*20*	*5*	*28*	*7*	*53*
Cold wave	7	60	7	20	6	40	7	50
Snow/ice	7	50	8	20	6	10	8	50
Freezing	6	10	7	20	4	20	7	50
Hail	–	–	–	–	5	50	–	–
Heatwave	–	–	–	–	2	20		–
Wind	–	–	–	–	–	–	6	60
Meteorological risk	*267*		*147*		*129*		*368*	
Climatological category—Averages	*6*	*30*	*6*	*10*	*3*	*20*	*6*	*40*
Forest fire	6	30	6	10	–	–	6	30
Brush fire	–	–	–	–	3	20		
Flooding	–	–	–	–	–	–	5	50
Climatological risk	*180*		*60*		*60*		*220*	
Anthropogenic category—Averages	*5*	*15*	*6*	*10*	*5*	*10*	*5*	*10*
Traffic accident	6	10	6	10	5	10	5	10
Boating accident	4	20	5	10	5	10	4	10
Aircraft accident	–	–	–	–	–	–	5	10
Anthropogenic risk	*75*		*55*		*50*		*47*	
Sub destination average	**6**	**33**	**6**	**21**	**5**	**27**	**6**	**39**
Sub destination risk	**189**		**126**		**128**		**219**	

Notes: L = Likelihood score, scale of 0–10; S = Severity score, scale of 0–100

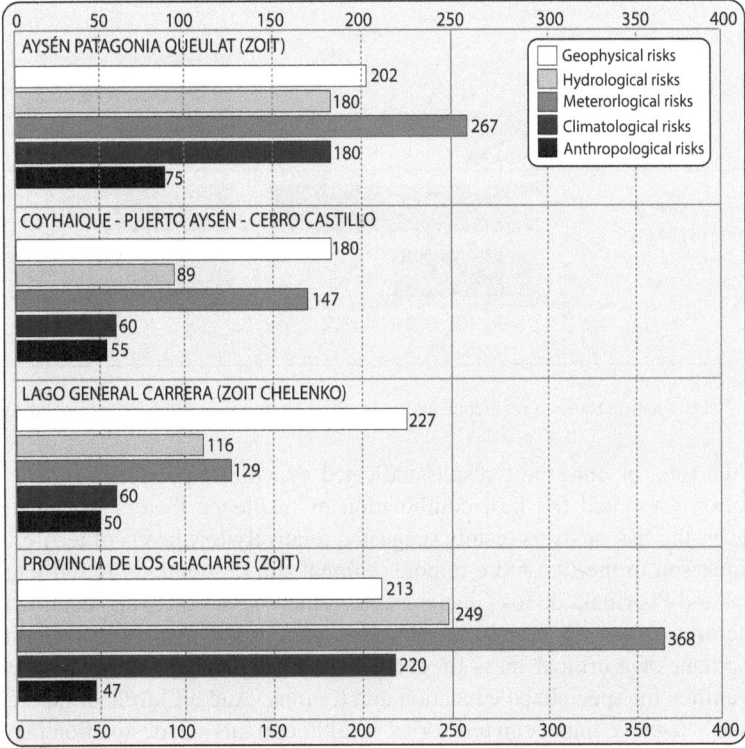

Fig. 14.6 Territorial risks by category for the four sub destinations in Aysén

Based on the data, *Anthropogenic* risks received the lowest ratings, with an average score of 57, which was influenced primarily by the high score within the Aysén Patagonia Queulat sub destination (75) and the low risk score within the Provincia de los Glaciares (47).

14.5   Discussion

14.5.1   Integrating Resilience Factors with Territorial Risks

This study investigated the synergy between community resilience factors for tourism and territorial risk to provide a more integrated view of tourism destination resilience. Integrating the resilience factors and risks for the four sub destinations in the Aysén Region of Chilean Patagonia provides a simple scorecard that may help tourism actors within these territories understand their current situations and how tourism in their territories is likely to respond in the face of natural adversities (Fig. 14.7).

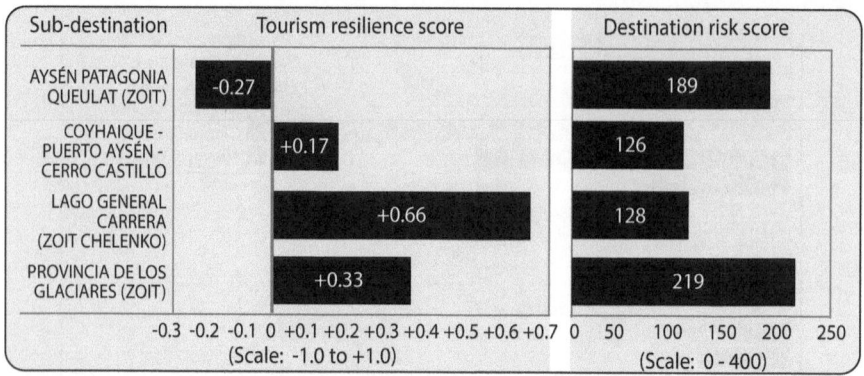

Fig. 14.7 Destination tourism resilience and risk scores for the four sub destinations in Aysén

At the time of our study, results indicated that the Lago General Carrera sub destination presented the best combination of resilience factors for facing risk. Moreover, the risk analysis results suggested relatively low levels of territorial risk in comparison to the two more remote regional sub destinations Aysén Patagonia Queulat and Provincia de los Glaciares. Nevertheless, our analysis identified some specific areas that will require continued attention and resources, including the development of a critical mass of good quality lodging providers and increased opportunities for specialized education and training. And all levels of natural risks can have a negative impact on territories, so although this sub destination fared well in our analysis, this does not mean that they are enough to positively overcome adversity. In general, favorably overcoming adversity requires that destinations have: a strong commitment to efficient financial support, municipal alliances in tourism, private alliances, public-private coordination bodies in tourism, training and capacity building for the sector, and stable participation mechanisms for the design, planning, and implementation of tourism development plans.

The Provincia de los Glaciares sub destination had the second highest resilience index of the four sub destinations but scored highest of the four in terms of risk factors. This sub destination includes three communes: Cochrane, Tortel, and O'Higgins, each with its own political and administrative teams and capacities; building tourism resilience will require strong collaborations between these groups. Efforts are required for the generation of public and private coordination bodies in tourism, as well as alliances between the different municipalities involved. Analysis of this sub destination also suggested the need for more and higher quality lodging, tourism support services, and public transport options. Also, our analysis suggested the importance of expanded education and training for the tourism sector, and the need for formal protocols for the management of natural disasters, accompanied by periodic simulations to improve awareness, participation, and preparedness among sector actors and local communities. Tourists should also be made aware of risks and disaster response procedures. Relatedly, a number of deficiencies were identified with respect to communication mechanisms, including insufficient coverage and bandwidth for cellular and internet connectivity, and low use of communication technologies by tourism service providers.

The Coyhaique-Puerto Aysén-Cerro Castillo scored positively in terms of tourism resilience but lower than the previous two sub destinations. This territory scored lowest in terms of risk factors; however, its location, between important urban points of connection, makes it more vulnerable to the effects that natural hazards could have. Specifically, with urban centers such as Puerto Aysén, Mañihuales, Coyhaique, Balmaceda, and Villa Cerro Castillo, this destination represents some of the larger urban centers regionally, with greater anthropogenic intervention; thus, the impacts of natural disasters are more widespread, extending beyond the natural environment of PAs to affect infrastructure and safety for human within the cities and towns. For this reason, we recommend a focus on building tourism ownership levels, through joint development of public financial incentives, participatory tourism planning design and implementation, and training and capacity building, including prevention training. Efforts should span the different municipalities and private actors of this sub destination, building connections and networks of professionals and technicians that are prepared to collaborate.

The Aysén Patagonia Queulat sub destination had the only negative score, indicating that this sub destination is least prepared to handle adversity. Moreover, this sub destination scored high in terms of destination risk. Priority actions for this sub destination include programs to evaluate and improve the safety of tourism sites and mechanisms to extend the tourism season, reducing dependence on the relatively short Patagonia summer. Integrated destination planning and management arose as areas that needed increased emphasis within the Aysén Patagonia Queulat sub destination, with active participation from all of the affected municipalities and stakeholders. A number of negative tourism impacts surfaced in the data for Aysén Patagonia Queulat, including cleanliness, noise pollution, vehicle congestion, and an inflation of prices for goods, services, land, and housing. Addressing these issues will require coordination and participation beyond the tourism sector. Perhaps tourist dispersion can be improved, helping to address some of the crowding-related issues, if tourist resources that have not yet been leveraged are developed for safe and appropriate use. We recommend developing recreational and entertainment activities for the local population, with ancillary use by tourists. This may generate a greater disposition toward tourism among the local community and in parallel create additional cultural tourism products and experiences. It is tremendously important to have professionals in the area of tourism to support the different aspects of tourism development in the destination, from the provision of information, to planning and management, to disaster recovery. A number of other deficiencies were identified in our analysis, including a lack of sufficient, high-quality, tourist support services, public transport and health services, and the need for better management of social networks at the public and private sector levels so that they include a tourism focus.

14.6 Conclusions

This chapter employed the Gutiérrez (2013) model for measuring tourism resilience factors within four tourism sub destinations in the Aysén Region of Chilean Patagonia: Aysén Patagonia Queulat, Coyhaique-Puerto Aysén-Cerro Castillo, Lago General Carrera, and the Provincia de los Glaciares. Analysis of Gutiérrez's (2013) factors and indicators produced a set of valuable diagnostic snapshots for the destinations, providing important inputs for resilience strategies and plans (Figs. 14.2, 14.3, and 14.4). Moreover, applying the research process to the four Aysén sub destinations helped us identify current areas of strength as well as priorities for resilience development in the coming years.

Our analysis of the *Capacities* tourism resilience factor demonstrated positive results for all four sub destinations. Results were especially positive for the Lago General Carrera and Coyhaique-Puerto Aysén-Cerro Castillo sub destinations, where the factor scores were 0.420 and 0.304, respectively, on a scale of −0.5 to +0.5. This seems logical; these two sub destinations are the most consolidated within the region and include the largest urban centers, where there is a larger concentration of tourism infrastructure, operators, and secondary services. The *Connections* factor scores were highest for Lago General Carrera and the Provincia de los Glaciares (0.369 and 0.295, respectively, on a scale of −0.5 to +0.5). We attribute these high scores to the work that has been done over the past several years in the designation and management of the ZOITs. The Coyhaique-Puerto Aysén-Cerro Castillo sub destination scored much lower in this area, reflecting the lack of coordination and networking that has occurred to date between tourism stakeholders involved with the main sites of this destination that are dispersed within three separate communes: Aysén, Coyhaique, and Río Ibáñez. All four destinations received low scores for the *Ownership* resilience factor. We attribute these deficiencies to two main factors. First, throughout the Aysén Region, there is a scarcity of tourism professionals and technicians to manage and support the development of tourism, especially at a sub destination level. Second, tourism-related territorial planning, including risk management and disaster recovery, is lacking, especially in terms of coordinated plans between sectors and communes. Finally, a review of the factor scores for the Aysén Patagonia Queulat sub destination revealed deficiencies in all areas as compared with the other sub destinations.

Next, we conducted a risk analysis, building on prior work from the United Nations Office for Disaster Risk Reduction (2017a, b, 2021), to illuminate several common risks across the four sub destinations and provide them with comparisons that can inform both local and regional tourism sector planning and management. Through this analysis, we identified the increased risks associated with more remote sub destinations, such as Aysén Patagonia Queulat and the Provincia de los Glaciares. The biggest areas of risks were associated with meteorological risks, though geophysical risks were also of concern. Anthropogenic risks were evaluated as lowest in our models for all four destinations; nevertheless, some of the biggest anthropogenic risks, such as recent social movements and the COVID-19 pandemic, were not captured by our model. More research is needed to develop and fine-tune tourism risk models, capable of capturing the broad range of contemporary anthropogenic risk, so that we can consider these factors in future planning and management.

Finally, we integrated these two analyses to provide an easily navigable dashboard (Fig. 14.7) that provides valuable information for developing local destination-level resilience strategies, public policies, and strategic roadmaps. At a regional level, this dashboard is helpful for informing priorities and areas for needed investment. For example, the combinations of scores in Aysén Patagonia Queulat and Provincia de los Glaciares indicate a regional priority for investment in tourism resilience planning and capacity building, including disaster planning and simulation.

To support these efforts, it is essential that public and private actors in the territories involved conduct more in-depth research to further understand the indicators and factors of tourism risk and resilience. Research and practice must incorporate all decision-makers, public and private managers, representatives of the communities, and also the tourists who visit each of these sub destinations. Co-management frameworks, such as the Tourism Area Response Network (TARN) proposed by Pennington-Gray et al. (2014), provide interesting frameworks from which to build and should be considered in future research and practice. There is little doubt that preparing nature-based tourism destinations where there are an abundance of PAs and a blend of urban and remote areas to manage and recover from crises and disaster is of vital importance. We hope this study will inform and advance the consolidation of actors in these destinations and prepare them to be stronger in the face of adversity.

Acknowledgments The authors are grateful for the funding provided by the Chilean Government project: CONICYT-PAI79170056.

References

E. Aedo, A. Rovira, J. Boldt, G. Inostroza, G. Pacheco, P. Szmulewicz, E. Szmulewicz, A. Arriagada, A. Malla, D. Quintana, *Turismo sustentable y áreas silvestres protegidas en Patagonia, Chile (Sustainable Tourism and Protected Wilderness Areas in Patagonia, Chile)*, ed. by P. Szmulewicz (Kultrún, 2020)

A. Al-Khateeb, How global tourism can become more sustainable, inclusive and resilient. World Economic Forum, 12 Apr 2021. https://www.weforum.org/agenda/2021/04/how-global-tourism-can-become-more-sustainable-inclusive-and-resilient/

S. Becken, B. Khazai, Resilience, tourism and disaster, in *Tourism and Resilience*, ed. by R.W. Butler, (CABI Publishing, Wallingford, 2017), pp. 96–104

D. Biggs, C.M. Hall, N. Stoeckl, The resilience of formal and informal tourism enterprises to disasters: Reef tourism in Phuket, Thailand. J. Sustain. Tour. **20**(5), 645–665 (2012). https://doi.org/10.1080/09669582.2011.630080

A. Blake, M. Sinclair, Tourism crisis management: US response to September 11. Ann. Tour. Res. **30**(4), 813–832 (2003). https://doi.org/10.1016/S0160-7383(03)00056-2

E. Calgaro, K. Lloyd, D. Dominey-Howes, From vulnerability to transformation: A framework for assessing the vulnerability and resilience of tourism destinations. J. Sustain. Tour. **22**(3), 341–360 (2014). https://doi.org/10.1080/09669582.2013.826229

S.R. Carpenter, K.J. Arrow, S. Barrett, R. Biggs, W.A. Brock, A.S. Crépin, et al., General resilience to cope with extreme events. Sustainability **4**(12), 3248–3259 (2012). https://doi.org/10.3390/su4123248

S. Chang, Urban disaster recovery: A measurement framework and its application to the 1995 Kobe earthquake. Disasters **34**(2), 303–327 (2009). https://doi.org/10.1111/j.1467-7717.2009.01130.x

Chilean National Emergency Office of the Ministry of the Interior, *Plan regional para la reducción del riesgo de catástrofes (Regional Plan for Disaster Risk Reduction)* (Ministerio del Interior, 2018)

Chilean National Tourism Service, *Plan de acción Región de Aysén del General Carlos Ibáñez del Campo sector turismo 2014–2018 (Aysén Region of General Carlos Ibáñez del Campo Tourism Sector Action Plan 2014–2018)* (Sernatur, 2014). Retrieved 13 Jan 2023, from https://www.sernatur.cl/wp-content/uploads/2018/10/Plan-de-Accio%CC%81n-Aysen.pdf

Chilean National Tourism Service, *Plan de acción para la gestión participativa de Zonas de Interés Turístico (ZOIT): Aysén Patagonia Queulat (Action Plan for Participatory Management of Zones of Tourist Interest (ZOIT): Aysén Patagonia Queulat)* (Sernatur, 2017). Retrieved 12 Jan 2023, from http://www.subturismo.gob.cl/wp-content/uploads/2015/10/Plan-Accion-ZOIT-Patagonia-Queulat.pdf

Chilean National Tourism Service, *Barómetro Región de Aysén (Aysén Región Barometer)* (Sernatur, 2019). Retrieved 23 Jan 2023, from https://estadisticas.aysenpatagonia.cl/app-assets/pdf/barometro_v11.pdf

Chilean National Tourism Service & Guazzinni Consultores, *Plan de acción para la gestión participativa de Zonas de Interés Turístico (ZOIT): Zona de interés turístico Chelenko (Action Plan for the Participatory Management of Zones of Tourist Interest (ZOIT): Chelenko Area of Tourist Interest)* (Sernatur, 2017). Retrieved 20 Dec 2022, from http://www.subturismo.gob.cl/wp-content/uploads/2015/10/Plan-Accion-ZOIT-Chelenko.pdf

Chilean Subsecretary of Tourism, *Plan de acción para la gestión participativa de las Zonas de Interés Turístico (ZOIT), Zona de interés turístico Provincia de los Glaciares (Action Plan for the Participatory Management of the Zones of Tourist Interest (ZOIT): Glacier Province Zone of Tourist Interest)* (Sernatur, 2017). Retrieved 20 Dec 2022, from http://www.subturismo.gob.cl/wp-content/uploads/2015/10/PLAN-ACCI%C3%93N-PROVINCIA-DE-LOS-GLACIARES.pdf

P. Cuevas-Reyes, Importancia de la resiliencia biológica como posible indicador del estado de conservación de los ecosistemas: Implicaciones en los planes de manejo y conservación de la biodiversidad (Importance of biological resilience as a possible indicator of ecosystem conservation status: Implications for biodiversity management and conservation plans). Biológicas **12**(1), 1–7 (2010)

N. De Sausmarez, The potential for tourism in post-crisis recovery: Lessons from Malaysia's experience of the Asian financial crisis. Asia Pac. Bus. Rev. **13**(2), 277–299 (2007). https://doi.org/10.1080/13602380601045587

Ecoespaña, & World Resources Institute, *Recursos mundiales: Las raíces de la resiliencia. Aumentar la riqueza de los pobres. Propiedad, capacidad, conexión (Global Resources: The Roots of Resilience. Increasing the Wealth of the Poor. Ownership, Capacity, Connection)* (Ecoespaña, 2009)

B.H. Farrell, L. Twining-Ward, Reconceptualizing tourism. Ann. Tour. Res. **31**(2), 274–295 (2004). https://doi.org/10.1016/j.annals.2003.12.002

C. Folke, S. Carpenter, T. Elmqvist, L. Gunderson, C.S. Holling, B. Walker, Resilience and sustainable development: Building adaptive capacity in a world of transformations. AMBIO: J. Hum. Environ. **31**(5), 437–440 (2002). https://doi.org/10.1579/0044-7447-31.5.437

T. Gale, A. Adiego, A. Ednie, A 360° approach to the conceptualization of protected area visitor use planning within the Aysén Region of Chilean Patagonia. J. Park. Recreat. Adm. **36**, 22–46 (2018). https://doi.org/10.18666/JPRA-2018-V36-I3-8371

G. García, *Medición de la capacidad de carga de la población local y los turistas en un destino turístico (Measurement of the Carrying Capacity of the Local Population and Tourists in a Tourist Destination)* (Universidad de Valencia, 2005)

D. Glaesser, *Crisis Management in the Tourism Industry* (Routledge, London, 2003)

GOAL, *Herramienta para medir la resiliencia comunitaria ante desastres. Guía metodológica (Tool for Measuring Community Resilience to Disasters Methodological Guide)* (DG ECHO, 2015). Retrieved 6 Jan 2023, from https://dipecholac.net/docs/herramientas-proyecto-dipecho/honduras/Guia-Medicion-de-Resiliencia.pdf

C. Goeldner, J.R.B. Ritchie, *Tourism: Principles, Practices, Philosophies* (Wiley, Hoboken, 2009)

Y. Guo, J. Zhang, Y. Zhang, C. Zheng, Examining the relationship between social capital and community residents' perceived resilience in tourism destinations. J. Sustain. Tour. **26**(6), 973–986 (2018). https://doi.org/10.1080/09669582.2018.1428335

C.A. Gutiérrez, *La resiliencia como factor clave en la recuperación de destinos turísticos : Aplicación al caso de un desastre natural en Chile (Resilience as a Key Factor in the Recovery of Tourist Destinations: Application to the Case of a Natural Disaster in Chile)* (Universidad de Valencia, 2013) https://roderic.uv.es/handle/10550/32139#.Y4-V7udrGfk.mendeley

C.M. Hall, D.J. Timothy, D.T. Duval, *Safety and Security in Tourism: Relationships, Management, and Marketing* (Routledge, London, 2003)

J.C. Henderson, *Tourism Crises: Causes, Consequences and Management* (Butterworth-Heinemann, Oxford, 2007)

G. Herrera, G. Rodríguez, Resiliencia y turismo: El caso de la ciudad de Baños de Agua Santa, Ecuador (Resilience and tourism: The case of the town of Baños de Agua Santa, Ecuador). Holos **3**, 253–280 (2016). https://doi.org/10.15628/holos.2016.4303

P.J. Holladay, R.B. Powell, Social-ecological resilience and stakeholders: A qualitative inquiry into community based tourism in the Commonwealth of Dominica. Caribb. Stud. **44**(1–2), 3–28 (2016). https://doi.org/10.1353/crb.2016.0000

J. Kalawski, A. Haz, Y ¿Dónde está la resiliencia? Una reflexión conceptual (And where is resilience? A conceptual reflection). Int. J. Psychol. **37**(2), 365–372 (2003)

N. Kontogeorgopoulos, Sustainable tourism or sustainable development? Financial crisis, ecotourism, and the "Amazing Thailand" campaign. Curr. Issue Tour. **2**(4), 316–332 (1999). https://doi.org/10.1080/13683509908667859

O. Lerbinger, *The Crisis Manager: Facing a Risk and Responsibility* (Lawrence Erlbaum Associates, Mahwah, 1997)

S. Levin, *Fragile Dominion* (Perseus, Reading, 1999)

B. McKercher, I. Young, The potential impact of the millennium bug on tourism. Tour. Manag. **20**(4), 533–547 (1999). https://doi.org/10.1016/S0261-5177(99)00016-3

D.S. Miller, C. Gonzalez, M. Hutter, Phoenix tourism within dark tourism. Worldw. Hosp. Tour. Themes **9**(2), 196–215 (2017). https://doi.org/10.1108/WHATT-08-2016-0040

J. Min, K.C. Birendra, S. Kim, J. Lee, The impact of disasters on a heritage tourist destination: A case study of Nepal earthquakes. Sustainability **12**(15), 6115 (2020). https://doi.org/10.3390/su12156115

D. Muñiz, J.A. Brea, Gestión de crisis en el turismo: La cara emergente de la sostenibilidad (Crisis management in tourism: The emerging face of sustainability). Rev. Enc. Cient. Tur. Manag. Stud. **6**, 49–58 (2010)

P.E. Murphy, G.G. Price, Chapter 9 – Tourism and sustainable development, in *Global Tourism*, ed. by W.F. Theobald, 3rd edn., (Butterworth-Heinemann, Oxford, 2005), pp. 167–193

H. Nakanishi, J. Black, K. Matsuo, Disaster resilience in transportation: Japan earthquake and tsunami 2011. Int. J. Disaster Resil. Built Environ. **5**(4), 341–361 (2014). https://doi.org/10.1108/IJDRBE-12-2012-0039

D. Pearce, C. Guala, K. Veloso, S. Llano, J. Negrete, A. Rovira, et al., Destination management in Chile: Objectives, actions and actors. Int. J. Tour. Res. **19**(1), 50–67 (2016). https://doi.org/10.1002/jtr.2083

L. Pennington-Gray, A. Schroeder, T. Gale, Co-management as a framework for the development of a tourism area response network in the rural community of Curanipe, Maule Region, Chile. Tour. Plann. Dev. **11**(3), 292–304 (2014). https://doi.org/10.1080/21568316.2014.890124

S. Pike, *Destination Marketing Organizations* (Elsevier, Oxford, 2004)

R. Plummer, D. Armitage, Charting the new territory of adaptive co-management: A Delphi study. Ecol. Soc. **12**(2), 10 (2007)

B. Richardson, Crisis management and the management strategy: Time to "loop the loop". Disaster Prev. Manag. Int. J. **3**(3), 59–80 (1994). https://doi.org/10.1108/09653569410795632

J. Ritchie, G. Crouch, *The Competitive Destination: A Sustainable Tourism Perspective* (CABI Publishing, Wallingford, 2003)

M. Rutter, Psychosocial resilience and protective mechanisms. Am. J. Orthopsychiatry **57**(3), 316–331 (1987). https://doi.org/10.1111/j.1939-0025.1987.tb03541.x

M. Scoville, Wartime tasks of psychiatric social workers in Great Britain. Am. J. Psychiatr. **99**(3), 358–363 (1942). https://doi.org/10.1176/ajp.99.3.358

R.A. Smith, J.C. Henderson, Integrated beach resorts, informal tourism commerce and the 2004 tsunami: Laguna Phuket in Thailand. Int. J. Tour. Res. **10**, 271–282 (2008). https://doi.org/10.1002/jtr.659

C. Stadel, Vulnerability, resilience and adaptation: Rural development in the tropical Andes. Pyrenees **163**, 15–36 (2008). https://doi.org/10.3989/pirineos.2008.v163.19

J.K. Strickland-Munro, H.E. Allison, S.A. Moore, Using resilience concepts to investigate the impacts of protected area tourism on communities. Ann. Tour. Res. **37**(2), 499–519 (2010). https://doi.org/10.1016/j.annals.2009.11.001

A. Tanana, C. Rodriguez, V. Gil, Strategic tourism management to address natural hazards in coastal areas: Lessons from Buenos Aires, Argentina. Tour. Rev. **74**(3), 503–516 (2019). https://doi.org/10.1108/TR-04-2018-0047

K. Tierney, M. Bruneau, Conceptualizing and measuring resilience: A key to disaster loss reduction. TR News **250**, 14–17 (2007)

T.J. Tyrrell, R.J. Johnston, Tourism sustainability, resiliency and dynamics: Towards a more comprehensive perspective. Tour. Hosp. Res. **8**(1), 14–24 (2008)

United Nations, *2009 UNISDR Terminology on Disaster Risk Reduction* (UNDRR, 2009)

United Nations, *Resolution Adopted by the General Assembly on 3 June 2015* (UN General Assembly, 2015). Retrieved 2 Dec 2022, from https://disasterlaw.ifrc.org/sites/default/files/media/disaster_law/2021-03/A_RES_69_283.pdf

United Nations Office for Disaster Risk Reduction, *Herramienta de auto-evaluación para la resiliencia frente a desastres a nivel local (Self-Assessment Tool for Disaster Resilience at the Local Level)* (UNISDR, 2017a). Retrieved 29 Nov 2022, from https://eird.org/camp-10-15/docs/herramienta-evaluacion.pdf

United Nations Office for Disaster Risk Reduction, *How to Make Cities more Resilient. A Handbook for Local Government Leaders* (UNDRR, 2017b). Retrieved 29 Nov 2022, from https://www.unisdr.org/campaign/resilientcities/assets/toolkit/Handbook%20for%20local%20government%20leaders%20%5B2017%20Edition%5D_English_ed.pdf

United Nations Office for Disaster Risk Reduction, *Quick Risk Estimation (QRE) Tool* (UDRR, 2021). Retrieved 7 Dec 2022, from https://mcr2030.undrr.org/quick-risk-estimation-tool

J. Varghese, N. Krogman, T. Beckley, S. Nadeau, Critical analysis of the relationship between local ownership and community resiliency. Rural. Sociol. **71**(3), 505–527 (2006). https://doi.org/10.1526/003601106778070653

B. Walker, D. Salt, *Resilience Practice: Building Capacity to Absorb Disturbance and Maintain Function* (Island Press, Washington, DC, 2012)

B. Walker, L. Gundersob, A. Kinzing, C. Kolkes, C. Carpenter, L. Schultz, A handful of heuristic and some propositions for understanding resilience in social-ecological system. Ecol. Soc. **11**(1), 13 (2006)

World Economic Forum, *The Global Risks Report 2020*, 15th edn. (World Economic Forum, Marsh & McLennan, Zurich Insurance Group, 2020). Retrieved 4 Dec 2022, from https://www3.weforum.org/docs/WEF_Global_Risk_Report_2020.pdf

Chapter 15
Evaluating Scientific Tourism Potential for Nature-Based Destinations: Expert Validation and Field Testing of Criteria and Indicators in the Aysén Región of Chilean Patagonia

Katerina Veloso, Fabien Bourlon, and Pablo Szmulewicz

Abstract The evolution of tourists' motivations is generating new approaches to tourism. One of them is scientific tourism (ST), which involves travel experiences with a focus on participation in scientific studies of various disciplines. ST has evolved significantly over the last decade in Chilean Patagonia, driven by public policies and the interest of various academic and private actors. These actors argue that involving travelers in research initiatives taking place in Patagonian destinations allows them to develop lasting connections with the heritage and institutions of these territories. This chapter presents an innovative process, including stakeholder identification, semi-structured survey interviews, and document analysis, to develop a matrix of weighted criteria to assess the potential for sustainable ST development within a destination. The process involves weighing four criteria: scientific research possibilities, supply of services to support scientific travel, value chain coordination, and current demand for ST in the destination. Each of these criteria is operationalized through a set of indicators that make it possible to evaluate

K. Veloso (✉)
Universidad Austral de Chile, Austral Patagonia, Valdivia, Los Rios Region, Chile

Universidad Austral de Chile, Faculty of Economics and Administrative Sciences, Valdivia, Los Rios Region, Chile
e-mail: katerina.veloso@uach.cl

F. Bourlon
Centro de Investigación en Ecosistemas de la Patagonia (CIEP), Sustainable Tourism Research Line, Human-Environmental Interactions Group, Coyhaique, Chile

Université Grenoble Alpes, Institute of Urban Planning and Alpine Geography – UMR 5194, Grenoble, Isere, France
e-mail: fabienbourlon@ciep.cl; Fabien.bourlon@umrpacte.fr

P. Szmulewicz
Universidad Austral de Chile, Faculty of Economics and Administrative Sciences, Valdivia, Los Rios Region, Chile
e-mail: pszmulew@uach.cl

© The Author(s) 2023
T. Gale-Detrich et al. (eds.), *Tourism and Conservation-based Development in the Periphery*, Natural and Social Sciences of Patagonia,
https://doi.org/10.1007/978-3-031-38048-8_15

the strengths, weaknesses, and opportunities for the sustainable development of scientific tourism services in the destination. Then, the method is field tested in the Aysén Region of Chilean Patagonia, affording the opportunity to further evaluate its assessment capacity.

Keywords Patagonia · Scientific tourism · Tourism destination · Tourism resources · Protected areas

15.1 Introduction

The evolution of tourists' motivations and behaviors has promoted the emergence of new approaches to tourism, diverging from the traditional sun and beach, mass tourism approach (Araújo Pereira and de Sevilha Gosling 2017; Pearce 2009). For example, scientific tourism (ST) represents a market niche that attracts visitors whose main motivation for travel is expanding their scientific knowledge and, above all, participating or collaborating in scientific studies (Bourlon and Mao 2011; Bourlon et al. 2021; Kosiewicz 2014; West 2008). Successful ST contains all the attributes of ecotourism in addition to human capital as a key factor and, in some cases, high-level technical equipment (Laing 2010). Opportunities for ST have emerged across diverse disciplines and in the different stages of the methodological process of research, creating tourism experiences that attract a variety of interests.

Public policies that promote Chile as an excellent natural laboratory for its resources and for the advancement of some scientific disciplines such as astronomy, glaciology, oceanography, and ecology, among others, have spurred interest in developing ST across Chilean Patagonia (Aguilera and Larrain 2018). The creation of new research centers and the interest of various academic and private actors have allowed Chile to become a natural laboratory for the advancement of scientific disciplines (Szmulewicz and Veloso 2013). However, the concept of ST is new to the Chilean tourism industry, and in many cases, the initiation of ST operations has outpaced regional ST planning. Thus, additional research is required to support adequate planning and integration of ST within the series of consecutive steps required for the creation of finished ST products (i.e., ST value chain; Veloso 2021).

The main objective of this chapter was to apply a new methodological tool to evaluate the potential of tourist destinations for the development of ST. After identifying the actors involved in each of the links of the ST value chain, a matrix of criteria and indicators was applied to evaluate the potential of a destination for sustainable ST services and programs. Then, our study participants, a sample of national and international ST experts, weighed each criterion and the associated sub-criteria indicators according to their importance for evaluating the potential for sustainable ST development within a destination. Finally, the matrix was applied to the Aysén Region in Chilean Patagonia, which stands out for its high presence of protected areas and its progress in ST initiatives.

15.2 Theoretical Framework

While the first studies on ST date back to the early 1990s, most of the available research has occurred in the last 5 years. The progression of these studies allows us to identify advances in the conceptualization of ST as well as the development of ST typologies and tourism experiences. Fortunately, a certain consensus has been reached regarding the benefits and requirements of destinations for the operation of ST programs.

Bourlon and Mao (2011) clarified the components of ST, emphasizing the importance of tourist participation in structured scientific activities and the supervision of experts or researchers. They recommended a mandatory orientation and training session to train tourists with respect to data collection, following the scientific method. The main objective of ST participants is to improve their knowledge and understanding of the place they are visiting. They seek to enrich their education, develop experience complying with research protocols, or strengthen research networks, among other motivations. By acquiring new knowledge, travelers feel that they are living an out-of-the-ordinary experience. The duration of the activity—an average of 8 days—is generally longer as compared to other types of tourism, allowing time for the necessary training, adequate integration into the new environment and the research experience, or the fulfillment of precise scientific objectives (Bourlon et al. 2021). This longer duration supports efforts by the Chilean Subsecretary of Tourism to promote the development of new forms of special interest tourism that increase arrivals and spending through longer stays (Chilean Subsecretary of Tourism 2012).

Some benefits of ST are that it can contribute toward increasing the educational level of the local population and promote the care—and therefore conservation—of natural resources (Ilyina and Mieczkowski 1992; Suguio and De Almeida 2011). ST offers many other potential benefits. It contributes to the resilience of communities and territories through the creation of shared knowledge and understanding of the essential socio-ecological characteristics and dynamics of the site (Bourlon et al. 2021). It supports the dispersion of tourist destinations visited by scientists as they make new trips to the sector after their field research. And it attracts researchers, their undergraduate and graduate students, national and international amateurs, and additional visitors—beyond the scientific community—by mobilizing families, friends, students, and other networks associated with scientists to visit the territory to learn about the ST work (Bourlon and Mao 2011).

The concept of ST has ignited interest from a variety of actors. Scientists see the potential of generating income to support the funding and dissemination of their research through leading ST programs (Pino 2020; Rozzi and Schüttler 2015). Entrepreneurs see an opportunity to insert innovative tourism products in their offerings that are hard to replicate and, thus, protected from a competitive standpoint (Constabel and Veloso 2020). And social organizations see the possibility of promoting sustainable local development that is based on scientific knowledge.

ST provides opportunities to establish relationships between scientists, visitors, specialized interpreters and guides, and the community where the study will be carried out (Molokáčová and Molokáč 2011). The development of ST opportunities involves the transformation of a scientific resource (e.g., a scientific phenomenon or object of study) into a tourism product (Bourlon 2020), thereby linking researchers to the local stakeholders interested in safeguarding the heritage and the socio-ecological systems in which they live (Bourlon et al. 2021). Effective ST destinations should have a local community that is prepared with basic knowledge of the scientific discipline that is of interest to the ST participants (Kosiewicz 2014). In addition to the traditional tourism products and services, like lodging and accommodations, this requires the participation of scientists, specialized guides from host institutions, and logistical services for explorations and studies.

In Chile, ST has become linked to the implementation of public policies that seek to promote the country's unique natural characteristics and its potential for scientific research. One reason for this is that the development of ST opportunities has attracted respected researchers who, in turn, have helped install specialized research centers for the study of astronomy, and Antarctic and sub-Antarctic ecosystems, among others (Aguilera and Larrain 2018; Szmulewicz and Veloso 2013). The National Commission for Scientific and Technological Research has called Chile a *Natural Laboratory*, favorable to the practice of scientific disciplines linked to specific territorial elements, including the night skies of the Atacama Desert, the Mapuche culture, the Valdivian rainforest, Patagonia's freshwater reserves and glaciers, and Antarctica.

There are not many specialized and validated methodologies for a general evaluation of a territory's tourism potential. One existing method, designed by the Inter-American Center for Tourism Training (CICATUR) and the Organization of American States (OAS) in 1979, established the parameters to be considered to estimate tourism development potential by inventorying the tourism heritage, the definition of hierarchies, and the subsequent categorization of tourism resources through field reconnaissance. Specifically looking at the evaluation of a destination's potential for the development of new forms of tourism, Szmulewicz and Álvarez (2002) designed and applied a method using wine tourism and vineyards as a case study. Based on demand, they identified a set of criteria they could evaluate to place potential wine tourism operations on a scale ranging from the destination's current level of tourism to their ideal level of tourism development.

Previous reviews have identified four basic approaches to the evaluation of tourism resources (Arnandis-i-Agramunt 2018; Camara and Morcate Labrada 2014; Varisco et al. 2014; Leno Cerro 1990). The first is analytical and considers the presence or absence of certain components. Intrinsic values of the resource are rated in order to evaluate natural areas, beaches, and, with some later modifications, other types of resources. A second approach aims to determine the value of possible alternative and compatible tourism resource-use opportunities that are not yet commercial in nature and therefore do not have a market value. A third approach is based on the evaluation of resource demand preferences expressed by tourists. This approach assumes that the higher the value placed on a given resource, the greater interest it will attract among potential or actual users. Finally, the last approach employs analytical methods to develop demand valuations from qualitative assessments, allowing researchers to weigh variables of interest.

To date, there is limited evidence of the application of any of the above approaches to evaluate ST resources and development opportunities. Within Chilean Patagonia, the methodology that comes closest is the approach designed by the Center for Investigation in Ecosystems of Patagonia (CIEP) (Bourlon 2020; Bourlon et al. 2021). Three criteria have been identified as important for supporting an ST offering: (1) a scientific topic of national or international relevance; (2) relevance of the scientific topic to local actors, including its relevance for inclusion in their tourism activities; and (3) capacity of local actors to generate a tourism offer with adequate services (specialized guides, logistic and transfer services, among others).

15.3 Methods

This study establishes and applies a tool to evaluate the potential of tourist destinations for the development of ST. The chapter results reflect a three-stage study: (1) a survey interview developed to evaluate ST competence and to collect stakeholder weightings of four criteria and underlying indicators for evaluating the potential for ST development; (2) stakeholders across the ST value chain were identified and asked to complete the survey and the survey results were then validated through focus groups; and (3) the study authors completed a document analysis process and replicated the process of criteria weighting using data from existing documents in combination with their own local expertise. The following sections outline the study process in terms of the local ST context, the participant survey, the survey data analyses, and the document analysis/author validation.

15.3.1 Scientific Tourism Context: The Aysén Region of Chilean Patagonia

The Aysén Region is located in Chilean Patagonia, south of the Lake District and north of the Magallanes Region. It is bordered by the Pacific Ocean to the west and Argentina to the east. Aysén is the third largest geographic region of Chile, yet one of the least populated. It is considered an extreme zone by the Chilean government due to its level of isolation, low population density, reduced presence of public services, and limited level of socioeconomic development (Gale et al. 2013). Landscapes within the Aysén Region vary greatly, from the Patagonian steppe to the fjords, passing through the Andes Mountain range, with its glaciers, lakes, rivers, forests, and pastures. The region has ecosystems whose values are internationally recognized and critical for biodiversity conservation (Mittermeier et al. 2003). Due to the presence of the Southern Ice Fields, the region contains one of the largest freshwater reserves on the planet.

Eighteen of Chile's 100 SNASPE Protected Areas are distributed throughout the Aysén Region, covering approximately 53,000 km^2 and representing more than a third of Chile's total protected lands (Gale et al. 2018). Various initiatives have been undertaken in an effort to develop and enhance ST in this region since 2007. Although Aysén is considered an excellent location for ST and numerous individual ST initiatives have occurred (Bourlon and Mao 2016), the formalized supply of ST remains very low.

15.3.2 Participant Survey Interviews

Based largely on the work of Bourlon and Mao (2016) and a supporting literature review, the ST value chain was characterized, identifying five categories of ST actors: (1) producers (tour guides, etc.); (2) providers (equipment and raw material providers); (3) intermediaries (travel agencies, etc.); (4) ST support institutions (research institutions, etc.); and (5) ST demand group (ST travelers). Snowball sampling was employed to develop a list of ST experts within the five categories (Naderifar et al. 2017).

A group of 50 stakeholders from different countries (Chile, France, Canada, USA, Mexico, and Germany) was interviewed: 20 participants representing the ST producers' group, 10 participants from ST support institutions, 10 participants representing ST intermediaries, and 10 participants from the ST demand group of travelers. The provider's group was not explicitly represented as a sampling group, though the backgrounds of several participants did overlap in this area. A semi-directed interview was conducted to have each participant complete an evaluative survey where they were asked to weigh the importance of criteria and sub-criteria indicators for determining the potential of sustainable ST development within a region.

15.3.3 Participant Survey Components

The survey interviews consisted of two parts. First, the participant's level of ST competence was gauged, and second, the participant's perceived importance of criteria and indicators to consider in planning for ST development was evaluated. These criteria and indicators were developed through a bibliographic review and forum discussions with working groups within the Scientific Tourism Network (Veloso 2021).

Unique sets of questions were developed to evaluate the levels of ST competence of the participant groups. The questions centered around the prevalence of ST within participants' professional experiences, their self-assessed knowledge regarding ST, and their experiences with supporting and implementing ST experiences (Table 15.1). The average competence score was 14 out of 40, and only data from participants who scored above average were included in the results of this chapter. This requirement was to provide a sense that our data represent the perspectives of individuals who can be considered *experts* when it comes to evaluating the potential for sustainable development of ST.

Study participants were asked to weigh four criteria (scientific potential, the supply of support services, the coordination of stakeholders, and current demand) according to their perceived influence on the territory's potential to develop ST. The first criterion, *scientific potential*, was understood as the scientific character of a territory and indicated the diversity and uniqueness of the scientific elements that might be investigated in a territory. The region's scientific potential helped identify scientific disciplines with the greatest potential to connect with scientific travel programs in order to develop ST experiences.

Table 15.1 Participant-tailored survey questions used to assess scientific tourism competence

Unique surveys	Questions used to evaluate scientific tourism competence
Designed for scientific travelers	Experience in scientific travel within your country Experience in international travel for scientific purposes Level of use of tour operators for the logistical organization of your scientific explorations Level of the utilization of specialized services
Designed for researchers	Degree of theoretical knowledge you have about scientific tourism The extent to which the concept of scientific tourism is used in their research, projects, and publications Level of referencing to works by foreign authors in their research on scientific tourism Degree of knowledge about scientific tourism experiences in the national and international context
Designed for ST support institutions	Degree of knowledge you have about scientific tourism Experience in supporting the development of scientific tourism programs Level of support for the development of scientific tourism networks Degree of support for research projects, studies, and publications on the subject
Designed for ST producers (tour operators)	Experience in the operation of scientific tourism programs, scientific explorations, and/or excursions for scientists or similar Degree of diversity in the operation of the services of these programs: logistical aspects, transportation, lodging, guiding, etc. Degree of awareness of the demand for this type of travel Level of management of the different segments by scientific tourism

The second criterion corresponded to *the supply of support services for scientific travel*, which reflected the relevance of the facilities offered within the territory to welcome ST, including the specialized supply of lodging, food, transportation, and human resources capable of operating ST products and programs. This criterion also considered the general tourism offer that complements the visitor's experience, including programs before or after the scientific program, the availability of souvenirs, recreational activities, currency exchange offices, etc., and the level of alignment between the actual destination image and the development of ST.

The third criterion was *scientific tourism stakeholder coordination*, which refers to the linkages that exist between stakeholders across the ST value chain. In cases where a strong connection between stakeholders was perceived, the types of existing stakeholder relationships were analyzed (formal, informal, legal, or de facto, etc.), as well as the level of connection. Specifically, we sought to identify connections between actors belonging to the same level of the value chain (i.e., intra-link) and connections between groups of actors across the value chain (i.e., inter-link).

The fourth criterion, *current demand*, involves the analysis of the demand dynamics and the different types of tourists that were attracted by the range of ST products (e.g., students, affinity groups, families). An understanding of current demand assisted in the evaluation of ST initiative viability. Some of the information that was relevant for this criterion included information about the volume of visitors arriving, their level of motivation to see and understand its scientific resources, their nationality or region of origin, and their main scientific and recreational motivations.

Participants weighted each of these criteria by assigning each one a percentage (for a total of 100% across the four criteria). Participants were then asked to assign another weighted percentage to several sub-criteria indicators within each criterion (for a total of 100% per criterion). Participants had the opportunity to recommend new indicators (in addition to those provided); however, no new indicators were reported. The provided indicators are listed in Tables 15.1 and 15.2.

15.3.4 Participant Survey Data Analyses

To validate the criteria for evaluating a region's ST potential, the mean percentage weight of each criterion was calculated (Table 15.2). Similarly, a weighted average was applied to define the relative weight of each sub-criteria indicator for these criteria. Once the relative weight of the criteria and indicators had been established by the participants, the matrix was validated through a focus group composed of 30 representatives of professionals, public institution representatives, and tour operators. The responses of the experts in the focus group were very similar to the survey interview results.

Table 15.2 Criteria and indicators for evaluating the potential for scientific tourism

Criteria	Indicators	%
Scientific potential of the territory 40%	1. Natural and cultural elements and phenomena of scientific relevance	21
	2. Importance given to research by the local community	13
	3. Exoticism of natural and cultural elements and phenomena	11
	4. Scientific infrastructure and equipment	10
	5. Valuation of scientific research	10
	6. Scientific dissemination and popularization programs	9
	7. Institutions with disciplines related to relevant topics	9
	8. Presence of important scientific networks	8
	9. Recognition of researchers in the scientific community	6
	10. Recognized graduate programs	3
Offer of support services for scientific travel 25%	1. Hospitality facilities for scientific travelers	20
	2. Qualified human resources to connect science and tourism	20
	3. Destination tourism image	17
	4. Formal scientific products and programs offered	13
	5. Connectivity	12
	6. Tourist products and services before and after the scientific trip	10
	7. Promotion of scientific internationalization	8
Articulation of scientific tourism stakeholders 25%	1. Joint strategies to promote scientific tourism	28
	2. Linkage in the scientific tourism value chain	27
	3. Willingness to welcome scientific travelers	25
	4. Programs to promote research networks	20
Current demand 10%	1. Travelers' appreciation of the territory	35
	2. Reasons for the trip	26
	3. Number of tourist arrivals for scientific purposes	17
	4. Profile of scientific travelers	16
	5. Origin (nationality)	6

15.3.5 Application of the Criteria Within the Aysén Region

Once the criteria had been validated, the research team field tested their results by applying a similar evaluation process to the Aysén Region. Background information provided by the Aysén regional government, other relevant public services, and academic/research centers were systematized and identified through consultations and review of data. This information included evaluation reports of public policies and programs, studies by private organizations, scientific publications, and data from sources referenced in websites or press releases that report on the regional situation for each indicator of each criterion.

The authors of this study, who are knowledgeable about the region and public policies in the fields of science, education, and tourism, weighted each criterion individually on a Likert-type scale (from 1 to 5) to evaluate the potential for ST development specifically within the region of Aysén. Subsequently, a consultation meeting was held to engage in a triangulation process whereby the research team discussed each individual evaluation and corroborated agreements or disagreements in their conceptualization of how weights should be assigned. Once the group's evaluations were complete, the average weighted values were used to calculate the results given in Table 15.2.

15.4   Results

15.4.1   Criteria for Scientific Tourism

The survey participants were asked to weigh the four criteria presented in the survey for their impact on Chilean Patagonia's potential for ST development: (1) scientific potential, (2) the supply of support services, (3) stakeholder coordination, and (4) current demand. The criteria were weighted as follows: 40% for scientific potential, 25% for stakeholder coordination, 25% for the supply of support services, and 10% for current demand (Table 15.2). The weight for the *supply of support services* (25%) encompassed (a) the scientific value of the territory as the main motivation for the trip, and (b) how support services can be developed later, without (c) ignoring the importance of this criterion in the quality of the traveler's experience. The weight for the *stakeholder coordination* (25%) considered the uniqueness of ST offerings, as this criterion tends to be quite fundamental within the tourism industry as a whole (Szmulewicz and Veloso 2013). *Current demand* was weighted much lower in relation to the other criteria, at 10%. Some experts who were consulted explained that the current demand for ST was not indicative of ST potential; rather, they considered the presence of scientific resources, the actors, and their coordination as more important for generating demand.

15.4.2   Defining the Four Criteria for the Evaluation of ST Potential

Participants assigned weights to indicators to better explain and contextualize the four criteria for evaluating ST potential (Table 15.2).

Within the *scientific potential* criterion, the highest average weight was given to the *natural and cultural elements and phenomena of scientific relevance* indicator

(21%), suggesting these elements are capable of motivating the development of scientific projects in the territory. Another relevant aspect to be considered is the *importance given by the local community to research on ST topics* (13% average weight). Several indicators received similar weighting, including the degree of *exoticism of the natural and cultural elements and phenomena of scientific relevance* (11% average weight), the *infrastructure and scientific equipment* (10% average weight—laboratories, workstations, equipment), the *valuation of scientific research by public and/or private institutions* through support and financing programs (10% average weight), the *communication capacity* for scientific dissemination and popularization programs (9% average weight), *scientific institutions* with disciplines related to the relevant aspects of ST (9% average weight), and the presence of important *scientific networks* (8% average weight). Some aspects of lesser relevance that should also be considered include the *recognition of researchers* (6% average weight) by the scientific community, scholarships, awards, and *recognized graduate programs* (3% average weight) in terms of their being accredited, and their trajectory, in years.

For the second criterion, *supply of support services*, the evaluation of existing *specialized hospitality facilities* available for ST travelers, and *qualified human resources* were each attributed high importance with 20% average weights, that is, the identification of lodging services, food, transportation, and souvenirs, among others, that allow for good reception of the ST visitor was considered important, as was the identification of specialized guides and translators that connect researchers and the tourism industry. The third highest average weight was associated with the *tourist image of the destination* (17% average weight). Other considerations that received lower weights included formal *scientific products and programs that are currently offered*, such as scientific explorations and research stays (13% average weight), and the digital *connectivity* within the ST destination (12% average weight).

Four indicators within the *scientific tourism stakeholder's coordination* criterion were given similar importance, all with average weights between 20% and 30%. These included the *review of programs that promote research networks* (both public and private), analyzing how the *actors in the value chain are linked* (whether or not they communicate and carry out joint actions), *joint strategies* (public and private) for promoting ST, and *willingness to welcome ST* (acceptance by researchers, research centers, tourism business, and the community at large).

Finally, for the *current demand of ST* criterion, understanding the *valuation of ST travelers*, mainly in terms of the perception of tourism satisfaction within the territory, was considered particularly important with an average weight of 35%. Next were *visitor motivations* (both scientific and general), with an average weight of 26%. Other important considerations were the estimation of the *proportion of tourists who visit for ST-related reasons* (17% average weight) and understanding *traveler profiles for ST* (purchasing behavior, age, scientific discipline, etc.—16% average weight). Interestingly, *visitor nationality* was rated as a less important indicator, with an average weight of 6%, suggesting that it may not be crucial to understand when evaluating the potential of a territory for ST since every destination has the potential to attract visitors.

15.4.3   Assessing the Potential for Scientific Tourism Within the Aysén Region of Chile

To field test the participant results, we replicated the participants' assessment of the importance of the criteria and indicators of ST development potential with an analysis that was specifically focused on the Aysén Region in Chilean Patagonia. Our weightings were based on our review of local data and analyses of existing documentation. Our logic in conducting this field test was that if our analysis closely aligned with the participant survey data, then we would have obtained a measure of validity for the process of analyzing ST potential within a region and also a sense of the potential for ST within the Aysén Region.

Column A of Table 15.3 shows the same weightings for each of the ST criteria, as determined by our study participants (from Table 15.2). Column B outlines the indicators for each criterion, and Column C lists the average of our (the authors) weightings of each indicator in terms of their importance for the region of Aysén. Column D represents the actual presence of each indicator within the Aysén Region (scale of 1–5; 1 = not present, 5 = very present) that we (the authors) assigned to each indicator. Column E represents the product of the importance weight (Column C) and the presence rating (Column D) that we assigned to each indicator. The score in Column F represents the total importance/presence (Column E) score for each criterion, weighted according to the overall weights the participants had assigned to each criterion in Column A (see Table 15.2).

15.4.4   Scientific Potential of the Aysén Region

Several considerations emerged from our own analysis of Aysén's scientific potential. Within the region, the most prominent fields of research that involve ST focus either on flora and fauna or on land dynamics (Bourlon et al. 2021). These two fields of study attract the attention of regional, national, and international research groups. There are also a small number of community organizations and local non-governmental organizations that promote citizen science, demonstrating that regional communities value research and are interested in collaborating in work on scientifically relevant topics. Examples of these groups include the Pitipalena Añihue Marine Protected Area Foundation and the NGO Aysén Mira al Mar, among others (Bourlon 2020).

Patagonia's international recognition for housing some of the most diverse and pristine remaining landscapes and historical Indigenous communities has caused the region to attract a large number of domestic and international researchers (Gale et al. 2013). However, the infrastructure for science is limited and there are very few scientific field laboratories within the region. While the level of resource allocation

Table 15.3 Gauging the potential for scientific tourism within the Aysén Region of Chilean Patagonia

A	B	C	D	E	F
Criteria	Indicators	%	Rating (1–5)	Importance/ presence product	Score
Scientific potential of the territory 40%	1. Natural and cultural elements and phenomena of scientific relevance	21.3	4	0.84	1.3
	2. Importance given to research by the local community	13.1	4	0.52	
	3. Exoticism of natural and cultural elements and phenomena	10.8	5	0.55	
	4. Scientific infrastructure and equipment	9.7	2	0.2	
	5. Valuation of scientific research	9.7	4	0.4	
	6. Scientific dissemination and popularization programs	8.5	3	0.27	
	7. Institutions with disciplines related to relevant topics	9.1	2.3	0.21	
	8. Presence of important scientific networks	8.4	2.7	0.22	
	9. Recognition of researchers in the scientific community	6.4	2	0.12	
	10. Recognized graduate programs	2.5	1	0.03	
	Weighted average of the ten indicators	*100*		*3.35*	
Offer of support services for scientific travel 25%	1. Hospitality facilities for scientific travelers	20.1	2	0.4	0.7
	2 Qualified human resources to connect science and tourism	19.5	3	0.6	
	3. Destination tourism image	16.7	4.5	0.77	
	4. Formal scientific products and programs offered	13.3	3	0.39	
	5. Connectivity	12.3	1.5	0.18	
	6. Tourist products and services before and after the scientific trip	10.8	4	0.4	
	7. Promotion of scientific internationalization	9.1	2.3	0.18	
	Weighted average of the seven indicators	*100*		*2.92*	

(continued)

Table 15.3 (continued)

A	B	C	D	E	F
Criteria	Indicators	%	Rating (1–5)	Importance/ presence product	Score
Articulation of st stakeholders 25%	1. Joint strategies to promote scientific tourism	28.1	3.5	0.98	0.8
	2. Linkage in the scientific tourism value chain	26.6	2.2	0.59	
	3. Willingness to welcome scientific travelers	24.1	4.5	1.13	
	4. Programs to promote research networks	20	2	0.4	
	Weighted average of the four indicators	*100*		*3.1*	
Current demand 10%	1. Travelers' appreciation of the territory	34.7	3	1.04	0.3
	2. Reasons for the trip	26.3	3.3	0.87	
	3. Number of tourist arrivals for scientific purposes	17	3	0.51	
	4. Profile of scientific travelers	16	2.8	0.45	
	5. Origin (nationality)	6	5	0.3	
	Weighted average of the five indicators	*100*		*3.17*	

within the region to scientific research is low, it remains significant when compared with the national average and when considering the region's low population density (Lozano et al. 2010). Public funds have allowed a considerable level of researchers to work within the region, though few internationally recognized researchers have been involved to date. This would be an important step considering the literature on ST development remains of local origin. International recognition of local researchers who stay within the region would be helpful for the development of a niche ST industry.

To date, the scope of ST research has mainly been at the regional level (Rivera-Polo et al. 2018). There are still few academic institutions installed within the region, and the number of scientific networks is low. However, there are links with researchers from other parts of the world, such as the Scientific TourismNetwork (www.scientific-tourism.org) initiated in 2018, and an ST group of initiatives included in the strategic roadmap of the ANID agenda for science, technology, innovation, and knowledge creation for the southern macrozone of Chile (NODO CTCI Austral, https://nodocienciaaustral.cl/) that seeks to develop and promote tourism that is connected with science and collaborations between researchers.

15.4.5 Scientific Tourism Support Services Offering in the Aysén Region

There are very few specialized lodging facilities for the ST traveler within the region. Scientists use the usual tourism facilities and must be self-sufficient when it comes to working in the field, which can be difficult given the frequently harsh environmental conditions. Although there is a significant level of services provided by general (non-ST specific) tour operators and travel agencies (http://recorreaysen. cl/; Chilean National Tourism Service 2017), the region of Aysén has a low level of services when compared with other emblematic destinations in Chile (Chilean Subsecretary of Tourism 2012).

Connectivity in the region remains weak (Muñoz and Torres Salinas 2010; Aysén Regional Government 2005), which considerably increases the costs of research. Flight rotations are relatively infrequent, transport is focused mainly on the larger towns, and digital connections are weak and unstable, all of which make field work difficult. There are three main laboratories that support ST field activity: one laboratory in Caleta Tortel, one in Raúl Marín Balmaceda, and the old school/pioneer museum in Cerro Castillo. A similar situation exists surrounding specialized human capital for ST, where few specialized professionals have been identified and few professional guides offer services specific to ST (Bourlon et al. 2021).

When it comes to considering the region's tourism image, the Patagonia brand stands out, which is internationally recognized for its commitment to protecting natural resources, for its focus on remote and pristine settings, and for visitor safety to remote settings. Although ST products have been identified (e.g., Scientific Products Catalog) and there is motivation and interest on the part of operators and guides, there is no formal and continuous offer in the market (Bourlon 2020). Existing ST products have emerged through publicly funded projects and the support of the Center for Investigation in Ecosystems of Patagonia (CIEP; www.ciep. cl) and its partners. Although there are national programs, like the Chilean National Agency for Research and Development (ANID) Regional program, that promote the dissemination of the region's research, resources dedicated to this end are relatively low in proportion to GDP and the existing strategy does not focus on participatory sciences that could favor ST (Lozano et al. 2010).

15.4.6 Scientific Tourism Stakeholder Articulation in the Aysén Region

There are science and innovation programs in place that support ST, but no specific joint strategies for ST have yet been identified. The coordination of the ST value chain is still weak. Some initiatives have been led by CIEP with support from

Chilean or international public funds, including the creation of an ST network and the incubation of a dedicated ST operator in Aysén (Exploraysen S.A.) whose mission was to promote ST products provided by the actor/partners (Bourlon 2020; Bourlon and Mao 2016). However, once public funding ends, the continuity of these types of initiatives by private entities has been low. Other institutions, such as the Universidad Austral de Chile, the Universidad de Magallanes, the Chilean National Corporation for Development (CORFO), and the Chilean National Tourism Service (SERNATUR), have promoted and supported small ST initiatives (Bourlon et al. 2022). Furthermore, local interest in receiving scientific travelers has been documented in several publications (Bourlon 2020). There are some programs that promote ANID research networks and resources from the Regional Government of Aysén, but these resources are low in comparison to other development priorities such as infrastructure and promotion of traditional extractive activities.

15.4.7 Current Scientific Tourism Demand Within the Aysén Region

The demand for ST in the Aysén Region of Chile has not been investigated to the degree that it has in other regions of the world, including North American, European, Brazilian, and other Chilean markets (Chilean National Tourism Service 2017). Interviews and information in the press indicate that scientists who come and work at CIEP, the Universidad Austral de Chile, or other science and technology organizations are very interested in staying longer. There is an emotional affinity with the region on the part of visiting scientists (Bourlon and Mao 2016; Gale et al. 2013). There are no studies to date into the number of arrivals for scientific reasons, but it is estimated that in proportion to other segments, this is low, particularly with respect to the national level as in the Atacama area and in relation to the Astronomical observatories. The profile of the visitor to the Aysén Region, which is of an age group and income stratum higher than the national average, suggests that there is a potential interest in buying products of higher value and higher cost (Chilean National Tourism Service 2017; Chilean Subsecretary of Tourism 2012).

Overall, the Aysén Region has a strong potential for scientific tourism due to scientific relevance in certain fields, but weak growth prospects due to the lack of specific public and private scientific outreach and low current levels of participatory science programs. New incentive programs for scientific research, such as those included in the NODO CTCI Austral roadmap, should increase the relevance of the region for ST and help connect science and tourism in the territory.

15.5   Conclusions

In Chile, public policies that promote the advancement of science in fields related to the natural landscape have spurred interest in ST opportunities. As a relatively new concept, the initial ST programs and operations within the region have shown potential and sparked interest in further developing this niche within regional tourism planning initiatives. To continue to build on current ST offerings within the Aysén Region of Chilean Patagonia, an analysis of the regional potential for ST development is needed, as well as data that informs strategic priorities in terms of regional ST investment. This knowledge will help the region build an ST industry that maximizes sociocultural benefits to the communities and promotes sustainable economic profitability while minimizing environmental impacts.

Considering the importance of innovation in isolated territories and emerging nature tourism destinations, the development of ST appears as an interesting opportunity for social and economic development based on science and knowledge. Hence the importance of having methodologies to quantify or estimate the opportunity gaps that the region has for a sustainable ST industry that benefits the spectrum of actors in the value chain including local communities. Based on similar criteria and indicators to those outlined in this study—(1) scientific relevance; (2) the supply of support services for scientific travel; (3) coordination between the actors involved in this type of travel; and (4) the current demand for scientific travel in the territory—the strengths, weaknesses, and opportunities within a territory can be evaluated. ST planning can be further informed by engaging stakeholders who have local expertise in examining ST potential across such criteria. For each criterion, a set of indicators can be used to evaluate the potential of a territory.

Within the specific context of the Aysén Region of Chilean Patagonia, the method demonstrates the considerable potential for ST development. Our analysis demonstrated the region has good scientific potential (3.35 out of a maximum of 5), an average supply of support services for the scientific traveler (2.92), good evidence of articulation of actors to facilitate ST (3.10), and that current demand is incipient and significant (3.17). This qualitative evaluation highlights key attributes of the territory. Although the Aysén Region has an established trajectory of ST development in Chile, it seems important to continue applying the methodology to specify and establish programs to reduce the gaps identified for reaching the region's full level of sustainable ST operation. Throughout the Aysén Region and Chilean Patagonia more generally, it is necessary to improve land, air, and digital connectivity. The region would also benefit from the generation of specialized hospitality facilities for the scientific traveler, which will contribute to the increase in demand for this new form of tourism. In addition, it is important to advance coordination efforts between the actors of the value chain of ST with strategies that ensure current and future initiatives remain sustainable over time.

Although our work suggests that the proposed criteria and indicators may be generalizable, they should be reviewed and possibly adapted to destination contexts when applied to new regions. We recommend this methodology for local managers seeking to develop ST, especially where it is possible to collect information across each of the four indicators as outlined in this chapter. This method provides a guide for identifying the gaps and quantifying costs and efforts associated with successful and sustainable ST development. It is important to consider that once initiated, the evaluation of the potential for ST destinations is dynamic in nature. Thus, a periodic replication of the process is necessary, including revalidation of the weighted indicators with focus groups. And much will depend on those who lead the initiative to promote ST and the emphasis they place on intervention, development, and innovation criterion and indicators within the framework of their specific development strategy.

A territorial strategy, based on these guidelines, has the ultimate goal of contributing to greater objectivity when making decisions regarding this new form of tourism. In addition, researchers and scientific institutions can contribute to the territory's socioeconomic development by promoting initiatives that improve scientific infrastructure and equipment, strengthen national and international networks, and increase recognition of relevant discoveries within a territory. The creation of science-based undergraduate and graduate training programs can, in turn, foster the development of new knowledge and regional human capital and thus strengthen the capacities of local community actors.

Acknowledgments This work was supported by Chile's National Research and Development Agency (ANID) under ANID's Regional Program R17A10002 and the CIEP R20F0002 project.

References

J.M. Aguilera, F. Larrain, *Laboratorios naturales para Chile* (Natural laboratories for Chile). (Ediciones UC, 2018)

G. Araújo Pereira, M. de Sevilha Gosling, Los viajeros y sus motivaciones. Un estudio exploratorio sobre quienes aman viajar (Travelers and their motivations. An exploratory study on those who love to travel). Estudios y Perspectivas en Turismo **26**(1), 62–85 (2017)

R. Arnandis-i-Agramunt, Una revisión a la planificación de los recursos: sobre los enfoques de evaluación y los modelos de adaptación a uso turístico (A review of resource planning: on evaluation approaches and models of adaptation to tourism use). Investigaciones Turísticas **15**, 168–197 (2018). https://doi.org/10.14198/INTURI2018.15.08

Aysén Regional Government, *Atlas región de Aysén 2005* (Aysén region atlas 2005). (Gobierno Regional de Aysén, 2005)

F. Bourlon, La ciencia como recurso para el desarrollo turístico sostenible de los archipiélagos Patagónicos (Science as a resource for sustainable tourism development in the Patagonian archipelagos). PASOS, Revista De Turismo Y Patrimonio Cultural **18**(4), 795–810 (2020). https://doi.org/10.25145/j.pasos.2020.18.057

F. Bourlon, P. Mao, Las formas del turismo científico en Aysén, Chile (The forms of scientific tourism in Aysén, Chile). Gestión Turística **15**, 74–98 (2011). https://doi.org/10.4206/gest. tur.2011.n15-04

F. Bourlon, P. Mao, *La Patagonia Chilena: un nuevo El Dorado para el turismo científico* (Chilean Patagonia: a new El Dorado for scientific tourism). (Ñire Negro, 2016)

F. Bourlon, T. Gale, A. Adiego, V. Álvarez-Barra, A. Salazar, Grounding sustainable tourism in science – a geographic approach. Sustainability **13**(13), 7455 (2021). https://doi.org/10.3390/su13137455

F. Bourlon, Y. Vialette, P. Mao, Science as a resource for territorial and tourism development of mountainous areas of Chilean Patagonia. J. Alp. Res. **10**(1), 10398. Advanced online publication (2022). https://doi.org/10.4000/rga.10398

C.J. Camara, F.d.l.Á. Morcate Labrada, Metodología para la identificación, clasificación y evaluación de los recursos territoriales turísticos del centro de la ciudad de Fort-de-France (Methodology for the identification, classification and assessment of territorial tourist resources of the city center of Fort-de-France). Arquitectura y Urbanismo **35**(1), 48–67 (2014)

Chilean National Tourism Service, *Anuario de turismo región de Aysén* (Tourism yearbook Aysén región). (Sernatur Aysén, 2017)

Chilean Subsecretary of Tourism, *Estrategia nacional de turismo 2012–2020* (National tourism strategy 2012–2020). (Subsecretaría de Turismo, 2012). Retrieved 20 Dec 2022, from http://www.subturismo.gob.cl/wp-content/uploads/2015/10/Estrategia-Nacional-de-Turismo-2012-2020.pdf

S. Constabel, K. Veloso, Planning and managing a paleontological tourism destination: the case of Pilauco-Osorno, Chile, in *Pilauco: A Late Pleistocene Archaeo-Paleontological Site: Osorno, Northwestern Patagonia and Chile*, ed. by M. Pino, G.A. Astorga, (Springer International Publishing, 2020), pp. 317–332

T. Gale, K. Bosak, L. Caplins, Moving beyond tourists' concepts of authenticity: place-based tourism differentiation within rural zones of Chilean Patagonia. J. Tour. Cult. Chang. **11**(4), 264–286 (2013). https://doi.org/10.1080/14766825.2013.851201

T. Gale, A. Adiego, A. Ednie, A 360° approach to the conceptualization of protected area visitor use planning within the Aysén region of Chilean Patagonia. J. Park Recreat. Admi. **36**(3), 22–46 (2018). https://doi.org/10.18666/JPRA-2018-V36-I3-8371

L. Ilyina, Z. Mieczkowski, Developing scientific tourism in Russia. Tour. Manag. **13**(3), 327–331 (1992). https://doi.org/10.1016/0261-5177(92)90106-H

J. Kosiewicz, Scientific tourism, aspects, religious and ethics values. Phys. Cult. Sport Stud. Res. **62**(1), 83–93 (2014). https://doi.org/10.2478/pcssr-2014-0014

J.H. Laing, Science tourism: exploring the potential for astrobiology funding and outreach, in Astrobiology Science Conference 2010: Evolution and Life: Surviving Catastrophes and Extremes on Earth and Beyond, League City, Texas, April 26–20, 2010, https://ui.adsabs.harvard.edu/abs/2010LPICo1538.5047L

F. Leno Cerro, La evaluación de los recursos turísticos: El caso del Canal de Castilla (The evaluation of tourism resources: the case of the Canal de Castilla). Treballs de Geografía **43**, 135–138 (1990)

P. Lozano, G. Inostroza, C. Salgado, *Región de Aysén: Diagnóstico de las capacidades y oportunidades de desarrollo de la ciencia, la tecnología y la innovación* (Aysén region: diagnosis of capacities and opportunities for the development of science, technology and innovation). (Programa Regional CONICYT, 2010). Retrieved 13 Dec 2022, from https://www.conicyt.cl/regional/files/2013/06/Aysen.pdf

R.A. Mittermeier, C.G. Mittermeier, T.M. Brooks, J.D. Pilgrim, W.R. Konstant, G.A.B. da Fonseca, C. Kormos, Wilderness and biodiversity conservation. Proc. Natl. Acad. Sci. U. S. A. **100**(18), 10309–10313 (2003). https://doi.org/10.1073/pnas.1732458100

L. Molokáčová, Š. Molokáč, Scientific tourism – tourism in science or science in tourism? Acta Geoturistica **2**(1), 41–45 (2011)

M.D. Muñoz, R. Torres Salinas, Conectividad, apertura territorial y formación de un destino turístico de naturaleza. El caso de Aysén (Patagonia Chilena) (Connectivity, territorial openness and the formation of a nature tourism destination. The case of Aysén (Chilean Patagonia)). Estudios y Perspectivas en Turismo **19**(4), 447–470 (2010)

M. Naderifar, H. Goli, F. Ghaljaie, Snowball sampling: a purposeful method of sampling in qualitative research. Strides Dev. Med. Educ. **14**(3), e67670 (2017). https://doi.org/10.5812/sdme.67670

P.L. Pearce, The relationship between positive psychology and tourist behavior studies. Tour. Anal. **14**(1), 37–48 (2009). https://doi.org/10.3727/108354209788970153

M. Pino, The Pilauco and Los Notros sites: a final discussion, in *Pilauco: A late Pleistocene archaeo-paleontological site: Osorno, northwestern Patagonia and Chile*, ed. by M. Pino, G.A. Astorga, (Springer International Publishing, 2020), pp. 333–340

F. Rivera-Polo, P. Rivera-Vargas, C. Alonso-Cano, Una mirada territorial al Sistema Universitario Chileno. El caso de la Universidad Regional de Aysén (UAY) (A territorial look at the Chilean University System. The case of the Regional University of Aysén (UAY)). Estudios Pedagógicos **44**(1), 427–443 (2018). https://doi.org/10.4067/S0718-07052018000100427

R. Rozzi, E. Schüttler, Primera década de investigación y educación en la Reserva de la Biosfera Cabo de Hornos: El enfoque biocultural del Parque Etnobotánico Omora (First decade of research and education in the Cape Horn Biosphere Reserve: The biocultural approach of the Omora Ethnobotanical Park). Anales del Instituto de la Patagonia **43**(2), 19–43 (2015). https://doi.org/10.4067/s0718-686x2015000200002

K. Suguio, J.R. De Almeida, Ecoturismo científico en la planicie costera del extremo litoral sur del estado de São Paulo – Brasil (Scientific ecotourism in the coastal plain of the extreme south coast of the state of São Paulo – Brazil). Estudios y Perspectivas en Turismo **20**(5), 1196–1213 (2011)

P. Szmulewicz, K. Álvarez, *Evaluación del potencial turístico de las viñas de la Ruta del Vino Valles de Curicó* (Evaluation of the tourism potential of the vineyards of the Curicó Valley Wine Route). (2002)

P. Szmulewicz, K. Veloso, Diseño de rutas turísticas en áreas rurales y naturales, orientaciones metodológicas (Design of tourist routes in rural and natural areas, methodological orientations), in *Turismo rural y en áreas protegidas (Rural tourism in protected areas)*, ed. by M.M. González, C.J. León, J. De Leon, S. Moreno, (Editorial Síntesis, 2013), pp. 99–117

C. Varisco, D. Castellucci, M. González, M.J. Muñoz, N. Padilla, L. Campoliete, G. Benseny, El relevamiento turístico: de CICATUR a la planificación participativa (The tourism survey: from CICATUR to participatory planning), in VI Congreso Latinoamericano de Investigación Turística, Neuquén, Argentina, September 25–27, 2014. http://nulan.mdp.edu.ar/id/eprint/2052/1/varisco.etal.2014.pdf

K. Veloso, *Diseño de una propuesta metodológica para la evaluación del turismo científico* (Design of a methodological proposal for the evaluation of scientific tourism). (Universidad Austral de Chile, 2021)

P. West, Tourism as science and science as tourism. Curr. Anthropol. **49**(4), 597–626 (2008). https://doi.org/10.1086/586737

Chapter 16
Contributions of Nature Bathing to Resilience and Sustainability

Angel Custodio Lazo Álvarez, Andrea Ednie, and Trace Gale-Detrich

Abstract This chapter discusses the merits of harmonious relationships between people, society, and nature, and their potential to help address increasing societal vulnerabilities. In recent years, The Chilean National Forestry Corporation (CONAF) has developed the concept of *nature bathing*, through its *Nature for Everyone* program. Nature bathing draws from validated programs that have been found to strengthen the immune system and reduce anxiety, depression, and stress, all of which may contribute to greater psychological resilience. Specifically, CONAF's Nature Bathing program integrates elements of the cosmovision and practices associated with *forest bathing* (*Shinrin Yoku*, in Japanese), *grounding*, and Andean Indigenous and popular culture. Accredited experts facilitate a 2–3-hour experience in a PA, promoting a reflective meditation (*mindfulness*), that immerses participants in the environment by activating the senses. This chapter reviews the health benefits attributed to spending time in nature and developing direct relationships with nature. Next, we share some practices and traditions being employed around the world to purposefully rebuild human connections with nature. Then, we delve into CONAF's Nature Bathing initiative, as an example of a transformative

A. C. Lazo Álvarez
Chilean National Forestry Corporation (CONAF), Management of Protected Wildlife Areas, Santiago, Metropolitan Region, Chile
e-mail: angel.lazo@conaf.cl

A. Ednie
University of Wisconsin – Whitewater, College of Education & Professional Studies, Whitewater, WI, USA
e-mail: edniea@uww.edu

T. Gale-Detrich (✉)
Centro de Investigación en Ecosistemas de la Patagonia (CIEP), Sustainable Tourism Research Line, Human-Environmental Interactions Group, Coyhaique, Aysen Region, Chile

Cape Horn International Center (CHIC),
Puerto Williams, Magallanes and Chilean Antarctica Region, Chile
e-mail: tracegale@ciep.cl

© The Author(s) 2023
T. Gale-Detrich et al. (eds.), *Tourism and Conservation-based Development in the Periphery*, Natural and Social Sciences of Patagonia,
https://doi.org/10.1007/978-3-031-38048-8_16

program designed to strengthen the role of PAs as public health infrastructure and help visitors build resilience while rediscovering their interconnectedness with nature.

Keywords Patagonia · Mindfulness · Forest bathing · Grounding · Human-nature relationships · Resilience

16.1 Introduction

> All people need beauty as they need bread, places to play and pray where nature can heal and give strength to body and soul. (John Muir 1912, p. 256)

The COVID-19 pandemic has been linked with an intensification of mental health challenges across the globe, affecting all sectors of society and a range of mental disorders including post-traumatic stress, compassion fatigue, anxiety, functional exhaustion or burnout, depression, and somatization (Ertem et al. 2020; Real-Ramírez et al. 2020). A recent survey (undertaken from February to March 2021) for the World Economic Forum regarding the effects of the pandemic on health, economic, and behavioral patterns, which surveyed more than 21,000 adults worldwide under the age of 75, found that 45% of survey participants perceived a deterioration in their mental health (IPSOS 2021). Chile was the second highest rated among the 30 countries included in the survey in terms of mental health declines, surpassed only by Turkey. In total, 56% of Chilean participants perceived that their mental health had declined during the COVID-19 pandemic. Another study, by Urzúa et al. (2020), which surveyed Chilean primary healthcare workers (20% of the sample, $n = 125$) and secondary care workers (80% of the sample), showed that most of them reported some degree of symptoms of anxiety (74%), distress (56%), depression (66%), and insomnia (65%). These percentages far exceeded the results of other studies outside the pandemic context.

To cope with the stress associated with the COVID-19 pandemic and other mental health stressors, individuals must draw on a range of protective elements, individual traits, and healthy mechanisms to help manage the adversity they experience and find ways to respond, adapt and even thrive as time goes on. This capacity is referred to as psychological resilience in individuals and families (Ingulli and Lindbloom 2013; Liu et al. 2017; Satici 2016; Timmerman 1981), and social resilience in communities and societies (Adger 2000; Berkes and Ross 2013; Berkes et al. 2002). Although different sets of principles apply to each of these groups, researchers have recognized an interrelationship between individual and community resilience capacity (Berkes and Ross 2013; Berkes et al. 2002). Transformations that enhance psychological resilience at the individual and family levels can positively affect resilience at the community and societal levels (Berkes and Ross 2013).

In Chile, increasing psychological and social resilience capacity is particularly important. Urbanization and associated lifestyle changes, including the dominance of automobile use, availability and proximity of highly processed, calorie dense foods, increased reliance on computers, televisions, and other technology, and growing use of labor-saving appliances in home and work settings, have all been linked with declines in daily physical activity and rise of sedentary lifestyles (Al-Nuaim et al. 2012; Carrillo-Larco et al. 2016). This lifestyle transformation has been increasingly linked to chronic medical conditions and diseases, including high blood pressure, cardiovascular disease, high glucose blood levels, increased lipids, type 2 diabetes, and certain types of cancer, as well as increased risk of contracting many contagious diseases including COVID-19 (Al-Nuaim et al. 2012; Carrillo-Larco et al. 2016; Ramírez and Agredo 2012; Urzúa et al. 2020). In Chile, sedentary lifestyles are common across all age groups within the population, with insufficient levels of physical activity for one in every five Chilean adults (Cristi-Montero and Rodríguez 2014; Pan American Health Organization and World Health Organization 2016). Obesity affects more than 60% of the population between 15 and 64 years of age (Food and Agriculture Organization of the United Nations & Pan American Health Organization, 2017), with close to 10% of all children under 5 years of age overweight. Green spaces are in short supply in Chilean cities and considered critical public infrastructure for supporting urban health.

Urbanization is also linked with a range of mental health illnesses and disorders attributed to stressors arising from the influence of overcrowding, pollution, higher levels of violence, and lower levels of social support (Marzukhi et al. 2020; Núñez-González et al. 2020; Srivastava 2009; Pan American Health Organization and World Health Organization 2016). Some of the more prevalent illnesses include substance abuse, alcoholism, depression, alienation, anxiety, and severe mental disorders (Srivastava 2009). Chile faces one of the highest rates of prevalence of disease from psychiatric illnesses (23.2%) in the world (Vicente et al. 2016). Mental health disorders (anxiety, depression, substance abuse and dependence, disruptive disorders, attention-deficit, hyperactivity disorder, and post-traumatic stress disorder) are widespread in Chile; they have been the most frequent reasons for medically prescribed leave in the country for more than a decade (Pan American Health Organization and World Health Organization 2016). Researchers estimate that 36% of Chileans 15 years of age and older have experienced one or more psychiatric disorders, and one in five Chileans have experienced disorders within the past 6 months (Vicente et al. 2016). Moreover, almost 40% of Chileans have direct exposure with some form of trauma, with the most common traumas experienced including witnessing the injury or death of other persons, experiencing a sudden accident or injury, or being the victim of physical assault (Zlotnick et al. 2006).

This chapter focuses on the potential of nature to contribute to human mental health and resilience. We propose that current conservation strategies based on public and private protected areas (PAs) and even urban and semi-urban parks, can effectively contribute to human health by supporting public health policies and innovations related to tourism development through programs and infrastructure that contribute to well-being. The 107 Protected Areas (PAs) of Chile's national system of protected natural areas (SNASPE) offer important infrastructure to support these outcomes, including intentionally designed trails and mindfulness programs within PAs that apply practices to support human and nature interconnectedness. These installations and programs can contribute to all levels of resilience, public health, and human well-being. The following sections illustrate our proposal. We begin with a brief review of the health benefits attributed to spending time in nature and developing direct relationships with nature. Next, we share some of the practices and traditions being employed around the world to purposefully rebuild human connections with nature. Then, we delve into the CONAF *Nature Bathing* initiative as an example of a transformative program designed to strengthen the role of PAs as public health infrastructure and help visitors build resilience and stronger mental and physical health while rediscovering their interconnectedness with nature. This chapter concludes with a list of next steps, including possible connections between SNASPE PAs, nature bathing, and wellness tourism.

16.2 Theoretical Constructs

16.2.1 Health Benefits of Time in Nature/Relationships with Nature

Reconnecting our urban communities with nature, through more direct relationships with public and private protected areas (PAs), and even urban and semi-urban parks, may help contribute to improving mental health and resilience. A growing body of evidence has allowed us to become aware and more confident that it is not only a *feeling of well-being* that we have by immersion through sensory contact in nature, but also actual physiological, emotional, and mental health benefits that have been measured and monitored by various studies (Lemieux et al. 2016; Newton 2007; Puhakka et al. 2017; Taff et al. 2019; Thomsen et al. 2018; Townsend and Henderson-Wilson 2016; UK Sustainable Development Commission 2008). Exposure to nature has been linked with providing a number of physical fitness and wellness benefits, or ecosystem services, including better general health, reduced blood pressure and pulse rate, increased lifespans, and reduced exposure to pollution (Mitchell and Popham 2007, 2008; Pretty et al. 2007; Sandifer et al. 2015; Wells et al. 2007). Also, nature-based recreation and leisure experiences have been associated with a range of emotional and psychological ecosystem services, leading to mental wellness outcomes, including restorative benefits (Hartig et al. 1997; Kaplan 1995) stress reduction (Morita et al. 2007; Ulrich et al. 1991), and the improvement of cognition and affect for people suffering from depression (Berman et al. 2012; McMahan and Estes 2015; Sandifer et al. 2015). Nature has also been linked with ecosystem services contributing to social well-being. For example, Sandifer et al. (2015) found evidence of increased social interactions, reduced aggression, positive intercultural and interracial interactions, and enhanced social support and cohesion from time in nature.

Even relatively short immersive experiences in nature have been shown to have positive impacts on mental health. For example, Mayer et al. (2009) observed that a 15-minute nature walk increased participants' connection to nature, awareness of their immediate environment, level of attention, positive emotions, and their capacity for self-reflection. In addition to general restorative effects (Kaplan 1993; Kaplan 1995; Ulrich et al. 1991), researchers have suggested that exposure to nature has buffering, or stabilizing, effects for those who were not stressed prior to exposure, and positive benefits for self-regulation and the ability to control impulses (Beute and De Kort 2014). Alliances between PAs, parks, and other sectors like healthcare and tourism, can provide new mechanisms and therapies to help people and communities increase psychological resilience, improve connections between people and nature, and foster a more tangible social valuation of the importance of these special places and the role they play in improving our well-being.

16.2.2 Shinrin Yoku

The concept of *Shinrin Yoku,* sometimes referred to as *forest bathing*, has been gaining worldwide attention since it was established by the Japanese Ministry of Agriculture, Forestry, and Fisheries in 1982. Since then, a variety of agencies and organizations have begun to develop infrastructure, programming, and practice that follow the *Shinrin Yoku* concept, particularly in Europe, Asia, and North America, and more recently in Latin America (Li 2018; Miyazaki 2018). Initially, the concept of *Shinrin Yoku* was created as a promotional campaign for people to visit the beautiful forests of Japan. However, inspired by the realization that individuals somehow feel better when surrounded by nature, several scientists began studying the psychological and physiological benefits of nature—particularly forests—for human health and well-being (Miyazaki 2018).

Shinrin Yoku refers to an alternative form of medicine wherein one fully engages with nature through their senses (Miyazaki 2018), and has been defined as, "making contact with and taking in the atmosphere of the forest: a process intended to improve an individual's state of mental and physical relaxation" (Park et al. 2010, p. 18). Song et al. (2016) described the aim of *Shinrin Yoku* as being to, "induce preventive medical effects to improve weakened immune function and prevent diseases by achieving a state of physiological relaxation through exposure to forest-origin stimuli" (p. 2). Song et al. (2017) later described how the effectiveness of *Shinrin Yoku* as a preventive or alternative therapy has been demonstrated through research.

Researchers have identified several explanations for forest bathing's potential to promote good health and prevent the onset of disease (Ohtsuka et al. 1998; Antonelli et al. 2019). For example, Li et al. (2009) determined that phytoncides (aromatic volatile substances derived from trees), in combination with decreased stress hormone levels, likely play a role in increasing the body's natural production of *killer cells* (white blood cells with enzymes that kill tumor cells or virus-infected cells).

Song et al. (2017) found that for office workers who participated in their study, a day-long forest therapy session could produce a sustained decrease in systolic and diastolic blood pressure that lasted up to 5 days. Likewise, in their study of 24 forests in Japan, Park et al. (2010) found that walking through and viewing forests lowered cortisol, pulse rate, and blood pressure, heightened parasympathetic nerve activity (ability to relax), and lowered sympathetic nerve activity (fight or flight response). Psychological effects of *Shinrin Yoku* have also been identified. For example, Morita et al. (2007) found forest bathing to be an effective stress reduction method, especially among participants experiencing chronic stress.

Research has also demonstrated numerous health benefits associated with forest exposure, in general. Although these studies are not specific to *Shinrin Yoku*, they represent benefits that would apply to forest bathing programs. For example, López-Pousa et al. (2015) found that participants with fibromyalgia reported fewer days with perceived pain and insomnia, and more days of wellness, after participating in walks through a mature forest. Moreover, Lee and Lee (2014) compared forest walking to city walking to determine whether the antioxidant and anti-inflammatory effects of forest environments are superior to those of city environments and found significant differences between the two. It was found that forest walking improved arterial stiffness and pulmonary function for the participant group of elderly Korean women in comparison to city walking.

16.2.3 Grounding

The concept of grounding (also known as earthing) refers to the benefits of direct body contact with the Earth's natural electric charge (Menigoz et al. 2020). The process of grounding focuses on restoring an electric connection to the Earth, which proponents consider having been lost over time for many people because of increasingly urban, modern lifestyles, and technologies. Increased urbanization and usage of indoor entertainment alternatives (e.g., television, computers, electronic gaming devices) tend to reduce peoples' time outside in natural environments. Additionally, recent changes in the raw materials used for apparel, from shoes that were primarily constructed from traditional materials like leather and wood to shoes that are constructed from synthetic materials with more insulated soles of rubber and plastics, have augmented the barriers between the human body and the Earth's ground (Gruebner et al. 2017). Grounding theorists propose that these changes in lifestyle may challenge our immune systems. For example, Oschman et al. (2015) have proposed that increasing human disconnection from the Earth may be an important and overlooked contributor to the global rise in non-communicable, inflammatory-related chronic diseases. Selye (1984) described how built-up inflammation, that grounding may help to offset, can hinder the movement of antioxidants and regenerative cells and result in incomplete cellular repair. This can cause an inflammatory cycle that can persist for a long period of time, eventually promoting the development of chronic diseases.

Research on grounding began in the late ninetieth century, when a back-to-nature movement in Germany claimed many health benefits from being barefoot outdoors, even in cold weather. Chevalier et al. (2012) recount that in the 1920s, after learning of the practices and beliefs of some of his patients, George Starr White, M.D., investigated the practice of sleeping grounded, which referred to sleeping on the ground or sleeping connected to the ground through copper water, gas, or even radiator pipes. He reported improved sleeping using these techniques, though these ideas never caught on in mainstream society (Chevalier et al. 2012).

In recent years, grounding research has regained momentum. Modern research considers how the surface of the Earth may be affected electrically by lightning strikes, solar radiation, and other atmospheric dynamics, and that there exist possible relationships between Earth's electrical charge and human health. Their work builds on modern knowledge about the ground and bodies of water, which have a continuously renewed supply of subatomic particles called free electrons that give the Earth a natural negative electric charge (Menigoz et al. 2020), and a parallel knowledge of the human body, with its own charge or electrical system at the cellular level. Grounding theorists propose that the grounding of the human body keeps it at the natural electrical potential (voltage) of the Earth. They hypothesize that connecting the body to the Earth enables free electrons from the Earth's surface to spread over—and into—the body, where they can have antioxidant effects (Oschman et al. 2015).

Some recent studies support these theories (Chevalier et al. 2012; Menigoz et al. 2020; Oschman et al. 2015). For example, Menigoz et al. (2020) suggested that physical contact with the Earth's natural charge has a stabilizing effect, reduces inflammation, and improves blood flow. Although research evidence regarding the benefits of grounding is still developing, at least 20 studies to date have reported benefits to having a grounded versus ungrounded body, such as increased energy, better sleep, and reduced pain and stress (Brown and Chevalier 2015; Chevalier et al. 2012; Menigoz et al. 2020; Oschman et al. 2015). The physiological effects of grounding have perhaps most commonly been measured by cortisol levels and subjective information about sleep, pain, and stress (Ghaly and Teplitz 2004). However, some studies have suggested measurable biological effects, such as differences in white blood cell concentrations, cytokines, and inflammatory responses (Oschman et al. 2015).

Concerns over the health risks associated with being disconnected to the Earth align closely with the benefits of *Shinrin Yoku*, as outlined above. Initiatives related to these concepts in places across the world inspired CONAF to develop similar initiatives that combine grounding with forest bathing with the goal of providing visitors opportunities to re-establish their connections with the Earth and experience the benefits associated with exposure to nature. In addition to physiological and health benefits, these programs emphasize the alignment between grounding, forest bathing, and Chilean culture and traditions.

16.2.4   Relation of Shinrin Yoku and Grounding, to Chilean Culture and Traditions

The concepts and practices of *Shinrin Yoku* and grounding complement many Indigenous peoples' worldviews, in that they view nature and humans as a single inter-connected entity. Earth-based spirituality is the oldest recorded religious worldview and the central tenet of the South American Pachamama cosmovision (Crane-Seeber and Crane 2010). Most Andean Indigenous groups view the Earth as a mother both for humans and for other beings with whom they share a fraternal relation. For example, in both Indigenous and Chilean popular culture, the Earth is often referred to as *pachamamá*. *Pacha* derives from an Aymara and Quechua term, meaning *earth*, or *world*, or *universe*. *Mamá*, derives from mother. Thus, *pachamamá*, as understood by the Quechua and Aymara communities, and other Andean ethnic groups that have received Quechua-Aymara influences, refers to the concept of the Earth as a mother, and as a deity of sorts. Similarly, the Mapuche people of Chile refer to mother Earth in their language (*mapudungún*) as *ñuke mapu*. This conceptualization of the Earth does not refer to the soil, the geological Earth, or the planet Earth, but rather is more encompassing in scope. Unlike the *pachamamá*, the *ñuke mapu* is not considered a deity; instead, within Mapuche religious beliefs she is considered the representation of the Mapuche *world* in the cosmography and the interaction of the Mapuche people within it.

The harmonious, spiritual, and respectful relationship with nature reflected within the concepts of *pachamamá* and *ñuke mapu* is central for many belief systems and is common in the worldview of Indigenous peoples. For example, within the Mapuche worldview, the term *küme mogen* signifies a life that is lived with balance between society, nature, and spirituality, and describes this life as the foundation of human existence, often referred to as *buen vivir* (in Spanish), or *good living* (in English). Similar concepts are found within the conscience and culture of most—if not all—Indigenous peoples (Tórtora Aravena 2021). The *küme mogen* is sustained by the *itro fill mogen*, a concept that has been maintained for centuries within the Mapuche cosmovision that refers to a synchronization between all forms of life and being, and considers all as being part of the same, without exception (Weke 2017). *Itro fill mogen* posits that is referred to as *itro fill mogen*, as the recognition and contemplation of this synchronization may provide healing by awakening or activating the senses and connecting them with life and the vibrations of nature.

There is ample evidence of the efforts made by post-Columbian forces to impose western-based spirituality onto Andean and Latin American Indigenous and popular culture. However, there has been a growing recognition of the cultural permanence of Earth-based spiritual beliefs over the last decade, especially in light of a rising social movement oriented around the concept of *buen vivir*, and a fundamental sense of harmony between humans and nature (Coscieme et al. 2020; Ednie et al. 2020; Gudynas 2011; Nicoletti and Barelli 2019; Villavicencio Calzadilla and Kotzé

2018). As a social movement, *buen vivir* encompasses many of the same tenets as are found in Indigenous cosmovision. For example, Walsh (2010) noted, "In its most general sense, buen vivir denotes, organizes, and constructs a system of knowledge and living based on the communion of humans and nature and on the spatial-temporal-harmonious totality of existence" (p. 18). Beginning in Bolivia and Ecuador, contemporary popular culture manifestations of *buen vivir* have been adopted by other South American cultures and countries, including Brazil, Columbia, Peru, and Chile. These manifestations, also driven through social movements, have contributed to the proposal of political and economic alternatives to mainstream neoliberal development theory, with a focus on agrarian, collective, and sustainable use of natural resources. Importantly, they posit a more direct connection between nature and human well-being, and align strongly with ecological concerns, Indigenous worldviews, and rights (Altmann 2014; Bressa Florentin 2019; Bruckmann 2010; Gudynas 2011; Merino 2016).

As such, *buen vivir* has begun to inform and permeate wide-ranging aspects of modern Andean life and culture, including the CONAF Nature Bathing program, which draws on a range of concepts, including *Shinrin Yoku, grounding,* and *buen vivir,* to immerse and connect visitor consciousness and senses with nature. With respect to visitor consciousness, CONAF's Nature Bathing program is oriented around the understanding that we are part of nature and that our mutual relationship is one of respect and reciprocal benefit for all forms of life. From a *sensory* standpoint, the program employs a multiverse—or pluriverse—lens, seeking to provoke increased recognition of our coexistence with all other living beings and the plural meanings and connotations that this implies.

16.3 Applying Theoretical Constructs in Chile

16.3.1 Implementation of CONAF Nature Bathing in Chilean Protected Areas

Intrigued by growing international interest about the health benefits associated with Forest Bathing, in 2016 CONAF began to explore how such programs might further the agency's goals to benefit local communities, promote healthy living habits, provide universally accessible and socially inclusive recreation opportunities, and build partnerships with communities, focusing first on developing ideas and internal communication within the agency and its Corp of administrators, rangers, and professionals. Early on, agreement was reached to develop the concept of *nature bathing*—as opposed to forest bathing—to reflect the importance and potential for connecting with the diverse range of Chilean ecosystems protected through SNASPE. SNASPE's 107 PAs extend well beyond forests, to include grasslands, pampas, deserts, glacial ice fields, ocean shoreline, and much more.

The first CONAF Nature Bathing activities occurred in 2018 through a combination of in-house and partnership programs. For example, a key partnership for CONAF has been the European Forest Therapy Institute (FTI). FTI runs a scholarship program to train and certify park rangers to guide forest bathing programs, including two CONAF rangers based in the Coquimbo Región of Chile. Another initial partnership program was developed with SENAMA, the Chilean National Service for the Elderly. This partnership focused on strengthening SENAMA's *Vive tu Naturaleza* (Live your Nature) program, which helps older adults focus on active and positive aging. CONAF's Nature Bathing activities were added as one of three lines of action within the existing *Vive tu Naturaleza* program.

In 2019, after these initial partner programs, CONAF began to work toward a more systematic integration of nature bathing practices within SNASPE PAs through their *Nature for Everyone* program by developing broader technical tools and training for its park ranger staff. Some of the technical tools were aimed at the regional offices of CONAF and focused on efforts to standardize program implementation in their respective PAs. Others were geared toward directly assisting PA visitors who had a current interest in these practices. Through these programs, CONAF's nature bathing practices took on their modern form, which involves providing park visitors with a 2–3-hour immersive experience in the nature of a PA. Park rangers facilitate the opportunity for visitors to activate their senses and immerse themselves in nature using a number of reflection and attentive meditation (mindfulness) practices (Lazo et al. 2021). During 2019, 50 park rangers were trained through the *Bosques Para Ti* (Forests for You) program, a non-profit group with a variety of PA and land management partners. CONAF's involvement in leadership training quickly grew, and CONAF staff eventually began leading workshops in partnership with other agencies. CONAF also began promoting nature bathing programs along with other efforts related to nature accessibility and social inclusion within Chile and the larger Latin American region. A wide variety of strategies and outlets were employed, including the support of several local and regional media outlets for radio and TV interviews, as well as dissemination of information through national and international magazines and social media.

During the COVID-19 pandemic in 2020, virtual meetings and events were held to continue training efforts as well as promote CONAF's work in universal accessibility, social inclusion, and nature bathing. These events included seminars, workshops, talks, and continuing training courses for park rangers. In addition, CONAF, and the former Forest Therapy Institute, organized the American Congress on Health Practices Associated with Nature. The congress was attended by 2000 people and 26 international experts from around the world, including the United States, Spain, Colombia, Chile, Argentina, Portugal, South Africa, Peru, Awajun Nation (Amazon Peru), Wampis Naw Peru, Ireland, and Brazil. As of December 2021, CONAF's Nature Bathing program had applied *Shinrin Yoku* and grounding techniques within 21 of its PAs. A total of 100 park rangers had received basic training and acquired Nature Bathing program facilitation techniques.

16.3.2 Components of CONAF Nature Bathing Programs Within Chile's Protected Areas

Specific locations within PAs have been designated as nature bathing sites and trails, selected for their potential to foster meaningful experiences and connections with nature. These sites often include qualities known to align with visitor preferences, like native forests, streams, and places with the presence of a variety of birds (Lazo et al. 2021). The trails are often developed to be ADA accessible and include basic, natural infrastructure to facilitate the Nature Baths experience (tree stumps as stools, etc.). CONAF's rangers are trained to implement Nature Bathing programs that typically consist of four steps (Fig. 16.1), with program components ranging from experience preparation to post-experience reflection (Lazo et al. 2021).

Step 1 focuses on conscious breathing to help participants prepare their bodies and minds to fully absorb the nature bathing experience. Forest rangers lead participants in a conscious breathing exercise, giving them a few moments to reflect on their breathing and awaken their senses to nature.

Fig. 16.1 The four-step CONAF Nature Bathing process

Step 2 sets the experiential program in motion. Participants begin to move through nature, stopping in places they find attractive to engage their senses with the environment. By focusing their minds on experiencing nature with a particular sense, acuity is activated, and participants begin to take presence of small details and elements that may not have been initially noticed. Participants are encouraged to take their time and engage with as many senses as they are able, noting distinctive color variations, differences in texture, sound and vibrations, tastes of edible plants, and smells generated by rubbing and touching natural elements.

Step 3 involves continuing movement through nature, focusing on enhancing participants' contemplation of and connection (emotional, spiritual, mental, and physical) with nature. Similar to Step 2, participants are encouraged to find places to stop and purposely engage their senses with natural elements. They are encouraged to focus on Here and Now, letting themselves grow in awareness that they are part of nature. This step can also include holding natural elements and/or expressing their connections with nature through art. If participants are in a group, they are encouraged to occasionally engage in quiet conversation about their experiences and observations.

Step 4 involves a purposeful end to the nature bathing experience. Sometimes participants are encouraged to draw a line, choose the last tree along a nature bathing trail, or place a branch on the ground and move past this chosen feature to symbolize the end of their session. Herbal teas or other small snacks are often shared, ideally made with local materials and ingredients. Participants then reflect on and share their experiences as they enjoy the flavors of the natural bathing environment.

16.3.3 Expanding the Benefits of Nature Bathing in Chilean PAs, Through Wellness Tourism

Nature-based tourism, which involves travel to and engagement with natural environments through recreation and contemplation, can contribute to health and wellness in a variety of ways, including physical health, psychological wellness, psychosocial development, and spiritual upliftment (Curtin 2009; Quine Lackey et al. 2021; Von Lindern 2015). For example, it can support psychological wellness by helping people recover from stress and mental fatigue (Ulrich et al. 1991; Weiler and Davis 1993), cognitive wellness by helping them concentrate better (i.e., attention restoration; Kaplan 1995), psychosocial development by improving social contact opportunities that allay personal isolation (Dryglas and Salamaga 2016), and contribute to spiritual upliftment through development of new values, increased personal awareness, and empathy (Wolf et al. 2017).

Intentional programs like CONAF's Nature Bathing may strengthen the health benefits of nature-based tourism as well as help PA gateway communities and tourism destinations recover their tourism sector post-pandemic through the development of domestic and international wellness tourism programs and experiences.

According to Lu et al. (2020), in 2019, more than eight-billion people visited nature reserves around the world for relaxation and recovery before the pandemic. Wellness tourism, according to the Global Wellness Institute (2021), is "travel associated with the pursuit of maintaining or enhancing one's personal well-being" (p. 77). Wellness tourism is a growing sector of tourism, accounting for 6.5% of all tourism trips in 2020, and is typically more heavily engaged in by domestic tourists (Global Wellness Institute 2021).

Wellness may be the primary or secondary focus of travel for wellness tourism segment participants. When wellness is a secondary focus for travel, travelers' interests in maintaining good health and/or a wellness lifestyle during their travels affects some of their choices and activities (Global Wellness Institute 2021). Most wellness travel is secondary (92% of wellness tourism trips and 90% of wellness tourism expenditures in 2020), and this is expected to increase post-pandemic as travelers place greater emphasis on self-care. As noted by the Global Wellness Institute (2021), the COVID-19 pandemic has "expanded our understanding of self-care toward a more holistic concept encompassing healthy eating, exercise, social connections, sleep, creativity, nature, and much more" (p. 90). Supporting this claim, Qiu et al. (2021) found that after years of social isolation there is a renewed hunger for touch, human connections, travel, and experiences in nature.

Chile is not currently listed within the top 20 destination markets for wellness tourism; however, the country is well-positioned to support additional wellness tourism opportunities based on its PAs and other natural settings such as thermal and mineral springs. Wellness tourism programs, including nature bathing, can be developed within and around Chile's PAs to offer the opportunity for visitors to experience wellness benefits within nature. Incorporating a greater focus on wellness within Chile's primarily adventure-focused nature-based tourism industry would help to add relevance and accessibility for a broader audience, diversifying the current offering and boosting PA efforts to support resiliency and sustainability across the region. Thus, future collaboration between PAs and their PA gateway communities and destinations might focus on the development of domestic and international wellness tourism programs and experiences.

16.4 Future Implications and Needs

CONAF's leadership continues in Chile and Latin America with regard to work on universal accessibility, social inclusion, and nature bathing, as demonstrated by their ongoing service in this area. They continue to provide technical advice to public agencies, such as National Park Services in Latin America and other international entities including work with the National Health Institute (INS) of Colombia and the Latin American Academy of Social Studies (ALES). CONAF also frequently consults with doctors specialized in mental health to promote partnerships with public health initiatives in Chile, and with Chilean municipalities to encourage the implementation of the nature bathing practice in urban parks and greenspace settings. Other efforts to develop resources and promote nature bathing experiences include the establishment of a virtual library of information, tools, and a series of digital files containing sensorial nature tours with immersive visual and soundscape elements of PAs (see https://www.conaf.cl/parques-nacionales/visitanos/banos-de-naturaleza/). This resource was developed to bring nature closer to all people at a time when many were either confined to their homes by the COVID-19 pandemic or were unable to visit PAs or other natural environments for other reasons.

As initial efforts to develop and implement nature bathing programs have demonstrated success and grown exponentially, leaders of the project have recently shifted focus toward standardizing this work and incorporating it within formal agency plans. In 2021, the official national Nature Baths Program was developed in the SNASPE units with the following objective: To guide, systematize, and standardize the technical management and promotion of health practices (Nature Baths) associated with the nature contained in the protected areas of the State of Chile, for the benefit of the integral health of all people who visit them (Lazo et al. 2021).

The program has now become more institutionalized within CONAF and is also continuing to expand within local communities around SNASPE protected areas through collaborative initiatives. For example, in September 2022, the program partnered with the Austral Garden Route Cooperative, a local community-based initiative that promotes the interconnectivity of humans and nature within the northern reaches of the Aysén Region of Patagonia. Together, these organizations co-produced the first online international symposium on reconnecting with nature for better health. Presenters included CONAF administrators and park rangers with experience in nature bathing, international experts, and members of the cooperative. Topics ranged from nature bathing and grounding to discussions of the Mapuche worldviews and how to better connect children with nature. Intertwined efforts like this help to broaden awareness and interest in the program, connecting CONAF rangers with entities and persons in their local communities that share similar interests and knowledge.

This chapter has focused on the potential of nature to contribute to human mental health and resilience. The 107 Protected Areas (PAs) of Chile's national system of protected natural areas (SNASPE) offer important infrastructure that can contribute to all levels of resilience, public health, and human well-being. Through low-cost, intentionally designed, universal access trails and socially inclusive mindfulness programs within PAs, we believe that conservation can support human and nature interconnectedness. Moreover, these programs and infrastructure can be replicated in private PAs and even urban parks. Considering current trends of urbanization and rising health concerns, the continual awareness and growth of nature bathing opportunities within Chilean and Patagonian PAs could have a notable effect on community resilience.

CONAF's intent is to continue to build nature bathing within the 21 PAs where programming and infrastructure are already established, and to expand the practice to additional parks. To do so, the agency is continually working to train more park rangers as nature bathing guides. In addition, CONAF is calling for research to validate the mental and physiological benefits of the Chilean native forest, including clinical studies that investigate the potential benefits of nature bathing for human resilience and health. The quality and diversity of Chile's SNASPE PAs offer the opportunity for several lines of research. Studies that differentiate the volatile organic compounds across Chilean forest types and identify how particular compounds contribute to human health could help to design nature bathing programs that target the most beneficial locations and activities. Alternatively, research that develops a better understanding of the perceived and experienced physiological and mental effects of particular Chilean ecosystems would also help agencies such as CONAF prioritize new nature bathing programs.

In addition to research, advocacy and communication efforts will be critical for expanding nature bathing programs throughout the region. Gradually, doctors (especially alternative medicine doctors) have begun to prescribe time in nature as a compliment to other medical treatments. To encourage the expansion of this practice, advocacy efforts should focus on increasing free access to PAs for persons participating in these types of medical therapies. Similarly, partnerships with the medical community around health benefits associated with nature exposure might focus on developing greater accessibility to particular forest environments that provide the greatest potential for preventative healthcare and healing. Finally, more effective strategic communication strategies will be needed to promote nature bathing programs within Chile and Patagonia. Dissemination methods that optimize and increase sectoral alliances will help raise awareness and support for programming. They will also contribute to continued evolution of nature bathing and the development of new programs that strengthen conservation and human interconnectivity with nature and align with societal goals of improved accessibility, health, and resilience.

Acknowledgments This work was supported by Chile's National Research and Development Agency (ANID) under ANID's Regional Program R17A10002; the CIEP R20F0002 project; and the CHIC-ANID PIA/BASAL PFB210018.

References

W.N. Adger, Social and ecological resilience: are they related? Prog. Hum. Geogr. **24**(3), 347–364 (2000). https://doi.org/10.1191/030913200701540465

A.A. Al-Nuaim, Y. Al-Nakeeb, M. Lyons, H.M. Al-Hazzaa, A. Nevill, P. Collins, M.J. Duncan, The prevalence of physical activity and sedentary behaviours relative to obesity among adolescents from Al-Ahsa, Saudi Arabia: rural versus urban variations. J. Nutr. Metab. **2012**, 417589 (2012). https://doi.org/10.1155/2012/417589

P. Altmann, Good life as a social movement proposal for natural resource use: the indigenous movement in Ecuador. Consilience **12**(1), 82–94 (2014). https://doi.org/10.7916/D8GF0T6G

M. Antonelli, G. Barbieri, D. Donelli, Effects of forest bathing (Shinrin-yoku) on levels of cortisol as a stress biomarker: a systematic review and meta-analysis. Int. J. Biometeorol. **63**(8), 1117–1134 (2019). https://doi.org/10.1007/s00484-019-01717-x

F. Berkes, H. Ross, Community resilience: toward an integrated approach. Soc. Nat. Resour. **26**(1), 5–20 (2013). https://doi.org/10.1080/08941920.2012.736605

F. Berkes, J. Colding, C. Folke, *Navigating Social-Ecological Systems: Building Resilience for Complexity and Change* (Cambridge University Press, 2002)

M.G. Berman, E. Kross, K.M. Krpan, M.K. Askren, A. Burson, P.J. Deldin, et al., Interacting with nature improves cognition and affect for individuals with depression. J. Affect. Disord. **140**(3), 300–305 (2012). https://doi.org/10.1016/j.jad.2012.03.012

F. Beute, Y.A.W. De Kort, Natural resistance: exposure to nature and self-regulation, mood, and physiology after ego-depletion. J. Environ. Psychol. **40**, 167–178 (2014). https://doi.org/10.1016/j.jenvp.2014.06.004

D. Bressa Florentin, *The Political Process of Buen Vivir: Contentious Politics in Contemporary Ecuador* (University of Bath, 2019)

R. Brown, G. Chevalier, Grounding the human body during yoga exercise with a grounded yoga mat reduces blood viscosity. Open J. Prev. Med. **5**(04), 159–168 (2015). https://doi.org/10.4236/ojpm.2015.54019

M. Bruckmann, Alternative visions of the indigenous people's movement in Latin America. Soc. Change **40**(4), 601–608 (2010). https://doi.org/10.1177/004908571004000411

R.M. Carrillo-Larco, A. Bernabé-Ortiz, T.D. Pillay, R.H. Gilman, J.F. Sánchez, J.A. Poterico, et al., Obesity risk in rural, urban and rural-to-urban migrants: prospective results of the PERU MIGRANT study. Int. J. Obes. **40**(1), 181–185 (2016). https://doi.org/10.1038/ijo.2015.140

G. Chevalier, S.T. Sinatra, J.L. Oschman, K. Sokal, P. Sokal, Earthing: health implications of reconnecting the human body to the earth's surface electrons. J. Environ. Public Health **2012**, 291541 (2012). https://doi.org/10.1155/2012/291541

L. Coscieme, H. da Silva Hyldmo, Á. Fernández-Llamazares, I. Palomo, T.H. Mwampamba, O. Selomane, et al., Multiple conceptualizations of nature are key to inclusivity and legitimacy in global environmental governance. Sci. Environ. Policy **104**, 36–42 (2020). https://doi.org/10.1016/j.envsci.2019.10.018

J. Crane-Seeber, B. Crane, Contesting essentialist theories of patriarchal relations: evolutionary psychology and the denial of history. J. Mens Stud. **18**(3), 218–237 (2010). https://doi.org/10.3149/jms.1803.218

C. Cristi-Montero, R.F. Rodríguez, Paradoja: "Activo físicamente pero sedentario, sedentario pero activo físicamente". Nuevos antecedentes, implicaciones en la salud y recomendaciones (Paradox: "Physically active but sedentary, sedentary but physically active". New background, health implications and recommendations). Rev. Med. Chil. **142**(1), 72–78 (2014). https://doi.org/10.4067/S0034-98872014000100011

S. Curtin, Wildlife tourism: the intangible, psychological benefits of human-wildlife encounters. Curr. Issue Tour. **12**(5–6), 451–474 (2009). https://doi.org/10.1080/13683500903042857

D. Dryglas, M. Salamaga, Applying destination attribute segmentation to health tourists: a case study of Polish spa resorts. J. Travel Tour. Mark. **34**, 503–514 (2016). https://doi.org/10.1080/10548408.2016.1193102

A. Ednie, T. Gale, K. Beeftink, A. Adiego, Connecting protected area visitor experiences, wellness motivations, and soundscape perceptions in Chilean Patagonia. J. Leis. Res. **53**(3), 377–403 (2020). https://doi.org/10.1080/00222216.2020.1814177

M. Ertem, S. Capa, M. Karakas, H. Ensari, A. Koc, Investigation of the relationship between nurses' burnout and psychological resilience levels. Clin. Exp. Health Sci. **10**(1–9), 9–15 (2020). https://doi.org/10.33808/clinexphealthsci.600924

Food and Agriculture Organization of the United Nations, & Pan American Health Organization, *Aprobación de nueva ley de alimentos en Chile: Resumen del proceso, entrada en vigor Junio 2016 [Approval of New Food Law in Chile: Summary of the Process, Entry into Force June 2016]* (FAO & PAHO, 2017). Retrieved 16 Dec 2022, from https://iris.paho.org/bitstream/handle/10665.2/51643/PolicyBriefOPSFAO_spa.pdf?sequence=1&isAllowed=y

M. Ghaly, D. Teplitz, The biologic effects of grounding the human body during sleep as measured by cortisol levels and subjective reporting of sleep, pain, and stress. J. Altern. Complement. Med. **10**(5), 767–776 (2004). https://doi.org/10.1089/acm.2004.10.767

Global Wellness Institute, *The Global Wellness Economy: Looking Beyond COVID* (Global Wellness Institute, 2021). Retrieved 23 Feb 2023, from https://globalwellnessinstitute.org/wp-content/uploads/2021/11/GWI-WE-Monitor-2021_final-digital.pdf

O. Gruebner, M.A. Rapp, M. Adli, U. Kluge, S. Galea, A. Heinz, Cities and mental health. Dtsch Arztebl Int. **114**(8), 121–127 (2017). https://doi.org/10.3238/arztebl.2017.0121

E. Gudynas, Buen Vivir: today's tomorrow. Development **54**(4), 441–447 (2011). https://doi.org/10.1057/dev.2011.86

T. Hartig, K. Korpela, G.W. Evans, T. Gärling, A measure of restorative quality in environments. Scand. Hous. Plann. Res. **14**(4), 175–194 (1997). https://doi.org/10.1080/02815739708730435

K. Ingulli, G. Lindbloom, Connection to nature and psychological resilience. Ecopsychology **5**(1), 52–55 (2013). https://doi.org/10.1089/eco.2012.0042

IPSOS, *One Year of COVID-19: IPSOS Survey Conducted for the World Economic Forum* (IPSOS, 2021). Retrieved 16 Dec 2022, from https://www.ipsos.com/sites/default/files/ct/news/documents/2021-04/wef_-_expectations_about_when_life_will_return_to_pre-covid_normal_-final.pdf

R. Kaplan, The role of nature in the context of the workplace. Landsc. Urban Plan. **26**(1–4), 193–201 (1993). https://doi.org/10.1016/0169-2046(93)90016-7

S. Kaplan, The restorative benefits of nature: toward an integrative framework. J. Environ. Psychol. **15**(3), 169–182 (1995). https://doi.org/10.1016/0272-4944(95)90001-2

A. Lazo, F. Peralta, H. Velásquez, P. Correa, C. Correa, P. Layana, *Manual diseño e implementación de senderos para la práctica de baños de la naturaleza [Manual Design and Implementation of Trails for the Practice of Nature Baths]* (National Forestry Corporation, 2021). Retrieved 18 Dec 2022, from https://warnercnr.colostate.edu/wp-content/uploads/sites/2/2021/06/MANUAL-DISENO-SENDEROS-SHINRIN-YOKU-ENERO2021.pdf

J.Y. Lee, D.C. Lee, Cardiac and pulmonary benefits of forest walking versus city walking in elderly women: a randomized, controlled, open-label trial. Eur. J. Integr. Med. **6**(1), 5–11 (2014). https://doi.org/10.1016/j.eujim.2013.10.006

C.J. Lemieux, S.T. Doherty, P.F.J. Eagles, M.W. Groulx, G.T. Hvenegaard, J. Gould, et al., Policy and management recommendations informed by the health benefits of visitor experiences in Alberta's protected areas. J. Park. Recreat. Adm. **34**(1), 24–52 (2016). https://doi.org/10.18666/JPRA-2016-V34-I1-6800

Q. Li, *The Power of the Forest. Shinrin Yoku: How to Find Happiness and Health Through Trees* (Viking Press, 2018)

Q. Li, M. Kobayashi, Y. Wakayama, H. Inagaki, M. Katsumata, Y. Hirata, et al., Effect of phytoncide from trees on human natural killer cell function. Int. J. Immunopathol. Pharmacol. **22**(4), 951–959 (2009). https://doi.org/10.1177/039463200902200410

J.J.W. Liu, M. Reed, T.A. Girard, Advancing resilience: an integrative, multi-system model of resilience. Personal. Individ. Differ. **111**, 111–118 (2017). https://doi.org/10.1016/j.paid.2017.02.007

S. López-Pousa, G. Bassets Pagès, S. Monserrat-Vila, M. de Gracia Blanco, J. Hidalgo Colomé, J. Garre-Olmo, Sense of well-being in patients with fibromyalgia: aerobic exercise program in a mature forest—a pilot study. Evid. Based Complement. Altern. Med. **2015**, 614783 (2015). https://doi.org/10.1155/2015/614783

N. Lu, C. Song, T. Kuronuma, H. Ikei, Y. Miyazaki, M. Takagaki, The possibility of sustainable urban horticulture based on nature therapy. Sustainability **12**(12), 5058 (2020). https://doi.org/10.3390/su12125058

M. Marzukhi, N. Masyitah Ghazali, O. Ling Hoon Leh, H. Yakob, A bidirectional associations between urban physical environment and mental health: a theoretical framework. Environ. Behav. Proc. J. **5**(13), 167–173 (2020). https://doi.org/10.21834/e-bpj.v5i13.2048

F.S. Mayer, C.M. Frantz, E. Bruehlman-Senecal, K. Dolliver, Why is nature beneficial?: the role of connectedness to nature. Environ. Behav. **41**(5), 607–643 (2009). https://doi.org/10.1177/0013916508319745

E.A. McMahan, D. Estes, The effect of contact with natural environments on positive and negative affect: a meta-analysis. J. Posit. Psychol. **10**(6), 507–519 (2015). https://doi.org/10.108 0/17439760.2014.994224

W. Menigoz, T.T. Latz, R.A. Ely, C. Kamei, G. Melvin, D. Sinatra, Integrative and lifestyle medicine strategies should include Earthing (grounding): review of research evidence and clinical observations. Explore **16**(3), 152–160 (2020). https://doi.org/10.1016/j.explore.2019.10.005

R. Merino, An alternative to "alternative development"?: Buen Vivir and human development in Andean countries. Oxf. Dev. Stud. **44**(3), 271–286 (2016). https://doi.org/10.1080/1360081 8.2016.1144733

R. Mitchell, F. Popham, Greenspace, urbanity and health: relationships in England. J. Epidemiol. Community Health **61**(8), 681–683 (2007). https://doi.org/10.1136/jech.2006.053553

R. Mitchell, F. Popham, Effect of exposure to natural environment on health inequalities: an observational population study. Lancet **372**(9650), 1655–1660 (2008). https://doi.org/10.1016/ S0140-6736(08)61689-X

Y. Miyazaki, *Shinrin Yoku: The Art of Japanese Forest Bathing* (Octopus Publishing Group, 2018)

E. Morita, S. Fukuda, J. Nagano, N. Hamajima, H. Yamamoto, Y. Iwai, et al., Psychological effects of forest environments on healthy adults: Shinrin-yoku (forest-air bathing, walking) as a possible method of stress reduction. Public Health **121**(1), 54–63 (2007). https://doi.org/10.1016/j. puhe.2006.05.024

J. Muir, The Yosemite. Garden City, N.Y., Century (1912). ISBN 0991996216

J. Newton, *Wellbeing and the Natural Environment: A Brief Overview of the Evidence* (University of Bath, 2007)

M.A. Nicoletti, A.I. Barelli, Blessed among all women: the missionary virgin, identity and territory in Patagonia. Stud. Relig. **48**(2), 258–281 (2019). https://doi.org/10.1177/0008429819831942

S. Núñez-González, J.A. Delgado-Ron, C. Gault, A. Lara-Vinueza, D. Calle-Celi, R. Porreca, D. Simancas-Racines, Overview of "systematic reviews" of the built environment's effects on mental health. J. Environ. Public Health **2020**, 9523127 (2020). https://doi. org/10.1155/2020/9523127

Y. Ohtsuka, N. Yabunaka, S. Takayama, Shinrin-yoku (forest-air bathing and walking) effectively decreases blood glucose levels in diabetic patients. Int. J. Biometeorol. **41**, 125–127 (1998). https://doi.org/10.1007/s004840050064

J. Oschman, G. Chevalier, R. Brown, The effects of grounding (earthing) on inflammation, the immune response, wound healing, and prevention and treatment of chronic inflammatory and autoimmune diseases. J. Inflamm. Res. **8**, 83–96 (2015). https://doi.org/10.2147/JIR.S69656

Pan American Health Organization & World Health Organization, *Situación de salud en las Américas: Indicadores básicos 2016 [State of Health in the Americas: Basic Indicators 2016]* (Organización Panamericana de la Salud, Organización Mundial de la Salud, 2016). Retrieved 19 Dec 2022, from https://iris.paho.org/bitstream/handle/10665.2/31288/ IndicadoresBasicos2016-spa.pdf?sequence=1&isAllowed=y

B.J. Park, Y. Tsunetsugu, T. Kasetani, T. Kagawa, Y. Miyazaki, The physiological effects of Shinrin-yoku (taking in the forest atmosphere or forest bathing): evidence from field experiments in 24 forests across Japan. Environ. Health Prev. Med. **15**, 18–26 (2010). https://doi. org/10.1007/s12199-009-0086-9

J. Pretty, J. Peacock, R. Hine, M. Sellens, N. South, M. Griffin, Green exercise in the UK countryside: effects on health and psychological well-being, and implications for policy and planning. J. Environ. Plan. Manag. **50**(2), 211–231 (2007). https://doi.org/10.1080/09640560601156466

R. Puhakka, K. Pitkänen, P. Siikamäki, The health and well-being impacts of protected areas in Finland. J. Sustain. Tour. **25**(12), 1830–1847 (2017). https://doi.org/10.1080/0966958 2.2016.1243696

M. Qiu, J. Sha, N. Scott, Restoration of visitors through nature-based tourism: a systematic review, conceptual framework, and future research directions. Int. J. Environ. Res. Public Health **18**(5), 2299 (2021). https://doi.org/10.3390/ijerph18052299

N. Quine Lackey, D.A. Tysor, G.D. McNay, L. Joyner, K.H. Baker, C. Hodge, Mental health benefits of nature-based recreation: a systematic review. Ann. Leis. Res. **24**(3), 379–393 (2021). https://doi.org/10.1080/11745398.2019.1655459

R. Ramírez, R.A. Agredo, El sedentarismo es un factor predictor de hipertrigliceridemia, obesidad central y sobrepeso (Sedentary lifestyle is a predictor of hypertriglyceridemia, central obesity and overweight). Rev. Colomb. Cardiol. **19**(2), 75–79 (2012). https://doi.org/10.1016/S0120-5633(12)70109-2

J. Real-Ramírez, L.A. García-Bello, R. Robles-García, M. Martínez, K. Adame-Rivas, M. Balderas-Pliego, et al., Well-being status and post-traumatic stress symptoms in health workers attending mindfulness sessions during the early stage of the COVID-19 epidemic in Mexico. Salud Ment. **43**(6), 303–310 (2020). https://doi.org/10.17711/sm.0185-3325.2020.041

P.A. Sandifer, A.E. Sutton-Grier, B.P. Ward, Exploring connections among nature, biodiversity, ecosystem services, and human health and well-being: opportunities to enhance health and biodiversity conservation. Ecosyst. Serv. **12**, 1–15 (2015). https://doi.org/10.1016/j.ecoser.2014.12.007

S.A. Satici, Psychological vulnerability, resilience, and subjective well-being: the mediating role of hope. Personal. Individ. Differ. **102**, 68–73 (2016). https://doi.org/10.1016/j.paid.2016.06.057

H. Selye, *The Stress of Life* (McGraw-Hill, 1984)

C. Song, H. Ikei, Y. Miyazaki, Physiological effects of nature therapy: a review of the research in Japan. Int. J. Environ. Res. Public Health **13**(8), 781–798 (2016). https://doi.org/10.3390/ijerph13080781

C. Song, H. Ikei, Y. Miyazaki, Sustained effects of a forest therapy program on the blood pressure of office workers. Urban For. Urban Green. **27**, 246–252 (2017). https://doi.org/10.1016/j.ufug.2017.08.015

K. Srivastava, Urbanization and mental health. Ind. Psychiatry J. **18**(2), 75–76 (2009). https://doi.org/10.4103/0972-6748.64028

B.D. Taff, V. Peel, W.L. Rice, G. Lacey, B. Pan, *Healthy Parks Healthy People: Evaluating and Improving Park Service Efforts to Promote Tourists' Health and Well-being Introduction* (Travel and Tourism Research Association: Advancing Tourism Research Globally, 2019), p. 7. https://scholarworks.umass.edu/ttra/2019/research_papers/27

J.M. Thomsen, R.B. Powell, C. Monz, A systematic review of the physical and mental health benefits of wildland recreation. J. Park. Recreat. Adm. **36**(1), 123–148 (2018). https://doi.org/10.18666/jpra-2018-v36-i1-8095

P. Timmerman, *Vulnerability, Resilience and the Collapse of Society. A Review of Models and Possible Climatic Applications* (Institute for Environmental Studies, 1981)

H.G. Tórtora Aravena, El "Buen Vivir" y los derechos culturales de naturaleza colectiva en el nuevo constitucionalismo latinoamericano descolonizador (The "Buen Vivir" and the cultural rights of collective nature in the new decolonizing Latin American constitutionalism). Rev. Derecho **28**, e3712 (2021). https://doi.org/10.22199/issn.0718-9753-2021-00015

M. Townsend, C. Henderson-Wilson, Healthy parks, healthy people: Evidence from Australia, in *Green Exercise: Linking Nature, Health and Well-Being*, ed. by J.P. Jo Barton, R. Bragg, C. Wood, (Routledge, 2016), pp. 89–99

UK Sustainable Development Commission, *Health, Place and Nature. How Outdoor Environments Influence Health and Well-Being: A Knowledge Base* (Sustainable Development Commission, 2008). Retrieved 23 Feb 2023, from https://www.sd-commission.org.uk/data/files/publications/Outdoor_environments_and_health.pdf

R.S. Ulrich, R.F. Simons, B.D. Losito, E. Fiorito, M.A. Miles, M. Zelson, Stress recovery during exposure to natural and urban environments. J. Environ. Psychol. **11**, 201–230 (1991)

A. Urzúa, A. Samaniego, A. Caqueo-Urízar, A. Zapata Pizarro, M. Irarrázaval Domínguez, Salud mental en trabajadores de la salud durante la pandemia por COVID-19 en Chile (Mental health problems among health care workers during the COVID-19 pandemic). Rev. Med. Chil. **148**(8), 1121–1127 (2020). https://doi.org/10.4067/s0034-98872020000801121

B. Vicente, S. Saldivia, R. Pihán, Prevalencias y brechas hoy; salud mental mañana (Prevalence and gaps today; mental health tomorrow). Acta Bioeth. **22**(1), 51–61 (2016). https://doi.org/10.4067/s1726-569x2016000100006

P. Villavicencio Calzadilla, L.J. Kotzé, Living in harmony with nature? A critical appraisal of the rights of Mother Earth in Bolivia. Trans. Environ. Law **7**(3), 397–424 (2018). https://doi.org/10.1017/S2047102518000201

E. Von Lindern, Setting-dependent constraints on human restoration while visiting a wilderness park. J. Outdoor Recreat. Tour. **10**, 29–37 (2015). https://doi.org/10.1016/j.jort.2015.06.001

C. Walsh, Development as Buen Vivir: institutional arrangements and (de)colonial entanglements. Development **53**(1), 15–21 (2010). https://doi.org/10.1057/dev.2009.93

B. Weiler, D. Davis, An exploratory investigation into the roles of the nature-based tour leader. Tour. Manag. **14**(2), 91–98 (1993). https://doi.org/10.1016/0261-5177(93)90041-I

J. Weke, Itrofill mogen: toda la vida sin excepción (Itrofill mogen: all life without exception). Endémico, 23 Nov 2017. https://endemico.org/itrofill-mogen-toda-la-vida-sin-excepcion/

N.M. Wells, S.P. Ashdown, E.H.S. Davies, F.D. Cowett, Y. Yang, Environment, design, and obesity: opportunities for interdisciplinary collaborative research. Environ. Behav. **39**(1), 6–33 (2007). https://doi.org/10.1177/0013916506295570

I.D. Wolf, G.B. Ainsworth, J. Crowley, Transformative travel as a sustainable market niche for protected areas: a new development, marketing and conservation model. J. Sustain. Tour. **25**(11), 1650–1673 (2017). https://doi.org/10.1080/09669582.2017.1302454

C. Zlotnick, J. Johnson, R. Kohn, B. Vicente, P. Rioseco, S. Saldivia, Epidemiology of trauma, post-traumatic stress disorder (PTSD) and co-morbid disorders in Chile. Psychol. Med. **36**(11), 1523–1533 (2006). https://doi.org/10.1017/S0033291706008282

Chapter 17
(Re)Imagining the Relationship Between Society and Nature in Northern Chilean Patagonia: Encounters and (Mis) Encounters with the Modern World

Hugo Marcelo Zunino and Florencia Spirito

Abstract In this chapter, we examine the emergence of alternative ways for inhabiting the territory of Southern Chile which invite us to consider that other models of living are possible. First, we compare Indigenous and Western paradigms for sustainability. This serves as a framework through which we can then evaluate different alternative living projects. We document and describe three local community projects that are being developed in the mountainous area of the Araucanía region, which have raised awareness through a profound transformation of the way people relate to nature, to each other, and to themselves. The three projects are as follows: (1) a community project that recreates Mesoamerican Indigenous practices; (2) the *Waldorf* Educational Project, which represents a pedagogical counterproposal developed by the European spiritual thinker Rudolf Steiner (1861–1925); and (3) permaculture projects that seek other, new forms of food production through a close link with nature. We observe that although these projects open paths, they also present limitations and contradictions to re-imagining our relationship with the world. We suggest that in order to respond to the current multidimensional crisis we must construct new forms of living that break away from the dualistic ways of thinking that we have inherited from modernity.

Keywords Patagonia · Human-nature relationships · Indigenous practices · Waldorf pedagogy · Permaculture

H. M. Zunino (✉)
Department of Social Sciences/Social Sciences Nucleus, Universidad de La Frontera, Temuco, Araucanía Region, Chile
e-mail: hugo.zunino@ufrontera.cl

F. Spirito
Department of Natural Sciences and Technology, Universidad de Aysén, Coyhaique, Aysén Region, Chile
e-mail: florencia.spirito@uaysen.cl

© The Author(s) 2023
T. Gale-Detrich et al. (eds.), *Tourism and Conservation-based Development in the Periphery*, Natural and Social Sciences of Patagonia,
https://doi.org/10.1007/978-3-031-38048-8_17

409

17.1 Introduction

17.1.1 Spaces of Opposition in Northern Chilean Patagonia

Contemporary times are characterized by the confluence of multiple crises, with the ecological crisis gaining relevance as evidenced by environmental degradation and the increasingly dire consequences of climate change on ecosystems and their biodiversity (IPCC 2021; Magnan et al. 2021). This situation has led the scientific world to identify the *Anthropocene* as a new geological era characterized by the role of humans as a transforming force of nature of global and geological scope, influencing the dynamics, and stability of ecosystems and their components (Svampa 2019). The responses of the modern state to the crisis have been to incorporate sustainability criteria in its policies that allow adjusting economic development models considering the preservation of the environment, goods, and services (Reyes-Guillén et al. 2018). However, these institutionalized sustainability efforts, generally derived from international pronouncements and agreements, have been insufficient to respond to the current situation of the planet. For example, we have exceeded by 74% the capacity of ecosystems to regenerate the natural goods that sustain life, thus exacerbating multiple social and political tensions, pushing the global system toward unknown changes (Rockström et al. 2009).

The above-described planetary situation reverberates in the Patagonia of Southern Chile. Forestry plantations (mainly of exotic trees such as pine and eucalyptus) have been driven by public policies that favor big capital, triggering socio-ecological conflicts in the territory, such as lack of access and land rights for Indigenous communities, elimination of native forests, water scarcity, food insecurity, and migration of peasants to urban areas (e.g., Carte et al. 2021; McNamara et al. 2021). At the same time, Chilean Patagonia has received an important migratory flow from the country's main urban centers, a flow characterized by young adults seeking to reinvent their ways of life in spaces associated with natural amenities (Hidalgo and Zunino 2012): a process that is sometimes associated with a deeper reinvention of their being and their relationship with others (Zunino et al. 2016; Zunino and Huiliñir 2017; Mardones and Zunino 2019).

In this chapter, we examine the emergence of alternative ways for inhabiting the territory of Southern Chile which invite us to consider that other models of living are possible. First, we compare and contrast Indigenous and Western paradigms for sustainability. This serves as a framework through which we can then evaluate different alternative living projects. We will document and describe three local community projects that are being developed in the mountainous area of the Araucanía region, which have raised awareness through a profound transformation of the way people relate to nature, to each other, and to themselves. The three projects are as follows: (1) a community project that recreates Mesoamerican Indigenous practices; (2) the *Waldorf* Educational Project, which represents a pedagogical counterproposal developed by the European spiritual thinker Rudolf Steiner (1861–1925); and (3) permaculture projects that seek other, new, forms of food production through a close link with nature. We observe that although these projects open paths, they also present limitations and contradictions to re-imagine our relationship with the world. We argue that these community initiatives are commendable and provide a first step but are insufficient to transform broader social scales. We suggest that in order to respond to the current multidimensional crisis produced from modernity, we must construct new forms of living, based on a rupture with the dualistic ways of thinking that commodify nature, generating the illusion of having an infinite economy in a finite world. This mechanistic illusion intertwines a model of domination and technical exploitation, dissociated from society and culture, which serves the reproduction of this hegemonic model and its mechanisms of power, knowledge, and being.

17.2 Theoretical Framework

17.2.1 The Depth of the Current Crisis and the Limits of Institutionalized Efforts

Contemporary times are characterized by a growing concern for the degradation of nature: forest fires of unprecedented size, extreme temperatures, mass extinction of species, melting glaciers, to name a few disasters that face us to an uncertain and catastrophic future (IPCC 2021). This ecological crisis has gone beyond the sphere of universities and academia to colonize everyday spaces, where climate change and global warming are recurring topics of conversation. The English newspaper *The Guardian* graphs this situation in all its intensity:

The signs of change are becoming more frequent in Sicily, where in August a monitoring station in the south-eastern city of Syracuse recorded a temperature of 48.8 °C, the highest ever set in Europe. Data collected by the Balkans and Caucasus observatory put the average temperature rise on the island over the past 50 years at almost 2 °C, rising to 3.4 °C in Messina on the north-east coast (Tondo 2021, para. 13).

The crisis we face as humanity is not only ecological, but interferes with four fundamental dimensions of community life: it is a *social crisis,* in that it manifests itself as a fracturing of society through racism, oppression, and marginalization of people; it is a *cultural crisis,* represented by the deployment of individualism, consumerism, and materialism in different spheres of our lives; it is an *economic* crisis, in that only a few people are owners of the means of production and global wealth; and it is a *political crisis,* marked by the elimination of spaces for deliberation, participation, self-determination, and emancipation of people and societies (Giraldo 2014; Segato 2013). This multidimensional crisis crosses all spheres of reproduction of life, thus transforming itself into a civilizational crisis with its root in modern logic. Under this paradigm, the *Self* is reduced to a point that it reflects on an external world that is alien to it, forgetting that the existence of the *Self* is only possible through an inter-corporal relationship with everything else: that is, in the link with other plant-bodies, other animal-bodies, other water-bodies (Giraldo 2012). Accordingly, several works define this hegemonic model as epistemi-cidal, eco-cidal, agro-cidal, and communitari-cidal, insofar as it proposes a homogeneous global society disconnected from local community logics, breaking the singularity of the local, of culture, to establish a (single) hegemonic and homogeneous way of life that can be controlled globally (Agoglia 2012; Giraldo 2012).

Since the 1970s and 1980s, concepts such as sustainability and resilience have been constructed as an institutional response to the crisis; multi-scale strategies designed and applied by governments and international organizations in the search for economic development models that can bring together elements of environmental and social equity (Reyes-Guillén et al. 2018). The accepted concept of Sustainable Development (SD) aims to enable the satisfaction of the needs of this generation without compromising the satisfaction of the needs of future generations (World Commission on Environment and Development 1987). The notion of resilience has emerged in parallel to the concept of sustainability and focuses on the amount of change that a system can undergo without losing its functions and structure. When considering social-ecological systems, the definition of resilience broadens to incorporate the capacity of the social system to organize and increase its ability to adapt to changing conditions in the socio-economic and natural environment (Folke et al. 2010, 2021).

The present circumstances we are experiencing, with accelerated climatic and environmental collapse, show the limitations of these institutionalized strategies. Between 1970 and 2016, in tandem with the emergence of the concepts of sustainability and SD, a global average of 68% of mammals, birds, amphibians, reptiles, and fish became extinct, mainly in Latin America and the Caribbean. So, after more than 50 years of the SD paradigm and a model based on indefinite economic growth, the concept of sustainability is being questioned for its inability to respond to the problems associated with the style of development that prevails in the world today (Herrera 2020). From these traditional sustainability discourses, problems can be solved only within the current scheme of thinking, with an exacerbated reliance on technology as the only solution to environmental and climate problems (e.g., Østergaard et al. 2020). Recent research aims to explore actions and conceptual tools that are radically different from the classical approaches to SD. This exploration becomes relevant when we recognize that the continuity of our civilization, based on capitalist forms of appropriating nature and human beings, has become unsustainable due to the lack of resilience of social and ecological systems (Bennett 2016; Ceballos et al. 2017).

17.2.2 Beyond Sustainability

From the perspective of Servigne and Stevens (2015), new adaptive actions must be advanced in anticipation of collapse. This implies an extraordinary effort to holistically understand and interpret, based on evidence, creativity, and imagination, in order to emancipate ourselves from our current capitalist dependence and embrace a holistic, radical, and systemic socio-technical transition. From a critical perspective of sustainability, in order to think of alternative ways of relating to each other and to our environment, it is central to overcome the cognitive separation between theory and practices, between the object and the subject, shifting attention to human beings and the way we construct realities from our experiences and stories. In this way, we recognize social movements, disputes, and actions as alternatives to modern hegemony, which question and build new ways of living, amalgamating the material, the human, and the natural with the spiritual, validating other frameworks of production and reproduction of life (Escobar 2011, 2012, 2019).

In particular, the Latin American decolonial perspective proposes the implementation of methodologies that allow us to value and consider *other ways of knowing*, encouraging a political model that seeks to change the terms and conditions of conversation between European knowledge, local knowledge, and Indigenous knowledge (Mignolo 2003; Robbins 2006; Restrepo and Rojas 2010). In recent years, a significant number of initiatives have arisen aimed at constructing spaces for thought and practice based on the wisdom of the Indigenous peoples of the continent, inviting us to think about the "pluriverse," where multiple forms of knowledge are possible (Escobar 2019; De La Cadena 2010; De Sousa 2010; Gudynas 2011). These efforts are oriented toward the construction of a relational ontology that allows us to overcome the abyss that separates us from the world, reconnecting us with nature and other human beings. Indigenous groups do not artificially separate nature and people; instead, they focus on holistic thinking and the relations and interdependencies between people and the environment. Social-ecological systems approaches attempt to do the same, but from a Western perspective.

This interweaving of different kinds of knowledge is hampered by the difficulties we modern researchers have in experiencing connections with our own natural and social environments. For more than 500 years, scientific thinking has colonized our practices, positioning us as passive observers of a reality that is alien to us. For Indigenous peoples, knowledge is revealed through an inner journey in which the person relates to a cosmic whole, based on his or her experience with their place and territory. This inner journey toward knowledge is part of a collective identity that is articulated around Pachamama (or *Ñuke Mapu* for Mapuche people) which enables the recognition of spiritual forces or energies that govern nature and grant powers to the elements it contains (Neira Ceballos et al. 2012). For Cajete (2004), the knowledge of Indigenous people about nature derives from direct phenomenological experience with the lived space, practices from which cosmologies are derived that connect human beings with the universe, and they, as participants, are an active and creative part of that totality. Similarly, Ñanculef Huaiquinao (2016) described that the Mapuche people, the original inhabitants of Chilean Patagonia, employed phenomenological observation to understand nature, based on a sense of unity with nature. This unity is also reflected in their language: Mapuzugun is the onomatopoeia of the primordial sound that emerges from nature and communicates its essence and state, the manner in which nature speaks to human beings, and in which nature and human beings converse through a universal, indivisible, indecipherable language.

Thus, efforts toward a pluriversal understanding of the world must overcome the shortcomings of Western research in order to fully comprehend the Indigenous relationship with the world. More often than not, these limitations have resulted in proposals that are largely reflective and theoretical, lacking in lived experience. It is precisely for this reason that we have chosen to consider the practice-based efforts of human communities that are trying to foster other ways of relating to the social and natural environment. These social experiments aim to re-imagine; to construct spaces for deliberation, participation, self-determination, and emancipation from the ethics of nurturance. They seek to provide new spaces to reconsider our relationship with nature, by emphasizing the interdependence between people and nature (Battson 2018). For Knierbein and Viderman (2018), relational ontologies have grounded many of the communal initiatives that have been developed, where they have generated spaces that attempt to interrelate human beings with the non-human worlds that surround them. In this way, from the multidimensional crisis that we have characterized, new spaces have emerged for rethinking human relationships with nature that center their focus on the interdependence between people and nature (Gilligan 2017).

Agroecology is one of the relevant practices associated with a new relational paradigm that is evolving as a counter-hegemonic proposal capable of transforming the agri-food system, based on other ways of inhabiting and relating to nature, in its totality. Agroecology breaks with modernity by defining itself as a plural and pluri-epistemological science, which coordinates the contributions of diverse disciplines and forms of local, peasant, or traditional knowledge, with the aim of developing and promoting sustainable, resilient, and locally based agri-food governance systems (Carlile and Garnett 2021). Thus, Agroecology develops horizontal and egalitarian ways of connecting people, establishing bonds with the past and the future by rescuing agricultural management practices that have been shaped over thousands of years, adapting themselves to changing climatic conditions and thereby providing resilience to these productive systems. These concepts have been fostered by peasant and Indigenous movements, claiming their ancestral culture, and are based on organizing work and production around principles of solidarity and cooperation. They relativize the capitalist economy by recognizing, valuing, and promoting non-capitalist forms of production and consumption, based on collective action; thus, establishing sovereignty and ensuring the right to food—which is inextricably linked to the right to life—as an exercise in transforming the world (Holt-Gimenez and Altieri 2013).

In the northern part of Chilean Patagonia, which we will refer to as *Norpatagonia*, various initiatives and community practices have emerged that offer alternatives to experience other social orders and ways of life. Many of these initiatives are being articulated through migratory processes of idealistic young people who have ventured into the reconstruction of their own lives and sense of identity (Huiliñir-Curío and Zunino 2017; Mardones and Zunino 2021). In this chapter, we will reflect on three active community projects that will allow us to observe and discuss the possible ways to respond and add meaning to our contemporary world and the multiple crises we face as humanity. Through our observations, we will reflect on the enormous challenges facing our society, to overcome the abyss between people and nature that our way of thinking has generated, making us strangers in the world we have created (Sanhueza-Céspedes 2018).

17.3   Case Study Methods

In this chapter, we will focus on the territory along the northern shore of Villarrica Lake, in the northern reaches of Chilean Patagonia (*Norpatagonia*). The city of Pucón is the epicenter of this territory and the setting for an interesting migratory process. Although there are no statistical surveys that show the magnitude of this migration, qualitative evidence suggests a strong migration during the period between 1992 and 2002, during which the total population of the Pucón Commune increased from 14,000 to 22,000 inhabitants (Chilean National Statistics Institute 1992, 2002). This wave of migration has been characterized through the term *lifestyle migrants*, which refers, in this case, to young adults who are relatively well-off (i.e., persons with graduate degrees, or entrepreneurs with significant social and cultural capital), and choose to migrate to places that—for various reasons—offer them opportunities to achieve what can be broadly defined as a better quality of life (Benson and O'Reilly 2009). In Pucón, the impact of this migrant population on the socio-cultural fabric of the commune has been remarkable. The arrival of lifestyle migrants has contributed to expanded social and cultural options, as expressed in numerous tourist ventures, including restaurants, hostels, bars, and non-traditional services featuring a variety of alternative products. This process of social transformation has unfolded in parallel with the consolidation of the tourism industry and the actions of public-private agents that have *touristified* the city, through infrastructure and services that have been designed solely for tourists at the cost of local culture and presence, resulting in early forms of socio-residential segregation (Huiliñir-Curío et al. 2019).

Our research focuses on three amenity-migrant driven initiatives unfolding in the mountainous area of the Araucanía region. Generally, these three initiatives have been undertaken by people who have intentionally decided to migrate to the Pucón Commune from large urban centers. The three initiatives of interest included a neo-Shamanic-inspired community, an educational project, and permaculture movements taking place in the territory. Specifically, we sought to characterize the initiatives that are driving lifestyle change, to reflect on the possibilities of social transformation they are provoking, and to visualize internal and external tensions that emerge from participants' realization of their life projects. We will work with three sources of information. First, we will use secondary sources to systematize the initiatives that are being deployed in the territory, drawing from social networks and recent research that has studied community initiatives from the perspective of life-style migration. Second, we will incorporate data from a set of in-depth interviews undertaken during the last 7 years with lifestyle migrants in these three alternative communities. Third, through self-reflective work, we will integrate our own direct first-hand experiences with people who are participating in these initiatives, during their daily deliberations. In analytical terms, we privilege the practical dimension of these experiences, emphasizing the deep observation of the phenomenon, moving away from abstraction to dialogue with *the phenomenon itself.*

17.4 Results

In this section we present a summary of each of the three amenity-migrant driven initiatives unfolding in the mountainous area of the Araucanía region, with exam-ples of the data that informed the subsequent discussion section in which we focus on the possibilities and limits of social transformation and the larger process of rethinking our relationship with the world.

17.4.1 The Community for Life Neo-Shamanic Community

Our study indicates that the Community for Life (CFL) neo-shamanic community offered a locus for people of diverse origins to discuss and adapt traditional knowl-edge to a contemporary context. Although CFL was not an Indigenous community, it characterized itself as a depository of original knowledge that has been passed down from ancestral Indigenous communities. Spiritual leaders of CFL identified themselves as *mestizos*, a Spanish language term which signifies "mixed," in English, and is used to refer to persons of mixed Spanish and Indigenous heritage. They considered themselves to be descendants of one original community that shared in brotherhood with all beings on Earth, forming a great *family of life.*

Similar to other neo-shamanic communities that have formed in Latin America, we observed the CFL adding a contemporary context to traditional symbols and rituals. For example, members of CFL placed Mother Earth at the center of their cosmovision, referring to the community as follows:

> [CFL] is there… to be able to serve the spirit… to be able to honor Pachamama [Mother Earth], to be able to honor nature, to honor life from a spiritual sense, and to follow ances-tral traditions, that is it. There is nothing new, nothing new is invented, ancestral traditions are being followed, that is the substance, that is the basis, that is what motivated me [to be here]. (Interview 1)

Mother Earth is conceived as the divine element permeating all that is perceived and is related to the sensitivity and intuition that flourishes in the South American continent.

Another example of the adaptation of traditional symbols to the contemporary context lies in one of the core aspects of the CFL's belief system, the Prophecy of the Eagle and the Condor. In the prophecy, the condor represents a force streaming from the south and is complemented by the eagle, a symbol of logical thought coming from the north. According to the prophecy, the union of the forces of both hemispheres will harmonize life and create knowledge to protect the Earth and all sentient beings (LaDuke 1992). This prophecy plays a central role in symbolizing the spiritual drive of CFL and is frequently referred to in ceremonies and rituals. One community member explained the prophecy, saying:

> ... the "Prophecy of the Eagle and the Condor" is to join the poles, join the two halves of the Earth, and from there create a better humanity, which is in tune with nature, with Earth. Religion is re-linking with the Earth, it is going back to the natural roots... it has to be guided by the stars, to be guided by the moon, to respect, to honor what is there, what it gives us, that is, how to see again.

The power of the cosmos was behind the impulse to seek a better, more united humanity. While nature, earth, and the moon are understood as the female elements of the cosmos, the stars symbolize the male element, the Great Spirit. This prophecy shows the duality of the world and the need to unite the opposites. The drive for balance and the harmonization of opposites also appeared in the impulse to unite people of diverse origins and overcome the detrimental aspects they perceive in society. Our interviews helped illuminate CFL members' dissatisfaction with the world they left behind and their hopes for the future:

> And suddenly this space, this spiritual current, based on ancestral traditions, gives people better satisfaction, gives them more answers to their concerns, whereas traditional religions do not give them those answers. I consider that it is because of the spiritual need of people, to fill that emptiness they feel, which leads them to approach [the community], and many discover an affinity with these ancestral traditions, and begin their journey.

17.4.2 The Waldorf-Pucón Education Community

At the time of this research, the Waldorf-Pucón education community occupied 40,000 m² in the Los Riscos sector of the Pucón Commune, about 5 kilometers from the city of Pucón, along the shores of Villarrica Lake, one of the main tourist destinations in the territory. The land for this community was donated by a private individual with the explicit purpose of building an anthroposophical-oriented school. Contributions from the education community were used to finance the school complex, consisting of three classroom modules with a total of approximately 220 m² of floor space. One of the modules is used primarily for community activities and administrative purposes.

The school began to operate in 2006 in a private, urban residence in the city of Pucón, and in 2012, moved to its current location. The neighborhood is characterized by rapid residential-tourist expansion and constitutes one of the most expensive and exclusive sectors of the city. Following the relocation, school enrollments grew rapidly between 2012 and 2014, going from 18 students in 2007, to 104 students in 2015 (Waldorf Pucón 2015). The school's website (www.waldorfPucón.org) describes its own community, saying,

> Many of us are immigrants to this area, either from other parts of Chile or from other parts of the world. We feel the call of nature, of simple living, and the need for time to share with family and to form a human community that supports one another. We have diverse spiritual visions of the world; we accept all religions and beliefs, and we follow the principles of Waldorf pedagogy given by Rudolf Steiner. (Waldorf Pucon 2015)

The school's architecture was a central tenet in their education philosophy, which put education in action through bioconstruction. Classrooms were made from straw bales and mud, with the intention of integrating an educational experience with the natural environment. They included uneven walls, floors, and ceilings, along with the use of irregular geometric shapes for door and window frames; all characteristic of buildings informed by the anthroposophical vision applied to Waldorf education. Anthroposophical architecture strove to create and protect places for socializing, which were used for student performances and as a meeting place for the educational community. This is consistent with research that has documented the natural environment as a central element in anthroposophical architecture, which stimulates contact with the external world and defines spaces suitable for social interaction (Oliveira and Imai 2015).

The Waldorf-Pucón community was organized around the school, with about 20 residences situated in the vicinity. Similar to the results that emerged from interviews with the (CFL) Neo-Shamanic Community, interview participants from the Waldorf-Pucón community described being guided by the drive to construct a world that was different from the one they left behind, specifically noting desires leave behind the individualistic approach of our current societies and instead foster a society that values community support. One of the school's directors described how this drive manifested within the project, suggesting intentional deviation from current educational norms:

> [The motivations for this project mark a] ... contrast to the Western world; traditional education is not educating the whole being; it is only transferring knowledge and filling-in with knowledge. Children need to have other types of learning and knowledge instances...

The Waldorf Pucón community was internally divided between those who dogmatically follow Steiner's model of thinking and those who mix different practices and ways of knowing. The search for balance was a constant challenge for the directors of the establishment, and tensions between these two points of view were frequent in meetings and deliberative assemblies. Nevertheless, the community has maintained a certain cohesion, forming a colorful suburban space that brought together a range of community activities, including barter fairs, commemoration ceremonies, various seminars, and recreational activities. Nevertheless, recent research conducted by Vergara et al. (2019) indicated the Waldorf Pucón community was culturally segregated from the surrounding population, which they characterized as mostly low-income Chilean peasants and Mapuche with a traditional way of life. The researchers also documented that contact between the two groups was restricted to sporadic meetings in the local neighborhood council and/or salaried work of the peasants either in the Waldorf Pucón school or in the homes of Waldorf-Pucón community members.

17.4.3 The Permaculture Movement in the Pucón Commune of Norpatagonia

The first approaches to permaculture in Chile date back to 2008, when the first certified course in permaculture design took place, held in the country's central zone. In the Pucón Commune, there are records of environmental initiatives with a focus on permaculture dating back to the early 1990s, linked to ecotourism ventures that sought to contribute to local development in a more sustainable way and thus enhance the conservation commitment of the people living in the Commune.

These initiatives were mainly driven by European and North American migrants, who reported moving to the region in search of greater contact with the pristine ecosystems that characterized the territory at that time. Vestiges of these early permaculture initiatives are still visible today. For example, one of the ecolodges in the center of Pucón offers its food services through the positioning of inviting visitors to, "...make food an instance to share and savor what the earth offers us today, to become aware of the present and enjoy our community and its locality."

One of the oldest projects observed through our research was located in the Huirilil sector, 48 km east of the city of Pucón. This community defined themselves as "custodians of a natural valley, lovers of nature and the development of consciousness." The community provides tourist services such as hiking and trekking, camping spaces, and amenities for events and ceremonies, with the objective of providing, "learning spaces that help to build meaning, aligned with the care and protection of the environment, in an attempt to perpetuate life for our future generations."

One of the more recent examples of intentional communities associated with permaculture design was recorded in the village of Ñancul, 65 km from the city of Villarrica, where a couple of health professionals, originally from Chile's capital city of Santiago, bought 60,000 m² in 2016, to develop one of the most ambitious permaculture projects in the territory, including sustainable housing, spaces for activities, and enclosures for cultural and educational gardens and farms.

These examples have laid the foundation for current permaculture trends in the Pucón and Villarrica Communes, where an increase in permaculture experiences was noted by our research team, mainly in peri-urban and rural areas of the territory. Guided by philosophies and concepts deriving from *deep ecology*, a *culture of peace*, and a desire to *co-create the present we dream of* these initiatives proposed alternative forms of food and fiber production, in harmony with the natural systems where they were immersed. Several of these permaculture projects sold their surplus production in the agro-ecological farmers' market in Pucón, offering their organic products at what was considered *a fair price* (Adams 2005).

Every permaculture project we observed in the study area shared a vision of transforming its communities into sustainable and energetically self-sufficient places that are environmentally and socially resilient. Most of these permaculture projects also incorporated spiritual practices that went beyond food production. Participants sought to increase awareness of the planetary situation, under the precepts that society must understand and cultivate connections with life by contributing to sustainability and good living. Research participants described the need for society to recognize that we are not separate from nature, and as such, the way we treat the planet is actually a reflection of the way we treat ourselves.

17.5 Discussion and Conclusions

17.5.1 Possibilities and Limits of Social Transformation

Our study examined three options emerging in Chilean Norpatagonia for constructing new ways of life that rethink our relationship with nature (and move away from the human/nature duality). The practices outlined above offer spaces for collective experimentation that reclaim the production and reproduction of the commons by reconceptualizing our relationship with non-human nature. They represent promising alternatives to the dominant approach that organizes, normalizes, and homogenizes these territories. Nevertheless, they face a series of limitations and contradictions, and do not succeed in altering the hegemonic dynamics.

In the CFL Neo-Shamanic community, we observed a set of practices that proclaimed unity between the human and natural worlds and the development of *hybrid narratives* (De la Torre 2014), in which traditional symbols and rituals are being adapted for a contemporary context. The Waldorf-Pucón education community aspired similarly, working toward the goal of creating a community environment founded on a living relationship with the spiritual world. Meanwhile, The Permaculture Movement in the Pucón Commune of Norpatagonia has tested alternative forms of food production that contribute to the agro-ecological transition of territories. These three initiatives align with what Svampa (2012) proposed as an eco-territorial shift, based on these new forms of activism and citizen engagement that are being expressed in Latin American territories and are centered on the defense of natural resources, biodiversity, and the environment.

Nevertheless, our analysis also identified some contradictions in the community initiatives. In each case, the initiatives formed segregated and privileged microterritories, where exclusionary and privatizing practices were common, replicating capitalist tendencies. The permaculture movements we observed behaved as independent utopian communities: while they practiced and experimented with more sustainable ways of relating to nature, they did not foster social networks with surrounding communities, and are therefore unlikely to be effective in spurring meaningful societal change. Furthermore, such isolation inhibits connection to regional marketplaces and limits their potential to oppose industries—such as resource extraction and other commercial activities—that erode the very sustainable livelihoods they seek to protect. Without local organization, it seems almost impossible for agroecology to grow geographically and to achieve the scale necessary to become a dominant form of food production (Mier y Terán Giménez Cacho 2018).

Despite the existence of these and other initiatives in the region seeking alternative lifestyles, capitalist expansion has not slowed down in the study area. On the contrary, the cities of Pucón and Villarrica are rapidly expanding into rural areas through subdivisions and land development ventures, which inexorably lead to a disregard of non-human nature's right to existence and a rampant development of the countryside. Rising land values and levels of interest from investors and individuals during the confinements imposed by the pandemic have motivated many area landowners to subdivide and sell part of their land. This rural growth and transformation have been compounded by rural young adults from the region who are migrating to the city (Vergara et al. 2019). The area today appears saturated with people—migrants from nearby urban centers—with increasing levels of segregation, and congestion problems. The subdivision has also produced environmental problems, including the loss of native forest from land clearing, that has affected vulnerable species. Subdivisions are also an important factor in water scarcity problems; first, because as native forests are eliminated, so too are the related water retention ecosystem services they provide, and second, because of the increased resource demands generated by new housing centers in areas that are not prepared. The generation of these new projects in unplanned areas causes territorial segregation, and pressures local governments to provide basic services in areas where they do not have the capacity. Thus, while alternative structures to capitalism have emerged in the area, they have failed, so far, to challenge the predominant model and scale up to more general social and spatial levels.

The inability of these initiatives to bring about meaningful change and propose new sustainability and socio-ecological resilience frameworks is not surprising. Each project was inserted in an already established thought paradigm. This paradigm has its foundation in classical thinkers such as Descartes, Bacon, and Kant whose ideas drove the scientific process for producing knowledge. In these thinkers we find a matrix that enshrines an ontological separation between humans and the world. We ponder over an external world that appears to us as something preordained and alien to our consciousness. This way of thinking is what has shaped our entire institutional system and our various approaches to regulating the social, natural, and territorial spheres. Thus, government plans and programs, bureaucratic and administrative regulations, cultural intervention programs, economic incentive programs, and other forms of intervention are dominated by a hierarchical and linear approach to the development process. Under these conditions, initiatives tend to replicate the elements that make up the underlying paradigm, including the social, economic, and cultural pathologies that we have examined in this chapter. It seems an illusion to believe that change can be effective without overturning the underlying thought paradigm that anchors us to the world we know and inhibits any impulse to reconnect with other scenarios. Thus, in our opinion, the main challenge demanding our attention is to transform the thought paradigm that we have inherited from modernity, and that continues to act in full force even in those discourses that are more progressive and emancipatory.

17.5.2   Rethinking Our Relationship with the World

Norpatagonia provides a unique setting for the study of alternative approaches to individual and community life that challenge the supremacy of rationality. The communities we have studied offer a starting point for investigating new ways of thinking and relating to other forms of life. Nevertheless, even these approaches have their own tensions and contradictions. We have demonstrated that in some cases, many of the efforts being made to build alternative lifestyles end up replicating forms of social exclusion and harmful human-nature relationships typical of capitalist societies.

We continue to value these efforts to imagine—and put into practice—new ways of relating to one another. These experiments in alternative living offer important learning opportunities for understanding what alternative futures might look like and for critical reflection on how Western and capitalist paradigms often prevail, even in the most well-intentioned projects.

Proposing new ways for civilization is a slow process. It takes time to build living relationships that extend and expand, that embrace diversity, and that reinvent the world, the territories, and the governance of goods and food. We need new socio-environmental pacts and alternatives. Proposals like those that arise from agroecology, from the common good, from *buen vivir*, from radical ecological democracy, from ecovillages, from ecofeminism, from community-based economies, from civilizational transitions, and from cooperative ecosystems. These are just some of the concepts that have been proposed to confront, reinvent, and replace our current social and environmental structures, and move past the civilizational crisis of the twenty-first century with a new paradigm founded on an ethic of respect, care, and equality (Kothari et al. 2019).

Such transformation will require more than a purely reflexive and abstract academic endeavor. We researchers also have to transform the way we understand our relationship with the world. Without this revision, we will continue to find it difficult to affirm the transformative flow that crosses important segments of humanity in these times of multidimensional crisis. We must have the audacity to think differently. This is a relevant epistemological challenge that invites us to observe the world in a different way, redefine our relationship with nature, removing the human/nature dualism through more citizen engagement, increased social networking, and support that moves beyond the modernist thinking that has colonized the world for more than 500 years. This epistemological revision seems central to contribute to the pluriverse and relational alternatives that have been proposed. The way out of the crisis of civilization should not be sought within it, but outside of it.

References

D. Adams, *Organic functionalism: An important principle of the visual arts* (Waldorf School Crafts and Architecture, 2005) Retrieved January 19, 2023, from https://www.waldorfresearchinstitute.org/pdf/BACraftsArchtRev.pdf

O. Agoglia, El Marco categorial de la crisis ambiental en un contexto globalizado [The categorial framework of the environmental crisis in a globalized context]. Desbordes **3**, 25–40 (2012). https://doi.org/10.22490/25394150.1190

G. Battson, *Love and Ecology as an Integrative Force for Good and as Resistance to the Commodification of Nature and Planetary Harms: Introducing Fluminism* (University of Wales Trinity Saint David, 2018)

J. Bennett, Global ecological crisis: Structural violence and the tyranny of small decisions, in *Addressing Global Environmental Challenges from a Peace Ecology Perspective*, ed. by H. Brauch, U. Oswald Spring, J. Bennett, S. Serrano Oswald, (Springer International Publishing, 2016), pp. 55–75

M. Benson, K. O'Reilly, Migration and the search for a better way of life: A critical exploration of lifestyle migration. Sociol. Rev. **57**(4), 608–625 (2009). https://doi.org/10.1111/j.1467-954X.2009.01864.x

G. Cajete, Philosophy of native science, in *American Indian Thought*, ed. by A. Waters, (Blackwell, 2004), pp. 45–57

R. Carlile, T. Garnett, *What Is Agroecology? TABLE Explainer Series* (TABLE, University of Oxford, Swedish University of Agricultural Sciences and Wageningen University & Research, 2021)

L. Carte, Á. Hofflinger, M.H. Polk, Expanding exotic forest plantations and declining rural populations in La Araucanía, Chile. Land **10**(3), 283 (2021). https://doi.org/10.3390/land10030283

G. Ceballos, P.R. Ehrlich, R. Dirzo, Biological annihilation via the ongoing sixth mass extinction signaled by vertebrate population losses and declines. Proc. Natl. Acad. Sci. **114**(30), E6089–E6096 (2017). https://doi.org/10.1073/pnas.1704949114

Resultados oficiales censo de población 1992: Población total país, regiones, comunas, por sexo y edad [official results of the 1992 population census: Total population of the country, regions, communes, by sex and age], in *INE (National Institute of Statistics)*, (Chilean National Statistics Institute, 1992) Retrieved January 6, 2023, from https://bibliotecadigital.ciren.cl/bitstream/handle/20.500.13082/13354/MC0055470.pdf?sequence=2&isAllowed=y

Chilean National Statistics Institute, *Chile: Ciudades, pueblos y caseríos [Chile: Cities, towns and villages]* (INE (National Institute of Statistics), 2002)

M. De la Cadena, Indigenous cosmopolitics in the Andes: Conceptual reflections beyond "politics". Cult. Anthropol. **25**(2), 334–370 (2010). https://doi.org/10.1111/j.1548-1360.2010.01061.x

R. De la Torre, Los newagers: El efecto colibrí. Artífices de menús especializados, tejedores de circuitos en la red, y polinizadores de culturas híbridas [The new agers: The hummingbird effect. Artificers of specialized menus, weavers of circuits in the network, and pollinators of hybrid cultures]. Religião & Sociedade **34**(2), 36–64 (2014). https://doi.org/10.1590/S1984-04382014000200003

B. De Sousa Santos, *Refundación del Estado en América Latina: Perspectivas Desde Una epistemología del Sur [Refounding the State in Latin America: Perspectives from an Epistemology of the South]* (Plural Editors, 2010)

A. Escobar, ¿«Pachamámicos» versus «Modérnicos»? Tabula Rasa **15**, 265–273 (2011)

A. Escobar, Más allá del desarrollo: Postdesarrollo y transiciones hacia el pluriverso [Beyond development: Postdevelopment and transitions to the pluriverse]. Revista de antropología social **21**, 23–62 (2012). https://doi.org/10.5209/rev_RASO.2012.v21.40049

A. Escobar, Thinking-feeling with the earth: Territorial struggles and the ontological dimension of the epistemologies of the south, in *Knowledges Born in the Struggle*, ed. by B. de Sousa Santos, M.P. Meneses, (Routledge, 2019), pp. 41–57

C. Folke, S.R. Carpenter, B. Walker, M. Scheffer, T. Chapin, J. Rockström, Resilience thinking: Integrating resilience, adaptability and transformability. Ecol. Soc. **15**(4), 20 (2010). http://www.ecologyandsociety.org/vol15/iss4/art20/

C. Folke, S. Carpenter, T. Elmqvist, L. Gunderson, B. Walker, Resilience: Now more than ever. Ambio **50**(10), 1774–1777 (2021). https://doi.org/10.1007/s13280-020-01487-6

J. Gilligan, Nature of collaboration across disciplines, in *Pathways to Collaboration*, ed. by R. Holowinsky, A. Channell, O.J. Crocomo, J.P. Kreier, W.R. Sharp, (CreateSpace Independent Publishing, 2017), pp. 433–457

O.F. Giraldo, El discurso moderno frente al "pachamamismo": La metáfora de la naturaleza Como recurso y el de la tierra Como madre [The modern discourse facing "pachamamismo": The metaphor of nature as a resource and that of the earth as a mother]. Polis **33**, 1–13 (2012). http://journals.openedition.org/polis/8502

O.F. Giraldo, *Utopías en la era de la supervivencia: Una Interpretación del buen vivir [Utopias in the Age of Survival: An Interpretation of Good Living]* (Editorial Itaca, 2014)

E. Gudynas, Buen vivir: Germinando alternativas al desarrollo [Good living: Germinating alternatives to development]. América Latina en Movimiento **462**, 1–20 (2011)

H. Herrera, Pactos y alternativas socioambientales ante la crisis ambiental y civilizatoria del siglo XXI [Socio-environmental covenants and alternatives to the environmental and civilizational crisis of the 21st century]. Ideas Verdes **25** (2020) Retrieved January 18, 2023, from https://co.boell.org/sites/default/files/2021-01/IDEASVERDES_25_web.pdf

R. Hidalgo, H. Zunino, Negocio inmobiliario y migración por estilos de vida en la Araucanía lacustre: La transformación del espacio habitado en Villarrica y Pucón [Real estate business and lifestyle migration in the Araucanía lakeside: The transformation of the inhabited space in Villarrica and Pucón]. Revista Aus **11**, 10–13 (2012). https://doi.org/10.4206/aus.2012.n11-03

E. Holt-Giménez, M.A. Altieri, Agroecology, food sovereignty, and the new green revolution. Agroecol. Sustain. Food Syst. **37**(1), 90–102 (2013). https://doi.org/10.1080/10440046.2012.716388

V. Huiliñir-Curío, H.M. Zunino, Movilidad, utopías y lugares híbridos en Los Andes del sur de Chile [Mobility, utopias and hybrid places in The Andes of southern Chile]. Revista INVI **32**(91), 141–160 (2017)

V. Huiliñir-Curío, H.M. Zunino, L.F. De Matheus e Silva, Exclusión y desigualdad en localidades próximas a la Reserva Ecológica Privada Huilo-Huilo en el sur de Chile [Exclusion and inequality in localities near the Huilo-Huilo private ecological reserve in southern Chile]. Acme **18**(2), 335–363 (2019)

IPCC, *Climate change 2021: The physical science basis. Contribution of working group I to the sixth assessment report of the intergovernmental panel on climate change* ((In Press). Cambridge University Press, 2021)

S. Knierbein, T. Viderman, Space, emancipation and post-political urbanization, in *Public Space Unbound*, ed. by S. Knierbein, T. Viderman, (Routledge, 2018), pp. 3–19

A. Kothari, A. Salleh, A. Escobar, F. Demaria, A. Acosta, *Pluriverse: A Dictionary of Postdevelopment* (Icaria, 2019)

W. LaDuke, Indigenous environmental perspectives: A north American primer. Akwe: Kon J **9**(2), 52–71 (1992)

A.K. Magnan, H.-O. Pörtner, V.K.E. Duvat, M. Garschagen, V.A. Guinder, Z. Zommers, et al., Estimating the global risk of anthropogenic climate change. Nat. Clim. Chang. **11**(10), 879–885 (2021). https://doi.org/10.1038/s41558-021-01156-w

R.E. Mardones, H.M. Zunino, Repensando lo comunitario: Discursos de comunidades intencionales utópicas en Chile [Rethinking the communitarian: Discourses of utopian intentional communities in Chile]. Convergencia 26(81), 06 (2019). https://doi.org/10.29101/crcs.v0i81.10615

R.E. Mardones, H.M. Zunino, Emplazando la utopía. Reinvenciones del sujeto, la comunidad y el espacio habitado en Chile [Finding utopia. Reinventions of the individual, the community and the inhabited space in Chile]. Revista de Geografía Norte Grande 78, 49–69 (2021). https://doi.org/10.4067/S0718-34022021000100049

I. McNamara, A. Nauditt, M. Zambrano-Bigiarini, L. Ribbe, H. Hann, Modelling water resources for planning irrigation development in drought-prone southern Chile. International Journal of Water Resources Development 37(5), 793–818 (2021). https://doi.org/10.1080/0790062 7.2020.1768828

M. Mier y Terán Gimenez Cacho, O.F. Giraldo, M. Aldasoro, H. Morales, B.G. Ferguson, P. Rosset, et al., Bringing agroecology to scale: Key drivers and emblematic cases. Agroecol. Sustain. Food Syst. 42(6), 637–665 (2018). https://doi.org/10.1080/21683565.2018.1443313

W. Mignolo, Historias locales/diseños globales: Colonialidad, conocimientos subalternos y pensamiento fronterizo [Local histories/global designs: Coloniality, subaltern knowledge, and border thought] (Akal, 2003)

J. Ñanculef Huaiquinao, Tayiñ Mapuche kimün. Epistemología Mapuche – sabiduría y conocimientos [Mapuche Epistemology-Wisdom and Knowledge] (Universidad de Chile, 2016)

Z. Neira Ceballos, A.M. Alarcón, I. Jelves, P. Ovalle, A.M. Conejeros, V. Verdugo, Espacios ecológicos-culturales en territorio Mapuche de la región de la Araucanía en Chile [Ecological-cultural spaces in a Mapuche territory in the Araucanía region of Chile]. Chungará (Arica) 44(2), 313–323 (2012). https://doi.org/10.4067/S0717-73562012000200008

T.R. Oliveira, C. Imai, Identificação dos atributos da arquitetura escolar Waldorf: Un estudio de caso no interior paulista [Identification of Attributes of Waldorf School Architecture: A Case Study in the Countryside of São Paulo State] (IV Simpósio Brasileiro de Qualidade do Projeto no Ambiente Construído. Universidade Federal de Viçosa, Viçosa, 2015). https://doi.org/10.18540/2176-4549.6022

P.A. Østergaard, N. Duic, Y. Noorollahi, H. Mikulcic, S. Kalogirou, Sustainable development using renewable energy technology. Renew. Energy 146, 2430–2437 (2020). https://doi.org/10.1016/j.renene.2019.08.094

E. Restrepo, A. Rojas, Inflexión decolonial: Fuentes, conceptos y cuestionamientos [Decolonial inflexion: Sources, concepts and questions] (Universidad del Cauca, 2010)

I. Reyes-Guillén, X.F. Poblete Naredo, M.A. Villafuerte Franco, Historia del concepto desarrollo sustentable y su construcción en la población actual [History of the concept of sustainable development and its construction in the current population]. Espacio I+D, Innovación más desarrollo 7(17), 64–67 (2018). https://doi.org/10.31644/IMASD.17.2018.a05

P. Robbins, Research in theft: Environmental inquiry in a postcolonial world, in Approaches to Human Geography, ed. by S. Aitken, G. Valentine, (Sage, 2006), pp. 311–324

J. Rockström, W. Steffen, K. Noone, Å. Persson, F.S. Chapin III, E. Lambin, et al., Planetary boundaries: Exploring the safe operating space for humanity. Ecol. Soc. 14(2), 32 (2009)

F. Sanhueza-Céspedes, Extraños en un mundo que ellos han creado: El problema de la alienación en los textos del joven Marx (1843–1845) [Strangers in a World of their Own Making: The Problem of Alienation in the Texts of the Young Marx (1843–1845)] (Universidad de Chile, 2018)

R. Segato, La crítica de la colonialidad en ocho ensayos y una antropología por demanda [The Critique of Coloniality in Eight Essays and an Anthropology on Demand] (Editorial Prometeo, 2013)

P. Servigne, R. Stevens, Comment tout peut s'effondrer: Petit manuel de collapsologie à l'usage des générations présentes [How Everything Can Collapse: A Small Manual of Collapsology for the Use of Generations] (Le Seuil, 2015)

M. Svampa, Consenso de los commodities, giro ecoterritorial y pensamiento crítico en América Latina. [Commodity consensus, ecoterritorial turn and critical thinking in Latin America]. *Revista de Observatorio Social de América Latina* **13**(32), 15–38 (2012)

M. Svampa, El Antropoceno como diagnóstico y paradigma. Lecturas globales desde el Sur [The Anthropocene as diagnosis and paradigm. Global readings from the South]. Utopía y Praxis Latinoamericana **24**(84), 33–54 (2019)

L. Tondo. Southern Italy braced for rare Mediterranean hurricane. *The Guardian* (2021, October 27) https://www.theguardian.com/world/2021/oct/27/southern-italy-braced-for-rare-mediterranean-hurricane

L. Vergara, C. Sánchez, H.M. Zunino, Migración por estilo de vida: ¿Creando comunidades diversas y cohesionadas? El caso de Los Riscos, Pucón, Chile [Lifestyle migration: Creating diverse and cohesive communities? The case of Los Riscos, Pucón, Chile]. Revista Austral de Ciencias Sociales **36**, 47–67 (2019). https://doi.org/10.4206/rev.austral.cienc.soc.2019.n36-03

Waldorf Pucón. Asociación Educativa Waldorf Pucón [Waldorf Education Association Pucón] (2015) http://www.waldorfpucon.org/

World Commission on Environment and Development, *Our common future* (Alianza, 1987)

H. Zunino, V. Huiliñir, Utopías modernas y posmodernas en el sur de Chile: Rupturas y continuidades [Modern and postmodern utopias in southern Chile: Ruptures and continuities], in *Re-conociendo las geografías de América Latina y el Caribe [Re-Cognizing the Geographies of Latin America and the Caribbean]*, ed. by R. Sánchez, R. Hidalgo, F. Arenas, (Pontificia Universidad Católica de Chile, 2017), pp. 157–182

H.M. Zunino, L. Espinoza Arévalo, A. Vallejos-Romero, Los migrantes por estilo de vida como agentes de transformación en la Norpatagonia chilena [Lifestyle migrants as agents of transformation in Chilean Norpatagonia]. Revista de Estudios Sociales **55**, 163–176 (2016). https://doi.org/10.7440/res55.2016.11

Chapter 18
Catalyzing Holistic Conservation-Based Development Through Ethical Travel Experiences Rooted in the Bioculture of Patagonia's Subantarctic Natural Laboratories

Trace Gale-Detrich, Laura Sánchez Jardón, Andrés Adiego, Ricardo Rozzi, Pamela Maldonado, Matías Navarrete Almonacid, José Coloma Zapata, Diego Hernández Soto, Manuel Mora Chepo, Ronald Cancino Salas, Fabien Bourlon, Rodrigo Villa-Martínez, Lorna Moldenhauer Ortega, and Carla Henríquez V.

Abstract Synchronous losses of biological, linguistic, and cultural diversity are contributing to processes of *biocultural homogenization*, a persistent downgrading in how people perceive biodiversity, environmental, and cultural conditions, and what they consider as *normal*. Some have linked biocultural homogenization with neolib-

The original version of the chapter has been revised. A correction to this chapter can be found at https://doi.org/10.1007/978-3-031-38048-8_19

T. Gale-Detrich (✉)
Centro de Investigación en Ecosistemas de la Patagonia (CIEP), Sustainable Tourism Research Line, Human-Environmental Interactions Group, Coyhaique, Aysén Region, Chile

Cape Horn International Center (CHIC), Puerto Williams, Chile
e-mail: tracegale@ciep.cl

L. Sánchez Jardón
Cape Horn International Center (CHIC), Puerto Williams, Chile

Universidad de Magallanes, Coyhaique University Center, Coyhaique, Aysén Region, Chile
e-mail: laura.sanchez@umag.cl

A. Adiego
Centro de Investigación en Ecosistemas de la Patagonia (CIEP), Sustainable Tourism Research Line, Human-Environmental Interactions Group, Coyhaique, Aysén Region, Chile

Department of Geography and Territorial Planning, Universidad de Zaragoza, Zaragoza, Spain
e-mail: andres.adiego@ciep.cl

eral practices that emphasize standardization and efficiencies in order to enhance capital accumulation. In Chile, which is highly centralized, urban, and centered around neoliberal development, *biocultural homogenization* is especially concerning, as it may lower support for conservation. Thus, when thinking about conservation-based development in Patagonia, locally driven initiatives that foster biocultural reawakening, democratize science, and catalyze sustainable development, might help address biocultural homogenization and build conservation support. This mixed-methods, intrinsic case study explored three initiatives underway in the Aysén and Magallanes regions of Chile that address some of these outcomes: (1) Subantarctic Natural Laboratories; (2) 3-Hs Biocultural Ethic and FEP Cycle

R. Rozzi
Cape Horn International Center (CHIC), Puerto Williams, Chile

Department of Philosophy & Religion Studies, University of North Texas, Denton, TX, USA

Universidad de Magallanes, Sub-Antarctic Biocultural Conservation Program,
Puerto Williams, Chile
e-mail: rozzi@unt.edu

P. Maldonado
Universidad de Magallanes, GAIA Antarctica Investigation Center, Punta Arenas, Chile
e-mail: pamela.maldonado@umag.cl

M. Navarrete Almonacid · D. Hernández Soto
Department of Social Sciences, Universidad de la Frontera, Temuco, Araucanía Region, Chile
e-mail: m.navarrete17@ufromail.cl

J. Coloma Zapata
Universidad de la Frontera, Center for Social Research of the South,
Temuco, Araucanía Region, Chile

M. Mora Chepo · R. Cancino Salas
Universidad de la Frontera, Temuco, Araucanía Region, Chile
e-mail: manuel.mora@ufrontera.cl; ronald.cancino@ufrontera.cl

F. Bourlon
Centro de Investigación en Ecosistemas de la Patagonia (CIEP), Sustainable Tourism
Research Line, Human-Environmental Interactions Group, Coyhaique, Aysén Region, Chile

Université Grenoble Alpes, Institute of Urban Planning and Alpine Geography – UMR 5194,
Grenoble, Isère, France
e-mail: fabienbourlon@ciep.cl; Fabien.bourlon@umrpacte.fr

R. Villa-Martínez
Cape Horn International Center (CHIC), Puerto Williams, Chile

GAIA Antarctica Investigation Center, Universidad de Magallanes, Punta Arenas, Chile
e-mail: rodrigo.villa@umag.cl

L. Moldenhauer Ortega
Universidad de Magallanes, Coyhaique University Center, Coyhaique, Aysén Region, Chile
e-mail: lorna.moldenhauer@umag.cl

C. Henríquez V.
GAIA Antarctica Investigation Center, Universidad de Magallanes, Punta Arenas, Chile
e-mail: carla.henriquez@umag.cl

Approach; and (3) Scientific Tourism Collaborative Learning Networks. We sought to better understand their methods and implementations and explore how their integration might strengthen conservation-based development in Patagonia, through ethical travel experiences rooted in the bioculture of local communities. Results suggest promise for a combined approach; thus, additional research and consideration are merited.

Keywords Patagonia · Biocultural homogenization · Conservation-based development · Ethical travel · Natural laboratories · Biocultural ethic · Field environmental philosophy

18.1 Introduction

18.1.1 Biocultural Homogenization

Synchronous losses of biological, linguistic, and cultural diversity are contributing to increasing disconnections between human and non-human worlds. These losses have been associated with a persistent downgrading in how people perceive biodiversity, environmental and cultural conditions, plurality, and what they consider as *normal* (Gavin et al. 2015; Kesebir and Kesebir 2017; Maffi 2018; Rozzi et al. 2006; Rozzi 2012). This phenomenon, known as *biocultural homogenization*, has been linked to the progressive decline of human/human and human/non-human interactions (*extinction of experience*), which appears to diminish understanding and support for biodiversity protection (Soga and Gaston 2016, 2020). Specifically, scholars have noted that biocultural homogenization likely contributes to the general underestimation of the scope and extent of long-term environmental and cultural changes taking place, and accelerating loss of local biodiversity, languages, and culture (Rozzi et al. 2008; Rozzi 2012; Mackey and Claudie 2015).

Similar dynamics have provoked a loss of cultural diversity in much of the world; especially in terms of intangible cultural heritage, including oral traditions, rituals, festive events, social practices, performing arts, and knowledge and practices concerning nature, the universe, and traditional relationships that tie together land, flora, fauna, funga, and people (Bridgewater and Rotherham 2019; Mackey and Claudie 2015). Over the last century, intangible cultural heritage has eroded, as an increasingly hegemonic one-dimensional cultural lens has been evoked through popular media, standardized educational curriculums, and national food programs (e.g., fast food, standardized school lunches, plastic handicrafts, Disneyfication, English language use), at the expense of native species, ancient cultures, languages, medicines, religions, beliefs, and place-based ways of knowing (Rozzi 2018; Sõukand et al. 2022). These lenses pigeonhole our descriptions and understandings of the diversity of humans and other-than-humans, reducing diversity to a few quantitative indicators and syllogisms.

Apart from numerous health and well-being risks that arise from decreased human/human and human/non-human interactions that occur along with rural to urban transitions (Barreau et al. 2019; Chen et al. 2015; Commission on Social Determinants of Health 2008; Cox et al. 2018; Dalzell 2018; Kjellstrom 2006; Ledent 2012; Morris 2016; Senese and Wilson 2013; Weiss et al. 2020), authors have concluded that biocultural homogenization may drive significant biodiversity loss (Bridgewater and Rotherham 2019; Cox et al. 2018, Gavin et al. 2015; Rozzi et al. 2006; Stokes 2006). Stokes (2006) pointed out that in an increasingly urban world, most surviving biodiversity is conserved because people have decided its protection is important to them. Yet, as people become more isolated from nature, and culture becomes more homogenized, knowledge and support for environmental issues and nature conservation will likely decline (Bridgewater and Rotherham 2019; Cox et al. 2018).

18.1.2   Conservation-Based Development

To date, the concept of conservation-based development (CBD) in Patagonia has largely been advanced through private conservation initiatives that largely align with preservationist approaches that have been proposed globally. For example, Wilson's (2016) Half-Earth initiative called for protecting 50% of the world's land and sea, from human intervention or activity (Wilson 2016). While Wilson's Half-Earth proposal sought hardline preservation and protection, the more recent 30 × 30 Movement (Campaign for Nature 2021) proposes that ambitious protection goals will only be achieved by recognizing and incorporating a mixture of public and private lands with varying degrees of sustainable use (Campaign for Nature 2021; Wilson 2016). Differences in these and other global conservation proposals rest on the dilemma between integration of human and non-human worlds—*land sharing*—and/or the separation—*land sparing*—of these worlds (Fischer et al. 2014). Nature-based tourism is one of the most common uses of protected areas (PAs) to date, under a land-sharing framework, that combines conservation with the sustainable use of natural attractions and existing territorial infrastructure, including roads, PAs, trails, and ferry routes (Baum et al. 2017).

When developed and executed in a sustainable manner, these forms of CBD can help contribute to the UN Sustainable Development goals, especially with respect to biodiversity conservation through PAs and social/economic development through nature-based tourism (Carius and Job 2019; Deutsches Nationalkomitee MAB and UNESCO 2005; Hall 2019; Oldekop et al. 2016). Nevertheless, these forms of CBD largely remain rooted in the dominant neoliberal economic frameworks and paradigms, which have been criticized for overly narrow conceptions of nature that lead to its commodification, increased urbanization, worsening social inequities, and a homogenization of goods, services, and arts, driven by standardization and a constant focus on increased efficiencies and capital accumulation (Baldwin et al. 2019; Beer 2022; Butler 2018; Conway and Heynen 2006; Flores 2013; Gjorgjioska and Tomicic 2019; Igoe and Brockington 2007; Kashwan et al. 2019; Khan 2015; Kumi et al. 2014; Ranta 2018; Shreve 2012; Sims 2017; Slocum et al. 2019; Stahler-Sholk

2007; Wolters 2022). In essence, dominant neoliberal models have reinforced the one-dimensional practices that have led to the biocultural homogenization we currently experience (Flores 2013; Gjorgjioska and Tomicic 2019; Ranta 2018; Shreve 2012; Sims 2017; Stahler-Sholk 2007). Ranta (2018) posed, "...we are experiencing a crisis of humanity: a situation in which our social relations, our bodies, and our minds are being commoditized at an increasing pace. Development, as we know it, has reached its limits economically, environmentally, and socially" (p. 20).

Acknowledging these trends, conservation leaders have begun to recognize that neoliberal driven science and conservation models are not the only (or perhaps best) answer for addressing biodiversity declines. Increasingly, they are including "public education, policy, and other nonbiological dimensions in their work" (Stokes 2006, p. 6). Concretely, Rozzi et al. (2012) called for biological and cultural conservation initiatives—including ecotourism and other forms of tourism that seek to develop in a sustainable manner—to consider the sociocultural and ecological contexts of the ecoregions in which they work. These contexts must take a central role in programs, tourism offerings, and other efforts to connect humans with each other and with other forms of nature, facilitating transformative personal experiences that draw on a biocultural ethic, linking humans and nature within their common habitat (Rozzi et al. 2006; Rozzi 1999, 2012). For example, Rozzi et al. (2006) presented ten inter-related principles to link biocultural conservation to nature conservation, calling for (1) inter-institutional cooperation; (2) participatory approaches; (3) interdisciplinary conservation approaches that bridge humanities and sciences; (4) networking and international cooperation; (5) communication through the media; (6) flagship species; (7) outdoor education; (8) economic sustainability and ecotourism; (9) administrative sustainability; and (10) conservation research and conceptual sustainability, which they described as a continuously evolving process for perceiving, understanding, and co-existing with biocultural diversity.

Much has been written about the conflicts and tensions that have evolved in Patagonia when neoliberal development models and practices of nature-based tourism have been inserted, at a large scale, from *above* or *outside* (e.g., Aliste et al. 2018; Bachmann Vargas and van Koppen 2020; Borrie et al. 2020; Inostroza and Cànoves 2014; Núñez et al. 2021; Schweitzer 2020). Chilean policy, law, and practice have long-favored neoliberal development driven by the transnational private sector, through open, free markets, government guarantees, weak government regulation, and authority, especially in peripheral territories like Patagonia (Borrie et al. 2020; Khan 2015; Kashwan et al. 2019; Latta and Aguayo 2012; Miranda Cabaña 2016; Núñez et al. 2018, 2019, 2020; Orellana Calderón 2020).

Yet, there are several initiatives currently underway *within* and *from* Patagonia that seek to foster biocultural reawakening, democratize science, and catalyze more holistic approaches to sustainable development through ethical travel experiences. For example, a biocultural ethic approach, combining transdisciplinary science, environmental philosophy, and a series of replicable tools (i.e., Field Environmental Philosophy, 3-Hs Model) with education and ecotourism, has evolved within the UNESCO Cape Horn Biosphere Reserve (CHBR), in the southernmost Magallanes region of Chile. In parallel, for almost two decades, communities and researchers in

Aysén have worked to define models for combining science and tourism, through bottom-up, collaborative approaches (Bourlon et al. 2021, 2022; Bourlon and Mao 2011). Finally, a fairly recent initiative has taken a pluralistic approach to evolve and expand on the concept of *natural laboratories* (Aguilera and Larrain 2018, 2021; Rozzi 2018).

We believe a more conscious integration of these concepts and initiatives might impulse a path forward, enabling a locally driven evolution of CBD, which permits Patagonian territories to begin to evolve dominant neoliberal economic frameworks and paradigms, moving toward new models of sustainable development that encompass a more holistic view of human and non-human well-beings. Thus, we employ this chapter to explore each of these three initiatives and understand their methods, trajectories, strengths, and potential to contribute to our vision of CBD, through approaches that can be developed *within* and *from* Patagonia. Our guiding question is: How might we integrate current practices to achieve holistic CBD through ethical travel experiences rooted in subantarctic bioculture? (Fig. 18.1).

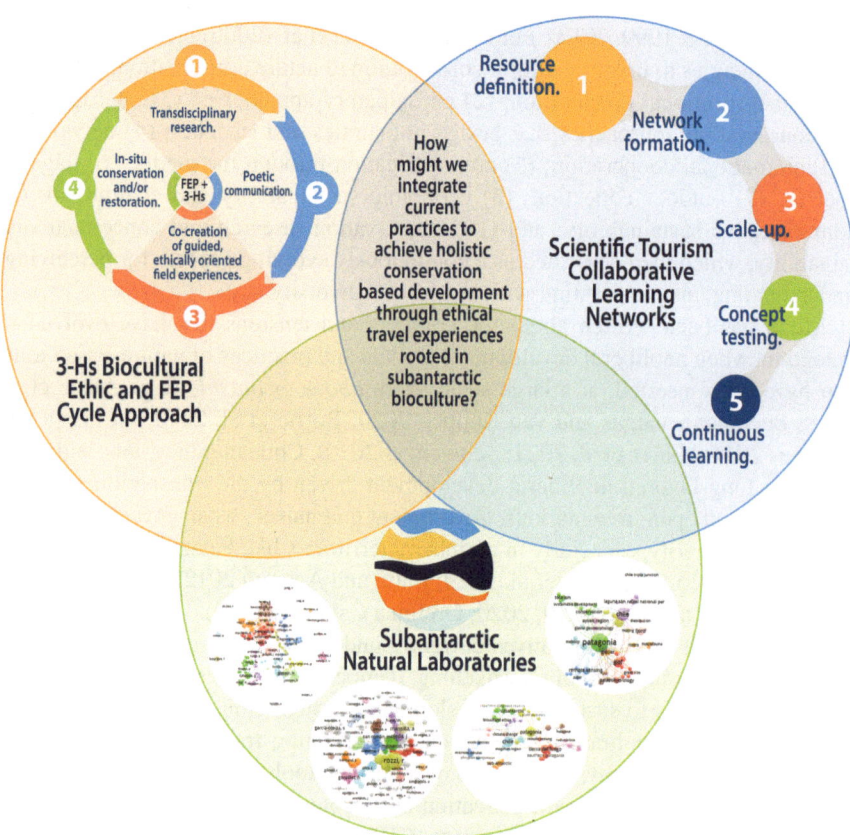

Fig. 18.1 Conceptual map for holistic conservation-based development through ethical travel experiences rooted in subantarctic bioculture

18.2 Methods

This chapter presents a mixed-methods intrinsic case study of three initiatives taking place in Chilean Patagonia—(1) the Collaborative network for the development of subantarctic natural laboratories: Strengthening scientific research and its socio-environmental and economic impacts in the Austral Macrozone project; (2) the pluralist, transdisciplinary, co-constructed educational and ecotourism programs within Omora Ethnobotanical Park (OEP); and (3) the collaborative learning networks that have been fostered around Scientific Tourism (ST) in Aysén—as possible components of an integrative post-neoliberal approach to CBD (Fig. 18.1). We chose these specific initiatives as our case study based on the particularities they involved, their trajectory within the territory, and our desire to better understand their potential for integration (Boblin et al. 2013; Stake 2003).

Qualitative inquiry focused on understanding the methods and strategies each initiative used to foster sustainable development through ethical travel experiences. For each of the three initiatives, we assembled a collection of primary and secondary data, including project documents, reports, and published academic papers; open-ended interviews with initiative collaborators, and participant observation during training initiatives, project work, and field visits (Ebneyamini and Sadeghi Moghadam 2018; Harrison et al. 2017; Rashid et al. 2019).

In addition, for the subantarctic natural laboratories (SNLs) initiative, a systematized geo-literature review and mixed methods geobibliometric analysis helped us evaluate and situate the scientific production that has occurred in two pilot SNLs during recent decades (Gale et al. 2022; Zakaria et al. 2021). These studies employed mixed methods, with a descriptive cross-sectional design. We began with a systematized literature review of peer-reviewed scientific output within the SNLs, during the period between 2000 and 2022, based on a consensual list of toponyms. Next, we invited project collaborators to nominate scientific output they felt to be important that did not surface in our search. At times, their submissions extended the original timeframe we had considered. The articles were georeferenced based on the sampling sites specified in the research. This allowed us to consider spatial aspects of the research and output in subsequent analysis. For the geobibliometric analysis, we employed VOSviewer software (version 1.6.15) to visualize bibliometric networks within the dataset, performing co-authorship and co-occurrence analyses to create graphics that would allow us to understand trends with respect to the lines of specialization and scientific collaboration in the pilot sites under study (van Eck and Waltman 2020). Clustering and relative link strength (collaboration) were indicated through differences in color, circle size, and the thickness of connecting lines (van Eck and Waltman 2020).

While data gathering and processing differed for each of the three initiatives, there were many similarities. For example, each of the three initiatives involved periods of time in which public funding was received. During these times, a fairly standard format of reporting occurred detailing objectives, beneficiaries, process, and outcomes. These commonalities allowed us to follow the recommendations of case study methodology experts (Ebneyamini and Sadeghi Moghadam 2018; Stake

2003) with respect to the holistic treatment of the intrinsic links and subjective realities of the data. A data saturation strategy guided decisions about the amount of data that was collected: the research team continued to explore and collect data until we agreed that no significant new information remained (Harrison et al. 2017; Rashid et al. 2019).

Qualitative data were processed and analyzed through an iterative approach of examining, categorizing, and tabulating the evidence gathered for each initiative to understand their processes, methods, and outcomes, with respect to our research interests: fostering biocultural reawakening, democratizing science, and catalyzing more holistic approaches to sustainable development through ethical travel experiences (Ebneyamini and Sadeghi Moghadam 2018). Our process involved a collaborative approach to data reduction, data display, and conclusion drawing and verification, in which the lead researcher conducted an initial open coding of the data to draw out primary themes, and then member checking and triangulation were employed to ensure agreement among the research team about core conclusions. Doubts and questions were resolved through an iterative process of engagement between the data and key informants who had directly participated within the three initiatives (Harrison et al. 2017; Rashid et al. 2019).

18.3 Results: Components of a Biocultural Conservation-Based Development System for Patagonia

The following section describes the results of the mixed-methods intrinsic case study with sub-sections dedicated to each of the three initiatives taking place in Chilean Patagonia: (1) Subantarctic Natural Laboratories; (2) 3-Hs Biocultural Ethic and FEP Cycle Approach; and (3) Scientific Tourism Collaborative Learning Networks.

18.3.1 Subantarctic Natural Laboratories: Fostering Transdisciplinary Research and Stewardship in Patagonia

Over the past decade, the concept of *natural laboratories* has emerged in Chile, in recognition of the potential to root science and learning within territories and harness the potential for positive spillover for local communities and economies (Aguilera 2013; Aguilera and Larraín 2021; Audretsch and Keilback 2007; Chilean National Innovation Council for Competitiveness 2013; Guridi 2018; Guridi et al. 2020). In this context, natural laboratories represent unique places that offer biophysical singularities and conditions, making them advantageous for scientific research and knowledge advancement (Aguilera 2013; Aguilera and Larraín, 2018).

These processes can occur at different spatial scales, from the micro-scale (e.g., in the collagen of a fossil bone), locality, country or continent level (Australia, Antarctica), to the macro-scale like the planet Earth. The areas of knowledge covered by these natural laboratories include earth sciences, medical, social, political, psychological, ethology, and communications.

An initial case of application of the natural laboratories concept in Chile is the one that was used to promote and internationalize astronomical research in the Atacama Desert: in Chile's northern regions, unique geologic, ecologic, climatic and atmospheric conditions conform some of the clearest, darkest, and driest skies in the world, favoring astronomic research (Aguilera and Larraín 2021; Barandiaran 2015). Chile began forming international agreements for astronomical research in the 1960s, and has since prioritized the development of policy, infrastructure, and capacities needed to ensure that this geography remains one of the best in the world for collaborative astronomical science. Moreover, since the late 1990s, strategic efforts have ensured that Chileans have sufficient access and opportunities to participate and benefit from the science taking place (Aguilera and Larraín 2021; Barandiaran 2015; Guridi 2018; Guridi et al. 2020).

Aguilera and Larraín (2021) posited the importance of local settings, interests, and actions regarding natural laboratories, recognizing these aspects as being central for enabling collaborative scientific advances and positive spillover effects. Specifically, they highlighted the need to cultivate local interest in the science taking place, suggesting the participation and support of local communities; both in terms of fostering transdisciplinary perspectives and mutual learning, and in the potential for science to create positive spillover effects for social and entrepreneurial innovation. They also emphasized the importance of protecting the unique conditions of natural laboratories, calling for policy, law, and practice that would ensure their territorial protection and conservation (Aguilera and Larraín 2021).

In 2021, Chile's National Agency for Research and Development (ANID) published a national call for proposals among actors in the Science, Technology, Knowledge and Innovation ecosystem, to promote the development of natural laboratories related to Chilean ocean, mountain, sky, desert, subantarctic, and Antarctic territories. They sought roadmaps for,

> …the implementation of scientific research in Natural Laboratories in harmony with the actors that coexist in these territories, considering those factors that support the development of scientific activity in them and the effects and impacts that such activity imposes on them. (anid.cl/concursos/concurso/?id=741)

18.3.1.1 Locally Led Conceptualization and Implementation of Subantarctic Natural Laboratories

Responding to this call, a group of scientific institutions working in the subantarctic regions of Aysén and Magallanes (i.e., University of Magallanes, CIEP, Pontificia Universidad Católica de Chile, and the Universidad de la Frontera) were awarded the funding to advance their proposal, *Collaborative network for the development of subantarctic natural laboratories: Strengthening scientific research and its socio-environmental and economic impacts in the Austral Macrozone* (Aysén and Magallanes). During 2022 and 2023, the project has established and coordinated a transdisciplinary collaborative network to co-define what subantarctic natural laboratories (SNLs) should encompass within the territory, including geographies, scientific, social, and ethical dimensions. Work has focused on identifying and prioritizing specific needs or areas for the Austral Macrozone, including the stakeholders and/or sectors that might benefit from purposeful involvement in science and related development, in accordance with pre-existing initiatives in the southern territory. A series of conceptualization workshops and participatory mapping exercises produced the following consensual working definition of SNLs:

SNLs comprise sites or geographic areas within the subantarctic Chilean region between 43° and 60° South, where processes and/or phenomena of scientific interest occur that, from a regional and global perspective, are essential to understand. The LNSs will promote forms of scientific development that integrate social dimensions, facilitating impact within the national and international scientific community, as well as in local communities. It is also fundamental that the science developed in SNLs encompasses activities and multiple scientific disciplinary perspectives. SNLs should be linked with public policies that enable the continuous development and improvement of place-based scientific capacities (i.e., disciplinary-related capacities and facilitating elements, such as access, infrastructure, and equipment), which enable the decentralized development of science, technology, knowledge, and innovation. Hence, these LNSs contribute to sustainable development of the local territory.

Working from this definition, transdisciplinary processes and collaborators defined four pilot LNSs: Palena River Delta (PRD), San Rafael Lagoon (SRL), Madre de Dios Archipelago (MDA), and Cape Horn (CH). Then, among other initiatives that have advanced work to operationalize the LNS concept within the Austral Macrozone territory, a geobibliographic characterization of scientific knowledge production, interests, and actors was developed for the two larger pilot LNSs—SRL and CH—to understand how science has been concentrated and evolved over the past few decades.

18.3.1.2 Geobibliographic Characterization of the Subantarctic San Rafael Lagoon Natural Laboratory Pilot Area

A total of 225 articles, published between 2000 and 2022, surfaced within the geobibliographic characterization of the SRL-SNL pilot area. These articles involved a total of 699 co-authors (Fig. 18.2). Articles were concentrated most heavily around Caleta Tortel, with several studies related to the Martinez, Baker, and Steffen Fjords. The eastern reaches of the Northern Patagonia Ice Field represented another research hotspot, with several projects situated around Cochrane, and the area of the Nef and Colonia Glaciers. The rest of the Northern Patagonia Ice Field and the areas surrounding the San Rafael Lagoon and Glacier were also represented within the

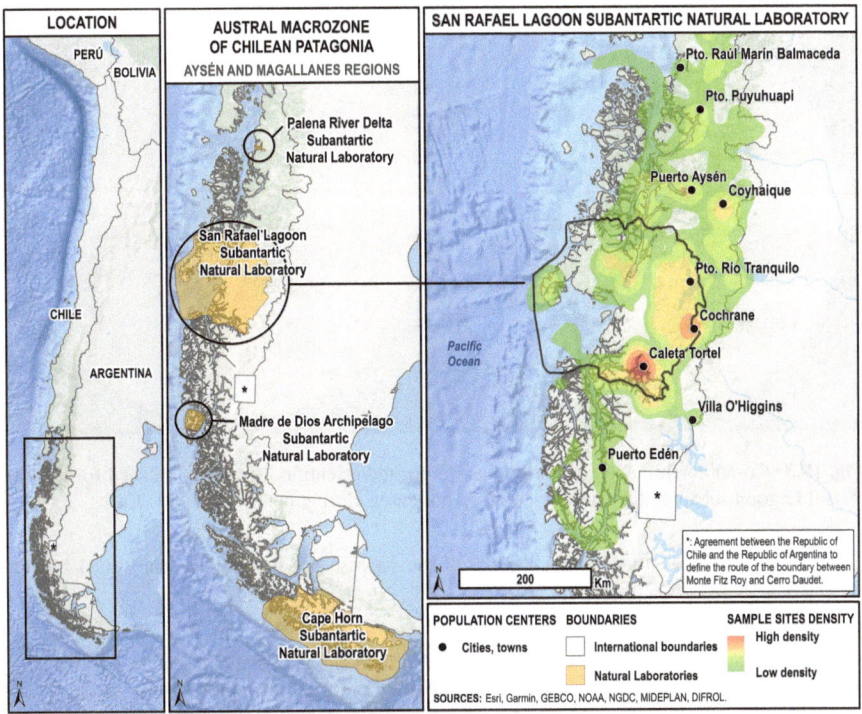

Fig. 18.2 Scientific hotspots of the San Rafael Lagoon subantarctic natural laboratory pilot area

research as well as isolated points in the surrounding communities of Coyhaique, the Aysén Fjord near Puerto Aysén, and the Puyuhuapi Fjord near Puerto Puyuhuapi.

Figure 18.3 demonstrates the co-authorship networks that emerged within the scientific literature characterized for the SRL-SNL pilot site, during the period 2000–2022. A total of 699 authors were identified. Of note are Glasser, N. (13 articles), Bertrand, S. (12 articles), Harrison, S. (12 articles), Casassa, G. (10 articles), and Reid, B. (10 articles), who were co-authors with 10 or more publications each. Most authors were involved in less than three publications. Algorithmic clustering approaches were used to construct article-level classifications based on the 70 keywords that were identified within the SRL-SNL articles, returning a total of 184 links between them. The links between keywords were analyzed using the Leiden algorithm (Traag et al. 2019), identifying 13 independent clusters. The first six of these clusters helped us infer the main lines of research that have occurred within the SRL-SNL pilot site, during the period 2000–2022. These lines have been described as: *Glaciology and climate change in Patagonia*, *Patagonian flora and fauna*, *Tourism, territories, and landscapes*, *Consequences of glacial melting in the Baker River*, *Photogrammetry applied to glaciological studie*s, and *Hydrochemistry and marine biology*.

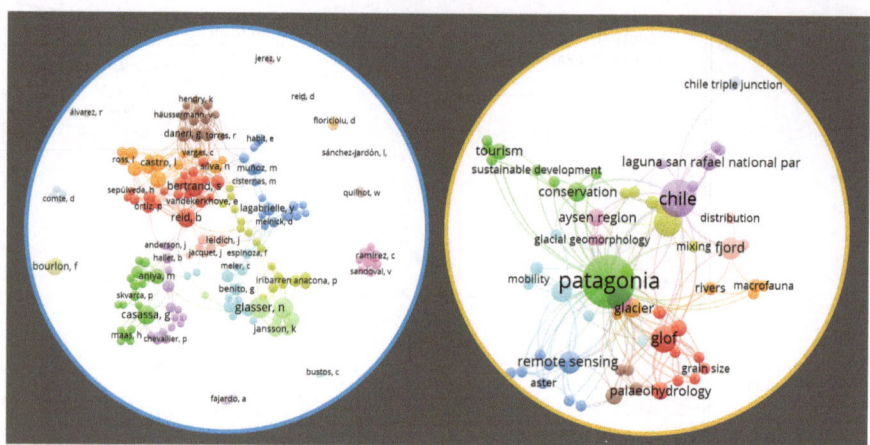

Fig. 18.3 Co-authorship and keyword networks for the scientific literature produced for the San Rafael Lagoon subantarctic natural laboratory pilot area, during the period 2000–2022

Table 18.1 Keywords associated with each of the six research lines in the San Rafael Lagoon subantarctic natural laboratory pilot area

Research line	Associated keywords
Glaciology and climate change in Patagonia	Patagonia, Chile, GLOF, North Patagonian Icefield, remote sensing, climate change, South America, fjord, Aysén region, conservation, geomorphology, glacier, mountain glaciers, palaeohydrology, aster, cachet 2 lake, glacier velocity, holocene, late holocene, neoglaciation, passive microwave, "little ice age," floodplain, glacier mapping, Tierra del Fuego, turbidite, grain size, Laguna san Rafael National Park, little ice age, rivers
Patagonian flora and fauna	Chilean Patagonia, biodiversity, biogeography, endemism, flora, mixing, lichens
Tourism, territories, and landscapes	Tourism, territory, actors, landscape
Consequences of glacier melting in the Baker River	Baker river, glacial retreat, mobility
Photogrammetry applied to glaciological studies	Disaster monitoring, image sequence processing, photogrammetric network
Hydrochemistry and marine biology	Macrofauna, *Percichthys trucha*, stable isotopes

Glaciology and climate change in Patagonia was the most developed research line that emerged, while the *Tourism, Territories, and Landscapes,* and *Photogrammetry applied to glaciological studie*s lines showed high levels of specialization and low volumes of output. The *Patagonian flora and fauna* and *Hydrochemistry and marine biology* lines were less developed, suggesting that these areas represent emergent interest within the SRL-SNL. Table 18.1 provides the primary keywords associated with the research in each of the six research lines of the SRL-SNL.

18.3.1.3   Geobibliographic Characterization of the Cape Horn Subantarctic Natural Laboratory Pilot Area

A total of 150 articles, published between 1980 and 2022, surfaced within the geo-bibliographic characterization of the CH-SNL subantarctic natural laboratory pilot area (Fig. 18.4). These articles involved a total of 529 co-authors. Articles were concentrated most heavily around Puerto Williams, with studies located in and around the OEP on Navarino Island, the Beagle Channel, and the Yendegaia Fjord. A little further west, we found another hotspot, related to the Darwin Mountain Range and Pía Bay. Other areas of research concentration were identified around Punta Arenas and in the Cape Horn Archipelago, in the southernmost part of the CH-SNL limits. Some terrestrial sectors like the Isla Grande de Tierra del Fuego north of Puerto Williams, the Brunswick Peninsula near Punta Arenas, and the area around Inútil Bay near Porvenir were also reflected within the dataset, as were some maritime sectors, like the Strait of Magellan and other fjords of the Magellanic archipelago.

Figure 18.5 demonstrates the co-authorship networks that emerged within the scientific literature characterized for the CH-SNL pilot site, during the period 1980–2022. For this pilot SNL, a total of 529 authors were identified. The most

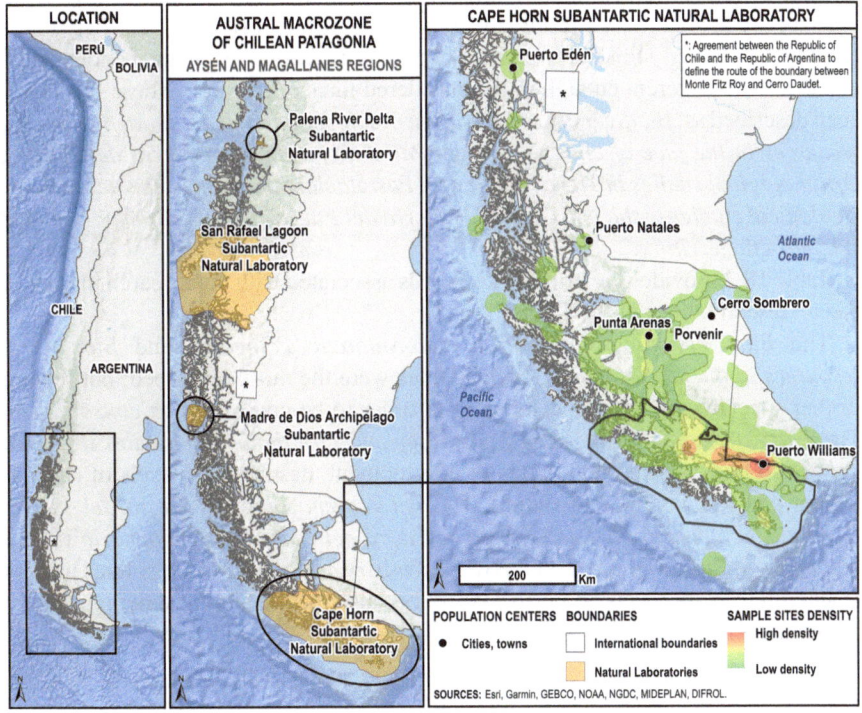

Fig. 18.4 Scientific hotspots of the Cape Horn subantarctic natural laboratory pilot area

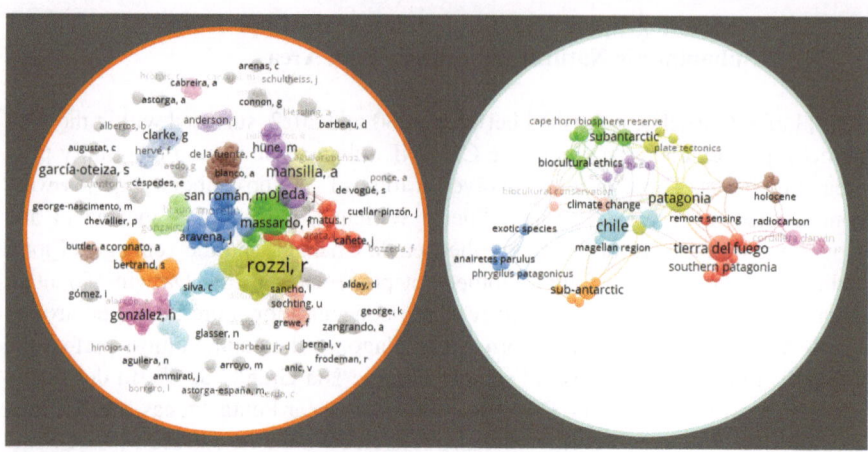

Fig. 18.5 Co-authorship and keyword networks for the scientific literature produced for the Cape Horn subantarctic natural laboratory pilot, during the period 1980–2022

productive co-authors were Rozzi, R. with 23 publications, Mansilla, A. with eight publications, Mackenzie, R., and Ojeda, J. with seven publications each, and García-Oteiza, S. with six publications. The remaining authors have five or fewer publications. A total of 61 keywords were identified within the CH-SNL articles, with 163 links between them. As before, the Leiden algorithm (Traag et al. 2019) was employed to identify links between the keywords, returning seven independent clusters that were coherent enough to be considered lines of research. These lines have been described as: *Biodiversity conservation in sub-Antarctic Patagonia, Biocultural resources in the face of climate change, Paleo-archaeology of Tierra del Fuego , Ornithological studies in Tierra del Fuego, Postglacial South American palynology, Geological studies in the Isla Grande of Tierra del Fuego,* and *Endozoochoric in the subantarctic territory.*

Table 18.2 provides the primary keywords associated with the research in each of the six research lines of the CH-SNL.

The *Biodiversity conservation in sub-Antarctic Patagonia* and *Biocultural resources in the face of climate change* lines were the most developed; but require further specialization and development in order to be considered as *consolidated.* The *Postglacial South American palynology* and *Ornithological studies in Tierra del Fuego* lines showed good rates of development, despite high levels of specialization and low volumes of output. The *Paleo-archaeology of Tierra del Fuego, Geological studies in the Isla Grande de Tierra del Fuego, Endozoochoric studies in the sub-Antarctic territory,* and *Hydrochemistry and marine biology* lines had not reached important levels of performance, because they deal with emerging topics within the CH-LNS.

Table 18.2 Keywords associated with each of the seven research lines in the Cape Horn subantarctic natural laboratory pilot

Research line	Associated keywords
Biodiversity conservation in sub-Antarctic Patagonia	Chile, Patagonia, conservation, subantarctic, biogeography, biodiversity, drake passage, glacier, Antarctica, limnology, archaea, bacteria, plate tectonics, protected areas, tectonics
Biocultural resources in the face of climate change	Environmental ethics, biocultural ethics, climate change, field environmental philosophy, bryophytes, Cape Horn biosphere reserve, ecology, lichens, sub-Antarctic Magellanic ecoregion, biocultural conservation, biosphere reserves, Cape Horn
Paleo-archaeology of Tierra del Fuego	Tierra del Fuego, southern Patagonia, little ice age, remote sensing, beagle channel, hunter-gatherers, neoglacial, southernmost South America, zooarchaeology
Ornithological studies in Tierra del Fuego	Exotic species, *Elaenia albiceps, Anairetes parulus, Navarino, Phrygilus patagonicus, Turdus falcklandii, Zonotrichia capensis*
Postglacial South American palynology	South America, holocene, pollen analysis, pollen preservation, southern westerly winds
Geological studies in the Isla Grande of Tierra del Fuego	Cordillera Darwin, in situ monazite, pseudosection, radiocarbon, thermocalc
Endozoochoric studies in the sub-Antarctic territory	Sub-Antarctic, diet, distribution, endozoochoric, mosses

18.3.2 Tools for Rebuilding Connections: The Habits, Co-inHabitants, Habitats (3-Hs) Biocultural Ethic and the Field Environmental Philosophy Cycle

Over the past 20 years, the OEP has developed and practiced a novel methodological approach to break down the barriers created by biocultural homogenization and reconnect visitors and students with other people, culture, and nature. The *Field Environmental Philosophy* (FEP) Cycle Approach (Fig. 18.6) helps participants overcome the physical and conceptual barriers that catalyze processes of biocultural homogenization through a purposeful cycle involving (1) transdisciplinary ecology and philosophy research, (2) composition of metaphors and communication through simple narratives, (3) design of field activities guided with an ecological and ethical orientation, and (4) the identification and implementation of in situ conservation areas. The FEP Cycle Approach is based on a systemic approach of the biocultural ethic (i.e., 3-Hs) that values the links that have co-evolved between life *Habits* of *co-inHabitants* (humans and other-than-humans) who share a common *Habitat* (Rozzi 2012). The FEP Cycle Approach represents a philosophical practice for epistemological and ethical reasons. It is e*pistemological* because participants not only investigate biological and cultural diversity, but they also investigate the methods, languages, and worldviews through which scientific and other forms of ecological knowledge are forged. It is *ethical* because the aim is not only to research and learn about biological and cultural diversity but, foremost, to learn to respectfully co-inhabit within it.

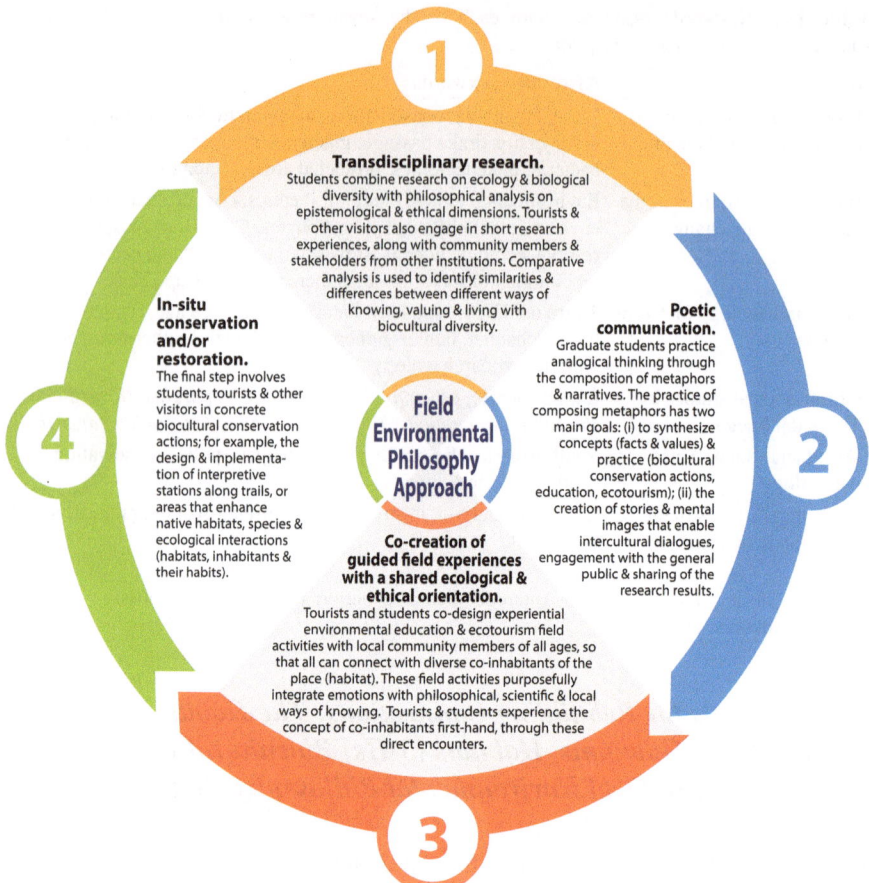

Fig. 18.6 The four-step cycle of the field environmental philosophy methodological approach

Whether or not conservation has an ethical obligation to benefit local communities is a question of values that should be negotiated and debated from the community level to transnational conservation forums (Igoe and Brockington 2007; Kashwan et al. 2019; Khan 2015; Kumi et al. 2014; Ranta 2018; Shreve 2012; Sims 2017). But such negotiations and debates are better informed by considering the role that humans from diverse cultures have played for centuries in the maintenance of biodiversity in different ecosystems, and the current role they play as custodians of biocultural rights (Bavikatte and Bennett 2015; Rozzi et al. 2018). A foundational principle of the biocultural ethic's 3-Hs framework is that conservation takes place in social-cultural-political contexts which are influenced by local biophysical and culturally particular conditions (Rozzi 2015b).

Thus, the FEP Cycle Approach recognizes that multiple forms of knowledge are necessary to solve social-environmental problems. For this reason, findings and data generated through this process are co-produced with and for the community, including decision makers and government authorities, to yield practices and actions that address problems at distinct scales. This integration of theory and practice fosters an ethic of responsibility, community solidarity, and concern for the well-being of the ecosystems, including their human and other-than-human co-inhabitants (Rozzi et al. 2012).

Applying the 3-Hs Biocultural Ethic and FEP Cycle in Omora Ethnobotanical Park The OEP, created in 2000, near Puerto Williams, is one of the core PAs within the Cape Horn Biosphere Reserve, and Chile's southernmost continental Long-Term Socio-Ecological Research (LTSER) site. With more than 20 years of ongoing research, monitoring, and community outreach, the OEP helps protect a representative mosaic of the Cape Horn archipelago's sub-Antarctic habitats found south of Tierra del Fuego, including the Róbalo River watershed and the Dientes de Navarino Mountain Range. Pluralist, transdisciplinary, co-constructed educational and ecotourism programs, based on the FEP Cycle Approach and the 3-Hs model, have promoted direct encounters with biocultural diversity for more than 20 years in OEP, forming an integral and long-standing priority within the OEP approach (Figs. 18.7 and 18.8).

These programs have been developed to help participants recover an awareness of their coexistence with a multiplicity of human and non-human beings. They help the OEP remediate the reduction of biocultural diversity taking place within the context of global and climate change, by rebuilding connections between humans and nature (Rozzi et al. 2010a, b, 2014). As such, they are a central part in the OEP approach to linking biocultural conservation with nature conservation, through the ten interrelated principles mentioned above.

These, and other FEP Cycle Approach experiences, have demonstrated the capacity of this methodology to transform socio-ecological relationships (Rozzi et al. 2014). They can involve an endless range of non-human inhabitants and a wide range of human participants in their creation and lived experiences: scientists, university students, administrators, policymakers, politicians, families, local holders of knowledge and experience, teachers, and schoolchildren, to name a few. They help people to "change the lenses" with which they normally experience bioculture. For example, the poetic communication and co-created field experiences help them appreciate values and symbolic-linguistic realities, usually reserved for philosophers (Rozzi et al. 2014). And, these first-hand experiences, driven by transdisciplinary research, transform their understanding of ecosystems, permitting them to approach and value new biophysical dimensions, usually studied by ecologists (Rozzi et al. 2014).

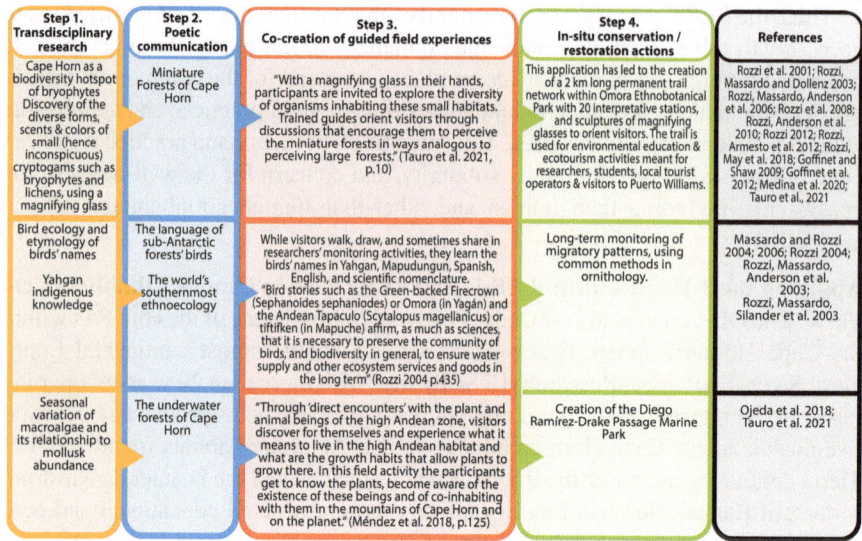

Fig. 18.7 Field environmental philosophy cycle approach: Bryophytes; Bird ecology and etymology; and Microalgae examples from Omora ethnobotanical park

Step 1. Transdisciplinary research	Step 2. Poetic communication	Step 3. Co-creation of guided field experiences	Step 4. In-situ conservation / restoration actions	References
Phenology of *Edwardsina dispar* (Insect: Diptera)	The River as a Community of Life	Submerged with a magnifying glass. Organized in three phases that help visitors to discover, appreciate and ethically value the submerged inhabitants, their habits and habitats. Phase 1: Observation, identification, and connecting with macro and micro habitats inside and outside the river; Phase 2: Face-to-face encounters with the submerged co-inhabitants. Phase 3: Participants leave the rock in the exact place where they found it and with it, all its inhabitants... " (Contador et al. 2018, p. 198).	Construction of a special circuit with 5 interpretative stations for environmental education and ecotourism activities, including platforms that extend over the edge of the stream to enable direct encounters with freshwater invertebrates using a magnifying glass.	Contador et al. 2012; 2018; Contador Mejias et al. 2022; Rendoll et al. 2020
Etymology of bryophytes and lichens' names	Cultivating a garden of names: Extending biocultural conservation and ethics to little perceived living beings	"Visitors are invited to observe, draw, and create names for bryophytes and lichens by using: (i) a magnifying glass or hand-lens to amplify the biophysical features of small plants, and (ii) the conceptual lenses of the biocultural ethics to broaden their understanding about how to interpret and respect the natural world." (Lewis et al. p.104)	Establishment of an interpretive station on the Cape Horn Miniature Forests trail. Printable resources, including a brochure are available for the people leading this activity.	Lewis et al. 2018
Spatial association in high Andean plant communities	High-Andean subantarctic gardeners	"Through 'direct encounters' with the plant and animal beings of the high Andean zone, visitors discover for themselves and experience what it means to live in the high Andean habitat and what are the growth habits that allow plants to grow there. In this field activity the participants get to know the plants, become aware of the existence of these beings and of co-inhabiting with them in the mountains of Cape Horn and on the planet." (Méndez et al. 2018, p.125)	Adaptation of areas so that they are apt for the observation of high Andean habitats	Méndez and Rozzi 2017; Méndez et al. 2018

Fig. 18.8 Field environmental philosophy cycle approach: Phenology; Entomology of bryophytes and lichen; and Spatial association examples from Omora ethnobotanical park

18.3.3   Collaborative Learning Networks: Scientific Tourism Research and Development in Aysén

Around 2006, the concept of *Scientific Tourism* (ST) began to evolve within the Aysén Region of Chilean Patagonia, influenced by growing recognition of the importance science could play in Western Patagonia for understanding the social-cultural and natural processes associated with climate change. Discussions advanced the idea of positioning tourism in Aysén in relation to areas of science, and in 2007, the Center for Investigation in Ecosystems of Patagonia (CIEP) was awarded funding for an initial regional innovation project that would develop concrete proposals around how to develop and promote ST. This project enabled early research to advance around the objectives and forms of ST.

From the offset, ST in Aysén focused on supporting socio-ecological systems sustainability while augmenting territorial connections, competitiveness, and concern (Bourlon 2020; Bourlon et al. 2021, 2022; Mao and Bourlon 2011, 2016; Mao et al. 2016). Actors defined ST as, "an activity where visitors participate in the generation and dissemination of scientific knowledge being developed by research and development centers" (scientific-tourism.org). It differed from learning tourism in that it was (and is) grounded in the scientific process and focused on knowledge generation and dissemination (Bourlon and Torres 2016). Four main segments for ST experiences have been identified that can overlap and intersect within destinations and projects: (1) exploration and adventure tourism, with a scientific dimension; (2) cultural tourism of scientific interpretation, which is close to ecotourism or, also, to industrial tourism; (3) scientific eco-volunteering; and (4) scientific research tourism (Bourlon and Mao 2011).

A second project, from 2009 to 2012, complemented project research with actions designed to create the structure and capacities for a ST Collaborative Learning Network. The project established the Consortium for ST in Patagonia, made up of nine Chilean and five international companies. CIEP coordinated the group of participating institutions through a management model called the Center for ST in Patagonia (CTCP) with the objective of supporting public activities, articulating pilot public-private ST initiatives, and knowledge sharing with organizations linked to the conservation and care of natural and cultural heritage. A broad group of experts was convened within a *Council for Science and Tourism in Aysén*, including representatives of the private sector (e.g., Coyhaique Chamber of Tourism, Patagonian Guides School, Conservación Patagónica, Grande Traversée des Alpes in France), the public sector (Chilean National Forestry Corporation; National Environment Commission; regional councils of culture, public lands, agricultural services, tourism, and hydric resources; representatives of municipalities and the regional government), and academic and research institutions (CIEP; Trapananda Center of the Universidad Austral de Chile; Department of Sociology and Anthropology of the Universidad de Concepción; the National Museum of Natural History of Santiago; the Museum of Pre-Columbian Art of Santiago; the Society of History and Geography of Aysén; Aúmen Conservation; University of Grenoble in

France; the University of Montana; the University of West Virginia; and The Sonoran Institute in the United States). During the project, the Council contributed to regional socioeconomic development based on tourism, supporting regional operators and entrepreneurs by providing work methodologies and offering training in the field of science with the aim of bringing science and tourism together. As well, the Council advised the CTCP on strategic lines of action and best practices to ensure that the ST concept safeguarded the regional public interest in terms of environmental protection, local economic development, and dissemination of the natural and cultural heritage.

The collaborative, interdisciplinary nature of the project achieved through the ST Collaborative Learning Network approach, became a core aspect that has defined subsequent ST in Aysén and beyond. From 2013 to 2016, this approach was employed for ST development within the Patagonian Archipelagos, as part of a technical cooperation agreement between CIEP and the Multilateral Investment Fund of the Inter-American Development Bank. During this period, work focused on helping local communities and tourism operators, working within the Patagonian archipelagos, develops a world-class destination for ST. Through the integration of tourism and science, this destination would enhance the development of scientific knowledge of its fragile ecosystems and transform that science into the primary resource for the creation of economically viable, socially inclusive, and environmentally sustainable tourism products. Through the expansion of the ST alliance of public and private partners with new actors who were strongly committed to inclusive territorial development of the Aysén coast, the project sought to promote ST as a mechanism for improving the quality of life and socioeconomic growth of local communities within the coastal towns of Aysén.

Over the years, as researchers and communities have learned together, ST development has become increasingly focused on connecting communities with their territories through science (Veloso 2021). In fact, the ST product development process (Fig. 18.9), which is built through ST Collaborative Learning Networks, has been shown to provide benefits for communities, in and of itself. Working through this process can contribute to sustainable practices that support resource patrimonialization and territorial coherence (Bourlon 2020; Bourlon and Torres 2016). The ST development process works through knowledge and learning exchanges which strengthen the scientific process, local knowledge, and heritage through coordinated collaboration and research conducted by scientists, stakeholders, communities, and visitors. This resulting collaboration chain, "increases focus and awareness around natural and cultural resource management, incentivizing a more resilient form of tourism development" (Bourlon et al. 2021, p. 20).

In the first stage, scientific heritage resources are identified for the geography of focus. During the second stage, destination level actor networks are cultivated, by matching local actors with scientists who live and/or work in the area, tourism entrepreneurs, and technical organizations (e.g., local government, funders, tourism management institutions), based on mutual areas of focus. Once the network has begun to work together, focus shifts in the third stage to a collaborative identification and prioritization of sites and themes that would be apt for linking scientific

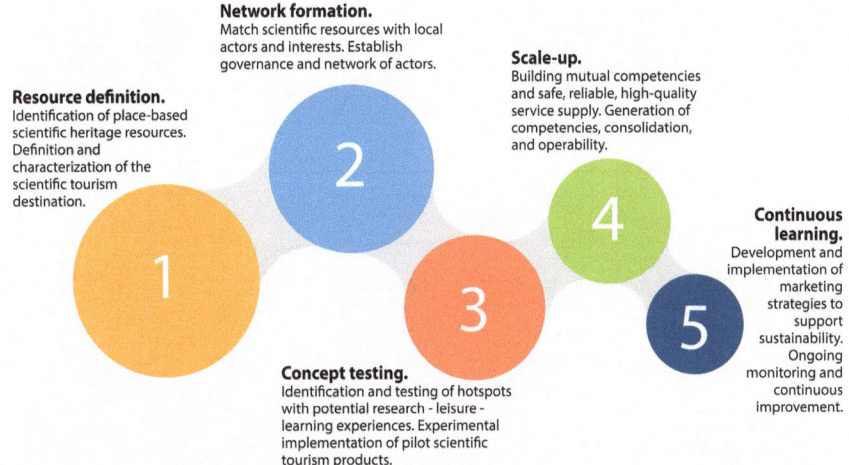

Resource definition.
Identification of place-based scientific heritage resources. Definition and characterization of the scientific tourism destination.

Network formation.
Match scientific resources with local actors and interests. Establish governance and network of actors.

Scale-up.
Building mutual competencies and safe, reliable, high-quality service supply. Generation of competencies, consolidation, and operability.

Continuous learning.
Development and implementation of marketing strategies to support sustainability. Ongoing monitoring and continuous improvement.

Concept testing.
Identification and testing of hotspots with potential research - leisure - learning experiences. Experimental implementation of pilot scientific tourism products.

Fig. 18.9 The five-phase Scientific Tourism Collaborative Learning Networks product development process

research with recreation and educational activities through pilot initiatives. This phase includes the process of developing the conceptual pilot initiatives, including plans for the activities, staffing needs, appropriate group sizes, protocols to ensure the resilience of resources and safety of participants, and initial pricing guides. The fourth phase implements the pilot, through a series of experiential learning tests that help the network build mutual competencies, refine a service supply-chain, and produce materials to support scientific dissemination and participatory science. During this phase, initial experiences are conducted with small groups of visitors who can provide feedback to help refine processes. Once the ST products have been thoroughly tested and supply-chains systems are consolidated, final pricing is set, and the fifth stage develops and implements strategies for communication, promotion, and market access to foster economic viability (Bourlon et al. 2021).

Collaborative Learning Networks Apply the Five-Phase Scientific Tourism Process in Aysén Since its creation, the ST process has been used to develop a range of products and destinations within local communities throughout the Aysén Region (Figs. 18.10 and 18.11). Thus, ST emphasizes the scientific heritage of places and people. It seeks to develop networks of local actors and scientists within a specific geography, so that they begin to learn together, in a process that joins collaborative transdisciplinary science, tourism, and traditional research practices. When communities decide to become ST destinations, they choose one or more long-term areas of science and begin a process of long-term study, experience, learning, and monitoring. Working together with the scientific community, they design tourism programs to integrate scientific fieldwork with tourism experiences. This process helps communities and territories build shared knowledge through its collaborative learning network approach.

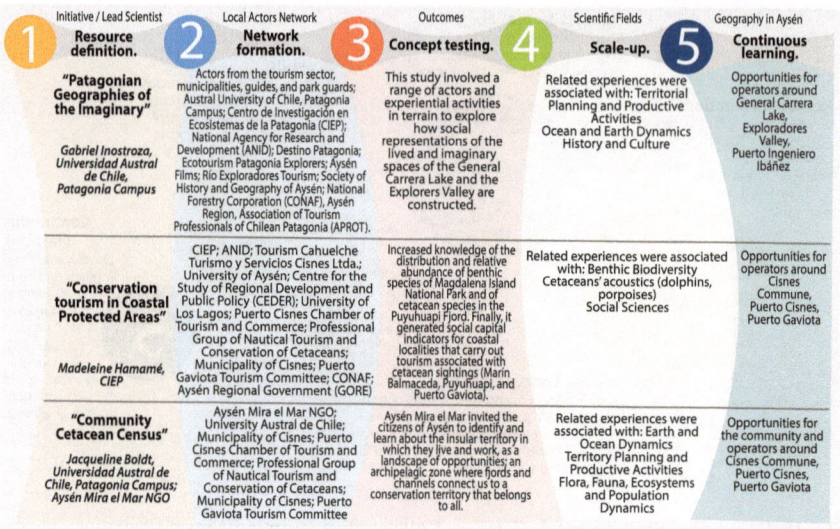

Fig. 18.10 Recent sustainable tourism initiatives in Aysén: Patagonian geographies of the imaginary; conservation tourism in coastal protected areas; and Community Cetacean Census

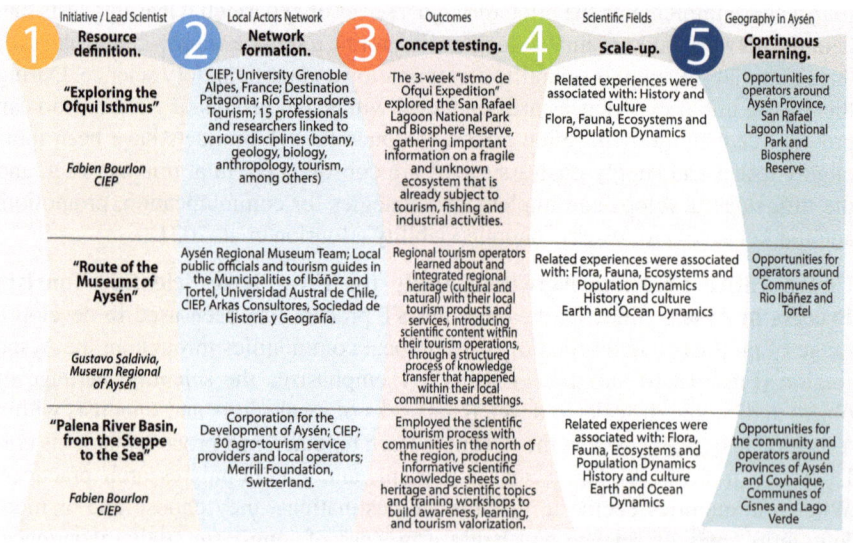

Fig. 18.11 Recent sustainable tourism initiatives in Aysén: Route of the Museums of Aysén; exploring the Ofqui Isthmus; and Palena River Basin, from the steppe to the sea

18.4   Discussion

This section discusses important considerations related to possible integration of the three initiatives developed within the Aysén and Magallanes regions of Chilean Patagonia: (1) 3-Hs Biocultural Ethic and FEP Cycle Approaches; (2) ST Collaborative Learning Networks; (3) Subantarctic Natural Laboratories, to promote biocultural conservation and more holistic forms of CBD (Hanspach et al. 2020). We begin by considering the potential to transfer the local experiences of the FEP Cycle Approach within the OEP and ST Collaborative Learning Networks within Aysén to other geographies, cultures, and contexts, as this would be an important consideration for building an integrated approach. Then, we discuss how the geobibliometric results can inform local strategies, employing an integrated approach.

18.4.1   Considerations for Transferability of the FEP Cycle Approach and Scientific Tourism Learning Networks

FEP's methodological approach seeks to integrate social, economic, and environmental dimensions of sustainability by generating new links between biological and cultural diversity at different spatial and social scales. By incorporating a philosophical foundation, the FEP Cycle Approach broadens both understanding and practices of sustainable tourism. What is more, recent experiences suggest strong potential to replicate and transfer the FEP Cycle Approach within other geographies, cultures, and contexts, including the Aysén Region.

Already, there are initiatives within the Aysén region that have incorporated the 3-Hs biocultural ethic. For example, social and environmental challenges for sustainable development drove an innovation based on wild and cultivated edible mushrooms, called *Hongusto* (from *hongo* = fungi and *gusto* = taste). Hongusto links key regional actors in education, tourism, and agricultural production with the scientific community, around basic knowledge of the ecology and biodiversity of edible fungi, offering a reference for regional initiatives for research, conservation, tourism, and sustainable production of mushrooms (Sánchez-Jardón et al. 2017). Before the project, fungi species diversity and ecological relevance were almost completely unknown; hence, local communities underestimated the long-term environmental consequences of potential biodiversity losses. The project incorporated ethical travel experiences through *fungal tourism* itineraries which continue to inspire local tour guides offering mycological tours. Rozzi's (1999) biocultural ethics was also employed in Aysén to inform the *"Subantarctic Sciences Open Laboratory,"* an initiative aimed at fostering the in situ connections between scientists and local communities. The initiative questioned why laboratories are typically reserved for scientists and viewed as the places from which the most valuable scientific knowledge is generated. Hence, the project employed open-spaces within the communities of Aysén to conduct scientific processes, including environmental knowledge exchange, research dissemination, and idea generation around solutions for environmental issues, and nature-based tourism concerns (Sánchez-Jardón et al. 2022).

The complete FEP Cycle Approach has been transferred to a range of other geographies, cultures, and biophysical contexts. It helped advance transdisciplinary research and experiences related to the starfish along the California Gulf, which are part of the rich biodiversity of southern Baja California. In this sea, starfishes have been widely studied from the natural sciences, but not from human disciplines. Nevertheless, native languages and traditions are still alive among fishermen and communities that inhabit the Baja California territory. Using the interdisciplinary approach of FEP, Moreno-Terrazas et al. (2022) investigated education and conservation processes that have occurred around the coastal and marine areas of Baja California Sur, combining marine biology, environmental history, and philosophy. With this approach, Moreno-Terrazas et al. (2022) designed a low-impact ecotourism practice oriented around the metaphor, *from the starfishes to the stars in the sky* and two field activities: *starfish viewing*, and *marine philosophical hiking*. These experiences facilitate face-to-face encounters with marine co-inhabitants that shine in the ineffable vermilion landscape of submerged ecosystems.

The ST Collaborative Learning Networks approach, as developed in Aysén, has also been replicated within other geographies, cultures, and contexts. For example, in 2018, existing networks were combined and expanded through the creation of the International Network for Scientific Tourism Research and Development (ISTN). The ISTN brings together institutions and actors in the fields of university education, scientific research, and tourism management, to promote mutual learning, collaboration, and support between researchers, trainers, entrepreneurs, public managers, and local communities that are interested in analyzing and promoting this form of tourism. ISTN members represent institutions around the world, including CIEP, the Universidad Austral de Chile, the Universitè Grenoble Alpes in France, the Université de Québec Trois Rivière in Canada, the Universidade de Caxias do Sul in Brazil, Chile's Pontificia Universidad Católica, the Victoria University of Wellington New Zealand, the Universidade de Algarve and the Poytechnic Institute of Tomar in Portugal, the Universidad Agraria Nacional La Molina of Peru, the Corporación Privada para el Desarrollo de Aysén, the Tecnológico de Estudios Superiores in Valle de Bravo Mexico, the Regional Museum of Aysén, the La Croisée des routes in the French Alps, the Aysén Mira al Mar Foundation, the Centro de Estudios Avanzados en Zonas Áridas (CEAZA), and the Centre de Recheches de Ecosystèmes d'Altitude (CREA) in Chamonix France.

A number of international ST projects have been developed through the local learning networks of the ISTN members. These include the Bravo Valley Monarch Butterfly Network in central Mexico; the Alpine Botanical Garden, Chalet-laboratory, and Alpe Gallery of Lautaret, France; ST training and support for the creation of a UNESCO Geopark in the Campos de Cima da Serra Gaúcha; a volunteering ST project with walking tours through Algarve's Mediterranean landscape and flora in Portugal; the Mont Blanc participatory science program, and Project RefLab, which involves high mountain ST based climate change monitoring in the Sentinel Shelters of the French Alps (Vialette et al. 2021).

18.4.2 Implications of the Geobibliometric Analysis of the Science Occurring in Pilot Subantarctic Natural Laboratories

While the 3-H framework of the biocultural ethic and the FEP Cycle Approach provide the methodological tools for co-constructing ethical travel experiences rooted in bioculture, and ST Collaborative Learning Networks offer the frameworks for building systems and commercialization, the development of subantarctic natural laboratories (SNLs) may provide an important ongoing policy and program mechanism to facilitate the evolution of subantarctic scientific research agendas and approaches. Results of the geobibliometric analysis developed within this intrinsic case study demonstrate some of the core themes and actors advancing scientific understanding within the SRL-SNL and CH-SNL and help identify gaps that require future concentration and support. For example, within the SRL there was a broad, well-developed line of research around glaciology and climate change that seems apt for incorporation within co-constructing ethical travel experiences. Other lines, which are less developed (e.g., *Consequences of glacier melting in the Baker River*; *Hydrochemistry and marine biology*), might find a way forward within the transdisciplinary approaches of the FEP Cycle Approach. And, while this area did indicate some social consideration, within the highly specialized work of the *Tourism, Territories, and Landscapes* line, the overall lack of socio-environmental approaches suggests the need for focused research agendas and capacity-building. The keywords that surfaced within the CH-SNL seemed much more connected to the local territory while, for example, many of the SRL-SNL keywords focused on broad concepts with global implications, but, with little specificity related to "place" (e.g., *Patagonia, Chile, remote sensing, climate change, fjord, glacier, biodiversity, landscape, mobility, macrofauna, stable isotopes*). In contrast, the CH-SNL keywords frequently (but not always) referenced place-based toponyms and phenomena (e.g., *subantarctic, Drake Passage, Antarctica, Cape Horn Biosphere Reserve, sub-Antarctic Magellanic ecoregion, Cape Horn, Tierra del Fuego, Beagle Channel, southernmost South America, southern westerly winds, Cordillera Darwin*). Species-based research also seemed to follow this trend. For example, within the Patagonian flora and fauna line of the SRL-SNL, keywords included broad concepts (i.e., Chilean Patagonia, biodiversity, biogeography, endemism, flora, mixing, lichens); yet, in the CH-SNL, the keywords for the Ornithological studies in Tierra del Fuego research line were very specific (i.e., *exotic species, Elaenia albiceps, Anairetes parulus, Navarino, Phrygilus patagonicus, Turdus falcklandii, Zonotrichia capensis*). These differences provoke a number of new research questions that merit future work. For example, has the place-based influence of the biocultural ethic that has been developed within CH-SNL provoked a more place-based long-term research agenda? How do these differences in focus affect local communities? How do they impact global research agendas? How would natural laboratory policies and programs best catalyze new capacities and strengths within these two, very distinct research agendas?

18.5   Conclusions

Rozzi (1999) coupled these 3-Hs of the biocultural ethic with three concepts of Earth Stewardship: the *habits* of stewardship, the communities of co-*inhabitants*, which include the *stewards*, and of course, the *habitats* of the Earth. By coupling the 3-Hs with Earth Stewardship concepts, the participation of stewards within habitats becomes explicit (Rozzi 2015b). Moreover, stewardship becomes an ethical imperative that should be incorporated into development policies as a matter of ecosocial justice (Rozzi 2015a). This approach markedly contrasts with the preservationist approach of the Half-Earth Project that calls for keeping half of the world's land and sea as wild and protected from human intervention or activity as possible (Wilson 2016). The Half-Earth initiative does not appropriately include the positive synergistic correlations found between biological and cultural diversity (Gorenflo et al. 2012; Maffi 2018).

Yet, as this chapter has demonstrated, there are several initiatives currently underway *within* and *from* Patagonia that seek to foster biocultural reawakening, democratize science, and catalyze more holistic approaches to sustainable development through ethical travel experiences. Based on the results of this intrinsic case study, we believe the three approaches that were analyzed (Fig. 18.1) offer the potential to move away from a preservationist approach and realize the potential of these positive synergistic correlations, through co-created scientifically based travel experiences dedicated to the conservation of nature and sociocultural aspects of local communities and its developed sustainably, integrating science, education, ethics, and economic profits (Rozzi et al. 2010a, b; Tauro et al. 2021).

We believe that the 3-Hs framework of the biocultural ethic and the FEP Cycle Approach can be linked with ST Collaborative Learning Networks, enhancing stages one through three of the existing ST Collaborative Learning Network product development process. This integration would enhance the potential of ST to link place-based scientific heritage resources with unforgettable experiences that could lead to increased well-being for those who visit and for those who are visited, potentially awakening in both a biocultural ethic for stewardship. And the integration of these two practices begins to formalize a process for locally driven CBD that moves from dominant neoliberal economic frameworks and paradigms, toward new models of sustainable development that encompass a more holistic view of human and non-human well-being. The SNL concept enhances these prospects by providing important resources, processes, policy, and oversight to support scientific agendas, processes, and governance within the unique and special places in which ethical biocultural travel experiences can occur, enabling collaborative scientific advances and positive spillover effects.

An integration of these three initiatives might lead to a new way of traveling or *touring* that better connects science and tourism by inviting visitors to appreciate the life *habits* of the co-in-Habitants (humans and other-than-humans) in their local shared *habitats* (Rozzi et al. 2012). Rozzi et al. (2010a, b) defined five essential attributes for developing ethical travel experiences rooted in local bioculture: (1) to value the uniqueness of each of the habitats visited, (2) to promote the conservation of habitats that support biological and cultural diversity (and their interrelationships), (3) to propose a framework of tourism practices that foster careful co-habitation relationships, (4) to seek the well-being of visitors and hosts who can maintain and promote their traditional practices, and (5) to consider the interrelationships between social, economic, and environmental sustainability (Rozzi et al. 2010a, b). These types of ethical travel experiences may represent a practice that invites both members of local communities and visitors to be guardians for the care of biocultural diversity. Moreover, tourism rooted in bioculture may catalyze experiences of ethical and aesthetic reconnection of society with nature. Additional research is warranted to test these hypotheses and the integration of these three initiatives.

Focused collaborative development of SNLs provide a socio-political framework for locally led strategies, providing them with important territorial monitoring information, improved scientific support and capacity, and the potential to influence and actively participate in place-based research agendas. When SNLs are developed to include the core habitats and areas protected through Chile's network of PAs (e.g., OEP, San Rafael Lagoon National Park, and Cape Horn National Park), and/or are protected under international accords and frameworks, like the UNESCO Man and the Biosphere (MAB) program, additional layers of oversight can further support locally led governance and stewardship. Specifically, this intergovernmental scientific program that operates under a land-sharing paradigm, supports a scientific basis for strengthening relationships between people and nature in their shared habitats, through socio-environmental approaches that promote bioculturally appropriate approaches to economic development (Deutsches Nationalkomitee MAB and UNESCO 2005; Fischer et al. 2014). Accordingly, these areas have systems for control and dialogue with scientists, which can be harnessed and improved through the development of collaborative governance and stewardship programs. According to the MAB website (en.unesco.org/mab), Biosphere Reserves are "learning places for sustainable development," where local communities and stakeholders actively participate in local level governance, stewardship, and monitoring, supported by tools, training, and experience sharing, through the MAB network of 701 Biosphere Reserves in 134 countries (Deutsches Nationalkomitee MAB and UNESCO 2005).

Rozzi et al.'s (2006) ten principles for linking biocultural conservation to nature conservation emphasized the importance of an integrated network for biocultural conservation, including locally led governance and stewardship, economic sustainability and ecotourism, national and international collaboration, and ethical outdoor education that helps humans connect with each other and with nature, within their shared habitat. To evolve toward post-neoliberal paradigms, these approaches must encounter legal and policy-based mechanisms that will allow them to overcome weak government regulation and authority, and obtain the resources and capacities needed to make locally led strategies that reject commodification and biocultural homogenization competitive (Khan 2015; Kashwan et al. 2019; Kumi et al. 2014; Orellana Calderón 2020).

The three initiatives analyzed within this case study seem to offer promising components for an integrated biocultural conservation network within the subantarctic Austral Macrozone of Chilean Patagonia. The FEP approach employed within the OEP has a long-standing record of integrating environmental ethics, arts, sciences, and education within the co-production of new educational and ecotourism programs and activities, as demonstrated within the six examples. The active Collaborative Learning Networks that have been fostered around ST in Aysén demonstrate ongoing interest amongst local communities to involve themselves with the science occurring in their territories through tourism. In addition, SNLs within Chilean PAs, and especially UNESCO Biosphere Reserves, provide innovative normative mechanisms for social linkage so that science of international importance can also support locally led strategies toward post-neoliberal forms of sustainable CBD (Deutsches Nationalkomitee MAB and UNESCO 2005).

Acknowledgments This work was supported by Chile's National Research and Development Agency (ANID) under the project, Collaborative network for the development of subantarctic Natural Laboratories: Strengthening scientific research and its socio-environmental and economic impacts in the Southern Macrozone (NODOSLN0002); ANID's Regional Program R17A10002; the CIEP R20F0002 project; ANID PAI79170138; ANID Vinculación Internacional FOVI220212; ANID FONDECYT Regular 1230020; and the CHIC-ANID PIA/BASAL PFB210018.

References

J.M. Aguilera, *Laboratorios naturales para una ciencia mundial [Natural laboratories for a global science]* (Comisión Nacional de Investigación Científica y Tecnológica, 2013) Retrieved January 4, 2023, from https://www.conicyt.cl/blog/2013/08/05/laboratorios-naturales-para-una-ciencia-mundial/

J.M. Aguilera, F. Larrain, *Laboratorios Naturales Para Chile [Natural Laboratories for Chile]* (Ediciones UC, 2018)

J.M. Aguilera, F. Larraín, Natural laboratories in emerging countries and comparative advantages in science: Evidence from Chile. Rev. Policy Res. **38**(6), 732–753 (2021). https://doi.org/10.1111/ropr.12450

E. Aliste, A. Núñez, M. Galarce, Geografías de lo sublime y el proceso de turistificación en Aysén-Patagonia. Turismo, territorio y poder [Geographies of the sublime and the process of touristification in Aysén-Patagonia. Tourism, territory and power], in *Araucanía-Norpatagonia II: La fluidez, lo disruptivo y el sentido de la frontera [Araucanía-Norpatagonia II: Fluidity, the disruptive and the sense of frontier]*, ed. by P. Núñez, A. Núñez, M. Tamagnini, (Universidad Nacional de Río Negro, 2018), pp. 249–269

D.B. Audretsch, M. Keilbach, The theory of knowledge spillover entrepreneurship. J. Manag. Stud. **44**(7), 1242–1254 (2007). https://doi.org/10.1111/j.1467-6486.2007.00722.x

P. Bachmann-Vargas, C.S.A. van Koppen, Disentangling environmental and development discourses in a peripheral spatial context: The case of the Aysén region, Patagonia, Chile. J. Environ. Develop. **29**(3), 366–390 (2020). https://doi.org/10.1177/1070496520937041

C. Baldwin, G. Marshall, H. Ross, J. Cavaye, J. Stephenson, L. Carter, C. Freeman, A. Curtis, G. Syme, Hybrid neoliberalism: Implications for sustainable development. Soc. Nat. Resour. **32**(5), 566–587 (2019). https://doi.org/10.1080/08941920.2018.1556758

J. Barandiaran, Reaching for the stars? Astronomy and growth in Chile. *Minerva* **53**(2), 141–164 (2015). https://doi.org/10.1007/s11024-015-9272-7

A. Barreau, J.T. Ibarra, F.S. Wyndham, R.A. Kozak, Shifts in Mapuche food systems in southern Andean forest landscapes: Historical processes and current trends of biocultural homogenization. Mt. Res. Dev. **39**(1), R12–R23 (2019). https://doi.org/10.1659/MRD-JOURNAL-D-18-00015.1

J. Baum, G.S. Cumming, A. De Vos, Understanding spatial variation in the drivers of nature-based tourism and their influence on the sustainability of private land conservation. Ecol. Econ. **140**, 225–234 (2017). https://doi.org/10.1016/j.ecolecon.2017.05.005

K.S. Bavikatte, T. Bennett, Community stewardship: The foundation of biocultural rights. J. Hum. Rights Environ. **6**(1), 7–29 (2015). https://doi.org/10.4337/jhre.2015.01.01

C.M. Beer, *Conservation, Development, and State Environment-Making in Chile* (University of California, 2022)

S.L. Boblin, S. Ireland, H. Kirkpatrick, K. Robertson, Using Stake's qualitative case study approach to explore implementation of evidence-based practice. Qual. Health Res. **23**(9), 1267–1275 (2013). https://doi.org/10.1177/1049732313502128

B. Borrie, T. Gale, K. Bosak, Privately protected areas in increasingly turbulent social contexts: Strategic roles, extent, and governance. J. Sustain. Tour. **30**(11), 2631–2648 (2020). https://doi.org/10.1080/09669582.2020.1845709

F. Bourlon, La ciencia como recurso para el desarrollo turístico sostenible de los archipiélagos patagónicos [Science as a resource for sustainable tourism development in the Patagonian archipelagos]. PASOS **18**(4), 795–810 (2020). https://doi.org/10.25145/j.pasos.2020.18.057

F. Bourlon, P. Mao, Las formas del turismo científico en Aysén, Chile [The forms of scientific tourism in Aysén, Chile]. Gestión Turística **15**, 74–98 (2011). https://doi.org/10.4206/gest.tur.2011.n15-04

F. Bourlon, R. Torres, Scientific tourism, a tool for tourism development in Patagonia. HAL, 1–12 (2016)

F. Bourlon, T. Gale, A. Adiego, V. Álvarez-Barra, A. Salazar, Grounding sustainable tourism in science – A geographic approach. Sustainability **13**(13), Article 7455 (2021). https://doi.org/10.3390/su13137455

F. Bourlon, Y. Vialette, P. Mao, Science as a resource for territorial and tourism development of mountainous areas of Chilean Patagonia. J. Alp. Res.. Advance Online Publication (2022). https://doi.org/10.4000/rga.10398

P. Bridgewater, I.D. Rotherham, A critical perspective on the concept of biocultural diversity and its emerging role in nature and heritage conservation. People Nat. **1**(3), 291–304 (2019). https://doi.org/10.1002/pan3.10040

R. Butler, Sustainable tourism in sensitive environments: A wolf in sheep's clothing? Sustainability (Switzerland) **10**(6), Article 1789 (2018). https://doi.org/10.3390/su10061789

Campaign for Nature (2021). *Why 30%?* Retrieved February 26, 2022, from https://www.campaignfornature.org/why-30-1

F. Carius, H. Job, Community involvement and tourism revenue sharing as contributing factors to the UN sustainable development goals in Jozani–Chwaka Bay National Park and Biosphere Reserve, Zanzibar. J. Sustain. Tour. **27**(6), 826–846 (2019). https://doi.org/10.1080/0966958 2.2018.1560457

F. Chen, H. Liu, K. Vikram, Y. Guo, For better or worse: The health implications of marriage separation due to migration in rural China. Demography **52**(4), 1321–1343 (2015). https://doi.org/10.1007/s13524-015-0399-9

Chilean National Innovation Council for Competitiveness (2013). *Orientaciones estratégicas para la innovación. Surfeando hacia el futuro. Chile en el horizonte 2025 [Strategic orientations for innovation. Surfing into the future. Chile on the 2025 horizon].* Retrieved January 20, 2023, from https://consejosociedades.files.wordpress.com/2013/08/orientaciones_estrategicas.pdf

Commission on Social Determinants of Health (CSDH) (2008). *Closing the gap in a generation: Health equity through action on the social determinants of health.* Retrieved January 16, 2023, from https://www.who.int/initiatives/action-on-the-social-determinants-of-health-for-advancing-equity/world-report-on-social-determinants-of-health-equity/commission-on-social-determinants-of-health

T. Contador Mejias, M. Gañan, J. Rendoll-Cárcamo, C.S. Maturana, H.A. Benítez, J. Kennedy, R. Rozzi, P. Convey, A polar insect's tale: Observations on the life cycle of Parochlus steinenii, the only winged midge native to Antarctica. Ecology **2022**, 1–9 (2022). https://doi.org/10.1002/ecy.3964

T.A. Contador, J.H. Kennedy, R. Rozzi, The conservation status of southern south American aquatic insects in the literature. Biodivers. Conserv. **21**(8), 2095–2107 (2012). https://doi.org/10.1007/s10531-012-0299-x

T. Contador, R. Rozzi, J. Kennedy, F. Massardo, J. Ojeda, P. Caballero, Y. Medina, R. Molina, F. Saldivia, F. Berchez, A. Stambuk, V. Morales, K. Moses, M. Gañan, G. Arriagada, J. Rendoll, F. Olivares, S. Lazzarino, Sumergidos con lupa en los ríos del Cabo de Hornos: Valoración ética de los ecosistemas dulceacuícolas y sus co-habitantes [Diving with a magnifying glass in the rivers of Cape Horn: Ethical assessment of freshwater ecosystems and their co-inhabitants]. Magallania (Punta Arenas) **46**(1), 183–206 (2018). https://doi.org/10.4067/S0718-22442018000100183

Conway, N. Heynen, *Globalization's Contradictions: Geographies of Discipline, Destruction and Transformation* (Routledge, 2006)

D.T.C. Cox, D.F. Shanahan, H.L. Hudson, R.A. Fuller, K.J. Gaston, The impact of urbanisation on nature dose and the implications for human health. Landsc. Urban Plan. **179**, 72–80 (2018). https://doi.org/10.1016/j.landurbplan.2018.07.013

S.E. Dalzell, R. College, *Bone Health in Gambian Women: Impact and Implications of Rural-to-Urban Migration and the Nutrition Transition*, vol 76 (University of Cambridge, 2018)

Deutsches Nationalkomitee MAB, & UNESCO, *Full of Life: UNESCO Biosphere Reserves, Model Regions for Sustainable Development* (Springer, 2005)

S. Ebneyamini, M.R. Sadeghi Moghadam, Toward developing a framework for conducting case study research. Int J Qual Methods **17**(1), 1–11 (2018). https://doi.org/10.1177/1609406918817954

J. Fischer, D.J. Abson, V. Butsic, M.J. Chappell, J. Ekroos, J. Hanspach, T. Kuemmerle, H.G. Smith, H. von Wehrden, Land sparing versus land sharing: Moving forward. Conserv. Lett. **7**(3), 149–157 (2014). https://doi.org/10.1111/conl.12084

N. Flores, The unexamined relationship between neoliberalism and plurilingualism: A cautionary tale. TESOL Q. **47**(3), 500–520 (2013). https://doi.org/10.1002/tesq.114

T. Gale, A. Adiego, A. Ednie, K. Beeftink, A. Báez, A systematized spatial review of global protected area soundscape research. Biodivers. Conserv. **31**, 2945–2964 (2022). https://doi.org/10.1007/s10531-022-02478-7

M.C. Gavin, J. McCarter, A. Mead, F. Berkes, J.R. Stepp, D. Peterson, R. Tang, Defining biocultural approaches to conservation. Trends Ecol. Evol. **30**(3), 140–145 (2015). https://doi.org/10.1016/j.tree.2014.12.005

M.A. Gjorgjioska, A. Tomicic, The crisis in social psychology under neoliberalism: Reflections from social representations theory. J. Soc. Issues **75**(1), 169–188 (2019). https://doi.org/10.1111/josi.12315

B. Goffinet, A.J. Shaw, *Bryophyte Biology*, 2nd edn. (Cambridge University Press, 2009)

B. Goffinet, R. Rozzi, L. Lewis, W. Buck, F. Massardo, *The Miniature Forests of Cape Horn: Ecotourism with a Hand-Lens – Los Bosques en Miniatura del Cabo de Hornos: Ecoturismo Con Lupa*, Bilingual English-Spanish edn. (University of Texas Press – Ediciones Universidad de Magallanes, 2012)

L.J. Gorenflo, S. Romaine, R.A. Mittermeier, K. Walker-Painemilla, Co-occurrence of linguistic and biological diversity in biodiversity hotspots and high biodiversity wilderness areas. Proc. Natl. Acad. Sci. **109**(21), 8032–8037 (2012). https://doi.org/10.1073/pnas.1117511109

J.A. Guridi, *Natural Laboratories as Policy Instruments for Technological Learning and Institutional Capacity Building: The Case of Chile's Astronomy Cluster* (Pontificia Universidad Católica de Chile, 2018)

J.A. Guridi, J.A. Pertuze, S.M. Pfotenhauer, Natural laboratories as policy instruments for technological learning and institutional capacity building: The case of Chile's astronomy cluster. Res. Policy **49**(2), 103899 (2020). https://doi.org/10.1016/j.respol.2019.103899

C.M. Hall, Constructing sustainable tourism development: The 2030 agenda and the managerial ecology of sustainable tourism. J. Sustain. Tour. **27**(7), 1044–1060 (2019). https://doi.org/10.1080/09669582.2018.1560456

J. Hanspach, L.J. Haider, E. Oteros-Rozas, A.S. Olafsson, N.M. Gulsrud, C.M. Raymond, et al., Biocultural approaches to sustainability: A systematic review of the scientific literature. People Nat. **2**(3), 643–659 (2020). https://doi.org/10.1002/pan3.10120

H. Harrison, M. Birks, R. Franklin, J. Mills, Case study research: Foundations and methodological orientations. Forum Qual. Sozialforschung **18**(1) (2017). https://doi.org/10.17169/fqs-18.1.2655

J. Igoe, D. Brockington, Neoliberal conservation a brief introduction. Conserv. Soc. **5**(4), 432–449 (2007)

G. Inostroza, G. Cànoves, Turismo sostenible y proyectos hidroeléctricos: Contradicciones en la Patagonia Chilena [Sustainable tourism and hydroelectric projects: Contradictions in Chilean Patagonia]. Cuad. Tur. **34**, 115–138 (2014)

P. Kashwan, L.M. MacLean, G.A. García-López, Rethinking power and institutions in the shadows of neoliberalism: (an introduction to a special issue of world development). World Dev. **120**, 133–146 (2019). https://doi.org/10.1016/j.worlddev.2018.05.026

S. Kesebir, P. Kesebir, A growing disconnection from nature is evident in cultural products. Perspect. Psychol. Sci. **12**(2), 258–269 (2017). https://doi.org/10.1177/1745691616662473

M.A. Khan, Putting 'good society' ahead of growth and/or 'development': Overcoming neoliberalism's growth trap and its costly consequences. Sustain. Dev. **23**(2), 65–73 (2015). https://doi.org/10.1002/sd.1572

T. Kjellstrom, Handbook of urban health: Populations, methods, and practice. Environ. Health Perspect. **114**(1) (2006). https://doi.org/10.1289/ehp.114-a64a

E. Kumi, A.A. Arhin, T. Yeboah, Can post-2015 sustainable development goals survive neoliberalism? A critical examination of the sustainable development-neoliberalism nexus in developing countries. Environ. Dev. Sustain. **16**(3), 539–554 (2014). https://doi.org/10.1007/s10668-013-9492-7

A. Latta, B.E.C. Aguayo, Testing the limits neoliberal ecologies from Pinochet to Bachelet. Lat. Am. Perspect. **39**(185), 163–180 (2012). https://doi.org/10.1177/0094582X12439050

J. Ledent, United Nations, Department of Economy and Social Affairs, population division. World urbanization prospects. The 1996 revision. New York, United Nations. (ST/ESA/SER. A/170), 1999. Cahiers Québécois de Démographie **29**(1), 179–185 (2012). https://doi.org/10.7202/010283ar

L. Lewis, C. Gottschalk-Druschke, C. Saldías, R. Mackenzie, J. Malebrán, B. Goffinet, R. Rozzi, Cultivando un jardín de nombres en los bosques en miniatura del cabo de hornos: Extensión de la conservación biocultural y la ética a seres vivos poco percibidos [Cultivating a garden of names in the miniature forests of Cape Horn: Extending biocultural conservation and ethics to little perceived living beings]. Magallania **46**(1), 103–123 (2018). https://doi.org/10.4067/S0718-22442018000100103

B. Mackey, D. Claudie, Points of contact: Integrating traditional and scientific knowledge for biocultural conservation. Environ. Ethics **37**(3), 341–357 (2015). https://doi.org/10.5840/enviroethics201537332

L. Maffi, Sustaining biocultural diversity, in *The Oxford Handbook of Endangered Languages*, ed. by K.L. Rehg, L. Campbell, (Oxford University Press, 2018), pp. 683–700

P. Mao, F. Bourlon, Le tourisme scientifique: Un essai de définition [Scientific tourism: An attempt at a definition]. Téoros **30**(2), 94–104 (2011). https://doi.org/10.7202/1012246ar

P. Mao, F. Bourlon, *Le tourisme scientifique en Patagonie Chilienne: Un essai géographique sur les voyages et explorations scientifiques [Scientific tourism in Chilean Patagonia: A geographic essay on scientific travel and exploration]* (L'Harmattan, 2016)

P. Mao, N Robinet, D Castro, *Creación de un destino de turismo científico: Análisis del proyecto, archipiélagos patagónicos, destino internacional para el turismo científico* [Creation of a scientific tourism destination: Project analysis, patagonian archipelagos, international scientific tourism destination]. (Centro de Investigación en Ecosistemas de la Patagonia (CIEP); International Development Bank—Multilateral Investment Fund (BID-FOMIN), 2016)

F. Massardo, R. Rozzi, Etno-ornitología Yagán y Lafkenche en los bosques templados de Sudamérica austral [Yagán and Lafkenche ethno-ornithology in the temperate forests of southern South America.]. Ornitol. Neotrop. **15**, 395–407 (2004)

F. Massardo, R. Rozzi, *The world's Southernmost Ethnoecology: Yahgan Craftsmanship and Traditional Ecological Knowledge* (Ediciones Universidad de Magallanes, 2006)

Y. Medina, F. Massardo, Rozzi, Educación, ecoturismo y conservación biocultural en los bosques en miniatura del Cabo de Hornos [Education, ecotourism, and biocultural conservation in the miniature forests of Cape Horn]. Magallania (Punta Arenas) **48**(2), 183–211 (2020). https://doi.org/10.4067/S0718-22442020000200183

M. Méndez, R. Rozzi, Jardineras subantárticas altoandinas en el Parque Etnobotánico Omora [High Andean sub-Antarctic gardeners in the Omora Ethnobotanical Park]. Revista Chagual **15**, 30–45 (2017)

M.O. Méndez, L. Cavieres, R. Rozzi, Jardineras subantárticas: Conocimiento y valoración de la flora altoandina [Sub-Antarctic gardeners: Knowledge and appreciation of high Andean flora]. Magallania **46**(1), 125–135 (2018)

F. Miranda Cabaña, Políticas del estado y la incorporación de espacios en la geografía del capitalismo: El Caso de Patagonia Aysén [State policy and the incorporation of spaces in the geography of capitalism: The case of Aysén Patagonia]. BEGEO **4**, 50–70 (2016)

R. Moreno-Terrazas, Z. Díaz-Gómez, H. González-Galván, M. Cariño-Olvera, M. Monteforte-Sánchez, Starfishes and sky stars: Field environmental philosophy education and ecotourism experiences in Baja California Sur, Mexico, in *Field Environmental Philosophy: Education for Biocultural Conservation*, ed. by R. Rozzi, A. Tauro, T. Wright, N. Avriel-Avni, R.H. May Jr., (Springer, 2022)

M. Morris, A statistical portrait of inuit with a focus on increasing urbanization: Implications for policy and further research. Aborig. Policy Stud. **5**(2), 4–31 (2016). https://doi.org/10.5663/aps.v5i2.27045

A. Núñez, E. Aliste, A. Bello, J. Astaburuaga, Eco-extractivismo y los discursos de la naturaleza en Patagonia-Aysén: Nuevos imaginarios geográficos y renovados procesos de control territorial [Eco-extractivism and nature discourses in Patagonia-Aysén: New geographical imaginaries and renewed processes of territorial control]. Revista Austral de Ciencias Sociales **35**, 133–153 (2018). https://doi.org/10.4206/rev.austral.cienc.soc.2018.n35-09

A. Núñez, F. Miranda, E. Aliste, S. Urrutia, Conservacionismo y desarrollo sustentable en la geografía del capitalismo: Negocio ambiental y nuevas formas de colonialidad en Patagonia-Aysén [Conservationism and sustainable development in the geography of capitalism: Environmental business and new forms of coloniality in Patagonia-Aysén], in *(Las) Otras geografías en Chile: Perspectivas Sociales Y Enfoques críticos [(the) Other Geographies in Chile: Social Perspectives and Critical Approaches]*, ed. by A. Núñez, E. Aliste, R. Molina, (Lom, 2019), pp. 23–46

A. Núñez, M. Benwell, E. Aliste, Interrogating green discourses in Patagonia-Aysén (Chile): Green grabbing and eco-extractivism as a new strategy of capitalism? Geogr. Rev. **112**(5), 688–706 (2020). https://doi.org/10.1080/00167428.2020.1798764

A. Núñez, G. Klier, E. Aliste, Ecoextractivismo y geopolítica de las periferias: El negocio de la diferencia en la Patagonia chilena/Argentina [Eco-extractivism and geopolitics of the peripheries: The business of difference in Chilean/Argentinean Patagonia], in *Araucania-Norpatagonia III*, ed. by A. Núñez, G. Kliear, E. Aliste, (Editorial UNRN, 2021), pp. 317–343

J. Ojeda, R. Rozzi, S. Rosenfeld, T. Contador, F. Massardo, J. Malebrán, J. González-Calderón, A. Mansilla, Interacciones bioculturales del pueblo Yagán con las macroalgas y moluscos: Una aproximación desde la Filosofía Ambiental de campo [Biocultural interactions of the Yagán people with seaweeds and mollusks: Field environmental philosophy approach]. Magallania (Punta Arenas) **46**(1), 155–181 (2018). https://doi.org/10.4067/s0718-22442018000100155

J.A. Oldekop, G. Holmes, W.E. Harris, K.L. Evans, A global assessment of the social and conservation outcomes of protected areas. Conserv. Biol. **30**(1), 133–141 (2016). https://doi.org/10.1111/cobi.12568

V. Orellana Calderón, In Chile, the post-neoliberal future is now. NACLA Rep. Am. **52**(1), 100–108 (2020). https://doi.org/10.1080/10714839.2020.1733239

E. Ranta, *Vivir Bien as an Alternative to Neoliberal Globalization: Can Indigenous Terminologies Decolonize the State?* (Taylor & Francis Group, 2018)

Y. Rashid, A. Rashid, M.A. Warraich, S.S. Sabir, A. Waseem, Case study method: A step-by-step guide for business researchers. Int. J. Qual. Method. Advance Online Publication **18**, 160940691986242 (2019). https://doi.org/10.1177/1609406919862424

J. Rendoll Cárcamo, T. Contador, M. Gañán, M. Houston, M. Troncoso, G. Arriagada, C. Saldías, P. Caballero, J. Malebrán, J. Kennedy, P. Convey, R. Rozzi, Filosofía Ambiental de Campo: Educación e investigación para la valoración ecológica y ética de los insectos dulceacuícolas [Environmental field philosophy: Education and research for the ecological and ethical valuation of freshwater insects]. Magallania (Punta Arenas) **48**(2), 213–228 (2020). https://doi.org/10.4067/S0718-22442020000200213

R. Rozzi, The reciprocal link between evolutionary ecological sciences and environmental ethics. Bio Science **49**, 911–921 (1999). https://doi.org/10.2307/1313650

R. Rozzi, Implicaciones éticas de narrativas Yaganes y Mapuches sobre las aves de los bosques templados de Sudamérica austral [Ethical implications of Yagán and Mapuche narratives about birds in the temperate forests of southern South America]. Ornitol. Neotrop. **15**, 435–444 (2004)

R. Rozzi, Biocultural ethics: Recovering the vital links between the inhabitants, their habits, and habitats. Environ. Ethic **34**(1), 27–50 (2012). https://doi.org/10.5840/enviroethics20123414

R. Rozzi, Implications of the biocultural ethic for earth stewardship, in *Earth Stewardship, Ecology and Ethics 2*, ed. by R. Rozzi, C. Palmer, D. Simberloff, E. Hargrove, F. Massardo, I.J. Klaver, J.B. Callicott, J.J. Armesto, K. Jax, S.T.A. Pickett, F.S. Chapin III, (Springer, 2015a), pp. 113–136

R. Rozzi, Earth stewardship and the biocultural ethic: Latin American perspectives, in *Earth Stewardship, Ecology and Ethics 2*, ed. by R. Rozzi, C. Palmer, D. Simberloff, E. Hargrove, F. Massardo, I.J. Klaver, J.B. Callicott, J.J. Armesto, K. Jax, S.T.A. Pickett, F.S. Chapin III, (Springer, 2015b), pp. 87–112

R. Rozzi, La Filosofía Ambiental de campo y la ecorregión subantártica de Magallanes Como un laboratorio natural en el Antropoceno [Environmental field philosophy and the Magellanic subantarctic ecoregion as a natural laboratory in the Anthropocene]. Magallania (Punta Arenas) **46**(1), 7–15 (2018). https://doi.org/10.4067/s0718-22442018000100007

R. Rozzi, C. Anderson, F. Massardo, J. Silander, Diversidad biocultural subantártica: El programa del Parque Etnobotánico Omora [Sub-Antarctic biocultural diversity: The Omora Ethnobotanical Park program]. Chloris chilensis **4**(2) (2001). https://www.chlorischile.cl/rozzi/rozzi.htm

R. Rozzi, F. Massardo, C. Anderson, S. McGehee, G. Clark, G. Egli, E. Ramilo, U. Calderón, C. Calderón, L. Aillapan, C. Zárraga, *Multi-Ethnic Bird Guide of the Austral Forests of South America* (Editorial Fantástico Sur—Universidad de Magallanes, 2003a)

R. Rozzi, F. Massardo, O. Dollenz, Un programa de investigación, educación y conservación biocultural a largo plazo en el extremo austral de América [A long-term biocultural research, education and conservation program in the far south of the Americas]. Austro Universitaria **14**, 50–59 (2003b)

R. Rozzi, F. Massardo, J. Silander, C. Anderson, A. Marin, Conservación biocultural y ética ambiental en el extremo austral de América: Oportunidades y dificultades para el bienestar ecosocial [Biocultural conservation and environmental ethics in the American far south: Opportunities and challenges for ecosocial wellbeing], in *Biodiversidad y globalización [Biodiversity and globalization]*, ed. by E. Figueroa, J. Simonetti, (Editorial Universitaria, 2003c), pp. 51–85

R. Rozzi, F. Massardo, C.B. Anderson, K. Heidinger, J.A. Silander, Ten principles for biocultural conservation at the southern tip of the Americas: The approach of the Omora Ethnobotanical Park. Ecol. Soc. **11**(1), 1–43 (2006). https://doi.org/10.5751/ES-01709-110143

R. Rozzi, X. Arango, F. Massardo, C. Anderson, K. Heidinger, K. Moses, Field environmental philosophy and biocultural conservation: The Omora Ethnobotanical Park educational program. Environ. Ethic. **30**(SUPPL. 3), 115–125 (2008). https://doi.org/10.5840/enviroethics200830supplement61

R. Rozzi, C. Anderson, C. Pizarro, F. Massardo, Y. Medina, A. Mansilla, J. Kennedy, J. Ojeda, T. Contador, V. Morales, K. Moses, A. Poole, J. Armesto, M. Kalin, Filosofía Ambiental de Campo y conservación biocultural en el Parque Etnobotánico Omora: Aproximaciones metodológicas para ampliar los modos de integrar el componente social ("S") en sitios de

Estudios Socio-Ecológicos a Largo Plazo (SESELP) [Field Environmental Philosophy and biocultural conservation in Omora Ethnobotanical Park: Methodological approaches to broaden the ways of integrating the social component ("S") in Socio-Ecological Long Term Study Sites (SESELP)]. Rev. Chil. Hist. Nat. **83**(1), 27–68 (2010a). https://doi.org/10.4067/S0716-078X2010000100004

R. Rozzi, F. Massardo, F. Cruz, C. Grenier, A. Muñoz, E. Mueller, J. Elbers, Galapagos and Cape Horn: Ecotourism or greenwashing in two emblematic Latin American archipelagoes? Environ. Phil. **7**(2), 1–32 (2010b)

R. Rozzi, J.J. Armesto, J.R. Gutiérrez, F. Massardo, G.E. Likens, C.B. Anderson, A. Poole, K.P. Moses, E. Hargrove, A.O. Mansilla, J.H. Kennedy, M. Willson, K. Jax, C.G. Jones, J.B. Callicott, M.T.K. Arroyo, Integrating ecology and environmental ethics: Earth stewardship in the southern end of the Americas. Bio Sci. **62**(3), 226–236 (2012). https://doi.org/10.1525/bio.2012.62.3.4

R. Rozzi, F. Massardo, T. Contador, R. Crego, M. Méndez, R. Rajan, L. Cavieres, J. Jiménez, Filosofía Ambiental de campo: Ecología y ética en las redes LTER-Chile e ILTER [Field Environmental Philosophy: Ecology and ethics in the LTER-Chile and ILTER networks]. Bosque(Valdivia) **35**(3), 439–447 (2014). https://doi.org/10.4067/S0717-92002014000300019

R. Rozzi, R.H. May, F.S. Chapin, F. Massardo, M.C. Gavin, I.J. Klaver, et al., From biocultural homogenization to biocultural conservation: A conceptual framework to reorient society toward sustainability of life, in *From Biocultural Homogenization to Biocultural Conservation*, ed. by R. Rozzi, R.H. May Jr., F.S. Chapin III, F. Massardo, M.C. Gavin, I.J. Klaver, et al., (Springer, 2018), pp. 1–17

L. Sánchez-Jardón, D. Soto, M. Torres, L. Moldenhauer, M. Solís Ehijos, J. Ojeda, B. Rosas, V. Salazar, C. Truong, *Hongusto, innovación Social en Torno a los Hongos Silvestres Y Cultivados en Aysén [Hongusto, Social Innovation around Wild and Cultivated Mushrooms in Aysén]* (Ediciones Universidad de Magallanes, 2017)

L. Sánchez-Jardón, R. Uribe-Paredes, D. Álvarez-Saravia, C. Aldea, V. Raimilla, E. Velázquez, S. Millán, J. Águila, Gestión regional de la información en biodiversidad: Fomentando la ciencia participativa en el sur de Chile [Regional biodiversity information management: Fostering participatory science in southern Chile]. Ecosistemas **31**(3), Article 2385 (2022). https://doi.org/10.7818/ECOS.2385

A. Schweitzer, Territórios cercados, territórios esvaziados e conservação da natureza no oeste da província de Santa Cruz, Patagônia Sul [Fenced territories, emptied territories and nature conservation in western Santa Cruz province, southern Patagonia]. Confins **47** (2020). https://doi.org/10.4000/confins.32551

L.C. Senese, K. Wilson, Aboriginal urbanization and rights in Canada: Examining implications for health. Soc. Sci. Med. **91**, 219–228 (2013). https://doi.org/10.1016/j.socscimed.2013.02.016

H. Shreve, Harmonization, but not homogenization: The case for Cuban autonomy in globalizing economic reforms. Indiana J. Glob. Leg. Stud. **19**(1), 365–390 (2012). https://doi.org/10.2979/indjglolegstu.19.1.365

M. Sims, Neoliberalism and early childhood. Cogent. Educ. **4**(1) (2017). https://doi.org/10.1080/2331186X.2017.1365411

S.L. Slocum, D.Y. Dimitrov, K. Webb, The impact of neoliberalism on higher education tourism programs: Meeting the 2030 sustainable development goals with the next generation. Tour. Manag. Perspect. **30**, 33–42 (2019). https://doi.org/10.1016/j.tmp.2019.01.004

M. Soga, K.J. Gaston, Extinction of experience: The loss of human-nature interactions. Front. Ecol. Environ. **14**(2), 94–101 (2016). https://doi.org/10.1002/fee.1225

M. Soga, K.J. Gaston, The ecology of human – Nature interactions. Proc. R. Soc. B: Biol. Sci **287**, 20191882 (2020). https://doi.org/10.1098/rspb.2019.1882

R. Sõukand, R. Kalle, A. Pieroni, Homogenisation of biocultural diversity: Plant ethnomedicine and its diachronic change in Setomaa and Võromaa, Estonia, in the last century. Biology **11**(2), Article 192 (2022). https://doi.org/10.3390/biology11020192

R. Stahler-Sholk, Resisting neoliberal homogenization: The Zapatista autonomy movement. Lat. Am. Perspect. **34**(2), 48–63 (2007). https://doi.org/10.1177/0094582X06298747

R.E. Stake, Case studies, in *Strategies of Qualitative Inquiry*, ed. by N.K. Denzin, Y.S. Lincoln, 2nd edn., (SAGE Publications Inc, 2003), pp. 134–164

D.L. Stokes, Conservators of experience. Bioscience **56**(1), 6–7 (2006). https://doi.org/10.164 1/0006-3568(2006)056[0007:COE]2.0.CO;2

A. Tauro, J. Ojeda, T. Caviness, K.P. Moses, R. Moreno-Terrazas, T. Wright, D. Zhu, A.K. Poole, F. Massardo, R. Rozzi, Field environmental philosophy: A biocultural ethic approach to education and ecotourism for sustainability. Sustainability (Switzerland) **13**(8), 1–22 (2021). https://doi.org/10.3390/su13084526

V.A. Traag, L. Waltman, N.J. van Eck, From Louvain to Leiden: Guaranteeing well-connected communities. Sci. Rep. **9**(1), 1–12 (2019). https://doi.org/10.1038/s41598-019-41695-z

N.J. van Eck, L. Waltman, *Manual for VOSviewer Version 1.6.15* (Univeristeit Leiden, 2020)

K. Veloso, *Diseño de una propuesta metodológica para la evaluación del turismo científico [Design of a methodological proposal for the evaluation of scientific tourism]* (Universidad Austral de Chile, 2021)

Y. Vialette, P. Mao, F. Bourlon, Scientific tourism in the French Alps: A laboratory for scientific mediation and research. Rev. Géogr. Alp. **109**, 2–15 (2021). https://doi.org/10.4000/rga.9189

B. Weiss, H.M. Dang, T.T. Lam, M.C. Nguyen, Urbanization, and child mental health and life functioning in Vietnam: Implications for global health disparities. Soc. Psychiatry Psychiatr. Epidemiol. **55**(6), 673–683 (2020). https://doi.org/10.1007/s00127-020-01838-4

E.O. Wilson, *Half-Earth: Our planet's Fight for Life* (WW Norton & Company, 2016)

T. Wolters, Why is ecological sustainability so difficult to achieve? An in-context discussion of conceptual barriers. Sustain. Dev. **30**(6), 2025–2039 (2022). https://doi.org/10.1002/sd.2326

R. Zakaria, A. Ahmi, A.H. Ahmad, Z. Othman, K.F. Azman, C.B. Ab Aziz, C.A.N. Ismail, N. Shafin, Visualising and mapping a decade of literature on honey research: A bibliometric analysis from 2011 to 2020. J. Apic. Res. **60**(3), 359–368 (2021). https://doi.org/10.108 0/00218839.2021.1898789

Correction to: Tourism and Conservation-based Development in the Periphery: Lessons from Patagonia for a Rapidly Changing World

Trace Gale-Detrich, Andrea Ednie, and Keith Bosak

Correction to:
T. Gale-Detrich et al. (eds.), *Tourism and Conservation-based Development in the Periphery*, Natural and Social Sciences of Patagonia, https://doi.org/10.1007/978-3-031-38048-8

On page 9 of Chapter 1, the name mentioned in the third line of the text was incorrect. It has now been changed from Carlos Menem to Raúl Alfonsin.

The original version of Chapters 11 and 18 was published with incorrect author names. Some of the authors' last names were incorrectly tagged as first names. The names have now been corrected.

Chapter 11:

Given Name	Last Name
Germaynee	Vela-Ruiz Figueroa

The updated version of these chapters can be found at
https://doi.org/10.1007/978-3-031-38048-8_1
https://doi.org/10.1007/978-3-031-38048-8_11
https://doi.org/10.1007/978-3-031-38048-8_18

Chapter 18:

Given Name	Last Name
Laura	Sánchez Jardón
Matías	Navarrete Almonacid
José	Coloma Zapata
Diego	Hernández Soto
Manuel	Mora Chepo
Ronald	Cancino Salas
Lorna	Moldenhauer Ortega

Index